UNIVERSITY OF NOTTINGHAM

10 0065289 2

KU-030-320

WITHDRAWN
FROM THE LIBRARY

UNIVERSITY OF NOTTINGHAM

WITHDRAWN
FROM THE LIBRARY

NOT TO
BE TAKEN
OUT OF
THE
LIBRARY

Compendium of Organic
Synthetic Methods

Compendium of Organic Synthetic Methods

Volume 6

MICHAEL B. SMITH

DEPARTMENT OF CHEMISTRY
UNIVERSITY OF CONNECTICUT
STORRS, CONNECTICUT

A Wiley-Interscience Publication

JOHN WILEY & SONS
New York • Chichester • Brisbane • Toronto • Singapore

Copyright © 1988 by John Wiley & Sons, Inc.

All rights reserved. Published simultaneously in Canada.

Reproduction or translation of any part of this work
beyond that permitted by Section 107 or 108 of the
1976 United States Copyright Act without the permission
of the copyright owner is unlawful. Requests for
permission or further information should be addressed to
the Permissions Department, John Wiley & Sons, Inc.

Library of Congress Catalog Card Number: 71-162800

ISBN 0-471-84896-4

Printed in the United States of America

10 9 8 7 6 5 4

UNIVERSITY OF NOTTINGHAM

1006052892

UNIVERSITY
LIBRARY
NOTTINGHAM

WITHDRAWN

FROM THE LIBRARY

PREFACE

It has been sixteen years since Ian and Shuyen Harrison first published the *Compendium of Organic Synthetic Methods* to facilitate the search for functional group transformations in the formidable body of the original literature. In Volume 2 this concept was expanded to include difunctional compounds. Louis Hegedus and Leroy Wade carried on this important compilation in Volume 3 and Wade continued with Volumes 4 and 5. Volume 6 is intended to continue the Harrisons' stated purpose of "a comprehensive one-volume listing of synthetic methods as an intermediary between the chemist and the literature."

Compendium of Organic Synthetic Methods, Volume 6, presents the functional group transformations and difunctional compound preparations of 1983, 1984, 1985, and 1986. The classification schemes of the first five volumes have been followed but a new chapter has been added. Each literature citation now includes all authors and the text concludes with an Author Index. The new Chapter 15 classifies the oxides of nitrogen, sulfur, and selenium with section headings identical to Chapters 1–14. The difunctional compounds now appear in Chapter 16. The experienced user of the *Compendium* will require no special instructions for the use of the new oxides chapter or the complete volume.

Author citations and the Author Index have been included to facilitate literature searches and to follow the current work of a particular author. The citations should not prove obtrusive and will hopefully assist the synthetic community.

I wish to express my gratitude to Professor Michael Edwards who first suggested this project to me. I also wish to thank Professor Leon Ghosez who provided the facilities for a pleasant sabbatical leave, which allowed completion of the research for this text. My thanks to Tae Woo Kwon, Jennline Sheu, Royce Menezes, Chung Jen Wang, Paul Keusenkothen, and Young Chan Son who proofread parts of the manuscript and offered many helpful suggestions. I want to thank my wife Sarah and son Steven for their patience and moral support throughout this work.

MICHAEL B. SMITH

Storrs, Connecticut
June 1987

CONTENTS

ABBREVIATIONS

Ac   Acetyl = $-\overset{\displaystyle O}{\overset{\|}{C}}CH_3$

acac   Acetylacetonate

AIBN   Azo-*bis*-isobutyronitrile

aq.   Aqueous

-B   9-Borabicyclo[3.3.1]nonylboryl

9-BBN   9-Borabicyclo[3.3.1.]nonane

Bn (Bz)   Benzyl = $-CH_2Ph$

Boc   *t*-Butoxycarbonyl = $-\overset{\displaystyle O}{\overset{\|}{C}}-O-t\text{Bu}$

bpy (bipy)   2,2'-Bipyridyl

Bu   *n*-Butyl = $-CH_2CH_2CH_2CH_3$

CAM   Carboxamidomethyl

CAN   Cerric ammonium nitrate = $(NH_4)_2Ce(NO_3)_6$

c-   Cyclo-

cat.   Catalytic

CBZ   Carbobenzoxy = $-\overset{\displaystyle O}{\overset{\|}{C}}OCH_2Ph$

COD   1,5-Cyclooctadienyl

Cp   Cyclopentadienyl

Cy   Cyclohexyl

DABCO   1,4-Diazabicyclo[2.2.2]octane

dba   Dibenzylidene acetone

DBE   1,2-Dibromoethane = $BrCH_2CH_2Br$

DBN   1,8-Diazabicyclo[5.4.0]undec-7-ene

DBU   1,5-Diazabicyclo[4.3.0]non-5-ene

DCC   1,3-Dicyclohexylcarbodiimide = $C_6H_{13}-N{=}C{=}N-C_6H_{13}$

DCE   1,2-Dichloroethane = $ClCH_2CH_2Cl$

DDQ   2,3-Dichloro-5,6-dicyano-1,4-benzoquinone

DEA   Diethylamine = $HN(CH_2CH_3)_2$

DEAD   Diethylazodicarboxylate

Dibal-H   Diisobutylaluminum hydride

Diphos-4   1,4-*bis*-(Diphenylphosphino)butane

DMAP   4-Dimethylaminopyridine

DME   Dimethoxyethane

DMF   *N,N*-Dimethylformamide = $H\overset{\displaystyle O}{\overset{\|}{C}}NMe_2$

dppe	1,2-*bis*-Diphenylphosphinoethane
dppf	*bis*-(Diphenylphosphine)ferrocene
dppp	1,3-*bis*-(Diphenylphospine)propane
dvb	Divinylbenzene
e$^-$	Electrolysis
ee	Enantiomeric excess
EE	Ethoxyethyl
Et	Ethyl = —CH_2CH_3
EDA	Ethylenediamine = $H_2NCH_2CH_2NH_2$
EDTA	Ethylenediaminetetraacetic acid
FMN	Flavin mononucleotide
fod	*tris*-(6,6,7,7,8,8,8)-Heptafluoro-2,2-dimethyl-3,5-octanedionate
Fp	Cyclopentadienyl*bis*carbonyliron
FVP	Flash vacuum pyrolysis
hv	Irradiation with light
1,5-HD	1,5-Hexadienyl
HMPA	Hexamethylphosphoramide = $(Me_3N)_3P{=}O$
HMPT	Hexamethylphosphorous triamide = $(Me_3N)_3P$
iPr	Isopropyl = —$CH(CH_3)_2$
LICA	Lithium cyclohexylisopropylamide
LDA	Lithium diisopropylamide = $LiN(iPr)_2$
LTMP	Lithium 2,2,6,6-tetramethylpiperidide
mcpba	*meta*-Chloroperoxybenzoic acid
Me	Methyl = —CH_3
MEM	β-Methoxyethoxymethyl = $MeOCH_2CH_2OCH_2$—
MOM	Methoxymethyl = $MeOCH_2$—
Ms	Methanesulfonyl = CH_3SO_2—
MTM	Methylthiomethyl = CH_3SCH_2—
NAD	Nicotinamide adenine dinucleotide
NADP	Sodium triphosphopyridine nucleotide
NBD	Norbornadiene
NBS	*N*-Bromosuccinimide
NCS	*N*-Chlorosuccinimide
Ni(R)	Raney nickel
Ⓟ	Polymeric backbone
PCC	Pyridinium chlorochromate
PDC	Pyridinium dichromate
PEG	Polyethylene glycol
Ph	Phenyl = —C_6H_5
Pip	Piperidine = HN—$\overline{CH_2CH_2CH_2CH_2}$
Pr	*n*-Propyl = —$CH_2CH_2CH_3$
Py	Pyridine = C_6H_5N
quant.	Quantitative yield
Red-Al	$[(MeOCH_2CH_2O)_2AlH_2]Na$
TBAF	*n*-$Bu_4N^+F^-$
TBHP	*t*-Butylhydroperoxide = Me_3COOH

tBu	*t*-Butyl = —C(CH$_3$)$_3$
TEMPO	Tetramethylpiperidinyloxy free radical
TFA	Trifluoroacetic acid = CF$_3$COOH
TFAA	Trifluoroacetic anhydride = CF$_3$COCCH$_3$
Tf	Triflate = —OSO$_2$CF$_3$
THF	Tetrahydrofuran
THP	Tetrahydropyran
TMEDA	Tetramethylethylenediamine
TMS	Trimethylsilyl = —Si(CH$_3$)$_3$
TMP	2,2,6,6-Tetramethylpiperidine
Tol	Tolyl = 4-C$_6$H$_4$CH$_3$
Tr	Trityl = —CPh$_3$
TRIS	Triisopropylphenylsulfonyl
Ts (Tos)	Tosyl = *p*-toluenesulfonyl
)))))	Sonication

INDEX, MONOFUNCTIONAL COMPOUNDS

Sections—heavy type
Pages—light type

PREPARATION OF →

FROM →

FROM \ PREPARATION OF	Acetylenes	Carboxylic acids, acid halides, anhydrides	Alcohols, phenols	Aldehydes	Alkyls, methylenes, aryls	Amides	Amines	Esters	Ethers, epoxides	Halides, sulfonates	Hydrides (RH)	Ketones	Nitriles	Olefins	Oxides
Acetylenes	**1**	**16** 8	**31** 19		**61** 67			**106** 139	**121** 165			**166** 205	**181** 239	**196** 248	
Carboxylic acids, acid halides, anhydrides	**2** 2	**17** 8	**32** 19	**47** 51	**62** 67	**77** 98	**92** 113	**107** 140		**137** 181	**152** 194	**167** 206		**197** 251	
Alcohols, phenols			**33** 20	**48** 52	**63** 68		**93** 113	**108** 143	**123** 165	**138** 182	**153** 195	**168** 209		**198** 252	**213** 272
Aldehydes	**4** 3	**19** 9	**34** 21	**49** 56	**64** 68	**79** 100	**94** 114	**109** 146	**124** 167	**139** 185	**154** 197	**169** 215	**184** 239	**199** 253	
Alkyls, methylenes, aryls		**20** 10	**35** 26	**50** 57	**65** 69		**95** 115	**110** 148	**125** 168	**140** 186		**170** 216		**200** 256	**215** 273
Amides			**36** 27			**81** 101	**96** 115	**111** 148				**171** 217	**186** 241		
Amines		**22** 11		**52** 58		**82** 104	**97** 117	**112** 149	**127** 169	**142** 187	**157** 198	**172** 217			**217** 274
Esters	**8** 4	**23** 11	**38** 28	**53** 58	**68** 69	**83** 106	**98** 125	**113** 149	**128** 169		**158** 198	**173** 218		**203** 257	
Ethers, epoxides			**39** 29	**54** 59	**69** 71			**114** 154	**129** 170	**144** 187	**159** 199	**174** 219		**204** 258	**219** 275
Halides, sulfonates, sulfates	**10** 5	**25** 12	**40** 33	**55** 59	**70** 71	**85** 107	**100** 125	**115** 155	**130** 171	**145** 188	**160** 200	**175** 220	**190** 243	**205** 260	
Hydrides (RH)		**26** 14		**56** 61	**71** 77		**101** 128	**116** 158		**146** 190		**176** 222	**191** 244	**206** 263	**221** 278
Ketones	**12** 5	**27** 14	**42** 35	**57** 61	**72** 79	**87** 108	**102** 128	**117** 158	**132** 173	**147** 191	**162** 203	**177** 223	**192** 245	**207** 264	
Nitriles		**28** 15			**73** 80	**88** 109	**103** 131	**118** 160			**163** 203	**178** 231	**193** 245	**208** 267	
Olefins	**14** 6	**29** 16	**44** 43	**59** 62	**74** 81	**89** 110	**104** 132	**119** 161	**134** 175	**149** 192		**179** 231		**209** 267	
Miscellaneous compounds	**15** 7	**30** 16	**45** 45	**60** 63	**75** 97	**90** 111	**105** 132	**120** 162	**135** 179	**150** 192	**165** 204	**180** 233	**195** 246	**210** 270	**225** 279

PROTECTION

	Sect.	Pg.
Carboxylic acids	**30A**	17
Alcohols, phenols	**45A**	46
Aldehydes	**60A**	64
Amines	**105A**	136
Ethers	**135A**	180
Ketones	**180A**	236

Blanks in the table correspond to sections for which no additional examples were found in the literature.

INDEX, DIFUNCTIONAL COMPOUNDS

Sections—heavy type
Pages—light type

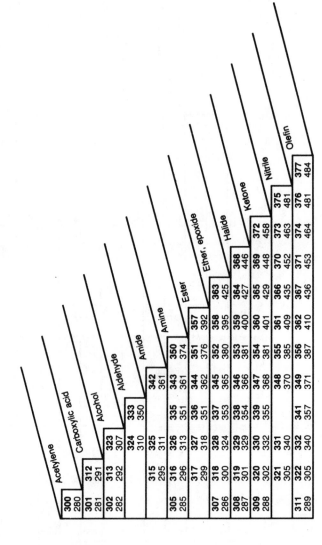

	Acetylene	Carboxylic acid	Alcohol	Aldehyde	Amide	Amine	Ester	Ether, epoxide	Halide	Ketone	Nitrile	Olefin
Acetylene	300 / 280											
Carboxylic acid	301 / 281	312 / 291										
Alcohol	302 / 282	313 / 292	323 / 307									
Aldehyde	303	314	324 / 310	333 / 350								
Amide	304	315 / 295	325 / 311	334 / 351	342 / 361							
Amine	305 / 285	316 / 296	326 / 313	335 / 351	343 / 361	350 / 374						
Ester	306	317 / 299	327 / 318	336 / 362	344 / 362	351 / 376	357 / 392					
Ether, epoxide	307 / 286	318 / 300	328 / 324	337 / 365	345 / 365	352 / 380	358 / 395	363 / 425				
Halide	308 / 287	319 / 301	329 / 329	338 / 366	346 / 366	353 / 381	359 / 400	364 / 427	368 / 446			
Ketone	309 / 288	320 / 302	330 / 332	339 / 368	347 / 368	354 / 381	360 / 401	365 / 429	369 / 448	372 / 458		
Nitrile	310	321 / 305	331 / 340	340 / 370	348 / 370	355 / 385	361 / 409	366 / 435	370 / 452	373 / 463	375 / 481	
Olefin	311 / 289	322 / 305	332 / 340	341 / 357	349 / 371	356 / 387	362 / 410	367 / 436	371 / 453	374 / 464	376 / 481	377 / 484

Blanks in the table correspond to sections for which no additional examples were found in the literature.

INTRODUCTION

Relationship between Volume 6 and Previous Volumes. *Compendium of Organic Synthetic Methods, Volume 6* presents about 1100 examples of published methods for the preparation of monofunctional compounds, updating the 7000 in Volumes 1 through 5. In addition, Volume 6 contains about 900 examples of preparations of difunctional compounds and various functional groups, updating the sections introduced in Volume 2. For Chapters 1 through 14 the same systems of section and chapter numbering are used as in previous volumes. Chapter 15 is new and classifies the preparation of oxides of nitrogen (nitro compounds, nitrones, *N*-oxides, etc.), sulfur (sulfoxides and sulfones), and selenium (selenoxides and Se dioxides) according to the scheme used for all other monofunctional compounds. Therefore, "Oxides from Acetylenes" appears as Section 211 and the chapter concludes with "Oxides from Miscellaneous Compounds", Section 225. The difunctional compounds now appear in Chapter 16 (Chapter 15 in Volumes 2 through 5) but the section numbers remain the same.

A minor change from previous volumes is the inclusion of thiols with all alcohol classifications and sulfides (thioethers) are classified with ethers. Both of these changes are reflected in the chapter titles, section titles, and citations. These changes do not alter the fundamental organization.

Each literature citation follows the notation used in previous volumes but is accompanied by a complete author list. The principal author(s) are denoted by an asterisk (*). Following Chapter 16 is an alphabetical listing of all authors (last name, initials).

Classification and Organization of Reactions Forming Monofunctional Compounds. Chemical transformations are classified according to the reacting functional group of the starting material and the functional group formed. Those reactions that give products with the same functional group form a chapter. The reactions in each chapter are further classified into sections on the basis of the functional group of the starting material. Within each section, reactions are loosely arranged in ascending order of year cited (1983–1986), although an effort has been made to put similar reactions together when possible. Review articles are collected at the end of each appropriate section.

The classification is unaffected by allylic, vinylic, or acetylenic unsaturation appearing in both starting material and product, or by increases or decreases

in the length of carbon chains; for example, the reactions t-BuOH →
t-BuCOOH, PhCH$_2$OH → PhCOOH, and PhCH=CHCH$_2$OH →
PhCH=CHCOOH would all be considered as preparations of carboxylic
acids from alcohols. Conjugate reduction and alkylation of unsaturated ketones, aldehydes, esters, acids, and nitriles have been placed in category
74 (alkyls from olefins).

The terms hydrides, alkyls, and aryls classify compounds containing reacting hydrogens, alkyl groups, and aryl groups, respectively; for example,
RCH$_2$—H → RCH$_2$COOH (carboxylic acids from hydrides), RMe → RCOOH
(carboxylic acids from alkyls), RPh → RCOOH (carboxylic acids from aryls).
Note the distinction between R$_2$CO → R$_2$CH$_2$ (methylenes from ketones) and
RCOR' → RH (hydrides from ketones). Alkylations involving additions across
double bonds are found in Section 74 (alkyls, methylenes, and aryls from
olefins).

The following examples illustrate the classification of some potentially con-
fusing cases:

RCH=CHCOOH → RCH=CH$_2$	Hydrides from carboxylic acids
RCH=CH$_2$ → RCH=CHCOOH	Carboxylic acids from hydrides
ArH → ArCOOH	Carboxylic acids from hydrides
ArH → ArOAc	Esters from hydrides
RCHO → RH	Hydrides from aldehydes
RCH=CHCHO → RCH=CH$_2$	Hydrides from aldehydes
RCHO → RCH$_3$	Alkyls from aldehydes
R$_2$CH$_2$ → R$_2$CO	Ketones from methylenes
RCH$_2$COR → R$_2$CHCOR	Ketones from ketones
RCH=CH$_2$ → RCH$_2$CH$_3$	Alkyls from olefins
RBr + CH CH → RC≡CR	Acetylenes from halides; also acetylenes from acetylenes
ROH + RCOOH → RCOOR	Esters from alcohols; also esters from carboxylic acids
RCH=CHCHO → RCH$_2$CH$_2$CHO	Alkyls from olefins
RCH=CHCN → RCH$_2$CH$_2$CN	Alkyls from olefins

**How to Use the Book to Locate Examples of the Preparation or Pro-
tection of Monofunctional Compounds.** Examples of the preparation of
one functional group from another are located via the monofunctional index
on p. xiii, which lists the corresponding section and page. Thus Section 1
contains examples of the preparation of acetylenes from other acetylenes;
Section 2, acetylenes from carboxylic acids; and so forth.

Sections that contain examples of the reactions of a functional group are
found in the horizontal rows of the index. Thus Section 1 gives examples of

the reactions of acetylenes that form acetylenes; Section 16, reactions of acetylenes that form carboxylic acids; and Section 31, reactions of acetylenes that form alcohols.

Examples of alkylation, dealkylation, homologation, isomerization, and transposition are found in Sections 1, 17, 33, and so on, lying close to a diagonal of the index. These sections correspond to such topics as the preparation of acetylenes from acetylenes; carboxylic acids from carboxylic acids; and alcohols, thiols, and phenols from alcohols, thiols, and phenols. Alkylations that involve conjugate additions across a double bond are found in Section 74 (alkyls, methylenes, and aryls from olefins).

Examples of name reactions can be found by first considering the nature of the starting material and product. The Wittig reaction, for instance is in Section 199 (olefins from aldehydes) and Section 207 (olefins from ketones). The aldol condensation can be found in the chapters on difunctional compounds in Section 324 (alcohol, thiol–aldehyde) and in Section 330 (alcohol, thiol–ketone).

Examples of the protection of acetylenes, carboxylic acids, alcohols, phenols, aldehydes, amides, amines, esters, ketones, and olefins are also indexed on p. xiii.

The pairs of functional groups alcohol, ester; carboxylic acid, ester; amine, amide; and carboxylic acid, amide can be interconverted by simple reactions. When a member of these groups is the desired product or starting material, the other member should, of course, also be consulted in the text.

The original literature must be used to determine the generality of reactions, although this is occasionally stated in the citation. This is only done in cases where such generality is stated clearly in the original citation. A reaction given in this book for a primary aliphatic substrate may also be applicable to tertiary or aromatic compounds. This book provides very limited experimental conditions or precautions and the reader is referred to the original literature before attempting a reaction. In no instance should the citation be taken as a complete experimental procedure. Not to refer to the original literature could be hazardous. The original papers usually yield a further set of references to previous work. Subsequent publications can be found by consulting the *Science Citation Index.*

Classification and Organization of Reactions Forming Difunctional Compounds. This chapter considers all possible difunctional compounds formed from the groups acetylene, carboxylic acid, alcohol, thiol, aldehyde, amide, amine, ester, ether, epoxide, thioether, halide, ketone, nitrile, and olefin. Reactions that form difunctional compounds are classified into sections on the basis of the two functional groups of the product. The relative positions of the groups do not affect the classification. Thus preparations of 1,2-aminoalcohols, 1,3-aminoalcohols, and 1,4-aminoalcohols are included in a single section. Difunctional compounds that have an oxide as the second group are found in the monofunctional sections for the nonoxide functional group. Therefore, the nitroketone product of oxidation of a nitroalcohol is found in Section 168 (ketones from alcohols and thiols). Conversion of an oxide to another functional group is generally found in the "Miscellaneous" section, so conversion of a nitroalkane to an amine is found in Section 105 (amines from miscellaneous compounds). The following examples illustrate the application of this classification system:

Difunctional Product	Section Title
$RC{\equiv}C{-}C{\equiv}CR$	Acetylene–acetylene
$RCH(OH)COOH$	Carboxylic acid–alcohol
$RCH{=}CHOMe$	Ether–olefin
$RCHF_2$	Halide–halide
$RCH(Br)CH_2F$	Halide–halide
$RCH(OAc)CH_2OH$	Alcohol–ester
$RCH(OH)CO_2Me$	Alcohol–ester
$RCH{=}CHCH_2CO_2Me$	Ester–olefin
$RCH{=}CHOAc$	Ester–olefin
$RCH(OMe)CH_2SO_2CH_2CH_2OH$	Alcohol–ether

How to Use the Book to Locate Examples of the Preparation of Difunctional Compounds. The difunctional index on p. xiv gives the section and page corresponding to each difunctional product. Thus Section 327 (alcohol, thiol–ester) contains examples of the preparation of hydroxyesters; Section 323 (alcohol, thiol–alcohol, thiol) contains examples of the preparation of diols.

Some preparations of olefinic and acetylenic compounds from olefinic and acetylenic starting materials can, in principle, be classified in either the monofunctional or difunctional sections; for example, $RCH{=}CHBr \rightarrow RCH{=}CHCOOH$, carboxylic acids from halides (Section 25, monofunctional compounds) or carboxylic acid–olefin (Section 322, difunctional compounds). In such cases both sections should be consulted.

Reactions applicable to both aldehyde and ketone starting materials are in many cases illustrated by an example that uses only one of them.

Many literature preparations of difunctional compounds are extensions of the methods applicable to monofunctional compounds. Thus the reaction RCl $\rightarrow$ ROH can be extended to the preparation of diols by using the corresponding dichloro compound as a starting material. Such methods are not fully covered in the difunctional sections.

The user should bear in mind that the pairs of functional groups alcohol, ester; carboxylic acids, ester; amine, amide; and carboxylic acid, amide can be interconverted by simple reactions. Compounds of the type RCH(OAc)CH$_2$OAc (ester–ester) would thus be of interest to anyone preparing the diol RCH(OH)CH$_2$OH (alcohol–alcohol).

Compendium of Organic
Synthetic Methods

CHAPTER 1
PREPARATION OF ACETYLENES

SECTION 1: Acetylenes from Acetylenes

$$HC \equiv C - \underset{\underset{CH_3}{|}}{\overset{\overset{CH_3}{|}}{C}} - OH \xrightarrow[\begin{array}{c} BnEt_3NCl/OH^-/RT \\ 2. \ PhBr, \ 80°C, \ 50 \ h \end{array}]{\begin{array}{c} 1. \ 2\text{-iodothiophene} \\ Pd[PPh_3]_4, \quad CuI \end{array}} Ph-C \equiv C \underset{S}{\diagdown\diagup}$$

80%

Carpita, A.; Lessi, A.; Rossi, R.[*]   **Synthesis**, (1984), 571

$$Ph-C \equiv CH \ + \ Ph-X \xrightarrow[NEt_3]{[PPh_3]_2PdX'_2} Ph-C \equiv C-Ph$$

$CHCl_3$-CuI; X = I; X' = Ph, I   77%

Bumagin, N.A.; Ponomarev, A.B.; Beletskaya, I.P.
                    Bull Acad Sci USSR, (1984), **33**, 1433

DMF; X = Tf; X' = Cl, Cl   91%

Chen, Q-Y.[*]; Yang, Z-Y. **Tetrahedron Lett**, (1986), **27**, 1171

$$\underset{Ph}{\overset{Ph}{\diagdown\diagup}}C=C=O \ + \ HC \equiv C-\underline{n}Pr \xrightarrow[THF, \ 120°C]{5\% \ Pd[PPh_3]_4} Ph_2CHC \equiv C-\underline{n}Pr$$

81%

Mitsudo, T.; Kadokura, M.; Watanabe, Y.[*]
                    Tetrahedron Lett, (1985), **26**, 3697

$$Me_3Si-C\equiv C-SiMe_3 \xrightarrow[\text{CH}_2\text{Cl}_2, \ -17°C]{\begin{array}{c} 2 \ \underline{t}Bu-Cl \\ AlCl_3 \end{array}} \underline{t}Bu-C\equiv C-\underline{t}Bu$$

70%

Capozzi, G.[*]; Ottana, R.; Romeo, G.; Marcuzzi, F.
Gazz Chim Ital, (1985), **115**, 311

$$\underline{n}Bu-C\equiv CH \xrightarrow{\begin{array}{l} 1. \ \underline{n}BuLi, \ RT \\ 2. \ I_2, \ -78°C \\ 3. \ R_2B-C_{Me}^{Et}{}_{,,,,}H \\ 4. \ NaOH/H_2O_2 \end{array}}$$

$$\underline{n}Bu-C\equiv C-C_{Me}^{Et}{}_{,,,,}H$$

57%

R = (structure)

(95%ee,R)

Brown, C.A.[*]; Desai, M.C.; Jadhav, P.K.
J Org Chem, (1986), **51**, 162

$$HC\equiv CCH_2OTHP \xrightarrow{\begin{array}{l} 1. \ MeLi, \ THF-Et_2O \\ \quad 0°C, \ 15 \ min \\ 2. \ \underline{n}C_{10}H_{25}Br \\ 3. \ DMSO, \ 0- \ RT, \ 3h \\ 4. \ HOH \end{array}} \underline{n}C_{10}H_{25}C\equiv CCH_2OTHP$$

84%

Chong, J.M.[*]; Wong, S.   **Tetrahedron Lett**, (1986), **27**, 5445

$$\text{(borane structure with } C_8H_{17} \text{ and } OMe) \xrightarrow{\begin{array}{l} 1. \ LiC\equiv CC_4H_9, \ 0°C \\ 2. \ I_2 \\ 3. \ ox. \end{array}} C_8H_{17}C\equiv CC_4H_9$$

75%

Sikorski, J.A.; Bhat, N.G.; Cole, T.E.; Wang, K.K.; Brown, H.C.[*]
J Org Chem, (1986), **51**, 4521

SECTION 2: Acetylenes from Acid Derivatives

$$\underset{\underset{Ph-C-Cl}{\overset{\|}{O}}}{} \quad \xrightarrow{\begin{array}{l}1.\ Li/Hg,\quad THF\\[6pt]2.\ Li/Hg\end{array}} \quad Ph-C\equiv C-Ph$$

51%

Horner, L.[*]; Dickerhof, K. **Chem Ber**, (1983), **116**, 1615

$$2\ Ph_3PEt^+\ Br^- \quad \xrightarrow{\begin{array}{l}1.\ 2\ \underline{n}BuLi\\[4pt]2.\ Et\underset{\overset{\|}{O}}{C}-Cl\\[4pt]3.\ FVP\ (750°C)\end{array}} \quad Et-C\equiv C-Et$$

80%

Aitken, R.A.[*]; Atherton, J.I. **JCS Chem Comm**, (1985), 1140

SECTION 3: Acetylenes from Alcohols and Thiols

No Additional Examples

SECTION 4: Acetylenes from Aldehydes

$$PhCH_2SO_2Ph \quad \xrightarrow{\begin{array}{l}1.\ \underline{n}BuLi,\ THF,\ -78°C\\[4pt]2.\ PhCHO\\[4pt]3.\ Ac_2O-Py\\[4pt]4.\ \underline{t}BuOK/THF\end{array}} \quad Ph-C\equiv C-Ph$$

86%

Mandai, T.[*]; Yanagi, T.; Araki, K.; Morisaki, Y.; Kawada, M.; Otera, J.

J Am Chem Soc, (1984), **106**, 3670

SECTION 5: Acetylenes from Alkyl, Methylenes, and Aryls

No Additional Examples

SECTION 6: Acetylenes from Amides

No Additional Examples

SECTION 7: Acetylenes from Amines

No Additional Examples

SECTION 8: Acetylenes from Esters

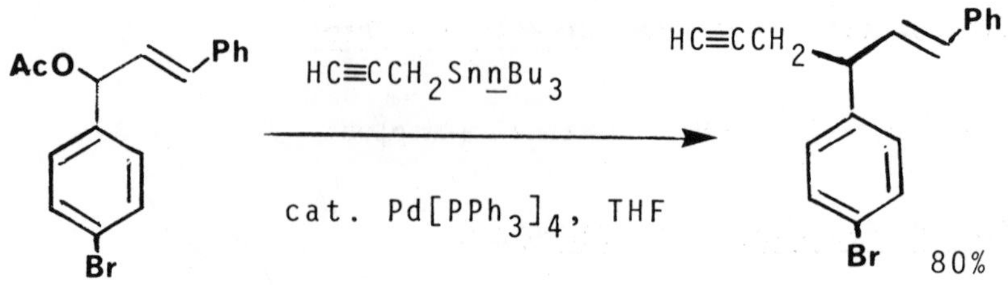

$HC≡CCH_2SnnBu_3$

cat. $Pd[PPh_3]_4$, THF

Keinan, E.[*], Peretz, M.   **J Org Chem**, (1983), **48**, 5302

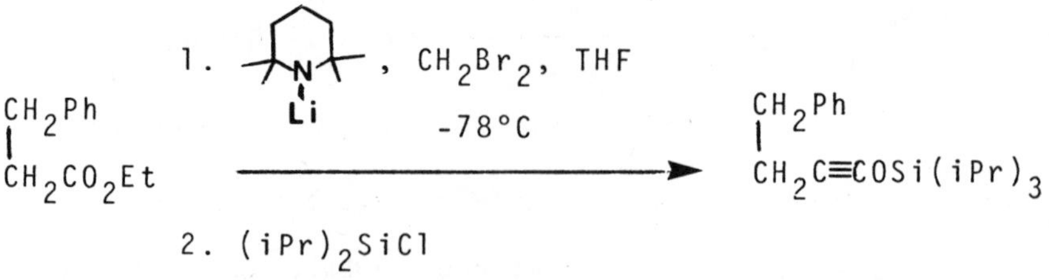

1. [piperidine structure], CH_2Br_2, THF
   $-78°C$

2. $(iPr)_2SiCl$

$\begin{array}{l} CH_2Ph \\ | \\ CH_2CO_2Et \end{array}$

$\begin{array}{l} CH_2Ph \\ | \\ CH_2C≡COSi(iPr)_3 \end{array}$

66%

Kowalski, C.J.[*], Lal, G.S., Haque, M.S.
                    J Am Chem Soc, (1986), **108**, 7127

SECTION 9: Acetylenes from Ethers, Epoxides, and Thioethers

No Additional Examples

SECTION 10: Acetylenes from Halides and Sulfonates

$$PhCH-CH_2Br \xrightarrow[\text{THF}]{2\ Ph_2P-Li} Ph-C \equiv CH$$

$$|$$
$$Br$$

71%

Gillespie, D.G.; Walker, B.J.[*]   JCS **Perkin I**, (1983), 1689

SECTION 11: Acetylenes from Hydrides

No Additional Examples

For examples of the reaction $RC \equiv CH \rightarrow RC \equiv C-C \equiv CR'$ see section 300 (Acetylene - Acetylene)

SECTION 12: Acetylenes from Ketones

$$2\ Ph_3P=CH\overset{\overset{O}{\|}}{C}Ph \xrightarrow[\substack{3.\ \underline{n}BuLi/THF \\ 4.\ CH_3I}]{\substack{1.\ PhSeBr \\ 2.\ 210°C}} Ph-C \equiv C-CH_3$$

45% overall

Braga, A.L.; Comasseto, J.V.[*]; Petragnani, N.
                Synthesis, (1984), 240

$$
\begin{array}{c}
O \\
\| \\
Ph-C-CH_3
\end{array}
\quad
\xrightarrow{\begin{array}{l} 1.\ LDA,\ THF \\ \quad\quad\quad O \\ \quad\quad\quad \| \\ 2.\ ClP(OEt)_2 \\ 3.\ 2.25\ LDA \\ 4.\ HOH \end{array}}
\quad
Ph-C\equiv CH
$$

80%

Negishi, E.[*]; King, A.O.; Tour, J.M. **Org Syn**, (1985), **64**, 44

SECTION 13: Acetylenes from Nitriles

No Additional Examples

SECTION 14: Acetylenes from Olefins

$$
CH_3CH_2CH=CHnBu \xrightarrow[\text{aq. EtOH, 70°C}]{Pd[OAc]_2 \cdot OPTA} Et-C\equiv C-nBu
$$

98%

OPTA = oligo-p-phenyleneterephthalamide

Cum, G.[*]; Gallo, R.; Ipasli, S.; Spadaro, A.
JCS Chem Comm, (1985), 1571

$$
CH_2=C=CH_2 \xrightarrow[\text{2. HOH}]{1.\ 2\ nBuLi} C_7H_{15}CH_2-C\equiv CH
$$

88%

Hooz, J.[*]; Calzada, J.G.; McMaster, D.
Tetrahedron Lett, (1985), **26**, 271

$$Me_2C=CH(CH_2)_{14}CH_3 \xrightarrow[\text{0 - 60°C}]{\text{NaNO}_2, \text{ AcOH}} CH_3(CH_2)_{14}C \equiv CCH_3$$

67%

Abidi, S.L.[*]   **Tetrahedron Lett**, (1986), **27**, 267

SECTION 15: Acetylenes from Miscellaneous Compounds

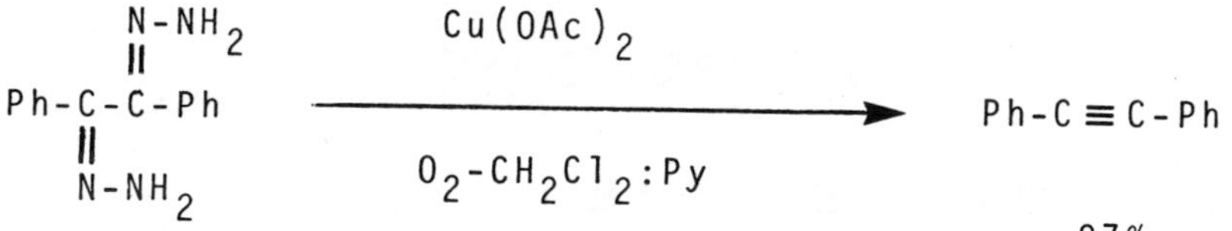

$$\underset{\overset{\displaystyle \|}{N-NH_2}}{\overset{\displaystyle N-NH_2}{Ph-C-C-Ph}} \xrightarrow[\text{O}_2-CH_2Cl_2:Py]{Cu(OAc)_2} Ph-C \equiv C-Ph$$

97%

Tsuji, J.[*]; Kezuka, H.; Toshida, Y.; Takayanagi, H.; Yamamoto, K.

Tetrahedron, (1983), **39**, 3279

SECTION 15A: Protection of Acetylenes

No Additional Examples

CHAPTER 2
PREPARATION OF CARBOXYLIC ACIDS, ACID HALIDES, AND ANHYDRIDES

SECTION 16: Acid Derivatives from Acetylenes

$$nC_6H_{11}-C{\equiv}CH \quad \xrightarrow{\begin{array}{l}1.\ nBuLi,\ THF \\ 2.\ PhSSPh \\ \\ 3.\ HOH \\ 4.\ HgSO_4,\ H_2SO_4 \\ \quad AcOH\end{array}} \quad nC_6H_{11}CH_2CO_2H$$

67%

Abrams, S.R.[*] **Can J Chem**, (1983), **61**, 2423

SECTION 17: Acid Derivatives from Acid Derivatives

$$2\ PhCO_2H \quad \xrightarrow[NEt_3,\ -30°\ \to\ 0°C]{Me_2N-\overset{\overset{\displaystyle Cl}{|}}{C}=NMe_2{}^+Cl^-} \quad Ph\overset{\overset{\displaystyle O}{||}}{C}-)_2O$$

88%

Fujisawa, T.[*]; Tajima, K.; Sato, T.
                    Bull Chem Soc Jpn, (1983), **56**, 3529

$$Ph\overset{\overset{\displaystyle O}{||}}{C}Cl \quad \xrightarrow[\text{2. KOTMS}]{\text{1. KOTMS, } Et_2O} \quad Ph\overset{\overset{\displaystyle O}{||}}{C}O^-K^+$$

(anhydrous)

Laganis, E.D.[*]; Chenard, B.L.
                    Tetrahedron Lett, (1984), **25**, 5831

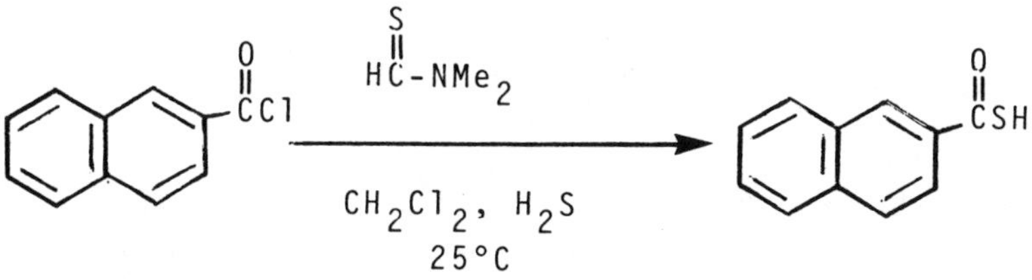

90%

Kobayashi, Y.[*]; Itabashi, K.[*]   **Synthesis**, (1985), 671

SECTION 18: Acid Derivatives from Alcohols and Thiols

No Additional Examples

SECTION 19: Acid Derivatives from Aldehydes

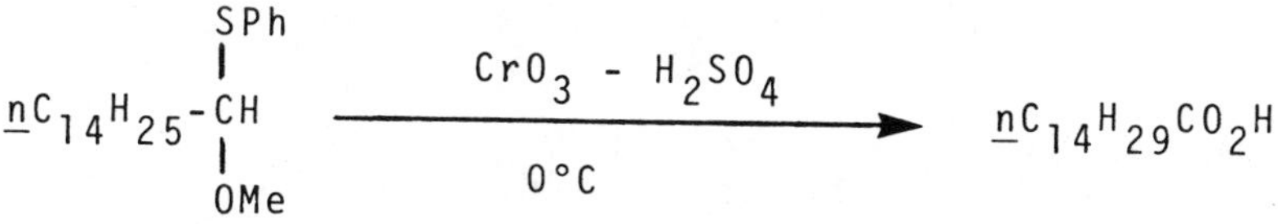

78%

Mandai, T.; Hara, K.; Nakajima, T.; Kawada, M.; Otera, J.[*]
                    Tetrahedron Lett, (1983), **24**, 4993

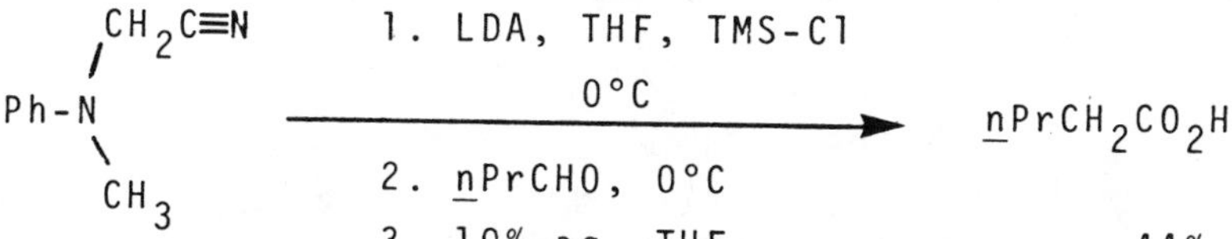

Takahashi, K.[*]; Shibasaki, K.; Agura, K.; Iida, H.
                    J Org Chem, (1983), **48**, 3566

nC_3H_7—CH=CH—CHO
$\xrightarrow[\begin{array}{c}2.\ Na_2SO_3 \quad 3.\ aq.\ HCl\end{array}]{\begin{array}{c}1.\ NaClO_2-H_2O_2,\\ CH_3CN,\ aq.\\ NaH_2PO_4\end{array}}$
nC_3H_7—CH=CH—CO_2H

88%

Dalcanale, E.[*]; Mantanari, F.[*] J Org Chem, (1986), 51, 567

$HCCH_2$⟨dioxolane⟩
$\xrightarrow[5\%\ NaH_2PO_4]{\begin{array}{c}KMnO_4,\ 10\ min\\ tBuOH\end{array}}$
$HOCCH_2$⟨dioxolane⟩

98%

Abiko, A.; Roberts, J.C.; Takemasa, T.; Masamune, S.[*]
Tetrahedron Lett, (1986), 27, 4537

$\begin{array}{c}CH_2Ph\\ |\\ CH_2CHO\end{array}$
$\xrightarrow[\begin{array}{c}2.\ piperidine,\ HOH,\ 90°C\\ 3.\ aq.\ HCl\end{array}]{\begin{array}{c}1.\ TrClO_4,\ tBuOOTMS\\ CH_2Cl_2,\ -45°C\end{array}}$
$\begin{array}{c}CH_2Ph\\ |\\ CH_2CO_2H\end{array}$

73%

Mukaiyama, T.; Miyoshi, N.; Kato, J.; Ohshima, M.
Chem Lett, (1986), 1385

Related methods: Carboxylic Acids from Ketones (Section 27)
    Also via: Esters (Section 109)

SECTION 20: Acid Derivatives from Alkyl, Methylenes, and Aryls

⟨structure with Ph⟩
$\xrightarrow[NaIO_4,\ 35°C]{\begin{array}{c}cat.\ cis-\\ [Rh(bpy)_2Cl_2]\\ 2\ HOH,\ aq.\ CH_3CN\end{array}}$
⟨structure with HOOC⟩

84%

Chakraborti, A.K.; Ghatak, U.R.[*] JCS Perkin I, (1985), 2605

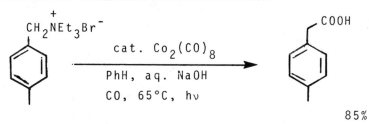

84%

Sasson, Y.[*]; Zappi, G.D.; Neumann, R.[*]
J Org Chem, (1986), **51**, 2880

SECTION 21: Acid Derivatives from Amides

No Additional Examples

SECTION 22: Acid Derivatives from Amines

85%

Brunet, J.J.; Sidot, C.; Caubere, P.[*]
J Org Chem, (1983), **48**, 1919

SECTION 23: Acid Derivatives from Esters

Me_2CHCO_2Me $\xrightarrow[\text{CH}_3\text{CN, reflux}]{\text{MeSiCl}_3, \text{ NaI}}$ Me_2CHCO_2H

70%

Olah, G.A.[*]; Husain, A.; Singh, B.P.; Mehrota, A.K.
J Org Chem, (1983), **48**, 3667

$$\underset{\bigcirc}{\overset{\overset{O}{\parallel}}{C-S\underline{t}Bu}} \xrightarrow[\text{Bu}_4\text{NBr, aq. CH}_3\text{CN}]{\overset{e^-}{\text{Pt electrodes}}} \underset{\bigcirc}{COOH}$$

94%

Kimura, M.; Matsubara, S.; Sawaki, Y.
                    JCS Chem Comm, (1984), 1619

$$\overset{O}{\underset{\parallel}{PhCOCH_2CH=CH_2}} \xrightarrow[\substack{EtOH, \ AcOH, \ 2h \\ reflux}]{Te/NaBH_4} \overset{O}{\underset{\parallel}{PhCOH}}$$

95%

Shobara, N.; Shanmugam, P.[*]   **Ind J Chem B**, (1986), **25B**, 658

Other reactions useful for the hydrolysis of esters may be found in Section 30A (Protection of Carboxylic Acids).

SECTION 24: Acid Derivatives from Ethers, Epoxides, and Thioethers

No Additional Examples

SECTION 25: Acid Derivatives from Halides and Sulfonates

$$\underset{Cl}{\overset{Cl}{CH_2=C}} \xrightarrow[\text{aq. } H_2SO_4, \ 2°C]{\underline{t}BuOH} \underline{t}BuCH_2CO_2H$$

90%

Randriamahefa, S.; Deschamps, P.; Gallo, R.[*]
                    Synthesis, (1985), 493

$$\begin{array}{c} R \\ | \\ PhCH \\ \backslash \\ Br \end{array} \xrightarrow{\;\;M(CO)_n,\; 1\ atm\ CO\;\;} \begin{array}{c} R \\ | \\ Ph-CH-CO_2H \end{array}$$

aq. benzene; R=H; M=Fe; n=5   75%

Tanguay, G.; Weinberger, B.; des Abbayes, H.[*]
                              Tetrahedron Lett, (1983), **24**, 4005

EtOH/NaOH; R=Me; M=Co; n=4   80%

Francalanci, F.; Gardano, A.; Foa, M.
                              J Organomet Chem, (1985), **282**, 277

$$\begin{array}{c} \text{Br} \\ \langle\!\!\!\bigcirc\!\!\!\rangle_O \end{array} \xrightarrow[\substack{Mg\ anode,\ 50°C \\ Bu_4NClO_4}]{\substack{CO_2,\ THF-HMPA \\ e^-}} \begin{array}{c} \text{COOH} \\ \langle\!\!\!\bigcirc\!\!\!\rangle_O \end{array}$$

                                               78%

Sock, O.; Troupel, M.; Perichon, J.
                              Tetrahedron Lett, (1985), **26**, 1509

$$PhI \xrightarrow[\substack{e^-,\ Pt\ electrode \\ DMF,\ Et_4NOTs,\ CO_2}]{\substack{cat.\ Pd[PPh_3]_4,\ PPh_3}} PhCO_2^-K^+$$

                                               92%

Torii, S.[*]; Tanaka, H.; Hamatani, T.; Morisaki, K.; Jutand, A.;
Pfluger, F.; Fauvarque, J.-F.
                    Chem Lett, (1986), 169

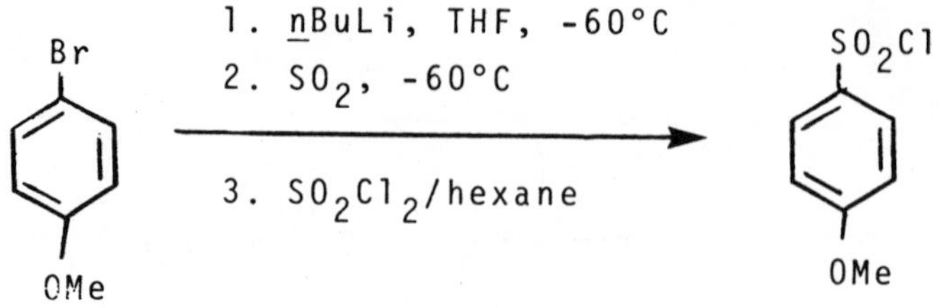
Br
1. nBuLi, THF, -60°C
2. SO2, -60°C
3. SO2Cl2/hexane
SO2Cl
OMe
OMe

95%

Hamada, T.*; Yonemitsu, O. **Synthesis**, (1986), 852

SECTION 26: Acid Derivatives from Hydrides

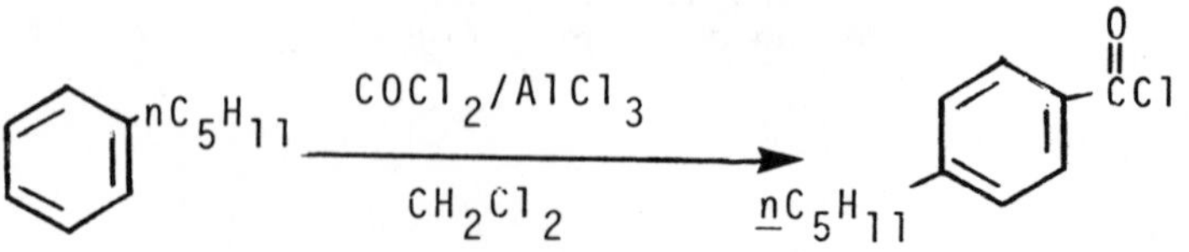
nC5H11
COCl2/AlCl3
CH2Cl2
O
CCl
nC5H11

55%

Neubert, M.E.; Fishel, D.L. **Org Syn**, (1983), **61**, 8

SECTION 27: Acid Derivatives from Ketones

$$PhC(=O)CH_2NEt_2 \xrightarrow[EtOH,\ 0°C]{10\ 30\%\ H_2O_2} PhCO_2H$$

89%

Wenkert, D.*; Eliasson, K.M.; Rudisill, D.
JCS Chem Comm, (1983), 392

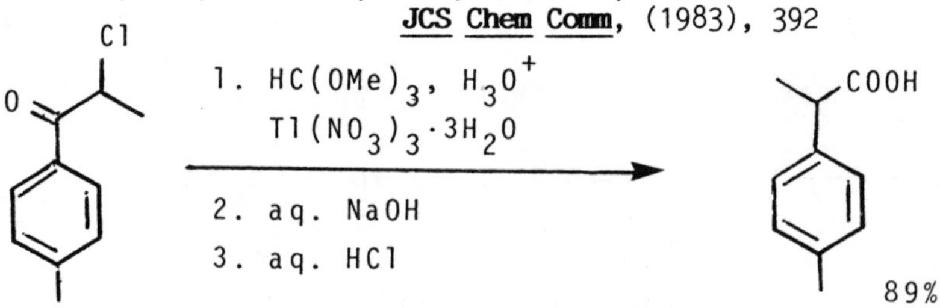
Cl
O
1. HC(OMe)3, H3O+
Tl(NO3)3·3H2O
2. aq. NaOH
3. aq. HCl
COOH

89%

Fujii, K.; Nakao, K.; Yamauchi, T.* **Synthesis**, (1983), 444

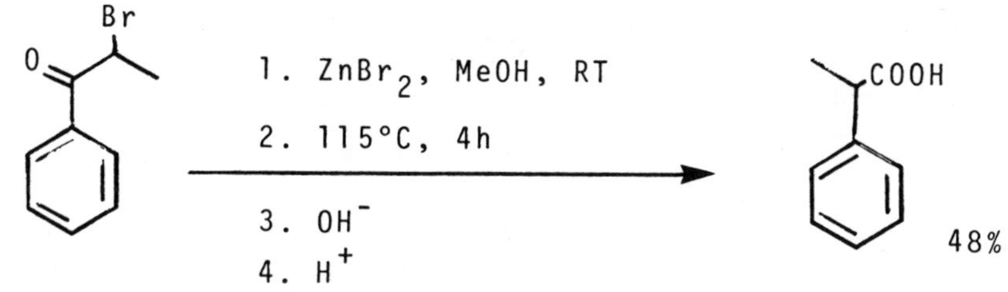
Br
O
1. ZnBr2, MeOH, RT
2. 115°C, 4h
3. OH⁻
4. H⁺
COOH
48%

Giordano, C.*; Castaldi, G.; Uggeri, F.; Gurzoni, F. **Synthesis**, (1985), 436

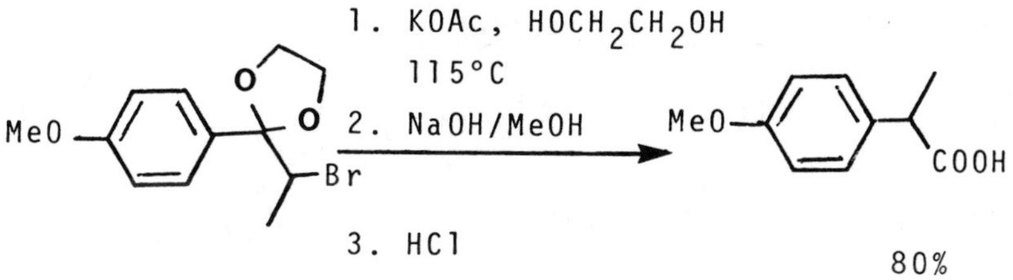
1. KOAc, HOCH2CH2OH
115°C
2. NaOH/MeOH
3. HCl
MeO
O
O
Br
MeO
COOH
80%

Castaldi, G.*; Giordano, C.; Uggeri, F. **Synthesis**, (1985), 505

$$Me_2CHCH_2C(=O)CH_3 \xrightarrow[\text{2. } H_3O^+]{\text{1. } NaBrO_2/NaBr,\ \text{aq. } NaOH,\ 0°C} Me_2CHCH_2CO_2H \quad 95\%$$

Kajigaeshi, S.*; Nakagawa, T.; Nagasaki, N.; Fujisaka, S. **Synthesis**, (1985), 674

SECTION 28: Acid Derivatives from Nitriles

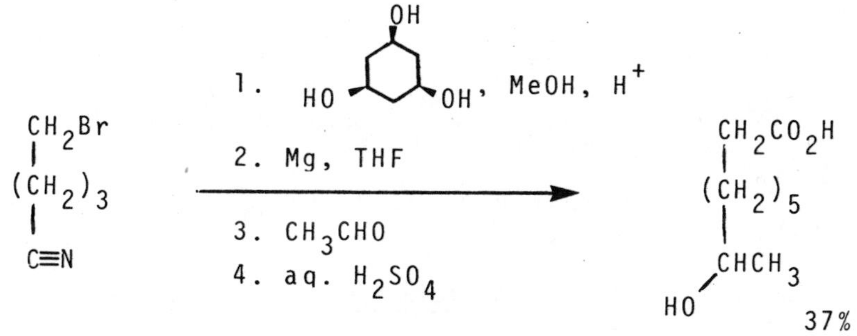
OH
1. HO OH, MeOH, H⁺
2. Mg, THF
3. CH3CHO
4. aq. H2SO4
CH2Br
(CH2)3
C≡N
CH2CO2H
(CH2)5
CHCH3
HO
37%

Voss, G.; Gerlach, H.* **Helv Chim Acta**, (1983), **66**, 2294

$$\underline{n}C_3H_7 \overset{C\equiv N}{\underset{C\equiv N}{C=C}} \xrightarrow[\substack{2. \text{ aq. } H_2SO_4 \\ 180°C}]{\substack{1. \ \underline{c}C_6H_{13} \quad NaBH_4 \\ \text{aq. } CH_2Cl_2, \ 20°C}} \overset{H_7C_3}{\underset{\underline{c}C_6H_{13}}{CHCH_2CO_2H}} \quad 45\%$$

Giese, B.[*]; Harnisch, H.; Lachhein, S.
                        Synthesis, (1983), 733

SECTION 29: Acid Derivatives from Olefins

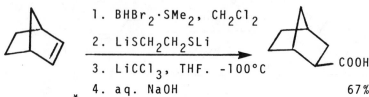

1. $BHBr_2 \cdot SMe_2$, CH_2Cl_2

2. $LiSCH_2CH_2SLi$

3. $LiCCl_3$, THF. $-100°C$

4. aq. NaOH   67%

Brown, H.C.[*]; Imai, T.   **J Org Chem**, (1984), **49**, 892

$$Cl_2C=CHSO_2\underline{n}Bu \xrightarrow{\text{aq. NaOH}} HO_2CCH_2SO_2\underline{n}Bu$$

                                        80%

Mirskova, A.N.; Kryukova, Yu.I.; Levkovskaya, G.G.; Guseva, S.A.; Voronkvo, M.G.
                        J Org Chem USSR, (1984), **20**, 545

SECTION 30: Acid Derivatives from Miscellaneous Compounds

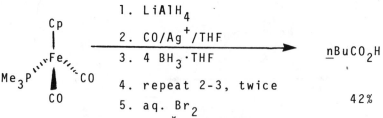

1. $LiAlH_4$

2. $CO/Ag^+/THF$

3. $4 \ BH_3 \cdot THF$

4. repeat 2-3, twice

5. aq. Br_2

$\underline{n}BuCO_2H$

42%

Brown, S.L.; Davies, S.G.[*]   **JCS Chem Comm**, (1986), 84

Review: "Synthesis of Dithiocarboxylic Acids"

Ramadas, S.R.; Srinivasan, P.S.; Ramachandran, J.; Sastry, V.V.S.K.

$$\text{Synthesis, (1983), 605}$$

SECTION 30A: Protection of Carboxylic Acid Derivatives

Kessler, H.[*]; Siegmeier, R.   **Tetrahedron Lett**, (1983), **24**, 281

Kunz, H.[*]; Waldmann, H.   **Angew Chem Int Ed Engl**, (1983), **22**, 62

85%

Martinez, J.[*]; Laur, J.; Castro, B.

$$\text{Tetrahedron Lett, (1983), 24, 5219}$$

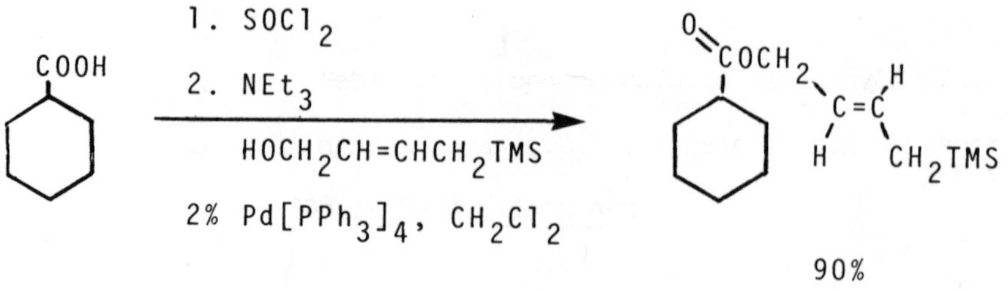
COOH
1. SOCl2
2. NEt3
HOCH2CH=CHCH2TMS
2% Pd[PPh3]4, CH2Cl2
O
COCH2
H
C=C
H
CH2TMS
90%

Mastalerz, H.* **J Org Chem**, (1984), **49**, 4092

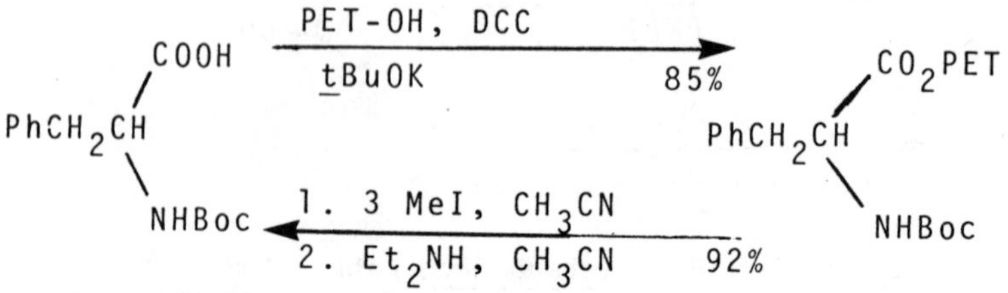
COOH
PhCH2CH
NHBoc
PET-OH, DCC
tBuOK
85%
CO2PET
PhCH2CH
NHBoc
1. 3 MeI, CH3CN
2. Et2NH, CH3CN
92%

PET = 2(2-pyridyl)ethyl

Kessler, H.*; Becker, G.; Kagler, H.; Wolff, M. **Tetrahedron Lett**, (1984), **25**, 3971

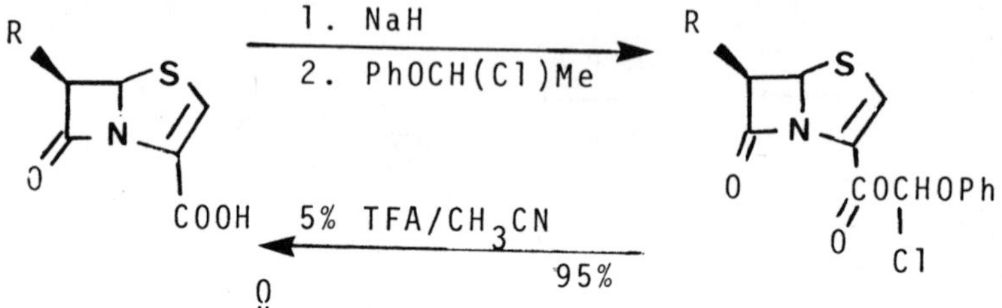
R
S
N
O
COOH
1. NaH
2. PhOCH(Cl)Me
5% TFA/CH3CN
95%
R
S
N
O
COCHOPh
O
Cl

R = $PhOCH_2C(=O)NH-$

Alpegiani, M.; Bedeschi, A.; Foglio, M.; Perrone, E. **Gazz Chim Ital**, (1984), **114**, 391

Other reactions useful for the protectio of carboxylic acids are included in Section 107 (Esters from Carboxylic Acids and Acid Halides) and Section 23 (Carboxylic Acids from Esters).

CHAPTER 3

PREPARATION OF ALCOHOLS, PHENOLS, AND THIOLS

SECTION 31: Alcohols and Thiols from Acetylenes

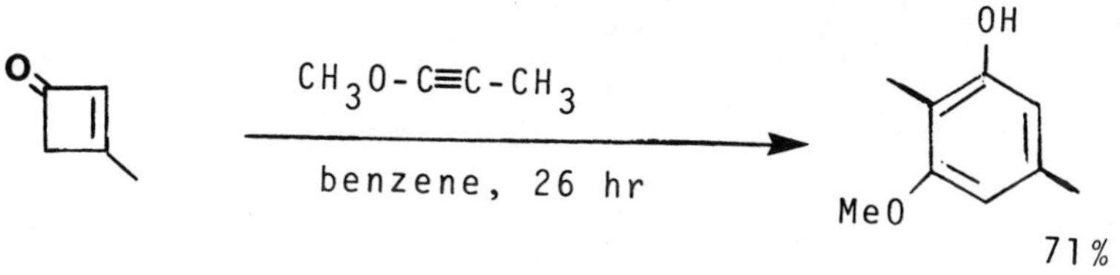

O
$CH_3O-C{\equiv}C-CH_3$
benzene, 26 hr
OH
MeO
71%

Danheiser, R.L.*; Gee, S.K. J Org Chem, (1984), 49, 1672

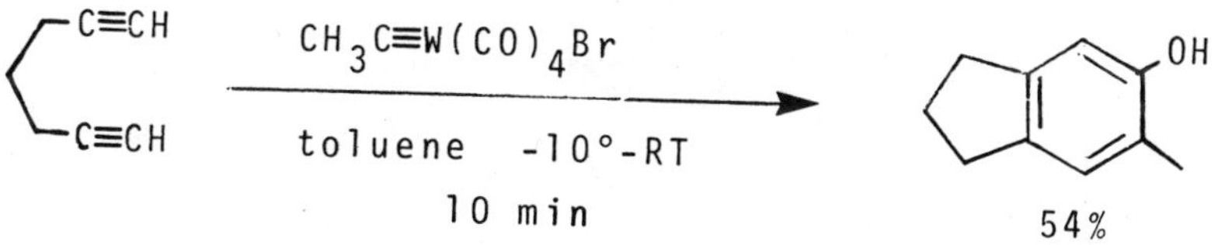

$C{\equiv}CH$
$C{\equiv}CH$
$CH_3C{\equiv}W(CO)_4Br$
toluene -10°-RT
10 min
OH
54%

Sivavec, T.M.; Katz, T.J.* Tetrahedron Lett, (1985), 26, 2159

SECTION 32: Alcohols and Thiols from Acid Derivatives

$$CH_3(CH_2)_4\overset{O}{\overset{\|}{C}}Cl \xrightarrow[MnCl_2\cdot 4H_2O,\ THF]{\underline{n}Bu_4N^+\ B_3H_8^-,\ RT} CH_3(CH_2)_4CH_2OH$$

B_3H_8 = octahydrotriborate 61%

Tamblyn, W.H.*; Aquadro, R.E.; DeLuca, O.D.; Weingold, D.H.; Dao, T.V.

Tetrahedron Lett , (1983), 24, 4953

$$\text{Br(CH}_2)_5\text{CO}_2\text{H} \xrightarrow[\substack{2. \text{ NaBH}_4, \ -18°C \\ 2 \text{ h}}]{\substack{1. \ \text{ClC-CCl/DMF} \\ \text{CH}_3\text{CN/THF}, \ -30°C}} \text{Br(CH}_2)_5\text{CH}_2\text{OH}$$

93%

Fujisawa, T.*; Mori, T.; Sato, T.   **Chem Lett**, (1983), 835

1. Zn(BH$_4$)$_2$, THF, 0°C
   TMEDA, 30 min

2. aq. HCl

93%

Kotsuki, H.*; Ushio, Y.; Yoshimura, N.; Ochi, M.
                    Tetrahedron Lett, (1986), **27**, 4213

SECTION 33: Alcohols and Thiols from Alcohols and Thiols

PhCHO

OH   62%

2 nBuLi

Et-X

92%

Braun, M.*; Ringer, E.   **Tetrahedron Lett**, (1983), **24**, 1233

$$\text{CH}_3(\text{CH}_2)_8\text{CH}_2\text{CH}_2\text{OH} \xrightarrow[\substack{200°C, \ 6 \text{ h} \\ 57 \text{ Bar}}]{\text{CH}_3\text{OH}, \ \text{Ni(R)}} \text{CH}_3(\text{CH}_2)_8\overset{\text{CH}_3}{\overset{|}{\text{C}}}\text{HCH}_2\text{OH}$$

72%

Sabadie, J.; Descotes, G.   **Bull Chem Soc Fr**, (1983), II253

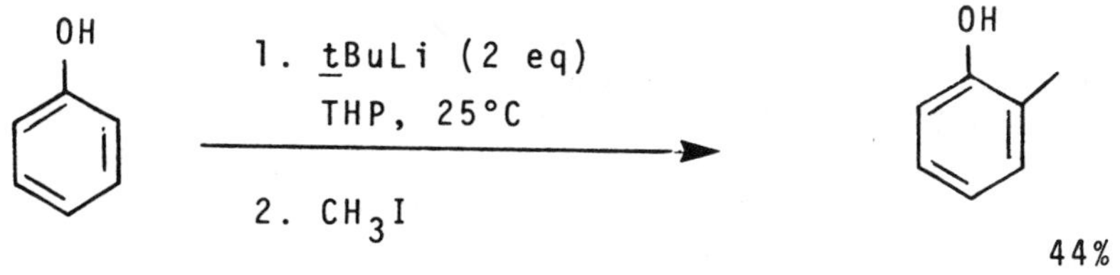
OH
1. tBuLi (2 eq)
THP, 25°C
2. CH3I
OH
44%

Posner, G.H.*; Canella, K.A. **J Am Chem Soc**, (1985), 107, 2571

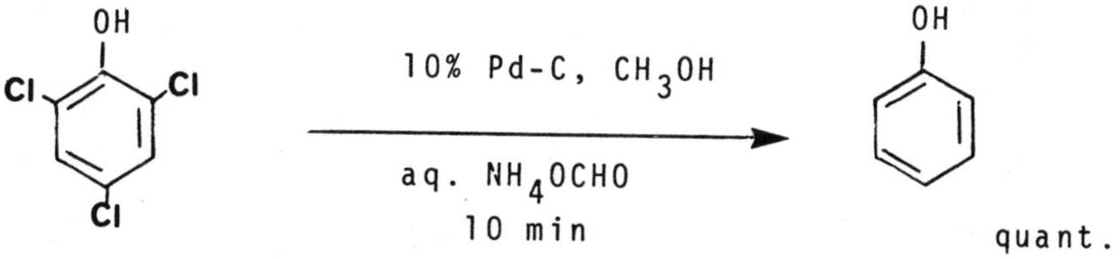
OH
Cl
Cl
Cl
10% Pd-C, CH3OH
aq. NH4OCHO
10 min
OH
quant.

Anwer, M.K.; Spatola, A.F. **Tetrahedron Lett**, (1985), 26, 1381

SECTION 34: Alcohols and Thiols from Aldehydes

The following reaction types are included in this section:

A. Reductions of Aldehydes to Alcohols.

B. Alkylation of Aldehydes, forming Alcohols.

Coupling of Aldehydes to form Diols is found in Section 323 (Alcohol - Alcohol).

Section 34A: Reductions of Aldehydes to Alcohols

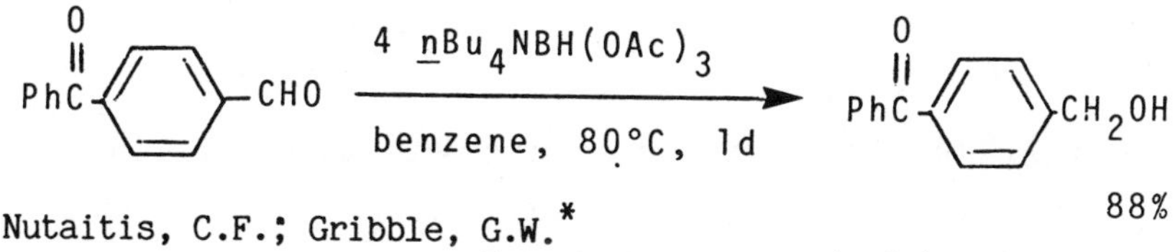
O
PhC
CHO
4 nBu4NBH(OAc)3
benzene, 80°C, 1d
O
PhC
CH2OH
88%

Nutaitis, C.F.; Gribble, G.W.*
Tetrahedron Lett, (1983), 24, 4287

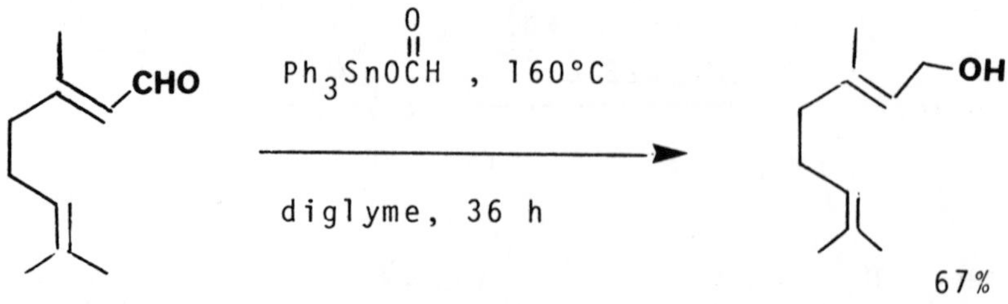
CHO
O
Ph_3SnOCH , 160°C
diglyme, 36 h
OH
67%

Wuest, J.D.*; Zacharie, B. J Org Chem, (1984), 49, 166

$$CH_3(CH_2)_4CHO \xrightarrow[\text{5 h}]{\text{BER, EtOH, 25°C}} CH_3(CH_2)_4CH_2OH$$

99%

BER = borohydride exchange resin

Yoon, N.M.*; Park, K.B.; Gyoung, Y.S.
Tetrahedron Lett, (1983), 24, 5367

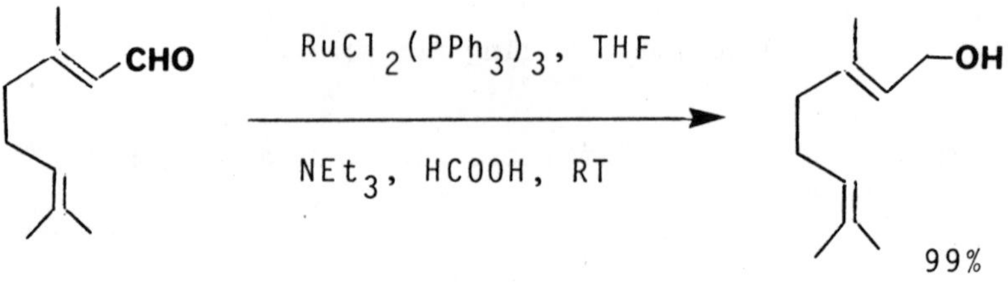
CHO
$RuCl_2(PPh_3)_3$, THF
NEt_3, HCOOH, RT
OH
99%

Khai, B.T.; Arcelli, A. Tetrahedron Lett, (1985), 26, 3365

$$CH_3(CH_2)_4CHO \xrightarrow[\text{HOH, iPrOH, reflux, 1 h}]{ZnOCl_2 \cdot 8H_2O} CH_3(CH_2)_4CH_2OH$$

83%

Matsushita, H.*; Ishiguro, S.; Ichinose, H.; Izumi, A.; Mizusaki, S.
Chem Lett, (1985), 731

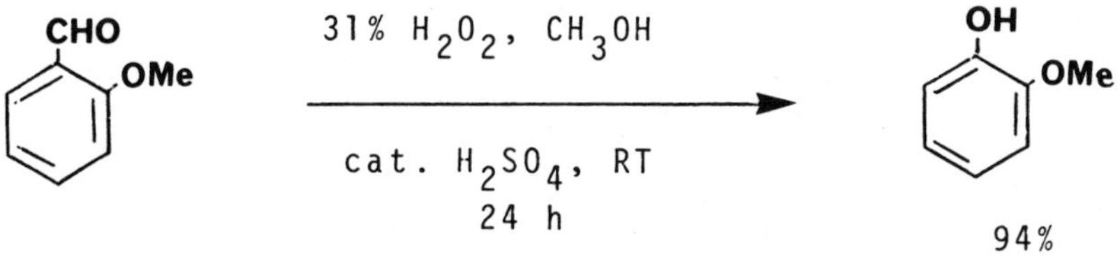

31% H_2O_2, CH_3OH

cat. H_2SO_4, RT
24 h

94%

Matsumoto, M.[*]; Kobayashi, H.; Hotta, Y.
J Org Chem, (1984), **49**, 4740

Review: "cis-Alkyl and cis-Acylrhodium and Iridium Hydrides:
Model Intermediates in Homogeneous Catalysis"

Milstein, D.[*]   **Accts Chem Res**, (1984), **17**, 221

Section 34B: Alkylation of Aldehydes, forming Alcohols

Aldol reactions are listed in : Section 324
(Aldehyde-Alcohol) and Section 330 (Ketone-Alcohol).

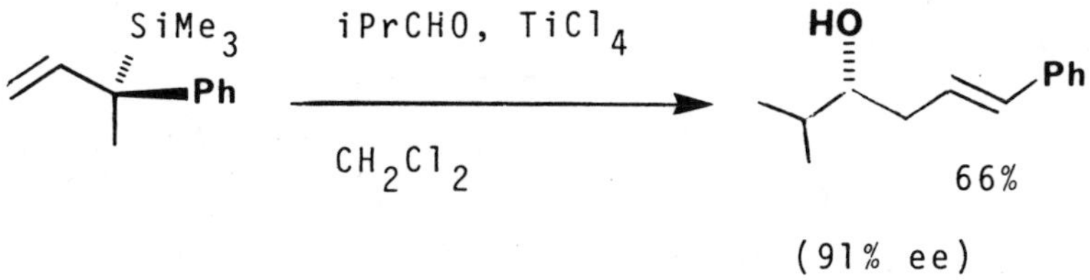

iPrCHO, $TiCl_4$

CH_2Cl_2

66%

(91% ee)

Hayashi, T.; Konishi, M.; Kumada, M.[*]
J Org Chem, (1983), **48**, 281

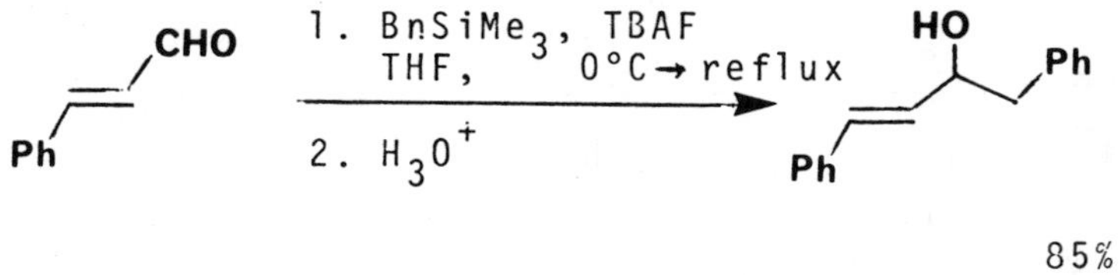

1. $BnSiMe_3$, TBAF
   THF, 0°C → reflux

2. H_3O^+

85%

Bennetau, B.; Dunogues, J.   **Tetrahedron Lett**, (1983), **24**, 4217

1. PhCHO, iPr_2NEt
   9BBN-OTf
   _____
   CH_2Cl_2/RT/17h
2. 30% H_2O_2/CH_3OH
   RT, 2h

78%

Hamana, H.[*]; Sugasawa, T.   **Chem Lett**, (1983), 333

$\underline{n}C_7H_{15}CHO$

1. BnBr, $2SmI_2$, THF
   2.5 min, RT
   _____
2. H_3O^+

OH
|
$\underline{n}C_7H_{15}CHBn$

86%

Souppe, J.; Danson, L.; Namy, J.L.; Kagan, H.B.[*]
                    J Organomet Chem, (1983), **250**, 227

1. CH_3CH_2CHO
   _____
2. H_2/Pt

(76% ee)

OH

+   [94]

OH

[6]

50%

Hoffmann, R.W.[*]; Landmann, B.
                    Angew Chem Int Ed Engl, (1984), **23**, 437

$\underline{n}C_5H_{11}CHO$

$CH_3Ti(OiPr)_3$, 0°C
_____
CH_2Cl_2, 15 h

OH
|
$\underline{n}C_5H_{11}CHCH_3$

95%

Reetz, M.T.[*]; Westermann, J.; Steinbach, R.; Wenderoth, B.;
Peter, R.; Ostarek, R.; Maus, S.
                    Chem Ber, (1985), **118**, 1421

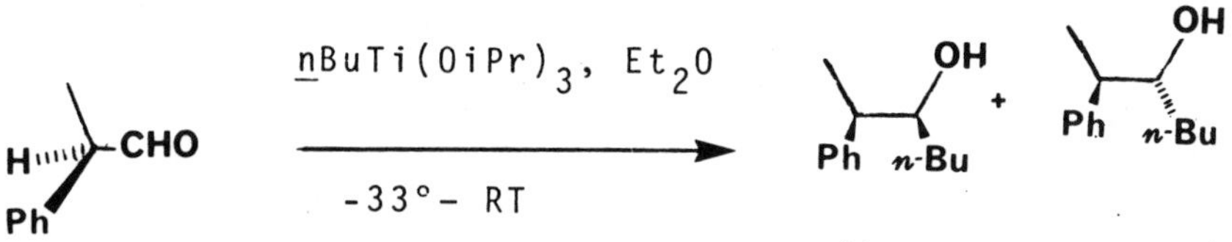

$$\underline{n}BuTi(OiPr)_3, \ Et_2O$$

-33° - RT

(88   :   12)

75%

Reetz, M.T.[*]; Steinbach, R.; Westermann, J.; Peter, R.; Wenderoth, B.

Chem Ber, (1985), **118**, 1441

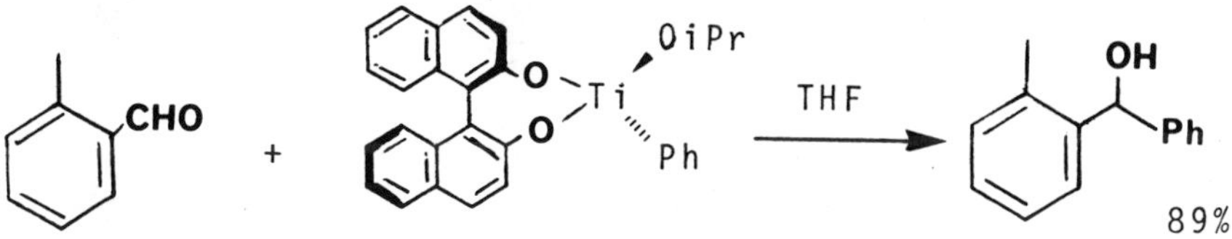

THF

89%

(85% ee)

Seebach, D.[*]; Beck, A.K.; Roggo, S.; Wonnacott, A.
Chem Ber, (1985), **118**, 3673

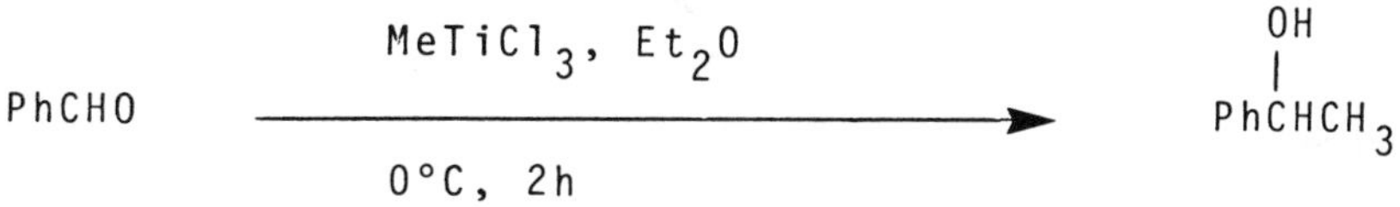

PhCHO

$$MeTiCl_3, \ Et_2O$$

0°C, 2h

$$\underset{PhCHCH_3}{\overset{OH}{|}}$$

98%

Reetz, M.T.[*]; Kyung, S.H.; Hüllmann, M.
Tetrahedron, (1986), **42**, 2931

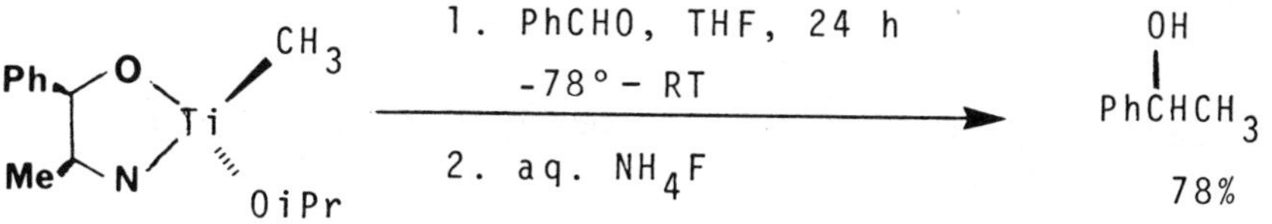

1. PhCHO, THF, 24 h
   -78° - RT

2. aq. NH$_4$F

$$\underset{PhCHCH_3}{\overset{OH}{|}}$$

78%

(85% ee)

Reetz, M.T.[*]; Kükenhöhner, T.; Weinig, P.
Tetrahedron Lett, (1986), **27**, 5711

1. PhLi, CuBr·Me$_2$S
   Me$_2$S, Et$_2$O, -30°C

2. tBuOK, THF
   60°C

77%

(97% ee, R)

Schreiber, S.L.*; Reagan, J.
Tetrahedron Lett, (1986), **27**, 2945

$CH_3(CH_2)_6CHO$

BnCl,   e⁻ ,   DMF

Mg anode, stainless
steel cathode, 0°C

Bu$_4$BBr

$CH_3(CH_2)_6CHBn$ with OH

95%

Sibille, S.; d'Incan, E.; Leport, L.; Perichon, J.
Tetrahedron Lett, (1986), **27**, 3129

Review: "Reactivity and Stereochemical Factors of the Reduction
of Carbonyl Functions with Metallic Borohydride"

Caro, B.; Boyer, B.; Lamaty, G.; Jaouen, G.
Bull Chem Soc Fr, (1983), II281

Related Methods: Alcohols from Ketones (Section 42)

SECTION 35: Alcohols and Thiols from Alkyl, Methylenes, and Aryls

No examples of the reaction RR' → ROH (R' = alkyl, aryl, etc.)
occur in the literature. For reactions of the type RH → ROH
(R = alkyl or aryl) see Section 41 (Alcohols and Phenols from
Hydrides).

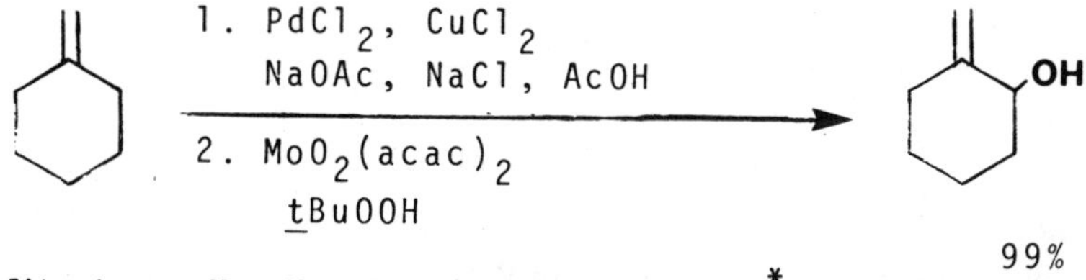

1. $PdCl_2$, $CuCl_2$
   NaOAc, NaCl, AcOH

2. $MoO_2(acac)_2$
   tBuOOH

99%

Jitsukawa, K.; Kaneda, K.; Teranishi, S.[*]
J Org Chem, (1983), 48, 389

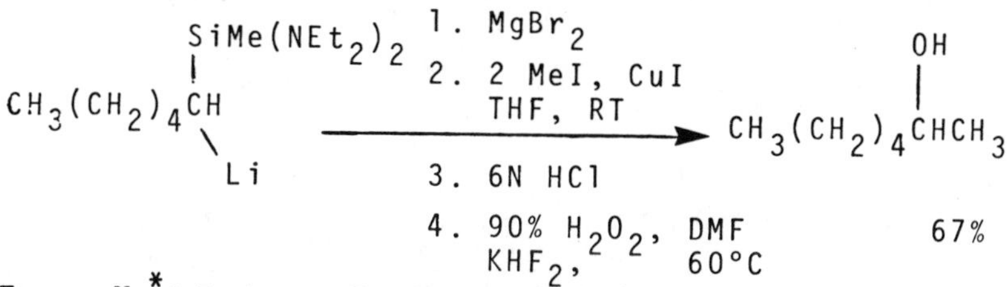

$SiMe(NEt_2)_2$
|
$CH_3(CH_2)_4CH$
\
Li

1. $MgBr_2$
2. 2 MeI, CuI
   THF, RT
3. 6N HCl
4. 90% H_2O_2, DMF
   KHF_2,   60°C

$CH_3(CH_2)_4CHCH_3$ with OH

67%

Tamao, K.[*]; Iwahara, T.; Kanatani, R.; Kumada, M.
Tetrahedron Lett, (1984), 25, 1909, 1913

SECTION 36: Alcohols and Thiols from Amides

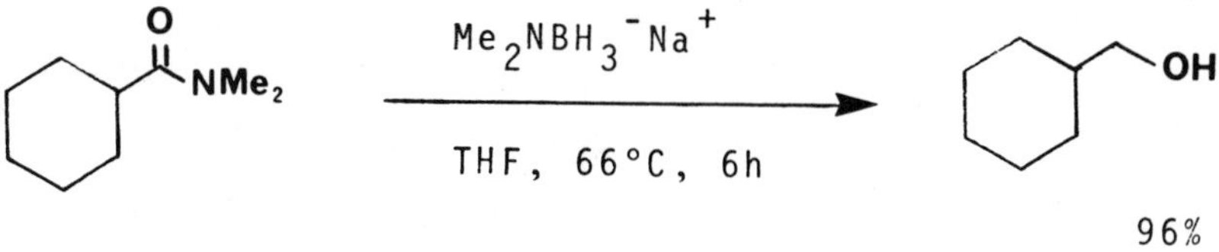

$Me_2NBH_3^-Na^+$

THF, 66°C, 6h

96%

Hutchins, R.O.[*]; Learn, K.; El-Telbany, F.; Stercho, Y.P.
J Org Chem, (1984), 49, 2438

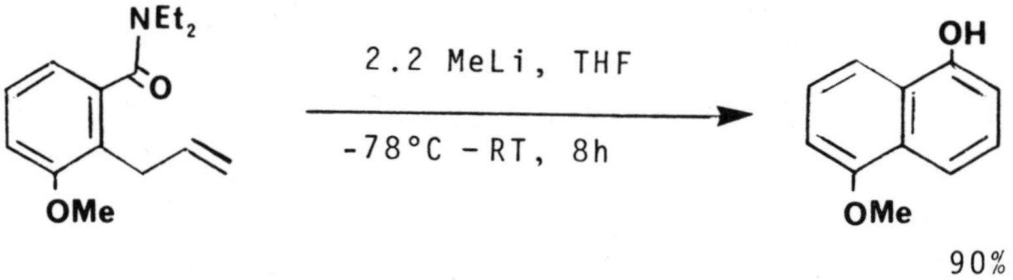

2.2 MeLi, THF

-78°C – RT, 8h

90%

Sibi, M.P.; Dankwardt, J.W.; Snieckus, V.[*]
J Org Chem, (1986), 51, 271

SECTION 37: Alcohols and Thiols from Amines

No Additional Examples

SECTION 38: Alcohols and Thiols from Esters

$$\begin{array}{c} CH_2 \\ \diagup \quad \diagdown \\ C\equiv N \qquad CO_2Me \end{array} \xrightarrow[\text{Et}_2\text{O, reflux, 3h}]{3 \text{ LiAlH}_4, \text{ SiO}_2} \begin{array}{c} CH_2 \\ \diagup \quad \diagdown \\ C\equiv N \qquad CH_2OH \end{array}$$

54%

Kamitori, Y.; Hojo, M.[*]; Masuda, R.; Inoue, T.; Izumi, T.
Tetrahedron Lett, (1983), **24**, 2575

$$\begin{array}{c} CH_2Ph \\ | \\ CH_2CO_2Et \end{array} \xrightarrow[\text{THF, } -20°\text{C, 24h}]{0.5 \text{ LiBH}_4-2\underline{n}\text{BuMgCl}} \begin{array}{c} CH_2Ph \\ | \\ CH_2CH\underline{n}Bu \\ | \\ OH \end{array}$$

74%

Comins, D.L.[*]; Herrick, J.J.
Tetrahedron Lett, (1984), **25**, 1321

$$\begin{array}{c} H \quad OSO_2\text{-pyridyl} \\ \vdots \\ CH_3C-(CH_2)_5CH_3 \end{array} \xrightarrow[\begin{array}{c} 2. \text{ H}_2, \text{ 10\% Pd-C} \\ \text{MeOH} \end{array}]{\begin{array}{c} 1. \underline{n}\text{Bu}_4\text{N}^+\text{NO}_3^- \\ \text{benzene, 0.5h} \end{array}} \begin{array}{c} OH \\ | \quad ,,,H \\ CH_3C-(CH_2)_5CH_3 \end{array}$$

60%

(100% ee)

Cainelli, G.; Manescalchi, F.; Martelli, G.; Panunzio, M.;
Plessi, L.
Tetrahedron Lett, (1985), **26**, 3369

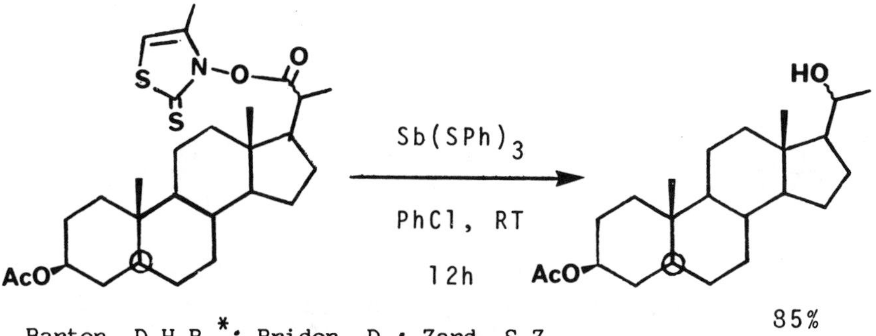

Barton, D.H.R.[*]; Bridon, D.; Zard, S.Z.
          JCS Chem Comm, (1985), 1066

85%

Related Methods: Carboxylic Acids from Esters (Section 23)
                 Protection of Alcohols (Section 45A)
                 Hydrolysis of Esters is covered in Section 23.

SECTION 39: Alcohols and Thiols from Ethers, Epoxides, and
            Thioethers

Additional examples of ether cleavages may be found in Section
45A (Protection of Alcohols and Phenols).

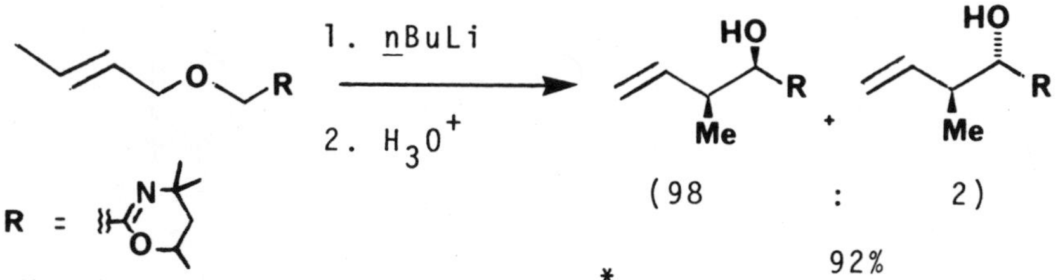

(98 : 2)

92%

Mikami, K.; Fujimoto, K.; Nakai, T.[*]
                 Tetrahedron Lett, (1983), 24, 513

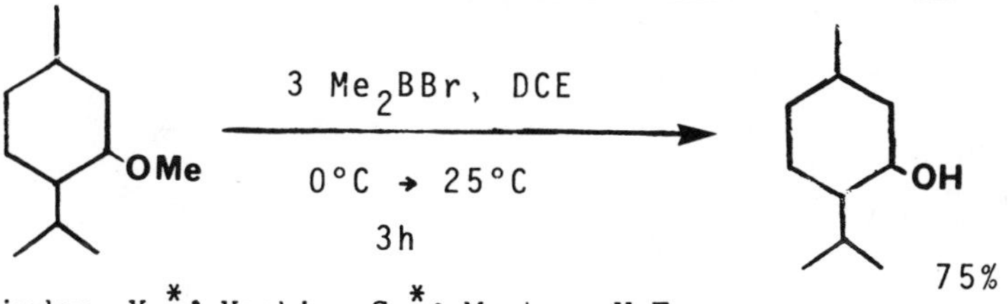

75%

Guindon, Y.[*]; Yoakim, C.[*]; Morton, H.E.
                 Tetrahedron Lett, (1983), 24, 2969

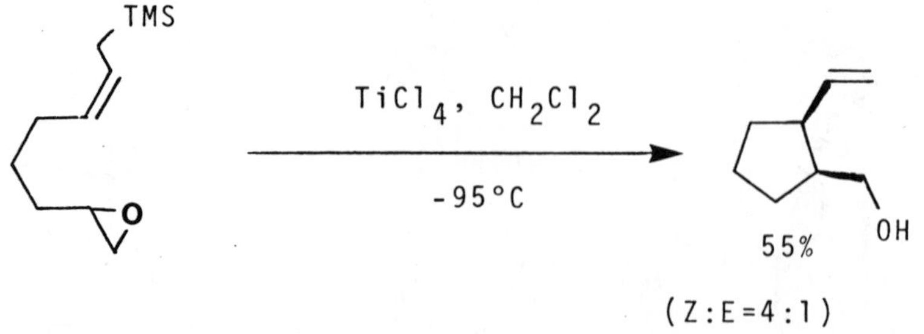

$TiCl_4$, CH_2Cl_2

-95°C

55%   OH

(Z:E=4:1)

Tan, T.S.; Mather, A.N.; Procter, G.; Davidson, A.H.
JCS Chem Comm, (1984), 585

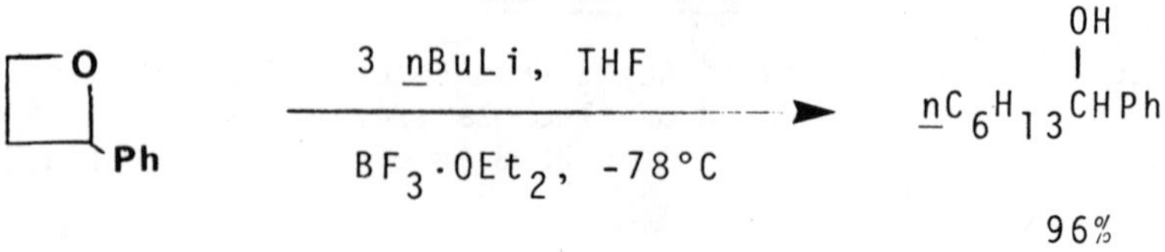

3 nBuLi, THF

$BF_3 \cdot OEt_2$, -78°C

$nC_6H_{13}CHPh$

OH

96%

Eis, M.J.; Wrobel, J.E.; Ganem, B.*
J Am Chem Soc, (1984), 106, 3693

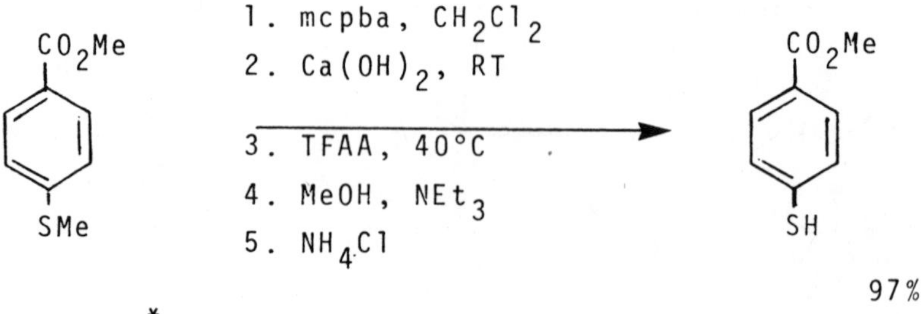

CO_2Me

1. mcpba, CH_2Cl_2
2. Ca(OH)$_2$, RT

3. TFAA, 40°C
4. MeOH, NEt$_3$
5. NH$_4$Cl

SMe

CO_2Me

SH

97%

Young, R.N.*; Gauthier, J.Y.; Coombs, W.
Tetrahedron Lett, (1984), 25, 1753

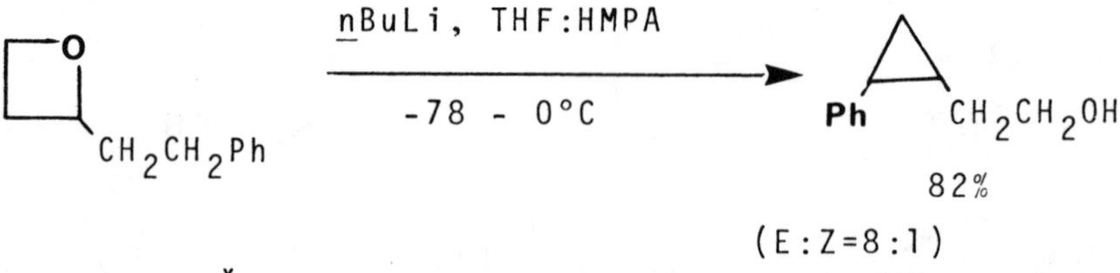

CH_2CH_2Ph

nBuLi, THF:HMPA

-78 - 0°C

Ph CH_2CH_2OH

82%

(E:Z=8:1)

Yamaguchi, M.*; Hirao, I.   Tetrahedron Lett, (1984), 25, 4549

$R = \underline{c}C_6H_{11}$; $R' = H$; $\underline{TiCl_4, Et_3SiH, -78°C}$ 85%

                                                  (96% ee)

Atsunori, M.; Ishihara, K.; Yamamoto, H.[*]

                    Tetrahedron **Lett**, (1986), <u>27</u>, 987

$R = \underline{n}C_8H_{17}$; $R' = H$ $\dfrac{1.\ MeMgBr}{2.\ PCC\quad 3.\ OH^-}$ 90%

                                                  (89% ee)

Lindell, S.D.; Elliott, J.D.; Johnson, W.S.[*]

                    Tetrahedron **Lett**, (1984), <u>25</u>, 3947

$R = \underline{n}C_9H_{19}$; $\underline{Zn/CH_2Cl_2/Me_2SiCl_2}$ 96%(69:31)

Vankar, Y.D.[*]; Arya, P.S.; Rao, C.T.

                    Syn Commun, (1983), <u>13</u>, 869

$R = \underline{n}C_6H_{13}$; $\dfrac{1.\ NaHTe/EtOH/reflux}{2.\ DBE\quad 3.\ Nickel}$ 75% (100:0)

                         boride

Barton, D.H.R.[*]; Fekih, A.; Lusinchi, X.

                    Tetrahedron **Lett**, (1985), <u>26</u>, 6197

                                                  84%

Eisch, J.J.[*]; Dua, S.K.; Behrooz, M.

                    J Org Chem, (1985), <u>50</u>, 3674

$$CH_3(CH_2)_3CH-CH_2 \quad \xrightarrow[\text{2. aq. } NH_4Cl]{\text{1. 2Ca, EDA}} \quad CH_3(CH_2)_3\overset{\overset{\displaystyle OH}{|}}{C}HCH_3$$

91%

Benkeser, R.A.[*]; Rappa, A.; Wolsieffer, L.A.
J Org Chem, (1986), **51**, 3391

$$\xrightarrow[\text{2. aq. } NH_4Vl]{\begin{array}{c}\text{1. 2.2Ca, } NH_3\\ \text{2h}\end{array}}$$

Hwu, J.R.[*]; Chua, V.; Schroeder, J.E.; Barrans Jr., R.E.;
Khoudary, K.P.; Wang, N.; Wetzel, J.M.
J Org Chem, (1986), **51**, 4731

$$\xrightarrow[\begin{array}{c}TiCl_4, CH_2Cl_2\\ -85°C, 5h\end{array}]{CH_2=CHCH_2SiMe_3} \quad CH_2=CHCH_2\overset{\overset{\displaystyle Ph}{|}}{\underset{\underset{\displaystyle Ph}{|}}{C}}(CH_2)_3OH$$

quant.

Oku, A.[*]; Homoto, Y.; Harada, T.   **Chem Lett**, (1986), 1495

$$CH_2=CH-CH-CH_2 \quad \xrightarrow[Et_2O, RT, 23h]{MeLi/Sm[N(TMS)_2]_3} \quad CH_2=CH\overset{\overset{\displaystyle CH_3}{|}}{C}HCH_2OH$$

81%

Mukerji, I.; Wayda, A.[*]; Dabbagh, G.; Bertz, S.H.[*]
Angew Chem Int Ed Engl, (1986), **25**, 760

Reviews:

"Cleavage of Ethers"

Bhatt, M.V.; Kulkarni, S.U.   **Synthesis**, (1983), 249

"Synthetically Useful Reactions of Epoxides"

Smith, J.G.[*]   **Synthesis**, (1983), 629

SECTION 40: Alcohols and Thiols from Halides and Sulfonates

$$\underline{n}C_6H_{13}Br \xrightarrow[\text{2. ox.}]{\text{1. } MeSBCH_2^{-}} \underline{n}C_6H_{13}CH_2OH$$

92%

Pelter, A.; Williams, L.; Wilson, J.W.; Singaram, B.
                        Tetrahedron Lett, (1983), **24**, 627, 631

1.$(iPrO)_2MeSiCH_2MgCl$
10% CuI

2. 90% H_2O_2

91%

Tamao, K.[*]; Ishida, N.; Kumada, M.[*]
                        J Org Chem, (1983), **48**, 2120

1. Mg, THF

2. $\begin{smallmatrix}O\\O\end{smallmatrix}$B-O-O-$\underline{t}$Bu
THF, overnite

3. 2N HCl, 0°C

91%

Hoffmann, R.W.; Ditrich, L.   **Synthesis**, (1983), 107

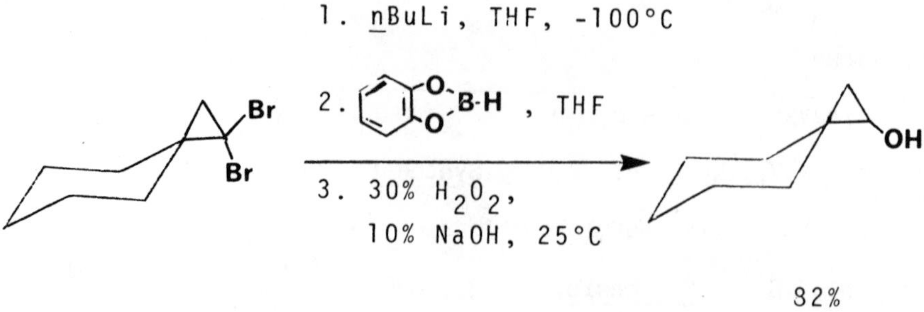

1. $\underline{n}BuLi$, THF, $-100°C$

2. [structure] $\overset{O}{\underset{O}{}}B\text{-}H$, THF

3. 30% H_2O_2,
   10% NaOH, 25°C

82%

Danheiser, R.L.*; Savoca, A.C.   J Org Chem, (1985), 50, 2401

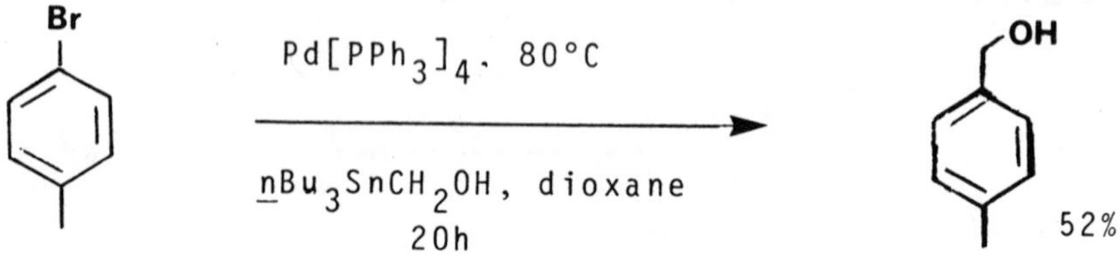

$Pd[PPh_3]_4$. 80°C

$\underline{n}Bu_3SnCH_2OH$, dioxane
20h

52%

Kosugi, M.; Sumiya, T.; Ohhashi, K.; Sano, H.; Migita, T.*
Chem Lett, (1985), 997

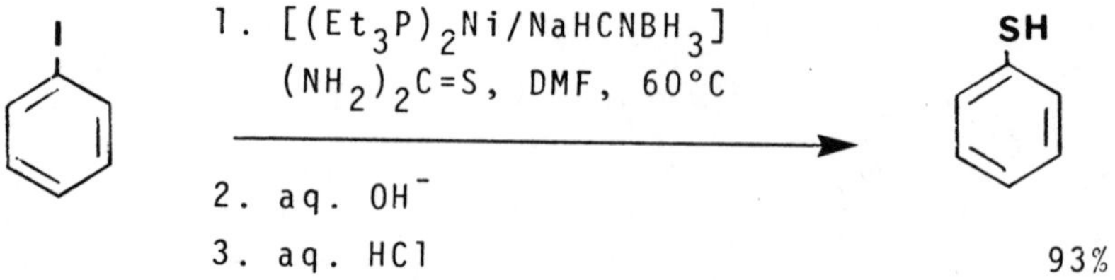

1. $[(Et_3P)_2Ni/NaHCNBH_3]$
   $(NH_2)_2C=S$, DMF, 60°C

2. aq. OH^-
3. aq. HCl

93%

Takagi, K.*   Chem Lett, (1985), 1307

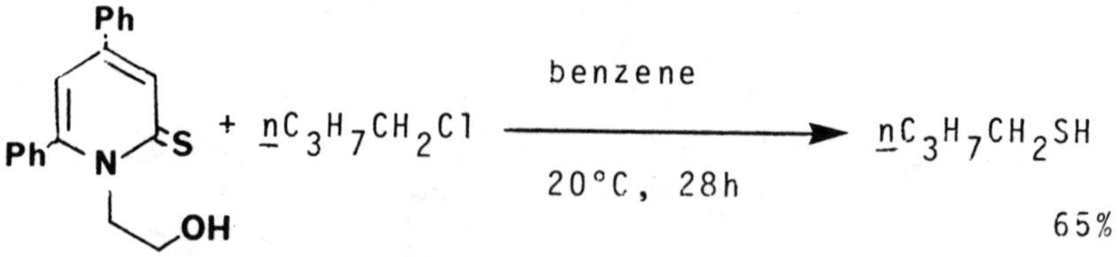

[structure] + $\underline{n}C_3H_7CH_2Cl$ benzene
20°C, 28h $\longrightarrow$ $\underline{n}C_3H_7CH_2SH$

65%

Molina, P.*; Alajarin, M.; Vilaplana, M.J.; Katritzky, A.R.*
Tetrahedron Lett, (1985), 26, 469

$$CH_3CH_2CH_2CHO \xrightarrow[\text{3. PhCH}_2\text{Br}]{\substack{\text{1. H}_2\text{S, TMS-Cl, Py} \\ \text{2. nBuLi}}} PhCH_2SH$$

50%

Harpp, D.N.[*]; Kobayashi, M.   **Tetrahedron** **Lett**, (1986), **27**, 3975

$$\underline{n}C_8H_{17}-Cl \xrightarrow[\text{2. aq. NaOH, RT}]{\substack{\text{1. NaOCH, } \underline{n}\text{Bu}_4\text{NBr} \\ \text{125°C, 1.5h}}} \underline{n}C_8H_{17}-OH$$

91%

Zahalka, H.A.; Sasson, Y.[*]   **Synthesis**, (1986), 763

SECTION 41: Alcohols and Thiols from Hydrides

No Additional Examples

SECTION 42: Alcohols and Thiols from Ketones

The following reactions types are included in this section:

A. Reductions of Ketones to Alcohols.

B. Alkylations of Ketones, forming Alcohols.

Coupling of ketones to give diols is found in Section 323 (Alcohol - Alcohol).

SECTION 42A: Reductions of Ketones to Alcohols

Asymmetric Reduction:

$$\underset{\text{RCR'}}{\overset{O}{\|}} \longrightarrow \underset{*}{\underset{\text{RCHR'}}{\overset{O}{|}}}$$

R = Pr
R' = Ph

$\xrightarrow[\text{30°C}]{BH_3 \cdot THF}$ quant.
(96% ee R)

Itsuno, S.[*]; Ito, K.; Hirao, A.; Nakahama, S.
JCS Chem Comm, (1983), 469
J Org Chem, (1984), 49, 555

R = Me
R'= Ph

$\xrightarrow{1. \quad , \text{HSiHPh}_2 \quad 2. \text{HOH}}$ 99%
(83.2% ee R)

Brunner, H.[*]; Riepel, G.; Weitzer, H.
Angew Chem Int Ed Engl, (1983), 22, 331

R = Me
R' = Bn

$\xrightarrow{\begin{array}{l}1. \ HSiHPh_2/[Rh(COD)Cl]_2 \\ 2. \ H_3O^+\end{array}}$ 91%
(52.7% ee R)

Brunner, H.[*]; Becker, R.; Riepel, G.
Organometallics, (1984), 3, 1354

R = Pr
R' = Ph

$\underset{}{S,S' \ [Ph\overset{O}{\overset{\|}{C}}NH-\overset{\text{COOH}}{\underset{|}{C}}H-CH_2-S]_2}$

$\xrightarrow{LiBH_4, \ \underline{t}BuOH/THF}$ 98%
(90% ee)

Soai, K.[*]; Oyamada, H.; Yamanoi, T.   JCS Chem Comm, (1984), 413

R = Bu
R' = Ph

$\xrightarrow[\text{30°C}]{THF}$ quant.
(73% ee R)

Itsuno, S.[*]; Hirao, A.; Nakahama, S.; Yamazaki, N.
JCS Perkin I, (1983), 1673

PhCH$_3$, -78°C, BF$_3$ 71%

R = iPr

R' = Ph   /NH$_3$/BH$_3$ (67% ee R)

Allwood, B.L.; Shahriari-Zavareh, H.; Stoddart, J.F.; Williams,
D.J.

JCS Chem Comm, (1984), 1461

R = Me   iPrOH. 37°C. NADP, TRI·HCl   61%

R' = iPr

TBADH, HSCH$_2$CH$_2$OH, pH 8.5   (86% ee)

Keinan, E.[*]; Hafeli, E.K.; Seth, K.K.; Lamed, R.
J Am Chem Soc, (1986), **108**, 162

R = Ph   iPrOH, Ir(COD)(acac), KOH
                                                                    87%
R' = Me
                (menthyl)Ph$_2$P, HBF$_4$, 83°C
                                                            (42% ee R)

Krause, H.W.; Bhatnagar, A.K.
J Organomet Chem, (1986), **302**, 265

R = Ph   1. PhCNH, Li, THF   71%

R' = Et

                2. aq. HCl   3. aq. NaHCO$_3$ (88% ee)

Soai, K.[*]; Yamanoi, T.; Oyamada, H.   **Chem Lett**, (1984), 251

R = Ph

R' = Me   2   -100°C   60%

                                                            (95% ee)

Noyori, R.[*]; Tomino, I.; Tanimoto, Y.; Nishizawa, M.
J Am Chem Soc, (1984), **106**, 6709

R = Ph   1. LiAlH$_4$; (-)-quinine, Et$_2$O   quant.

R' = Me   2. aq. H$_2$SO$_4$
                                                            (98% ee)

Červinka, O.; Fábryová, A.; Sablukova, I.
Coll Czech Chem Comm, (1986), **51**, 401

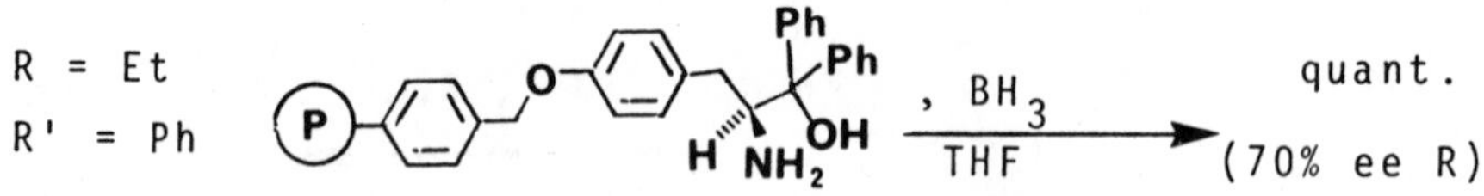

R = Et
R' = Ph

, BH$_3$ / THF

quant.
(70% ee R)

Itsuno, S.[*]; Nakano, M.; Ito, K.; Hirao, A.; Owa, M.; Konda, N.; Nakahama, S.

JCS Perkin I, (1985), 2615

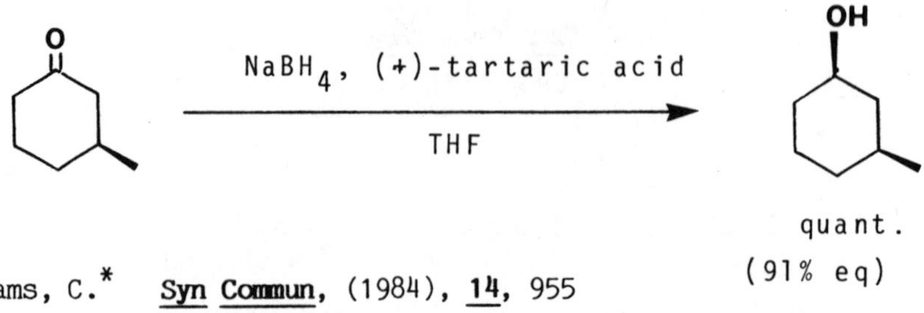

NaBH$_4$, (+)-tartaric acid

THF

OH

quant.
(91% eq)

Adams, C.[*]   Syn Commun, (1984), 14, 955

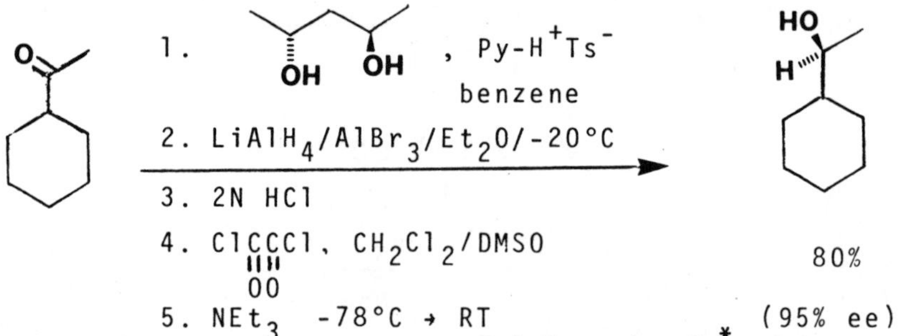

1. [structure with OH, OH], Py-H$^+$Ts$^-$
   benzene
2. LiAlH$_4$/AlBr$_3$/Et$_2$O/-20°C
3. 2N HCl
4. ClCCCl, CH$_2$Cl$_2$/DMSO
     00
5. NEt$_3$ -78°C → RT

HO
H

80%
(95% ee)

Mori, A.; Fujiwara, J.; Maruoka, K.; Yamamoto, H.[*]
Tetrahedron Lett, (1983), 24, 4581

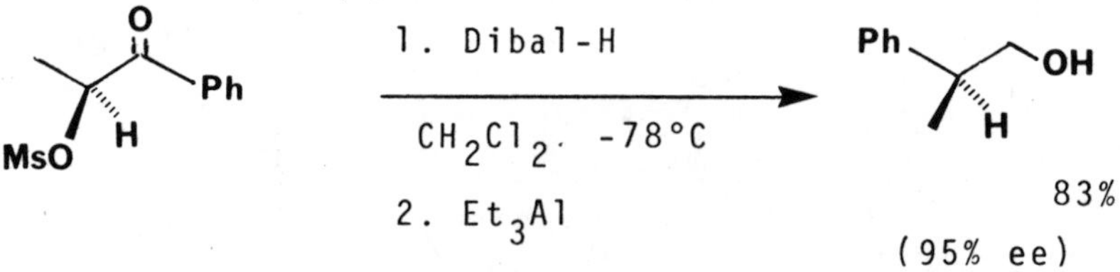

1. Dibal-H
   CH$_2$Cl$_2$. -78°C
2. Et$_3$Al

Ph [structure] OH
H

83%
(95% ee)

Suzuki, K.; Katayama, E.; Matsumoto, T.; Tsuchihashi, G.[*]
Tetrahedron Lett, (1984), 25, 3715

Non-Asymmetric Reduction:

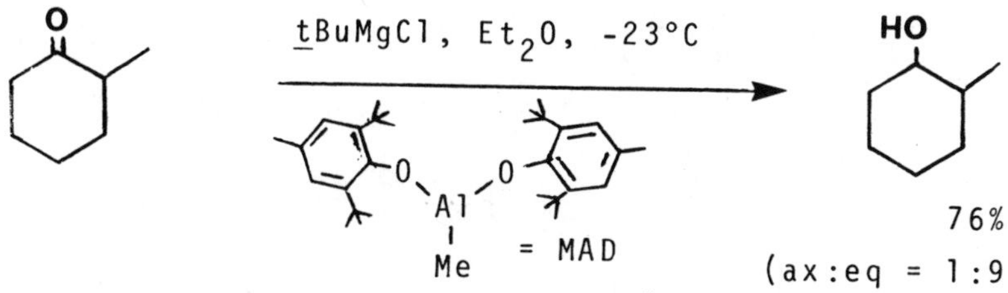

tBuMgCl, Et$_2$O, -23°C

= MAD

76%
(ax:eq = 1:9)

Maruoka, K.; Sakurai, M.; Yamamoto, H.[*]
Tetrahedron Lett, (1985), **26**, 3853

1. MAD, PhCH$_3$ 84%
2. MeLi   (ax:eq = 1:99)

Maruoka, K.; Itoh, T.; Yamamoto, H.[*]
J Am Chem Soc, (1985), **107**, 4573

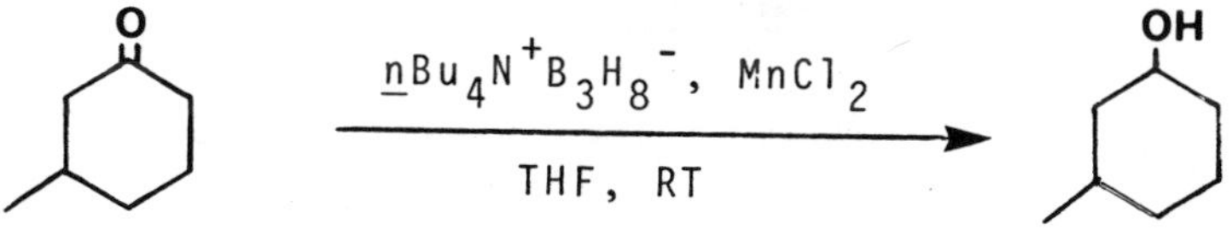

nBu$_4$N$^+$B$_3$H$_8^-$, MnCl$_2$

THF, RT

93%

B$_3$H$_8$ = octahydrotriborate

Tamblyn, W.H.[*]; Aquadro, R.E.; DeLuca, O.D.; Weingold, D.H.;
Dao, T.V.

Tetrahedron Lett, (1983), **24**, 4955

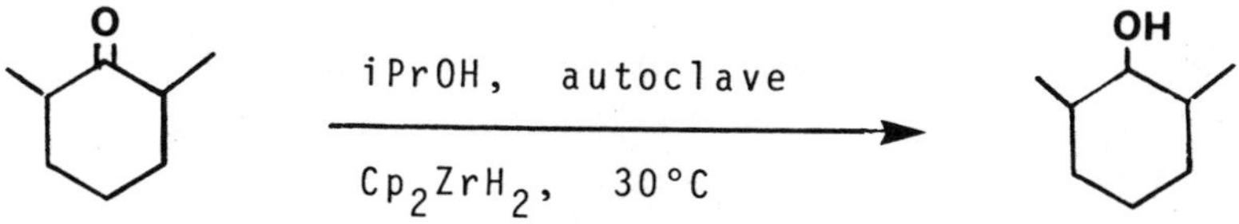

iPrOH, autoclave

Cp$_2$ZrH$_2$, 30°C

92%

Ishii, Y.[*]; Nakano, T.; Inada, A.; Kishigami, Y.; Sakurai, K.;
Ogawa, M.[*]

J Org Chem, (1986), **51**, 240

$$CH_3CH_2\overset{\displaystyle O}{\overset{\displaystyle \|}{C}}CH_3 \xrightarrow[\text{AcOH, } 25°C]{\text{PPN}^+\text{HCr(CO)}_3^-, \text{ THF}} CH_3CH_2\overset{\displaystyle OH}{\overset{\displaystyle |}{C}}HCH_3$$

PPN = <u>bis</u>(triphenylphosphine)imminium   70%

Gaus, P.L.[*]; Kao, S.C.[*]; Youngdahl, K.; Darensbourg, M.Y.[*]

J Am Chem Soc, (1985), **107**, 2428

$$CH_3(CH_2)_2\overset{\displaystyle O}{\overset{\displaystyle \|}{C}}CH_3 \xrightarrow[\text{iPrOH, reflux}]{\text{ZrOCl}_2 \cdot 8H_2O/H_2O} CH_3(CH_2)_2\overset{\displaystyle OH}{\overset{\displaystyle |}{C}}HCH_3$$

quant

Matsushita, H.[*]; Ishiguro, S.; Ichinose, H.; Izumi, A.; Mizusaki, S.

Chem Lett, (1985), 731

1. Ph_2SbH, THF, $AlCl_3$
   RT

2. aq. H_2SO_4

96%

Huang, Y.Z.; Shen, Y.C.; Chen, C.

Tetrahedron Lett, (1985), **26**, 5171

$\underline{n}Bu_3SnH$, MeOH, 55°C

1400 mPa, 24h

quant

Dequeil-Castaing, M.; Rahm, A.[*]; Dahan, N.

J Org Chem, (1986), **51**, 1672

1. $HSCH_2CH_2SH$, H^+
2. 3 $\underline{n}BuLi/Et_2O$
3. HOH

89%

Wilson, S.R.[*]; Georgiodis, G.M.   **Org Syn**, (1983), **61**, 74

SECTION 42B: Alkylation of Ketones, Forming Alcohols

Aldol reactions are listed in Section 330 (Ketone-Alcohol).

$$PhCCH_3 \quad (O)$$

1. $PhCH_2SiMe_3$, THF
   $\underline{n}Bu_4NF$, 0°C
2. H_3O^+

$$PhCCH_2Ph \quad (OH, CH_3)$$ 75%

Bennetau, B.; Dunogues, J.   **Tetrahedron Lett**, (1983), **24**, 4217

2 MI_2, THF, RT

cat. $Fe(dbm)_3$

R = $\underline{t}Bu$	M = Yb	95%	(Z:E = 1:1.3)
R = H	M = Sm	95%	(Z:E = 1.1.5)

Molander, G.A.[*]; Etter, J.B.   **J Org Chem**, (1986), **51**, 1778
Tetrahedron Lett, (1984), **25**, 3281

$$PHCH_2CCH_3 \quad (O)$$

$MeTiCl_3$, Et_2O

0°C, 2h

$$PhCH_2CCH_3 \quad (OH, CH_3)$$ 81%

Reetz, M.T.[*]; Kyung, S.H.; Hüllmann, M.
Tetrahedron, (1986), **42**, 2931

1. $\underline{n}BuCeCl_2$, THF
   -78°C

2. aq. NH_4Cl

88%

Imamoto, T.[*]; Sugiura, Y.; Takujama, N.
Tetrahedron Lett, (1984), 25, 4233

$\underline{t}BuLi$, Et_2O

-58°C

78%

Cooke Jr., M.P.[*]; Houpis, I.N.
Tetrahedron Lett, (1985), 26, 4987

$PhCCH=CH_2$

1. 57% HI

2. Zn, TMS-Cl, THF
   heat

68%

Narasimhan, N.S.[*]; Patil, P.A.
Tetrahedron Lett, (1986), 27, 5133

Reviews:

"Metal-Ammonia Reduction of Cyclic Aliphatic Ketones"

Huffman, J.W.[*]   Accts Chem Res, (1983), 16, 399

"An Introduction of Chiral Centers into Acyclic Systems Based On
Stereoselective Ketone Reduction"

Oishi, T., Nakata, T.   Accts Chem Res, (1984), 17, 338

"Microbial Asymmetric Catalysis - Enantioselective Reduction of
Ketones"

Sih, C.J.[*]; Chen, C.-S.
Angew Chem Int Ed Engl, (1984), 23, 570

"Stereoselective Acyclic Ketone Reduction"

Nakata, T.[*]; Fukui, M.; Ohtsuka, H.; Oishi, T.[*]
                    Tetrahedron, (1984), **40**, 2225

"Mechanism and Stereochemistry of Alkali Metal Reductions of Cyclic and Unsaturated Ketones in Protic Solvents"

Pradhan, S.K.[*]   **Tetrahedron**, (1986), **42**, 6351

Related Methods: Alcohols from Aldehydes (Section 34)

SECTION 43: Alcohols and Thiols from Nitriles

                    No Additional Examples

SECTION 44: Alcohols and Thiols from Olefins

For the preparation of diols from olefins see Section 323 (Alcohol - Alcohol).

(E:Z = 1:1)

Taylor, R.T.[*]; Flood, L.A.   **J Org Chem**, (1983), **48**, 5160

87%

TPP = meso-tetraphenylporphyrinato

Okamoto, T.[*]; Oka, S.   **J Org Chem**, (1984), **49**, 1589

$$CH_3$$
$$PhC=CH_2 \xrightarrow[\begin{array}{c}18\text{-crown-6, }30°C\\1h\end{array}]{NaBH_4, \ TiCl_3, \ THF} \quad \begin{array}{c}CH_3\\PhCHCH_2OH\end{array}$$

84%

Lee, H.S.; Isagawa, K.*; Toyoda, H.; Otsuji, Y.
Chem Lett, (1984), 673

$$CH_3(CH_2)_5-B-CH_3 \xrightarrow[\begin{array}{c}2. \ NaOMe/MeOH\\3. \ NaOH/H_2O_2\end{array}]{\begin{array}{c}1. \ LiCHCl_2, \ THF\\-100°C - 25°C\end{array}} \quad CH_3(CH_2)_5CH$$

OH

63%

Brown, H.C.*; Imai, Y.; Perumal, P.T.; Singaram, B.
J Org Chem, (1985), **50**, 4032

$$\xrightarrow[\begin{array}{c}2. \ HOCH_2CH_2OH, \ 6N \ NaOH/MeOH\\THF, \ 30\% \ H_2O_2, \ 50°C\end{array}]{\begin{array}{c}1. \quad BH_2 \ Li \cdot OEt_2, \ 9.5h\\MeI, \ Et_2O, \ RT\end{array}}$$

HO

89%

(97% ee)

Masamune, S.*; Kim, B.M.; Petersen, J.S.; Sato, T.; Veenstra, S.J.
J Am Chem Soc, (1985), **107**, 4549

$$MeO_2C-(CH_2)_8CH=CH_2 \xrightarrow[\begin{array}{c}2. \ Me_3N-O, \ DMF\\KHF_2, \ 130°C\end{array}]{\begin{array}{c}1. \ HSiMeEt_2\\RhCl(PPh_3)_3\end{array}} \quad MeO_2C-(CH_2)_{10}-OH$$

72%

Sakurai, H.*; Ando, M.; Kawada, N.; Sato, K.; Hosomi, A.*
Tetrahedron Lett, (1986), **27**, 75

Review: "The Use of Chiral Organoboranes in Organic Synthesis"

Matteso, D.S.[*]   **Synthesis**, (1986), 973

SECTION 45: Alcohols and Thiols from Miscellaneous Compounds

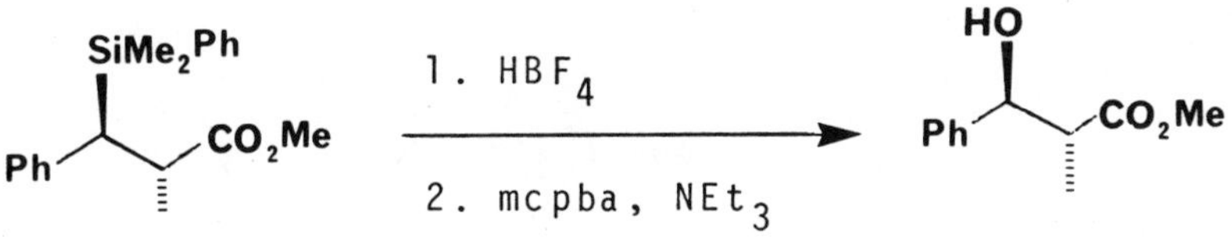

74%

Fleming, I.[*]; Henning, R.; Plaut, H.   **JCS Chem Comm**, (1984), 29

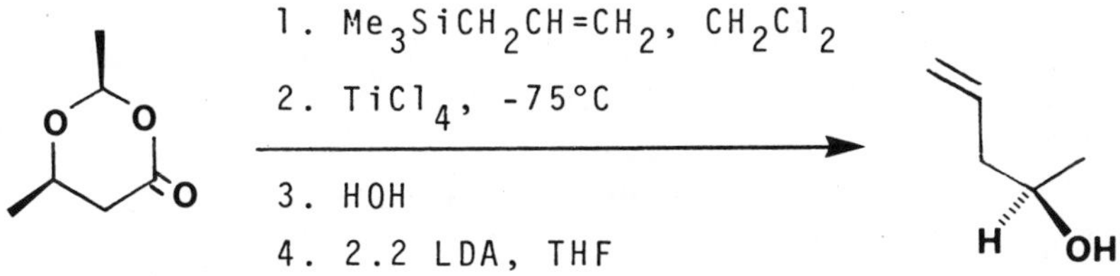

(94% ee R)   76%

Seebach, D.[*]; Imwinkelried, R.; Stucky, G.
                    Angew Chem Int Ed Engl, (1986), **25**, 178

PhSSPh → 1. NaBH$_4$, THF-MeOH reflux 2. aq. HCl → PhSH

quant.

Ookawa, A.; Yokoyama, S.; Soai, K.[*]
                    Syn Commun, (1986), **16**, 819

For conversions of boranes to alcohols (Section 44)

SECTION 45A: Protection of Alcohols and Thiols

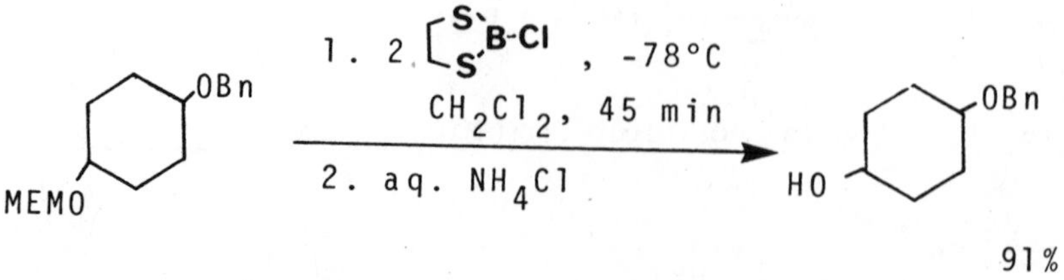

$$\text{1. 2} \left[\begin{array}{c} S \\ S \end{array} \right] B\text{-Cl} \ , \ -78°C$$
$$CH_2Cl_2, \ 45 \ min$$
$$\text{2. aq. } NH_4Cl$$

91%

Williams, D.R.[*]; Sakdarat, S.
Tetrahedron Lett, (1983), 24, 3965

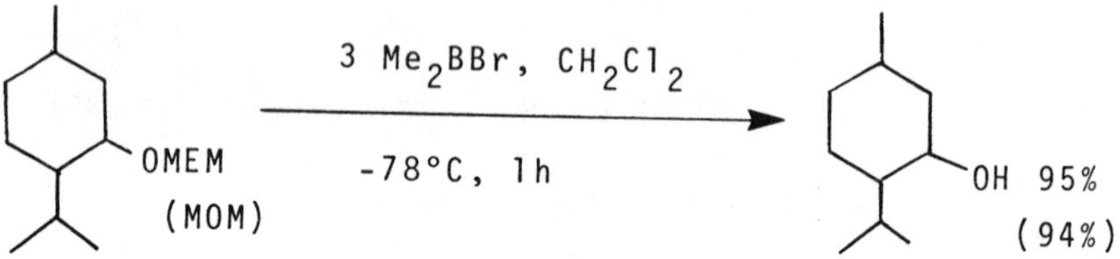

$$3 \ Me_2BBr, \ CH_2Cl_2$$
$$-78°C, \ 1h$$

OH 95%
(94%)

(MOM)

Guindon, Y.[*]; Yoakim, C.; Morton, H.E.
J Org Chem, (1984), 49, 3912
Tetrahedron Lett, (1983), 24, 3969

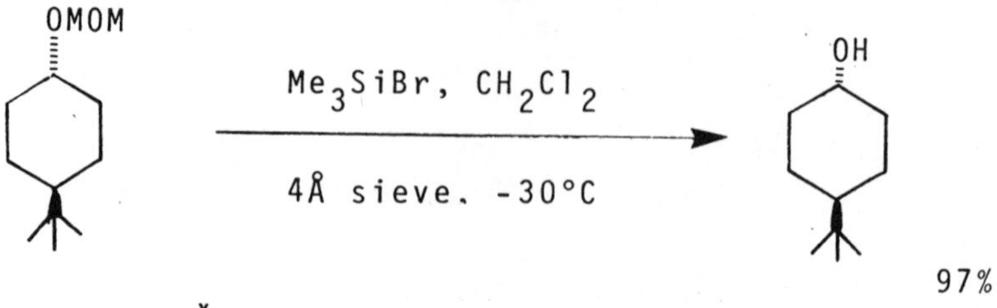

OMOM

$$Me_3SiBr, \ CH_2Cl_2$$
$$4Å \ sieve. \ -30°C$$

OH

97%

Hanessian, S.[*]; Delorme, D.; Dufresne, Y.
Tetrahedron Lett, (1984), 25, 2515

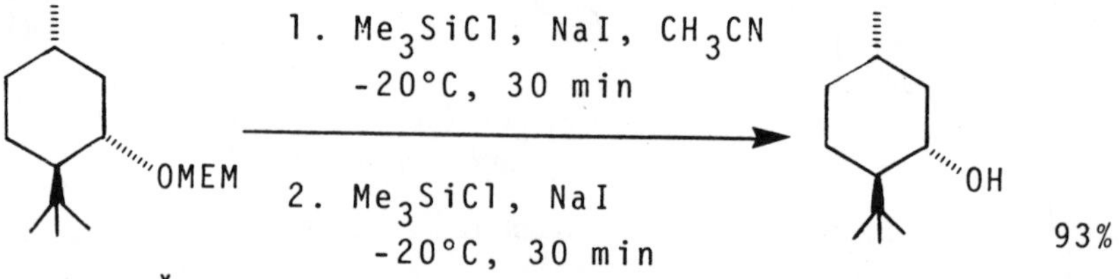

1. Me_3SiCl, NaI, CH_3CN
   $-20°C$, 30 min

2. Me_3SiCl, NaI
   $-20°C$, 30 min

OMEM

OH

93%

Rigby, J.H.[*]; Wilson, J.Z.   Tetrahedron Lett, (1984), 25, 1429

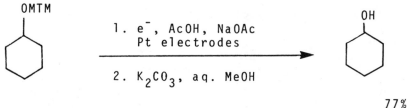

1. e⁻, AcOH, NaOAc
   Pt electrodes

2. K_2CO_3, aq. MeOH

77%

Mandai, T.; Yasunaga, H.; Kawada, M.; Otera, J.*
                    Chem Lett, (1984), 715

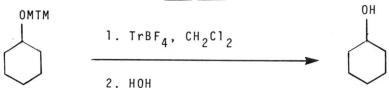

1. $TrBF_4$, CH_2Cl_2

2. HOH

80%

Chowdhury, P.K.; Sharma, R.P.*; Batuah, J.N.
                    Tetrahedron Lett, (1983), **24**, 4485

$MeOCH_2OMe$ (solvent)

TMS-I

1 h

95%

Olah, G.A.*; Husain, A.; Narang, S.C.   **Synthesis**, (1983), 896

$CH_2=CH$ $\overset{OBn}{\underset{CH_3}{}}$, cat. $PdCl_2(COD)$

$Ph(CH_2)_3OH$ 96% $Ph(CH_2)_3OCOBn$

H_2, 5% Pd-C

94%

Mukaiyama, T.*; Ohshima, M.; Murakami, M.
                    Chem Lett, (1984), 265

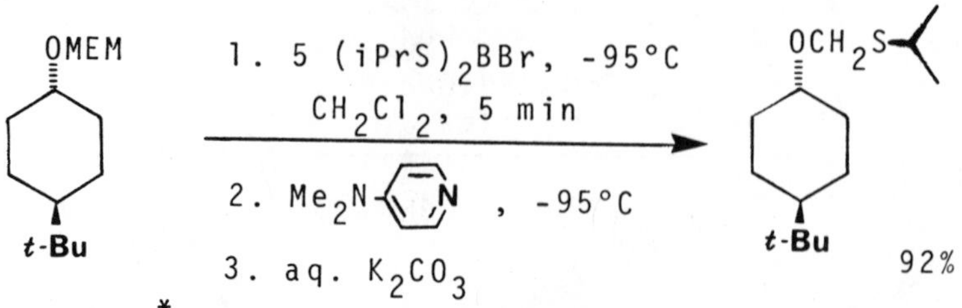
OMEM
t-Bu
1. 5 (iPrS)2BBr, -95°C
CH2Cl2, 5 min
2. Me2N-pyridine, -95°C
3. aq. K2CO3
OCH2S
t-Bu
92%

Corey, E.J.*; Hua, D.H.; Seitz, S.P.
Tetrahedron Lett, (1984), **25**, 3

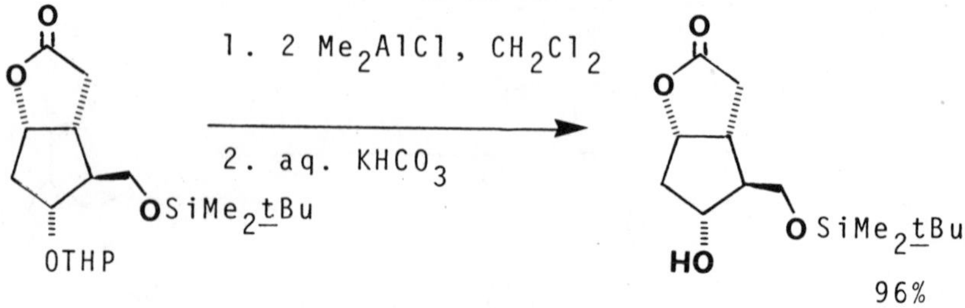
1. 2 Me2AlCl, CH2Cl2
2. aq. KHCO3
OSiMe2tBu
OTHP
HO
OSiMe2tBu
96%

Ogawa, Y.; Shibasaki, M.* **Tetrahedron Lett**, (1984), **25**, 663

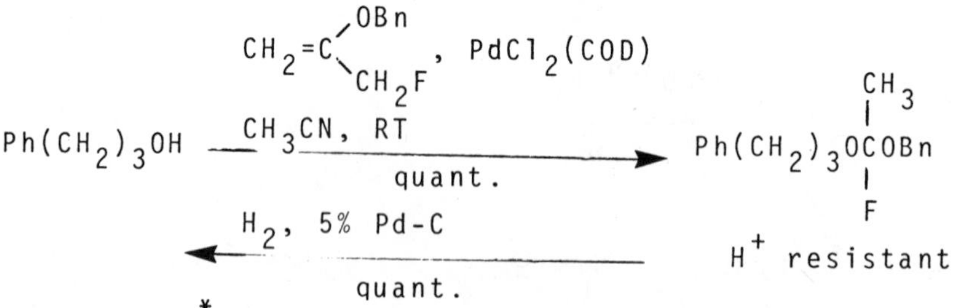
CH2=C(OBn)CH2F, PdCl2(COD)
Ph(CH2)3OH
CH3CN, RT
quant.
Ph(CH2)3OC(CH3)(F)OBn
H2, 5% Pd-C
quant.
H+ resistant

Mukaiyama, T.*; Ohshima, M.; Nagaoka, H.; Murakami, M.
Chem Lett, (1984), 615

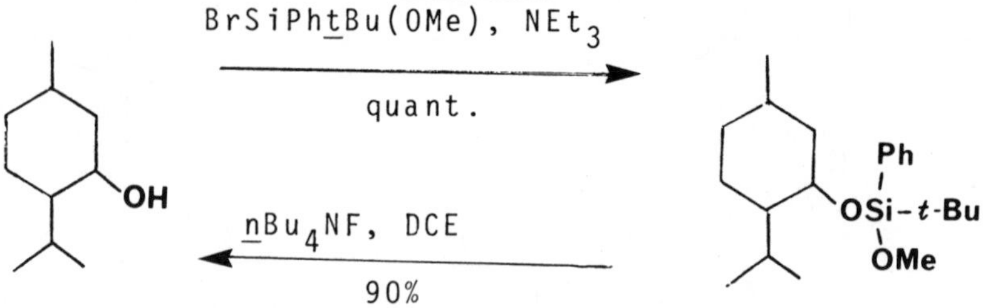
BrSiPhtBu(OMe), NEt3
quant.
OH
Ph
OSi-t-Bu
OMe
nBu4NF, DCE
90%

Guindon, Y.*; Fortin, R.; Yoakim, C.; Gillard, J.W.
Tetrahedron Lett, (1984), **25**, 4717

$$\text{, CuBr}_2/\text{THF}$$

RT, 20h

85%

HCl, EtOH/HOH
reflux

90%

Fetizon, M.*; Hanna, I. **Synthesis**, (1985), 806

MeO⟨⟩OH, PPh$_3$, CH$_2$Cl$_2$

DEAD, RT   99%

$\text{THPO(CH}_2)_4\text{OH}$ ⟶ $\text{THPO(CH}_2)_4\text{OAs}$

CAN, aq. CH$_3$CN

As = ⟨⟩-OMe

Fukuyama, T.*; Laird, A.A.; Hotchkiss, L.M.
                    Tetrahedron Lett, (1985), **26**, 6291

1. Me$_2$BBr, CH$_2$Cl$_2$, -78°C

2. CH$_3$SH, iPr$_2$NEt
   CH$_2$Cl$_2$, -78°C

OMOM ⟶ OMTM

91%

Morton, H.E.*; Guindon, Y.   **J Org Chem**, (1985), **50**, 5379

1. aq. HF, CH$_3$CN

2. aq. NaHCO$_3$

96%

1. nBu$_4$NF/THF, 0°C

2. aq. NH$_4$Cl

96%

Collington, E.W.; Flinch, H.*; Smith, I.J.
                    Tetrahedron Lett, (1985), **26**, 681

quant.

Boeckman Jr., R.K.[*]; Potenza, J.C.
                    Tetrahedron **Lett**, (1985), **26**, 1411

King, P.F.[*]; Stroud, S.G.   **Tetrahedron Lett**, (1985), **26**, 1415

54%

Braish, T.F.; Fuchs, P.L.[*]   **Syn Commun**, (1986), **16**, 111

Related Methods: Esters from Alcohols (Section 123)
                    Alcohols from Ethers (Section 39)
                    Esters from Alcohols (Section 108)
                    Alcohols from Esters (Section 38)

CHAPTER 4
PREPARATION OF ALDEHYDES

SECTION 46: Aldehydes from Acetylenes

No Additional Examples

SECTION 47: Adehydes from Acid Derivatives

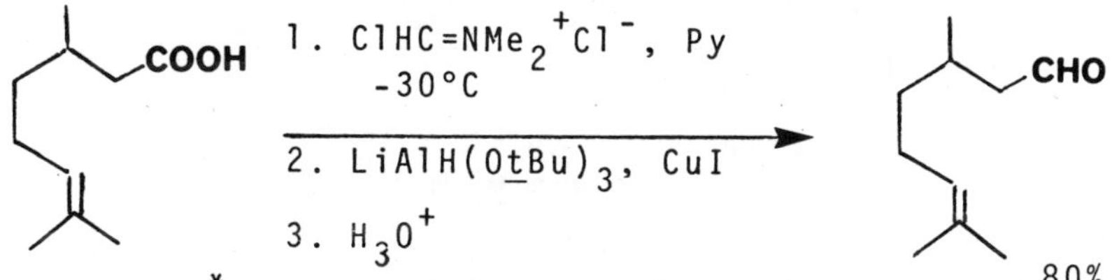

COOH
1. ClHC=NMe2+Cl−, Py
-30°C
2. LiAlH(OtBu)3, CuI
3. H3O+
CHO
80%

Fujisawa, T.*; Mori, T.; Tsuge, S.; Sato, T.
Tetrahedron Lett, (1983), 24, 1543

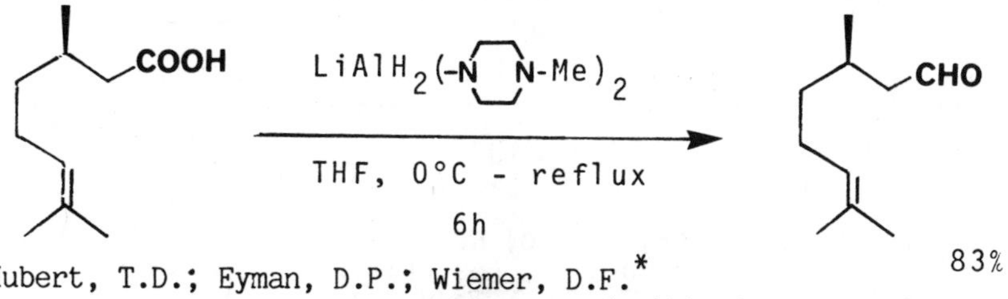

COOH
LiAlH2(-N N-Me)2
THF, 0°C - reflux
6h
CHO
83%

Hubert, T.D.; Eyman, D.P.; Wiemer, D.F.*
J Org Chem, (1984), 49, 2279

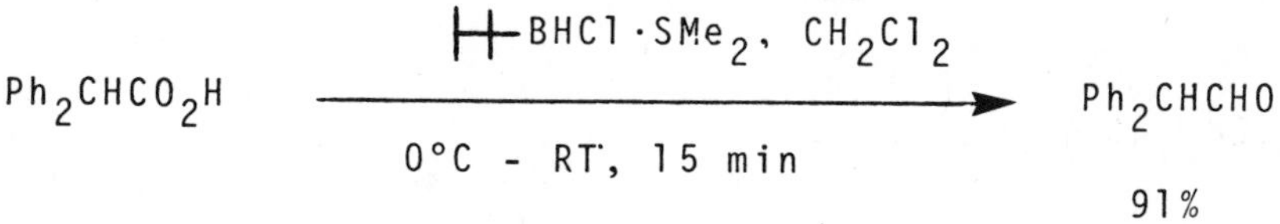

BHCl·SMe2, CH2Cl2
Ph2CHCO2H
Ph2CHCHO
0°C - RT, 15 min
91%

Brown, H.C.*; Cha, J.S.; Nazer, B.; Yoon, N.M.
J Am Chem Soc, (1984), 106, 8001

$$\underset{\overset{|}{PhCHCOOH}}{\overset{OH}{}} \xrightarrow[\substack{Et_2O/HOH \\ RT}]{NaClO, 2h} PhCHO$$

95%

Carlsen, P.H.J.[*]

Acta Scand B, (1984), 343

$$CH_3(CH_2)_{16}\overset{O}{\underset{||}{C}}Cl \xrightarrow[\text{hexane}]{\textbf{P}-NMe_3{}^+BH_4{}^-} CH_3(CH_2)_{16}\overset{O}{\underset{||}{C}}H$$

(P) = Amberlyst A-26

93%

Gordeev, K.Yu.; Serebrennikova, G.A.; Evstigneeva, R.P.
J Org Chem USSR, (1985), **21**, 2393

SECTION 48: Adehydes from Alcohols and Thiols

$$PhCH_2OH \xrightarrow[\substack{PhCH_3, DBU}]{PhSePh, NCS} PhCHO$$

93%

Takai, K.[*]; Yasumura, M.; Negoro, K.
J Org Chem, (1983), **48**, 54

$$CH_3(CH_2)_3CH_2OH \xrightarrow[\substack{RuCl_2(PPh_3)_2}]{\text{COOH} \atop \text{IO, } CH_2Cl_2} CH_3(CH_2)_3CHO$$

quant.

Müller, P.; Godoy, J.   **Helv Chim Acta**, (1983), **66**, 1790

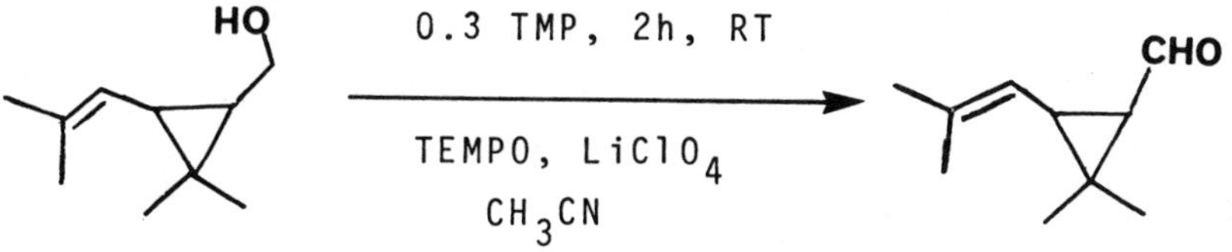
HO
0.3 TMP, 2h, RT
TEMPO, $LiClO_4$
CH_3CN
CHO

59%

Semmelhack, M.F.*; Chou, C.S.; Cortes, D.A.
J Am Chem Soc, (1983), **105**, 4492

$CH_2=C(CH_3)CH_2OH$ —[$Pd(OAc)_2$, PhI / NEt_3, CH_3CN, 100°C, 11h]→ $PhCH_2CH(CH_3)CHO$ 82%

Buntin, S.A.; Heck, R.F.* **Org Syn**, (1983), **61**, 82

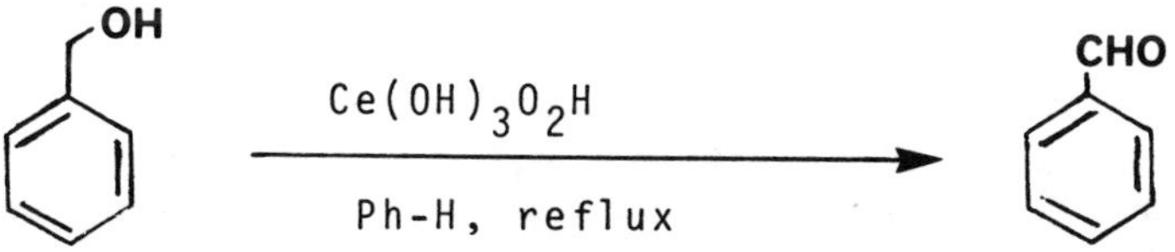
OH
$Ce(OH)_3O_2H$
Ph-H, reflux
CHO

95%

Firouzabadi, H.; Iranpoor, N. **Syn Commun**, (1984), **14**, 875

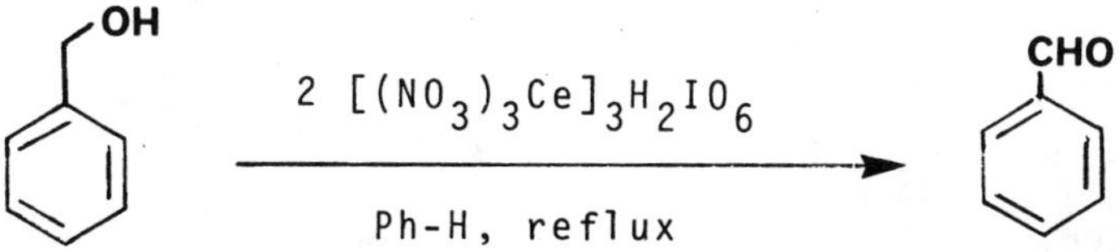
OH
2 $[(NO_3)_3Ce]_3H_2IO_6$
Ph-H, reflux
CHO

95%

Firouzabadi, H.; Iranpoor, N.; Hajipoor, G.; Toofan, J.
Syn Commun, (1984), **14**, 1033

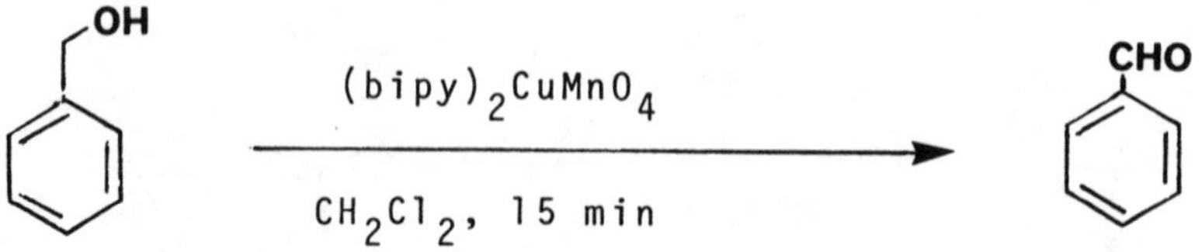

OH

$(bipy)_2CuMnO_4$

CH_2Cl_2, 15 min

CHO

95%

Firouzabadi, H.[*]; Sardarian, A.R.; Naderi, M.; Vessal, B.
Tetrahedron, (1984), **40**, 5001

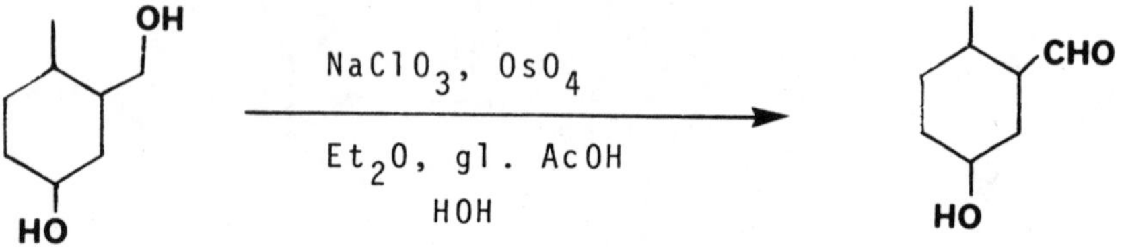

OH

$NaClO_3$, OsO_4

Et_2O, gl. AcOH
HOH

HO

CHO

HO

30%

Maione, A.M.; Romeo, A.[*]   **Synthesis**, (1984), 955

$CH_3CH_2CH_2CH_2OH$

$Pd(OAc)_2$, $NaHCO_3$

$\underline{n}Bu_4NCl$, PhI
DMF, RT, 48h

$CH_3CH_2CH_2CHO$

80%

Choudary, B.M.[*]; Reddy, N.P.; Kantam, M.L.; Jamil, Z.
Tetrahedron Lett, (1985), **26**, 6257

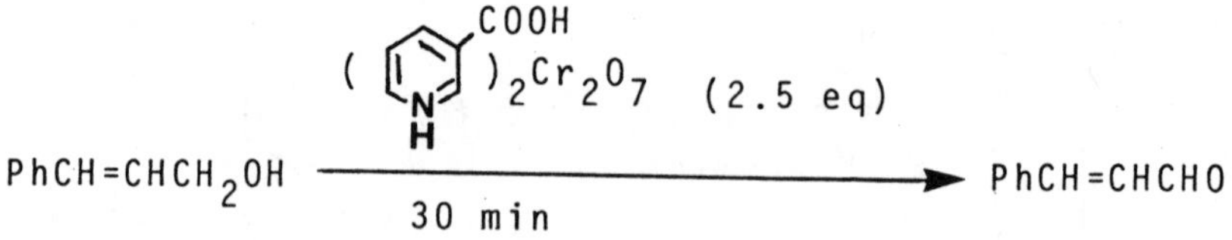

$\left(\begin{array}{c} \\ \end{array} \right)_2 Cr_2O_7$ (2.5 eq)

COOH

PhCH=CHCH$_2$OH

30 min

PhCH=CHCHO

80%

Lopez, C.; Gonzalez, A.; Cossio, F.P.; Palomo, C.[*]
Syn Commun, (1985), **15**, 1197

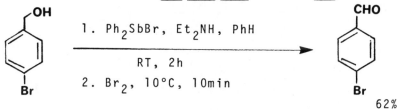

LAPA-EDA

130°C, 17h

74%

LAPA = lithium 3-aminopropanamide   (95% Z)

Hoffmann, H.M.R.[*]; Köver, A.; Pauluth, D.
JCS Chem Comm, (1985), 812

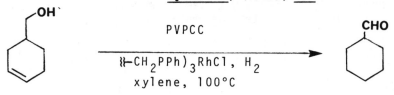

1. Ph$_2$SbBr, Et$_2$NH, PhH

RT, 2h

2. Br$_2$, 10°C, 10min

62%

Huang, Y.[*]; Shen, Y.; Chen, C.   Synthesis, (1985), 651

(NH$_4$)$_6$Mo$_7$O$_{24}$·4H$_2$O

6N HCl, dioxane, reflux
4h

63%

Sur, M.; Adak, M.M.; Pathak, T.[*]; Hazra, B.; Banerjee, A.
Synthesis, (1985), 652

PVPCC

$+$CH$_2$PPh)$_3$RhCl, H$_2$
xylene, 100°C

80%

PVPCC = poly(vinylpyridinium)chlorochromate
Bergbreiter, D.E.[*]; Chandran, R.
J Am Chem Soc, (1985), 107, 4792

$$DMSO-SbCl_5$$

benzene, reflux
2h

56%

Yamamoto, J.[*]; Ito, S.; Tsuboi, Ta.; Tsuboi, Ts.; Tsukihara, K.
Bull Chem Soc Jpn, (1985), **58**, 470

$$(Ph_3\overset{+}{P})_2CH_2Cr_2O_7^{-2}$$

$$CH_2Cl_2, \quad 25°C, \quad 12h$$

75%

Cristau, H.-J.[*]; Torreilles, E.[*]; Morand, P.; Christol, H.
Tetrahedron Lett, (1986), **27**, 1775

$$CH_3(CH_2)_5CH_2OH \xrightarrow[\substack{PhH, K_2CO_3 \\ reflux, 90\ min}]{Pb(OAc)_4, \ Mn(OAc)_2} CH_3(CH_2)_5CHO$$

68%

Mihailovic, M.L.[*]; Kostantinovic, S.; Vukicevic, R.
Tetrahedron Lett, (1986), **27**, 2287

Review: "Chromium (VI) Based Oxidants"

Firouzabadi, H.[*]; Iranpoor, N.; Kiaeezadeh, F.; Toofan, J.
Tetrahedron, (1986), **42**, 719

Related Methods: Ketones from Alcohols and Phenols (Section 168)

SECTION 49: Aldehydes from Aldehydes

Conjugate reductions and Michael alkylations of conjugated
aldehydes are listed in Section 74 (Alkyls from Olefins).

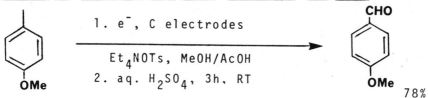

$\underline{n}C_4H_9CH_2CHO$

1. TMS-Cl
2. $CH_2I_2/Zn-Cu$
3. $Hg(OAc)_2/AcOH$
4. $CH_2=CHCN$, $NaBH_4$
5. KF

$\overset{O}{\overset{||}{HCCH(CH_2)_3C\equiv N}}$
$\underset{\underline{n}C_4H_9}{|}$ 60%

Giese, B.[*]; Horler, H. **Tetrahedron Lett**, (1983), **24**, 3221

CHO

e^-, Pt electrodes
────────────────────
MeOH, NaOTs
28°C

OMe
OMe
60%

Gora, J.; Smigielski, K.; Kula, J.   **Synthesis**, (1986), 586

Review: "Alkylation of Ketones and Aldehydes via Nitrogen
    Derivatives"

Whitesell, J.K.; Whitesell, M.A.   **Synthesis**, (1983), 517

Related Methods: Aldehydes from Ketones (Section 57)
                 Ketones from Ketones (Section 177)
                 Also via: Olefinic aldehydes (Section 341)

SECTION 50: Aldehydes from Alkyl, Methylenes, and Aryls

OMe

1. e^-, C electrodes
──────────────────────
Et$_4$NOTs, MeOH/AcOH
2. aq. H_2SO_4, 3h, RT

CHO

OMe
78%

Nishiguchi, I.[*]; Hirashima, T.   **J Org Chem**, (1985), **50**, 539

SECTION 51: Aldehydes from Amides

No Additional Examples

SECTION 52: Aldehydes from Amines

$$PhCH_2NH_2 \xrightarrow{\begin{array}{c} 1. \underline{m}TFMB, 6KOH, EtOAc \\ -78°C, 7h \\ \hline 2. 150°C, 2.5N HCl \end{array}} PhCHO$$

TFMB = (trifluoromethyl)benzenesulfonyl peroxide   77%

Hoffman, R.V.[*]; Kumar, A.   **J Org Chem**, (1984), **49**, 4011

Related Methods: Ketones from Amines (Section 172)

SECTION 53: Aldehydes from Esters

$$\begin{array}{c} 1. 2NaH/THF \\ MeSCH_2SO_2Tol \\ \hline 2. NaBH_4, MeOH \\ 3. K_2CO_3 \end{array}$$

91%

Ogura, K.[*]; Yahata, N.; Takahashi, K.; Iida, H.
Tetrahedron Lett, (1983), **24**, 5761

1. LTMP, CH_2Br_2, -90°C
   THF
2. $\underline{n}$BuLi, -90°C

3. ⬡   ruflux
4. TMS-Cl, -90 - 0°C   5. H^+ 68%

Kowalski, C.J.[*]; Haque, M.S.   **J Am Chem Soc**, (1986), **108**, 1325

3 $HSiEt_2Me$
50 atm CO

4% $Co_2(CO)_8$, ⬡

75%

Chatani, N.; Fujii, S.; Yamasaki, Y.; Murai, S.[*]; Sonoda, N.
J Am Chem Soc, (1986), **108**, 7361

SECTION 54: Aldehydes from Ethers, Epoxides, and Thioethers

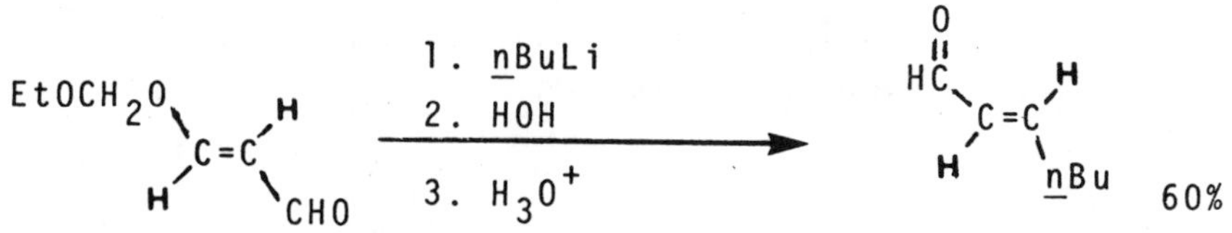

Maddaluno, J.; d'Angelo, J.* **Tetrahedron Lett**, (1983), **24**, 895

Related Methods: Ketones from Ethers and Epoxides (Section 174)

SECTION 55: Aldehydes from Halides and Sulfonates

$$CH_3CH_2CH_2Br \xrightarrow[\text{2. HOH}]{\text{1. } PhSCH_2SiMe_3\text{, nBuLi, TMEDA}} CH_3CH_2CH_2CHO \quad 70\%$$

Ager, D.J. **JCS Perkin I**, (1983), 1131

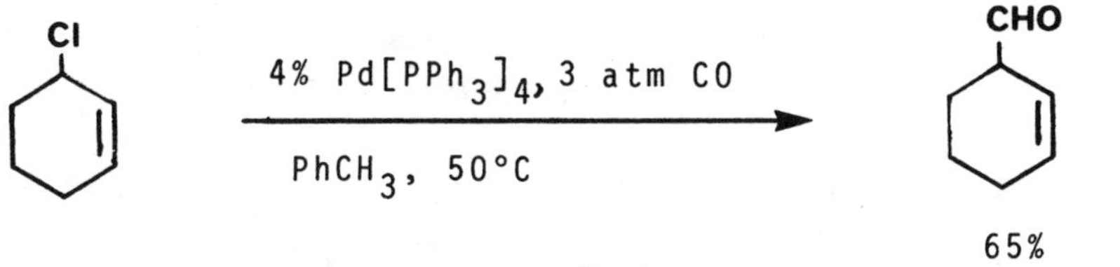

Baillargeon, V.P.; Stille, J.K.*
J Am Chem Soc, (1983), **105**, 7175

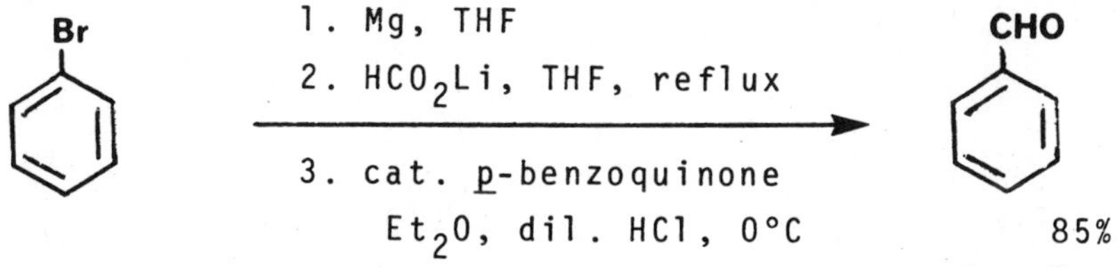

Bogavac, M.; Arsenijević, L.; Pavlov, S.; Arsenijević, V.*
Tetrahedron Lett, (1984), **25**, 1843

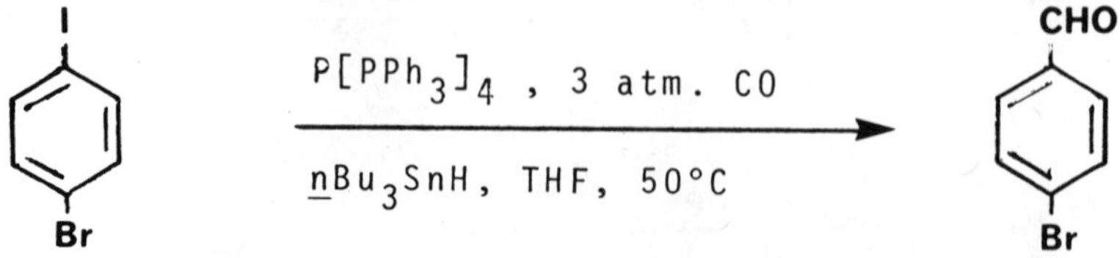

$$P[PPh_3]_4 \text{ , 3 atm. CO}$$
$$\underline{n}Bu_3SnH, \text{ THF, 50°C}$$

70%

Baillargeon, V.P.; Stille, J.K.*
J Am Chem Soc, (1986), **108**, 452

$CH_3(CH_2)_5I$

$$PtCl_2(PPh_3)_2, \text{ 120°C}$$
$$CO/H_2, \text{ } K_2CO_3$$
$$\text{dioxane, 9h}$$

$CH_3(CH_2)_5CHO$

86%

Takeuchi, R.; Tsuji, Y.; Watanabe, Y.*
JCS Chem Comm, (1986), 351

PhBr

$$\text{1. Li, DMF, 18°C, }\text{»))))}$$
$$\text{2. } H_3O^+$$

PhCHO

81%

Einhorn, J.; Luche, J.L.*
Tetrahedron Lett, (1986), **27**, 1791, 1793

$PhCH_2Br$

$$(\underline{n}Bu_4N)_2Cr_2O_7$$
$$\text{dioxane, reflux}$$
$$\text{3h}$$

PhCHO

85%

Ferraboschi, P.; Azadoni, M.N.; Santaniello, E.*; Trave, S.
Syn Commun, (1986), **16**, 43

$$CH_3(CH_2)_6CH_2Br \xrightarrow[\substack{115°C.\ NaI \\ 2h}]{DMSO.\ NaHCO_3} CH_3(CH_2)_6CHO$$

60%

Dave, P.; Byun, H.-S.; Engel, R.*   **Syn Commun**, (1986), **16**, 1343

SECTION 56: Aldehydes from Hydrides

$$\xrightarrow[\substack{Me_2S^+\text{-}SMe\ BF_4^- \\ RT.\ 2h}]{(PhS)_3CH,\ CH_2Cl_2}$$

70%

Smith, R.A.J.*; Bin Manas, A.R.   **Synthesis**, (1984), 166

SECTION 57: Aldehydes from Ketones

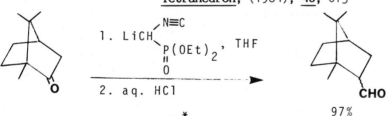

$$\xrightarrow[2h]{h\nu,\ PhH}$$

$$HCCH_2CH_2CH=C\begin{smallmatrix} CH_3 \\ \\ CH_2OH \end{smallmatrix}$$

39% E, 37% Z

Tortajada, J.; van Hemelryck, B.; Morizur, J.-P.
                              Tetrahedron, (1984), **40**, 613

$$\xrightarrow[2.\ aq.\ HCl]{1.\ LiCH\substack{N\equiv C \\ P(OEt)_2 \\ O},\ THF}$$

97%

Moskal, J.; van Leusen, A.M.*
                    Rec Trav Chim Pays Bas, (1986), **105**, 141

SECTION 58: Aldehydes from Nitriles

No Additional Examples

SECTION 59: Aldehydes from Olefins

$$CH_2=CH(CH_2)_3CH=CHCH_3 \xrightarrow[\text{cat.}]{\substack{2H_2/CO \\ PhH, \ RT}}$$

$$OHC(CH_2)_5CH=CHCH_3 \quad (12)$$

$$+ \quad 47\%$$

$$\underset{(1)}{CH_3\overset{\overset{CHO}{|}}{CH}(CH_2)_3CH=CHCH_3}$$

cat = $Rh(H)Co(PPh_3)_3$

Grigg, R.[*]; Reimer, G.J.; Wade, A.R.
JCS Perkin I, (1983), 1929

1. $BHBr_2 \cdot SMe_2$/pentane
2. $HOCH_2CH_2CH_2OH$
3. secBuLi/$MeOCH_2SPh$/THF
4. $HgCl_2$ 5. H_2O_2, pH 8

53%

Brown, H.C.[*]; Imai, T.   J Am Chem Soc, (1983), 105, 6285

1. e[-], Pt electrodes
   NaBr, CH_3CN/H_2O
2. 1% aq. H_2SO_4
3. e[-], Pt electrodes, 65°C
   aq. NaOH/PhH

97%

Torii, S.; Uneyama, K.; Ueda, K.   J Org Chem, (1984), 49, 1830

$$\xrightarrow[\substack{dioxane/H_2O \\ 3h}]{I_2/Ag_2O, \ RT}$$

92%

Kikuchi, H.[*]; Kogure, K.; Toyoda, M.
Chem Lett, (1984), 341

Related Methods: Ketones from Olefins (Section 179)

SECTION 60: Aldehydes Miscellaneous Compounds

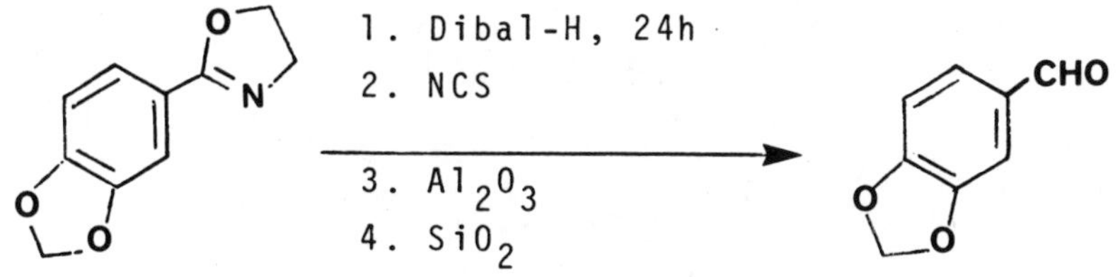

1. Dibal-H, 24h
2. NCS
3. Al_2O_3
4. SiO_2

62%

Meyers, A.I.[*]; Himmelsbach, R.J.; Reuman, M.
J Org Chem, (1983), 48, 4053

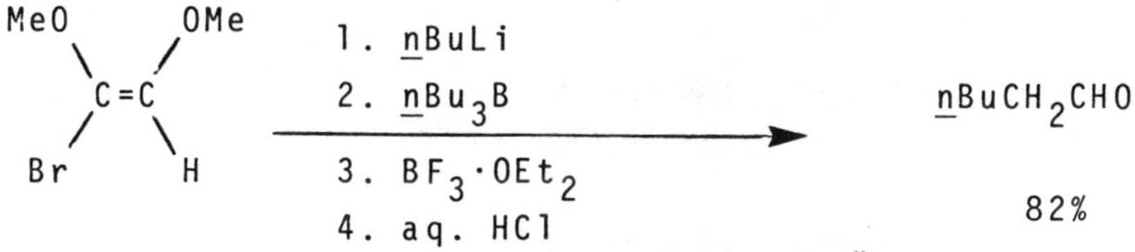

1. nBuLi
2. nBu₃B → nBu_3B
3. $BF_3 \cdot OEt_2$
4. aq. HCl

$nBuCH_2CHO$

82%

Koshino, J.; Sugawara, T.; Yogo, T.;Suzuki, A.[*]
Syn Commun, (1983), 13, 1149

$CH_3(CH_2)_6CH_2NO_2$

1. KOH, MeOH
2. conc. aq. $KMnO_4$
   $MgSO_4$, MeOH, 0°C

$CH_3(CH_2)_6CHO$

83%

Steliou, K.[*]; Poupart, M.A.   J Org Chem, (1985), 50, 4971

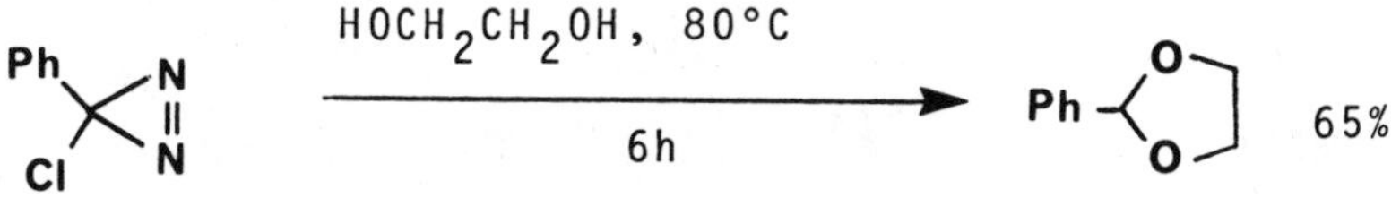

$HOCH_2CH_2OH$, 80°C

6h

65%

[+ PhCHO, 30%]

Liu, M.T.H.; Kokosi, J.   Heterocycles, (1985), 23, 3049

$$CH_3(CH_2)_5CH_2SO_2Ph \xrightarrow{\begin{array}{l}1.\ \underline{n}BuLi,\ THF,\ -78°C \\ 2.\ (MeO)_2BCl,\ -20°C \\ \quad hexane/THF \\ 3.\ mcpba,\ CH_2Cl_2 \\ \quad -20°C\end{array}} CH_3(CH_2)_5CHO$$

90%

Baudin, J.-B.; Julia, M.[*]; Rolando, C.
Tetrahedron Lett, (1985), **26**, 2333

$$\xrightarrow{\begin{array}{l}1.\ AlMe_3,\ DCE \\ \quad 25°C,\ 12h \\ 2.\ O_3/MeOH,\ -78°C\end{array}} \underline{n}C_6H_{13}CHCHO$$

85%
(S:R = 6.4:1)
(95% ee,S)

Maruoka, K.; Nakai, S.; Sakurai, M.; Yamamoto, H.[*]
Synthesis, (1986), 130

SECTION 60A: Protection of Aldehydes

CHO
|
$(CH_2)_8$
|
CH=CH_2

$\xrightarrow{\begin{array}{l}1.\ (P)\text{-}CH(OH)CH_2OH/PhH \\ \quad \underline{p}TsOH,\ heat \\ 2.\ B_2H_6 \\ 3.\ Me_3N \rightarrow O \\ 4.\ Py,\quad PhCCl \\ 5.\ aq.\ dioxane,\ \underline{p}TsOH\end{array}}$

CHO
|
$(CH_2)_9$
|
CH_2OCPh

65%

Hodge, P.[*]; Waterhouse, J.   **JCS Perkin I**, (1983), 2319

$$CH_3CH_2CHO \xrightarrow[-78°C\ -\ 0°C]{Si(OMe)_4} CH_3CH_2CH_2CH\begin{array}{l}OMe \\ OMe\end{array}$$

91%

Sakurai, H.[*]; Sassaki, K.; Hayashi, J.; Hosomi, A.
J Org Chem, (1984), **49**, 2808

CH$_3$(CH$_2$)$_7$CHO

$\xrightarrow[\text{(MeO)}_2\text{CMe}_2 \quad 85\%]{\text{CH}_2\text{OH} \;,\; \text{β-naphthol} \cdot \text{SO}_3\text{H}}$

CH$_3$(CH$_2$)$_7$

$\xleftarrow[\text{PhH, hν}]{}$
95%

Gravel, D.*; Murray, S.; Ladouceur, G.
JCS **Chem Comm**, (1985), 1828

nBuCHO

$\xrightarrow[\quad 87\% \quad]{\text{SH} \quad \text{NHMe}, \; \text{EtOH, reflux}}$

nBu

$\xleftarrow[\text{pH 7, NEt}_3 \qquad 86\%]{\text{AgNO}_3, \; \text{aq. CH}_3\text{CN, RT}}$

Chikashita, H.*; Ishimoto, N.; Kamazawa, S.; Itoh, K.
Heterocycles, (1985), **23**, 2509

CH$_3$(CH$_2$)$_4$CHO

$\xrightarrow[\text{quant.}]{\text{HSCH}_2\text{CH}_2\text{SH/SOCl}_2 \cdot \text{SiO}_2}$

CH$_3$(CH$_2$)$_4$

$\xleftarrow[\quad]{\text{SOCl}_2, \; \text{aq. SiO}_2}$

Kamitori, Y.; Hojo, M.*; Masuda, R.; Kimura, T.; Yoshida, T.
J Org Chem, (1986), **51**, 1427

nC$_6$H$_{13}$CH

SeMe

SeMe

$\xrightarrow[\text{RT, 90min}]{\text{Clayfen, pentane}}$

nC$_6$H$_{13}$CHO

Clayfen$_*$ = Fe^{+3}/K-10 Clay   91%
Laszlo, P.*; Pennetreau, P.; Krief, A.
Tetrahedron Lett, (1986), **27**, 3153

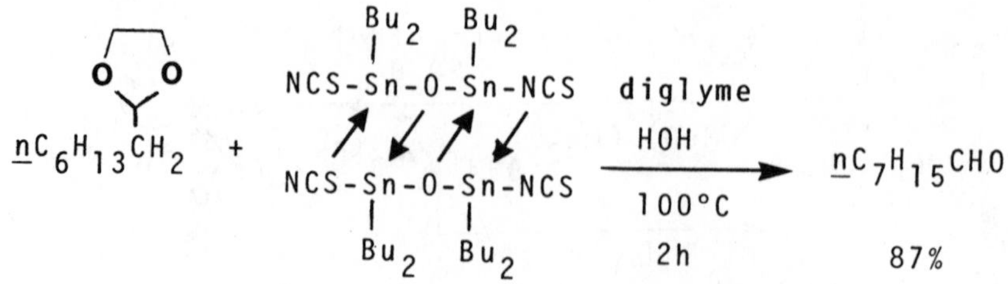

$$\underline{n}C_6H_{13}CH_2 \quad + \quad \begin{array}{cc} Bu_2 & Bu_2 \\ | & | \\ NCS-Sn-O-Sn-NCS \\ \\ NCS-Sn-O-Sn-NCS \\ | & | \\ Bu_2 & Bu_2 \end{array} \quad \xrightarrow[\substack{100°C \\ 2h}]{\substack{diglyme \\ HOH}} \quad \underline{n}C_7H_{15}CHO$$

87%

Otera, J.[*]; Nozaki, H.   **Tetrahedron Lett**, (1986), **27**, 5743

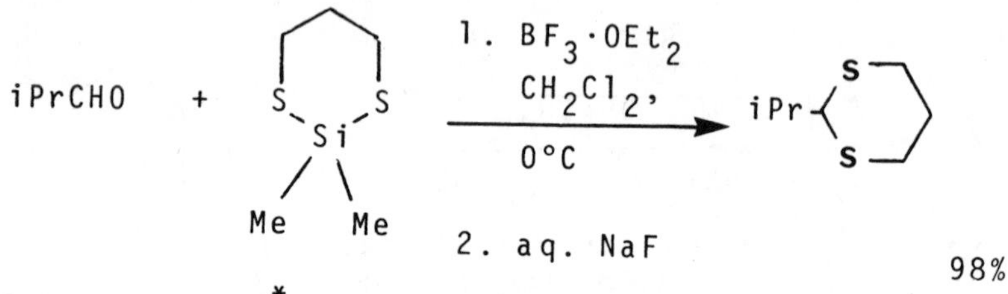

$$iPrCHO \quad + \quad \xrightarrow[\substack{2. \ aq. \ NaF}]{\substack{1. \ BF_3 \cdot OEt_2 \\ CH_2Cl_2, \\ 0°C}} \quad iPr$$

98%

Soderquist, J.A.[*]; Miranda, E.I.
                        Tetrahedron Lett, (1986), **27**, 6305

See Section 367 (Ether - Olefin) for
the formation of enol ethers. Many of
the methods in Section 180A (Protection
of Ketones) are also applicable to
aldehydes.

CHAPTER 5

PREPARATION OF ALKYLS, METHYLENES, AND ARYLS

This chapter lists the conversion of functional groups into Me, Et, ..., CH_2, Ph, etc.

SECTION 61: Alkyls, Methylenes, and Aryls from Acetylenes

PhC≡CH, 50°C

20% $H_3PO_4 \cdot BF_3$

25 min

82%

Zinov'eva, L.V.; Ryabov, V.D.   **J Org Chem USSR**, (1984), **20**, 540

Review: "Cobalt Mediated 2+2+2 Cycloadditions: A Maturing Synthetic Strategy"

Vollhardt, K.P.C.[*]   **Angew Chem Int Ed Engl**, (1984), **23**, 539

SECTION 62: Alkyls, Methylenes, and Aryls from Acid Derivatives

$Na_2S_2O_8$

$AgNO_3$

heat

46%

Fristad, W.E.; Klang, J.A.   **Tetrahedron Lett**, (1983), **24**, 2219

1. tBuLi, THF, -78°C
2. $PhCH_2Br$, -78°C - RT
3. pH 3, 0°C (10%HCl)
4. 170°C, 11 mmHg

90%

Muchowski, J.M.[*]; Solas, D.R.   **Syn Commun**, (1984), **14**, 453

<u>SECTION 63</u>: <u>Alkyls</u>, <u>Methylenes</u>, <u>and</u> <u>Aryls</u> <u>from</u> <u>Alcohols</u> <u>and</u>
<u>Thiols</u>

$(CH_3)_2C=CHCH_2OH$

1. $Me_2C=C$ 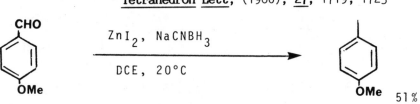 $\begin{smallmatrix}Cl\\N\end{smallmatrix}$ THF, 0°C

2. $PhCH_2CH_2MgBr$, CuI
   HMPA, -30°C

$PhCH_2CH_2CHCH=CH_2$
             CH_3 (91)

+

86% $Ph(CH_2)_3CH=CHCH_3$ (9)

Fujisawa, T.[*]; Iida, S.; Yukizaki, H.; Sato, T.
                    Tetrahedron Lett, (1983), **24**, 5745

$\begin{smallmatrix}Ph\\H\end{smallmatrix}C=C\begin{smallmatrix}H\\CH_2OH\end{smallmatrix}$

$\xrightarrow[\text{benzene, RT - reflux}]{(4-OMe)PhMgBr}$

$(Ph_3P)_2NiCl_2$

Ph—CH=CH—⟨OMe⟩

80%

Wenkert, E.[*]; Fernandes, J.B.; Michelotti, E.L.; Swindell, C.S.
                    Synthesis, (1983), 701

<u>SECTION 64</u>: <u>Alkyls</u>, <u>Methylenes</u>, <u>and</u> <u>Aryls</u> <u>from</u> <u>Aldehydes</u>

PhCHO

1. 2 MeSeH, CH_2Cl_2
2. nBiLi, THF, -78°C
3. nBuI, -78 - 20°C
4. nBuLi, THF, -78°C
5. $nC_6H_{13}Br$

$Ph-\overset{\underline{n}Bu}{\underset{H}{|}}-nC_6H_{13}$

63%

Clarembeau, M.; Krief, A.[*]
                    Tetrahedron Lett, (1986), **27**, 1719, 1723

CHO⟨⟩**OMe**

$\xrightarrow[\text{DCE, 20°C}]{ZnI_2, NaCNBH_3}$

⟨⟩**OMe**   51%

Lau, C.K.[*]; Dufresne, C.; Bélanger, P.C.; Piétré, S.; Scheigetz, J.

                    J Org Chem, (1986), **51**, 3038

Related Methods: Alkyls, Methylenes, and Aryls from Ketones
(Section 72)

SECTION 65: Alkyls, Methylenes, and Aryls from Alkyls, Methylenes, and Aryls

1. $\underline{n}$BuLi, TMEDA, RT, 4h

2. Me_3SiCl

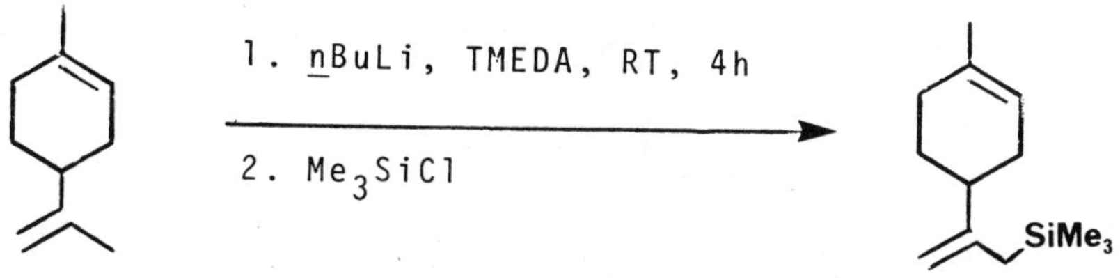

58%

Andrianome, M.; Delmond, B.[*]
Tetrahedron Lett, (1985), 26, 6341

SECTION 66: Alkyls, Methylenes, and Aryls from Amides

No Additional Examples

SECTION 67: Alkyls, Methylenes, and Aryls from Amines

No Additional Examples

SECTION 68: Alkyls, Methylenes, and Aryls from Esters

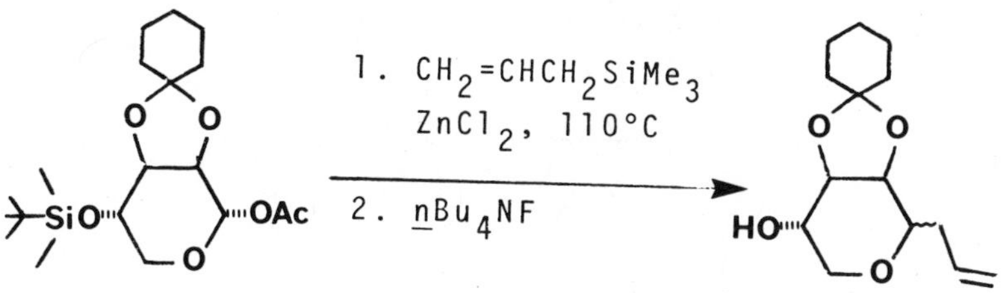

1. $CH_2=CHCH_2SiMe_3$
   $ZnCl_2$, 110°C

2. $\underline{n}Bu_4NF$

$(\alpha:\beta = 3:1)$ 90%

Kozikowski, A.P.[*]; Sorgi, K.L.; Wang, B.L.; Xu, Z.
Tetrahedron Lett, (1983), 24, 1563

$$CH_2=CHCH\underset{\overset{\displaystyle |}{CH_3}}{-}O\overset{\overset{\displaystyle O}{||}}{C}CF_3 \quad \xrightarrow[\substack{CF_3COOH \\ 80°C,\ 8h}]{benzene} \quad PhCH_2CH=CHCH_3$$

78%

Fujiwara, Y.*; Kuromaru, H.; Taniguchi, H.
J Org Chem, (1984), **49**, 4309

$$\underset{\substack{\displaystyle | \\ iPrCHCH=CHCH_3 \\ (\underline{E})}}{OAc} \quad \xrightarrow[\substack{THF-Et_2O-pentane \\ 0°C,\ 2h}]{[PhMe_2Si]_2CuLi} \quad \begin{array}{l} \underset{\substack{\displaystyle | \\ SiMe_2Ph}}{iPrCHCH=CHCH_3}\ (\underline{E}) \\ (26) \\ + \\ \underset{\substack{\displaystyle | \\ SiMe_2Ph}}{iPrCH=CHCHCH_3}\ (\underline{E}) \\ (74) \qquad 83\% \end{array}$$

Fleming, I.*; Thomas, A.P.   **JCS Chem Comm**, (1985), 411

NiCl$_2$, Zn, PPh$_3$

NaI, DMF
60°C)))))

85%

Yamashita, J.*; Inoue, Y.; Kondo, T.; Hashimoto, H.
Chem Lett, (1986), 407

CH$_3$MgI, 3% CuCN

Et$_2$O, 0°C

60%

Tseng, C.C.; Paisley, S.D.; Goering, H.L.*; Yen, S.-J.
J Org Chem, (1986), **51**, 2884, 2892

SECTION 69: Alkyls, Methylenes, and Aryls from Ethers, Epoxides, and Thioethers

The conversion ROR → RR' (R' = alkyl, aryl) is included in this section.

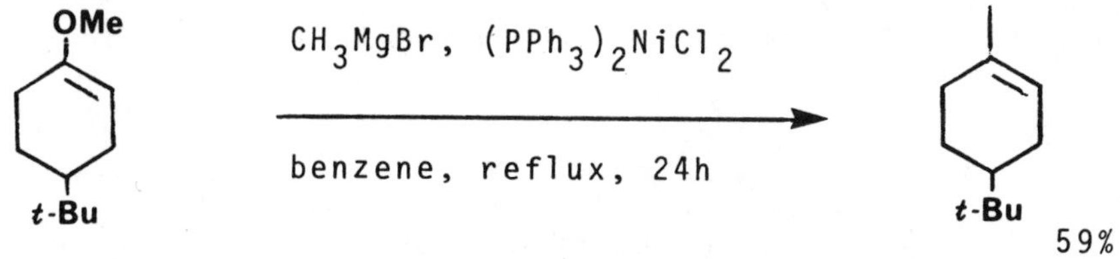

59%

Wenkert, E.*; Michelotti, E.L.; Swindell, C.S.; Tingoli, M.
J Org Chem, (1984), **49**, 4894

SECTION 70: Alkyls, Methylenes, and Aryls from Halides and Sulfonates

The replacement of halogen by alkyl or aryl groups is included in this section. For the conversion of RX → RH (X = halo) see Section 160 (Hydrides from Halides and Sulfonates).

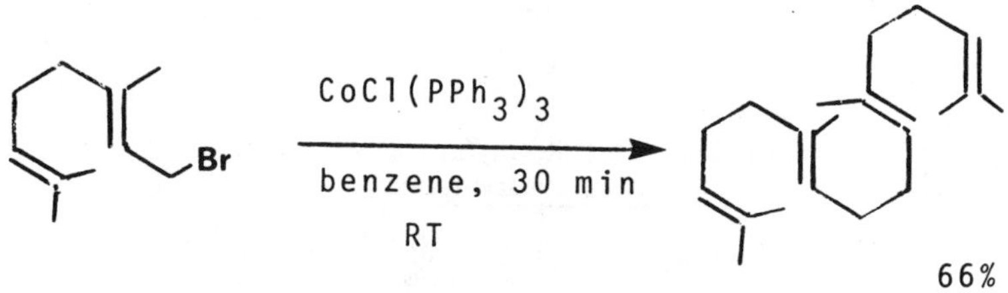

66%

Momosi, D.; Iguchi, K.; Sugujama, T.; Yamada, Y.*
Tetrahedron Lett, (1983), **24**, 921

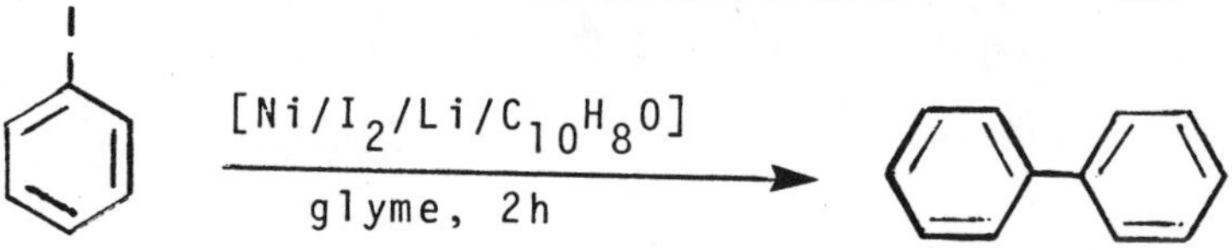

quant.

Matsumoto, H.; Inaba, S.; Rieke, R.D.*
J Org Chem, (1983), **48**, 840

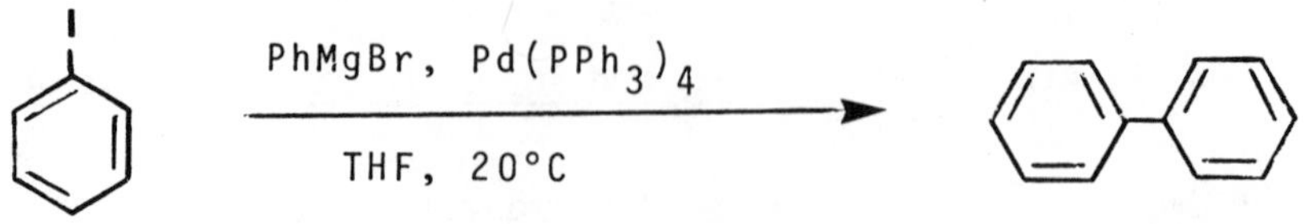

PhMgBr, Pd(PPh$_3$)$_4$

THF, 20°C

70%

Widdowson, D.A.[*]; Zhang, Y.-A.   **Tetrahedron**, (1986), **42**, 2111

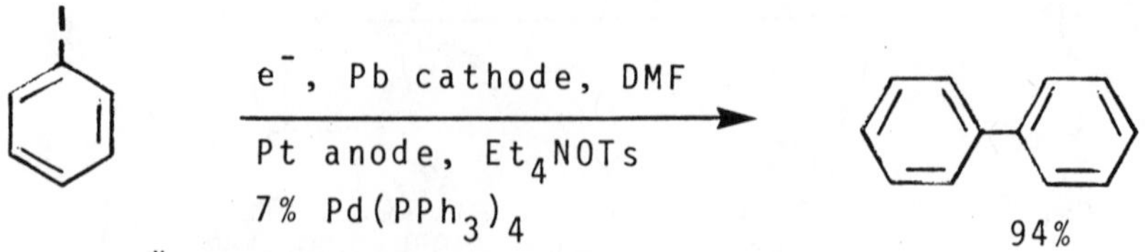

e$^-$, Pb cathode, DMF

Pt anode, Et$_4$NOTs

7% Pd(PPh$_3$)$_4$

94%

Torii, S.[*]; Tanaka, H.; Morisaki, K.
Tetrahedron Lett, (1985), **26**, 1655

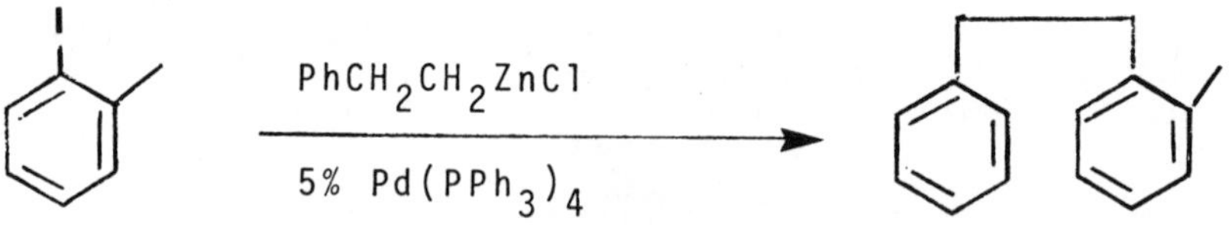

PhCH$_2$CH$_2$ZnCl

5% Pd(PPh$_3$)$_4$

73%

Negishi, E.[*]; Matsushita, H.; Kobayashi, M.; Rand, C.L.
Tetrahedron Lett, (1983), **24**, 3823

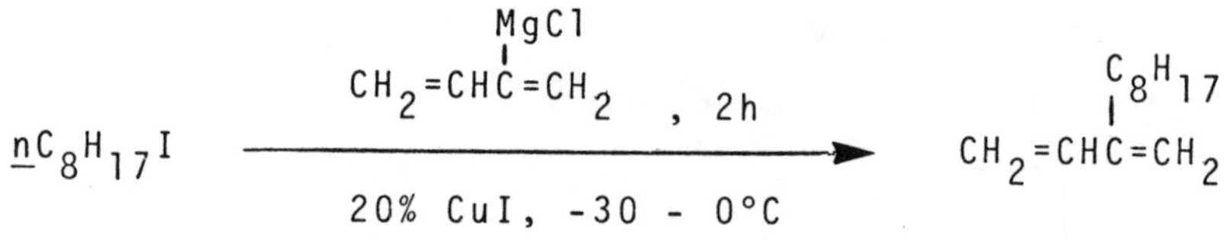

$\underset{|}{\overset{MgCl}{CH_2=CHC=CH_2}}$, 2h

$\underline{n}$C$_8$H$_{17}$I $\xrightarrow{\hspace{1cm}}$ $\underset{|}{\overset{C_8H_{17}}{CH_2=CHC=CH_2}}$

20% CuI, -30 - 0°C

60%

Nunomoto, S.[*]; Kawakami, Y.[*]; Yamashita, Y.
J Org Chem, (1983), **48**, 1912

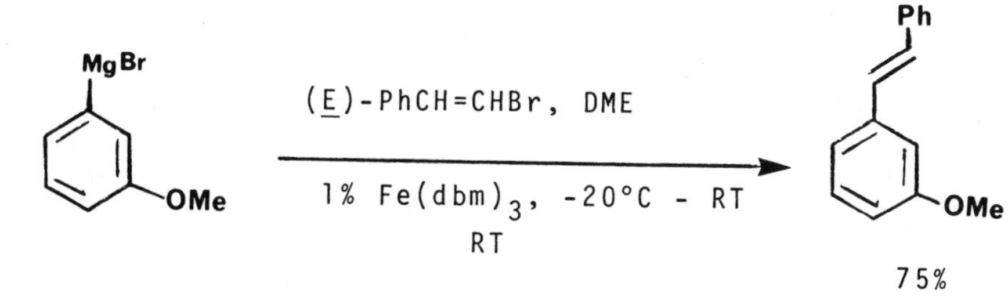

$(\underline{E})$-PhCH=CHBr, DME

1% Fe(dbm)$_3$, -20°C - RT
RT

75%

Molander, G.A.[*]; Rahn, B.J.; Shubert, D.C.; Bonde, S.E.
Tetrahedron Lett, (1983), **24**, 5449

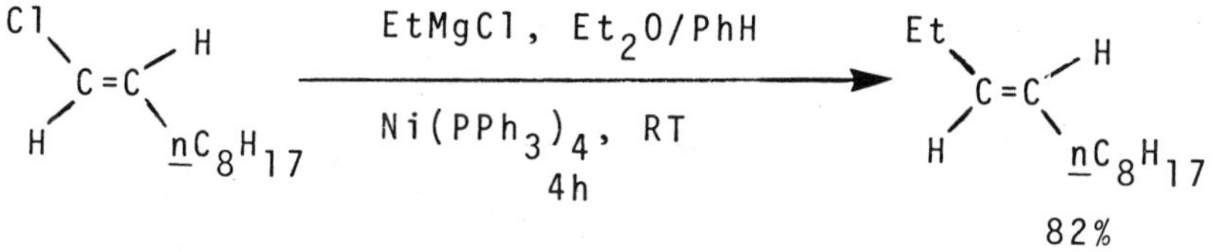

EtMgCl, Et$_2$O/PhH

Ni(PPh$_3$)$_4$, RT
4h

82%

Ratovelomanana, V.; Linstrumelle, G.
Syn Commun, (1984), **14**, 179

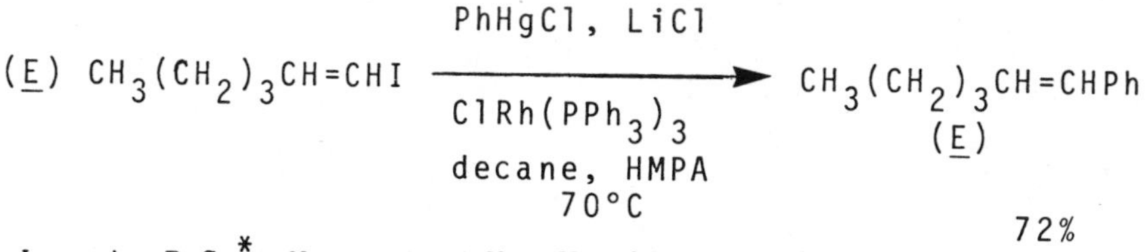

$(\underline{E})$ CH$_3$(CH$_2$)$_3$CH=CHI

PhHgCl, LiCl

ClRh(PPh$_3$)$_3$
decane, HMPA
70°C

CH$_3$(CH$_2$)$_3$CH=CHPh
$(\underline{E})$

72%

Larock, R.C.[*]; Narayanan, K.; Hershberger, S.S.
J Org Chem, (1983), **48**, 4377

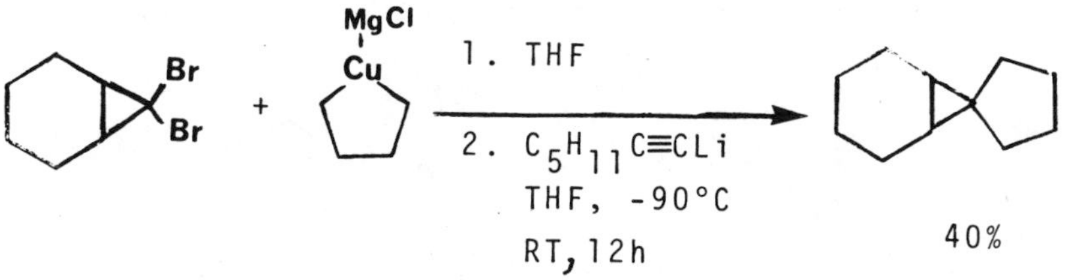

1. THF

2. C$_5$H$_{11}$C≡CLi
THF, -90°C
RT, 12h

40%

Scott, F.[*]; Mafunda, B.G.; Normant, J.F.; Alexakis, A.
Tetrahedron Lett, (1983), **24**, 5767

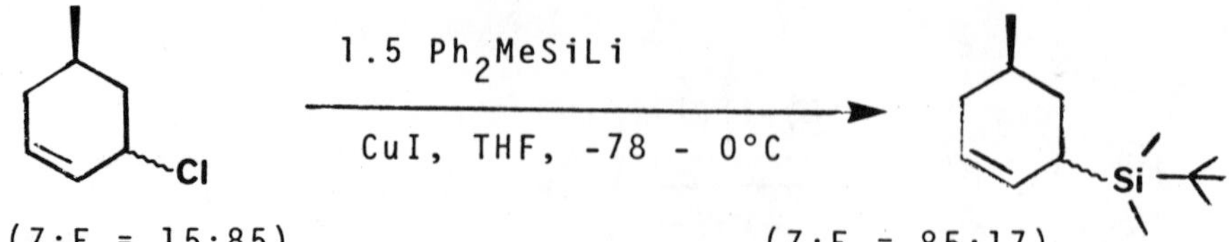

($\underline{Z}:\underline{E}$ = 15:85)

1.5 $Ph_2MeSiLi$

CuI, THF, -78 - 0°C

($\underline{Z}:\underline{E}$ = 85:17)

75%

Laycock, B.; Kitching, W.[*]; Wickham, G.
Tetrahedron Lett, (1983), **24**, 5785

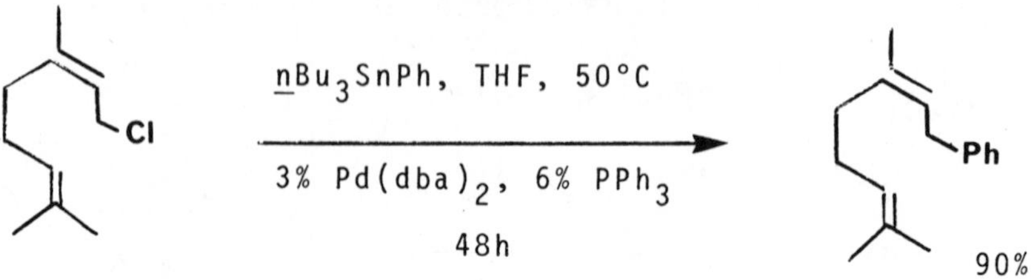

$\underline{n}Bu_3SnPh$, THF, 50°C

3% $Pd(dba)_2$, 6% PPh_3

48h

90%

Sheffy, F.K.; Stille, J.K.[*]   **J Am Chem Soc**, (1983), **105**, 7173

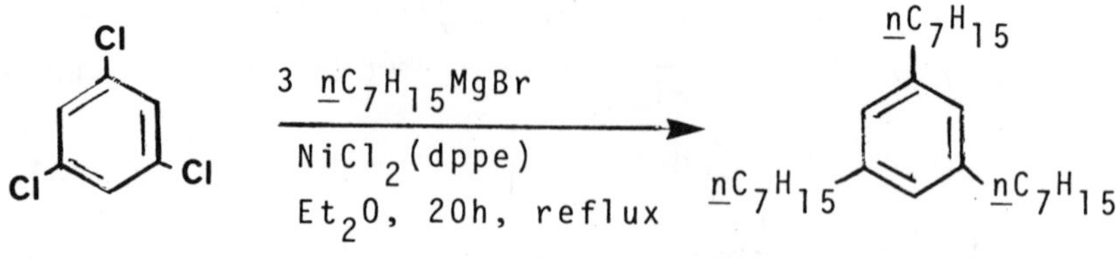

3 $\underline{n}C_7H_{15}MgBr$

$NiCl_2$(dppe)

Et_2O, 20h, reflux

87%

Eapen, K.C.; Dua, S.S.; Tamborski, C.[*]
J Org Chem, (1984), **49**, 478

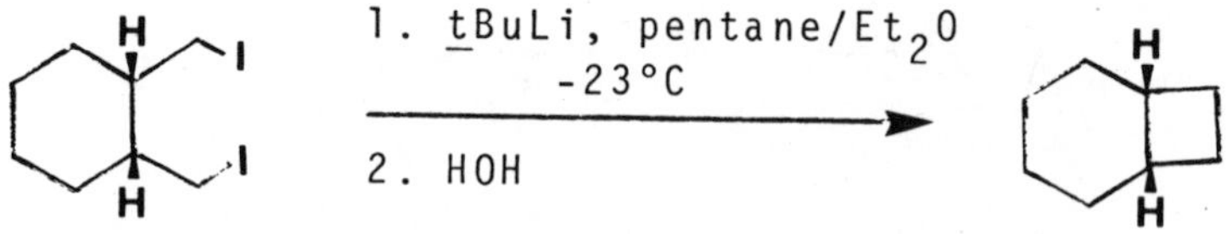

1. $\underline{t}BuLi$, pentane/Et_2O
   -23°C

2. HOH

97%

Bailey, W.F.[*]; Gagnier, R.P.; Patricia, J.J.
J Org Chem, (1984), **49**, 2098

$$(\underline{E})\ CH_3CH=CHCH_2Sn\underline{n}Bu_3$$
$$\xrightarrow{\hspace{2cm}}$$
$$CH_2Cl_2,\ 50°C$$
$$10\ KBar$$

88%

Yamamoto, Y.[*]; Muruyama, K.; Matsumoto, K.
JCS Chem Comm, (1984), 548

$$\underline{n}C_6H_{13}CH=CHCH_2Cl \xrightarrow[\text{TMPA, 5°C}]{\text{TMS-Cu}}$$

$$\underline{n}C_6H_{13}\underset{\underset{TMS}{|}}{C}HCH=CH_2\ (98)$$

+   87%

$$\underline{n}C_6H_{13}CH=CHCH_2THS\ (2)$$

Smith, J.G.[*]; Drozda, S.E.; Petraglia, S.P.; Quinn, N.K.; Rice,
E.M.; Taylor, B.S.; Viswanathan, M.
J Org Chem, (1984), **49**, 4112

$$CH_3CH_2\underset{\underset{CH_3}{|}}{C}HMgCl,\ Et_2O$$
$$\xrightarrow{\hspace{2cm}}$$
$$PdCl_2(dppf)$$

75%

Hayashi, T.[*]; Konishi, M.; Kobori, Y.; Kumada, M.; Higuchi, T.;
Hirotsu, K.
J Am Chem Soc, (1984), **106**, 158

$$CH_2=CHCH_2OAc,\ benzene$$
$$\underline{n}Bu_3SnSn\underline{n}Bu_3,\ reflux$$

86%

Yokoyama, Y.; Ito, S.; Takahashi, Y.; Murakami, Y.[*]
Tetrahedron Lett, (1985), **26**, 6457

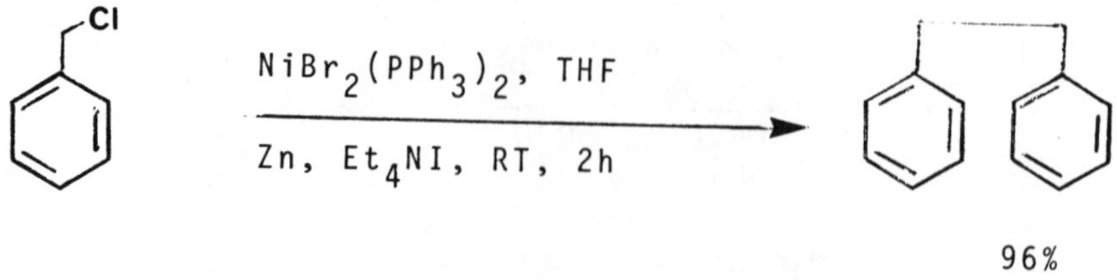

96%

Iyoda, M.[*]; Sakaitani, M.; Otsuka, H.; Oda, M.[*]
Chem Lett, (1985), 127

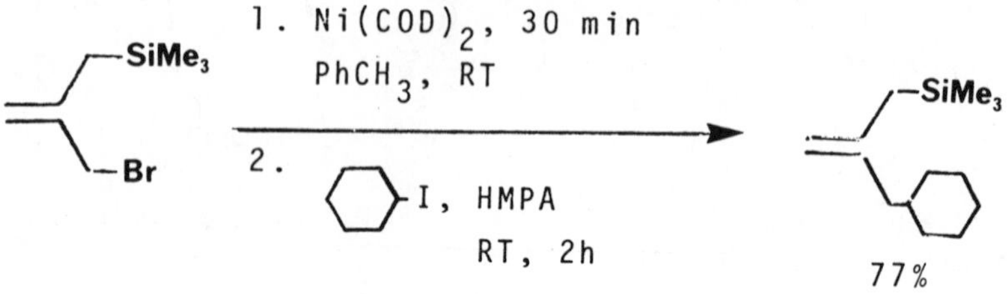

77%

Molander, G.A.[*]; Shubert, D.C.
Tetrahedron Lett, (1986), **27**, 787

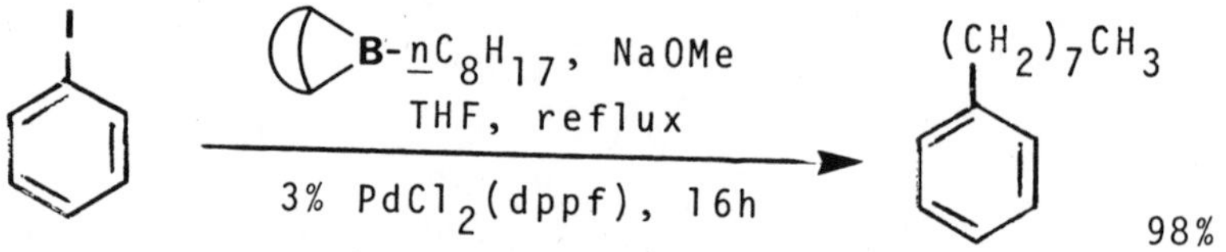

98%

Miyaura, N.; Ishiyama, T.; Ichikawa, M.; Suzuki, A.[*]
Tetrahedron Lett, (1986), **27**, 6369

Reviews:

"Copper Assisted Nucleophilic Substitution of Aryl Halogen"

Lindley, J.[*]   **Tetrahedron**, (1984), **40**, 1433

"The Chemistry of Higher Order Cuprates"

Lipshutz, B.H.[*]; Wilhelm, R.S.; Kozlowski, J.A.
Tetrahedron, (1984), **40**, 5005

SECTION 71: Alkyls, Methylenes, and Aryls from Hydrides

This section lists examples of the reaction RH → RR' (R,R' = alkyl or aryl). For the reaction C=CH → C=CR (R = alkyl or aryl) see Section 209 (Olefins from Olefins). For alkylations of ketones and esters, see Section 177 (Ketones from Ketones) and Section 113 (Esters from Esters).

$$CH_2=CHCO_2Me \xrightarrow[\substack{Co(CO)_8, \ 25°C \\ 3h}]{HSiMe_3, \ PhH} \overset{\displaystyle SiMe_3}{\underset{\displaystyle CH=CHCO_2Me}{|}}$$

65%

Takeshita, K.; Seki, Y.[*]; Kawamoto, K.; Murai, S.[*]; Sonoda, N.
JCS Chem Comm, (1983), 1193

$$PhCH=CHC=CH_2 \xrightarrow[\substack{Et_2O, \ RT}]{\substack{4 \ \underline{n}C_8H_{15}MgBr \\ 0.1 \ NiCl_2(dppp)}} PhCH=CHC=CH_2$$

with OP(OEt)$_2$‖O substituent → $\underline{n}C_8H_{17}$ substituent

78%

Sahlberg, C.; Quader, A.; Claesson, A.[*]
Tetrahedron Lett, (1983), **24**, 5137

1. $\underline{n}$BuLi
2. CuBr·SMe$_2$
3. Me$_2$C=CHCH$_2$Br

THF, -78°C - RT
2h

62%

Beswick, P.J.; Leach, S.J.; Masters, N.F.; Widdowson, D.A.[*]
JCS Chem Comm, (1984), 46

PhH, AcOH, 100°C

PhCH=CHCHO $\xrightarrow[\begin{array}{c}\text{Pd}(O_2CPh)_2,\ 3h\\ \underline{t}BuOO_2CPh\end{array}]{}$ Ph$_2$C=CHCHO

64%

Tsuji, J.[*]; Nagashima, H.   **Tetrahedron**, (1984), **40**, 2699

$\xrightarrow[\text{PhH, 40°C, 5h}]{\text{FeCl}_3,\ \text{K-10 Clay}}$

65%

Chalais, S.[*]; Cornelis, A.; Gerstmans, A.; Kolodziejski, W.;
Laszlo, P.[*]; Mathy, A.; Metra, P.
                    Helv Chim Acta, (1985), **68**, 1196

CH$_2$=CHPh $\xrightarrow[\begin{array}{c}\text{Pd(dba)}_2,\ \text{CH}_3\text{CN}\\ 40°C,\ 40\ \text{min}\end{array}]{}$

(97)

(3)

60%

Kikukawa, K.[*]; Naritomi, M.; He, G.-X.; Wada, F.; Matsuda, T.
                    J Org Chem, (1985), **50**, 299

$\xrightarrow[\text{2. Ni(R)}]{\begin{array}{c}1.\quad\quad,\ \text{SnCl}_4\\ -20°C,\ 30\ \text{min}\end{array}}$

90%

Ishibashi, H.[*]; Nakatani, H.; Umei, Y.; Ikeda, M.
                    Tetrahedron Lett, (1985), **26**, 4373

SECTION 72: Alkyls, Methylenes, and Aryls from Ketones

The conversions $R_2CO \rightarrow RR$, R_2CH_2, R_2CHR', etc. are listed in this section.

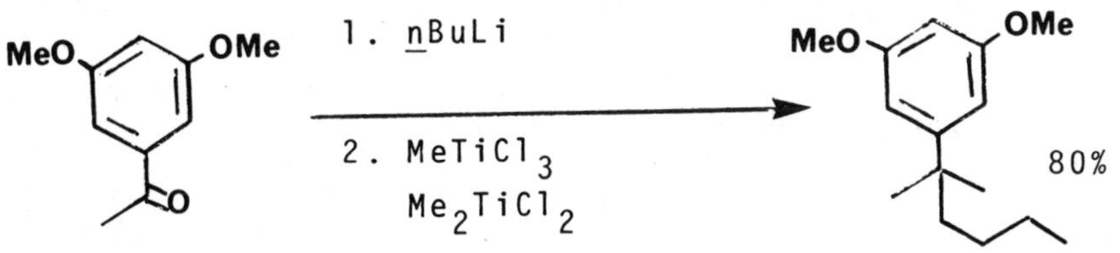

1. nBuLi

2. MeTiCl$_3$
   Me$_2$TiCl$_2$

80%

Reetz, M.T.[*]; Westermann, J.   **J Org Chem**, (1983), **48**, 254

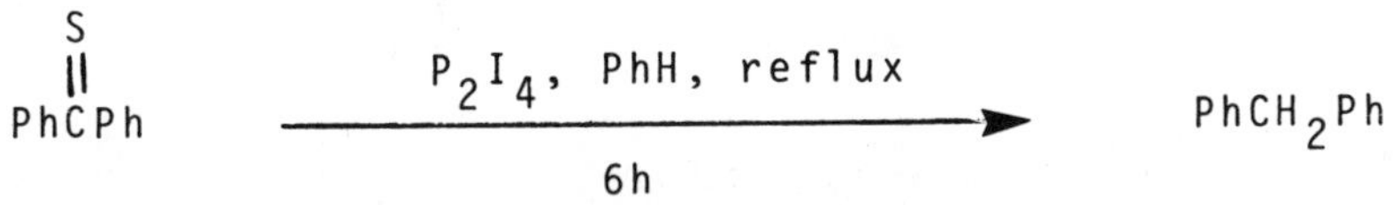

S
‖
PhCPh

P_2I_4, PhH, reflux

6h

PhCH$_2$Ph

85%

Suzuki, H.[*]; Tani, H.; Takeuchi, S.
                    Bull Chem Soc Jpn, (1985), **58**, 2421

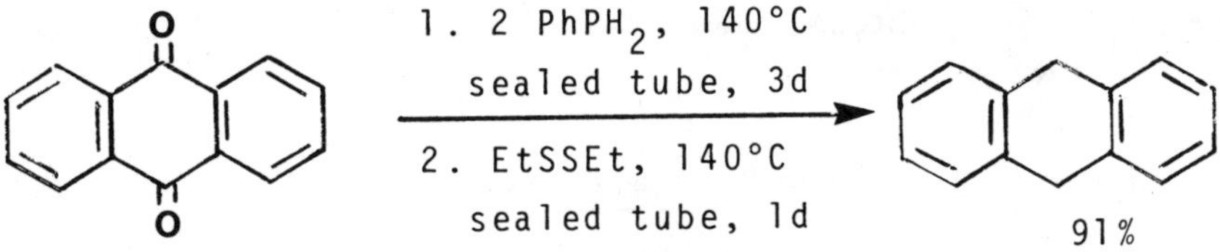

1. 2 PhPH$_2$, 140°C
   sealed tube, 3d

2. EtSSEt, 140°C
   sealed tube, 1d

91%

Tamano, M.; Koketsu, J.[*]   **Bull Chem Soc Jpn**, (1985), **58**, 2577

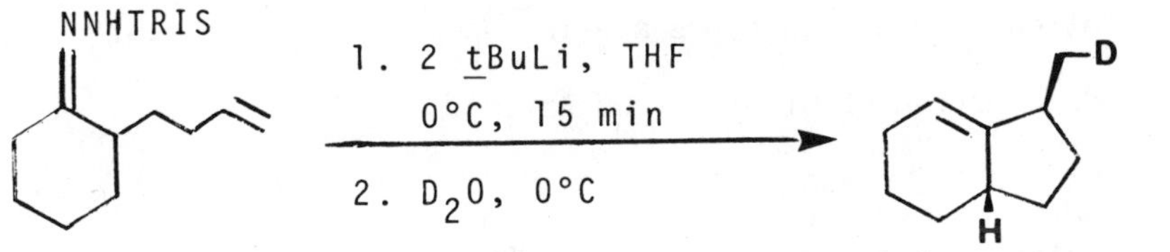

NNHTRIS

1. 2 tBuLi, THF
   0°C, 15 min

2. D$_2$O, 0°C

87%

Chamberlin, A.R.[*]; Bloom, S.H.
                    Tetrahedron Lett, (1986), **27**, 551

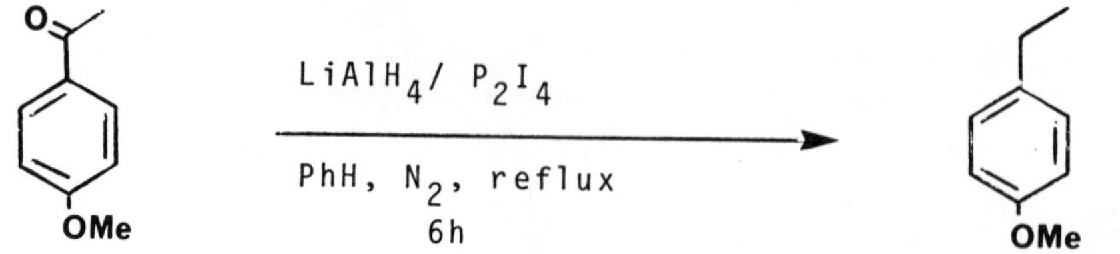
LiAlH4/ P2I4
PhH, N2, reflux
6h
OMe
OMe

65%

Suzuki, H.*; Masuda, R.; Kubota, H.; Osuka, A.
Chem Lett, (1983), 909

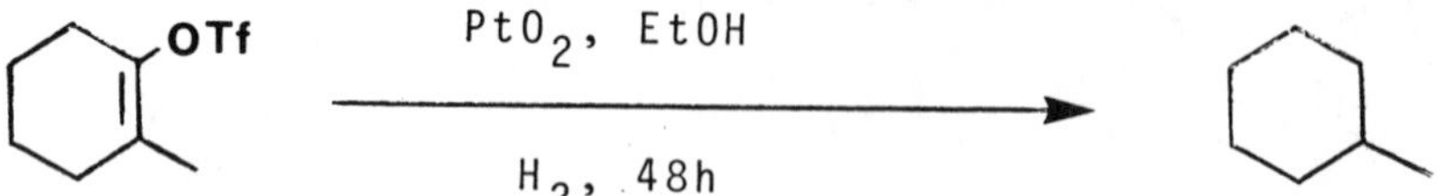
OTf
PtO2, EtOH
H2, 48h

81%

Subramanian, L.R.*; Garcia Martinez, A.*; Herrera Fernandez, A.; Martinez Alvarez, R.
Synthesis, (1984), 481

SECTION 73: Alkyls, Methylenes, and Aryls from Nitriles

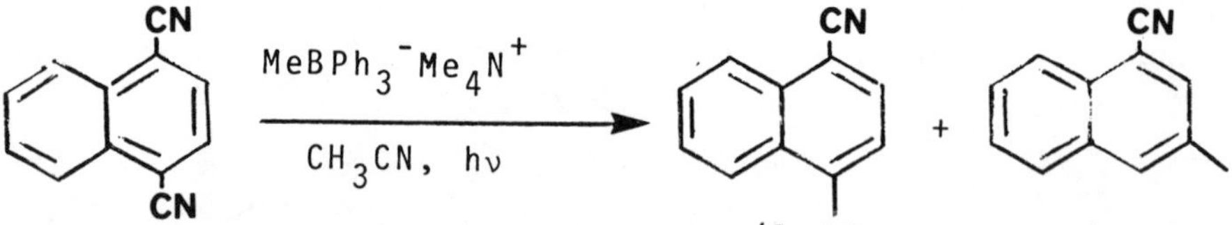
CN
CN
MeBPh3⁻Me4N⁺
CH3CN, hν
CN
CN
+

(1.25 : 1)
quant.

Lan, J.Y.; Schuster, G.B.* **Tetrahedron Lett**, (1986), **27**, 4261

SECTION 74: Alkyls, Methylenes, and Aryls from Olefins

The following reaction types are included in this section:

A. Hydrogenation of Olefins (and Aryls).

B. Formation of Aryls.

C. Alkylations and Arylations of Olefins.

D. Conjugate Reduction of Conjugated Aldehydes, Ketones, Acids, Esters, and Nitriles.

E. Conjugate Alkylations.

F. Cyclopropanations, including halocyclopropanations.

SECTION 74A: Hydrogenation of Olefins (and Aryls)

Reduction of aryls to dienes are listed in Section 377 (Olefin - Olefin).

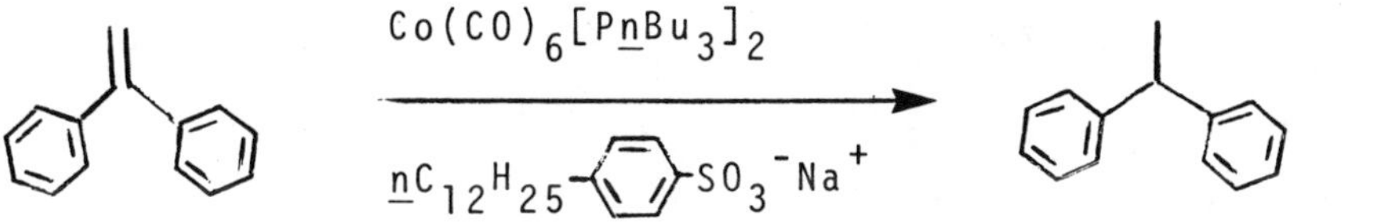

$Co(CO)_6[P\underline{n}Bu_3]_2$

$\underline{n}C_{12}H_{25}\!-\!\langle\!\rangle\!-\!SO_3{}^-Na^+$

97%

Alper, H.[*]; Heveling, J.   JCS Chem Comm, (1983), 365

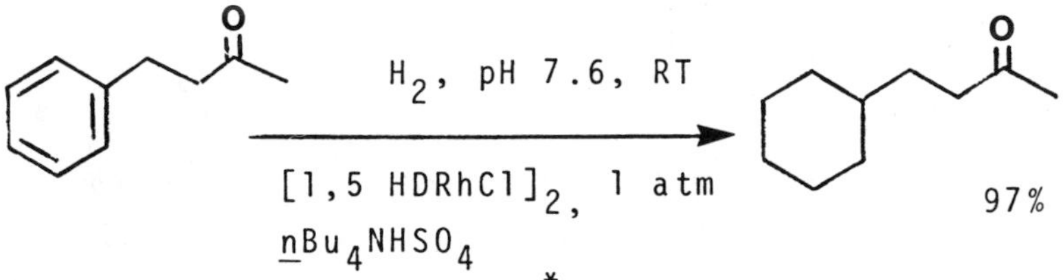

H_2, pH 7.6, RT

$[1,5\ HDRhCl]_2$, 1 atm
$\underline{n}Bu_4NHSO_4$

97%

Januszkiewicz, K.R.; Alper, H.[*]
                    Organometallics, (1983), 2, 1055

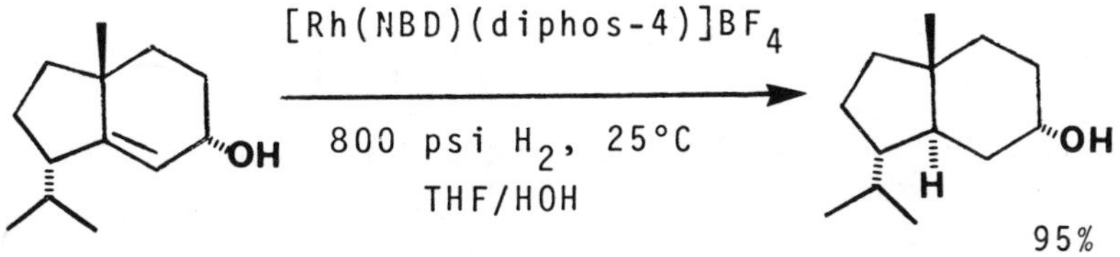

$[Rh(NBD)(diphos-4)]BF_4$

800 psi H_2, 25°C
THF/HOH

95%

Evans, D.A.[*]; Morrissey, M.M.   J Am Chem Soc, (1984), 106, 3866

$$Br(CH_2)_4CH=CH_2 \xrightarrow[\substack{C_6H_{14}, \ pH \ 7.6, \ 4h \\ 1 \ atm \ H_2}]{[1,5 \ HDRhCl]_2, \ RT} Br(CH_2)_5CH_3$$

78%

Januszkiewicz, K.R.; Alper, H.[*]   **Can J Chem**, (1984), **62**, 1031

$$CH_3(CH_2)_5CH=CH_2 \xrightarrow[\substack{autoclave, \ 145°C \\ 500 \ psi \ H_2, \ 10 \ min}]{[(\eta^4Ph_4C_4CO)(CO_2)Ru]_2} CH_3(CH_2)_6CH_3$$

quant.

Blum, Y.; Czarkie, D.; Rahamim, Y.; Shvo, Y.[*]
Organometallics, (1985), **4**, 1459

$$\xrightarrow[\substack{CH_2Cl_2, \ H_2, \ 0°C}]{[Ir(COD)(PCy)_3Py]PF_6}$$

99.9%

Crabtree, R.H.[*]; Davis, M.W.   **J Org Chem**, (1986), **51**, 2655

$$\xrightarrow[\substack{Me_3SiCl, \ DME, \ 45°C, \ 4h}]{NaH, \ \underline{t}AmONa, \ Ni(OAc)_2}$$

99%

Fort, Y.; Vanderesse, R.; Caubere, P.[*]
Tetrahedron Lett, (1986), **27**, 5487

Review: "cis-Akyl and cis-Acylrhodium and Iridium Hydrides: Model Intermediates in Homogeneous Catalysis"

Milstein, D.[*]   **Accts Chem Res**, (1984), **17**, 221

SECTION 74B: <u>Dehydrogenation</u> <u>to</u> <u>form</u> <u>Aryls</u>

No Additional Examples

SECTION 74C: <u>Alkylations</u> <u>and</u> <u>Arylations</u> <u>of</u> <u>Olefins</u>

$$PhCH=CHCH_2NMe_2 \xrightarrow[\text{2. MeOH}]{\substack{\text{1. } \underline{n}BuLi, \text{ hexane} \\ \text{reflux, 6h}}} PhCH_2\overset{\overset{\displaystyle nBu}{|}}{C}HCH_2NMe_2$$

52%

Richey Jr., H.G.*; Heyn, A.S.; Erickson, W.F.
    <u>J</u> <u>Org</u> <u>Chem</u>, (1983), <u>48</u>, 3821

$$CH_3(CH_2)_3CH=CH_2 \xrightarrow[\text{Cp}_2\text{ZrCl}_2, \text{ 1h}]{\text{MgEt}_2, \text{ Et}_2O, \text{ 20°C}} CH_3(CH_2)_3\overset{\overset{\displaystyle}{|}}{C}HCH_3$$
$$Et$$

93%

Dzhemilev, U.M.; Vostrikova, O.S.; Sultanov, R.M.
                    <u>Bull</u> <u>Acad</u> <u>Sci</u> <u>USSR</u>, (1983), <u>32</u>, 193

$$\xrightarrow[\text{CH}_2\text{Cl}_2, \text{ 3 min}]{\text{2 AlCl}_3, \text{ -78°C}}$$

61%

Trost, B.M.*; Ghadiri, M.R.   <u>J</u> <u>Am</u> <u>Chem</u> <u>Soc</u>, (1984), <u>106</u>, 7260

SECTION 74D: Conjugate Reductions

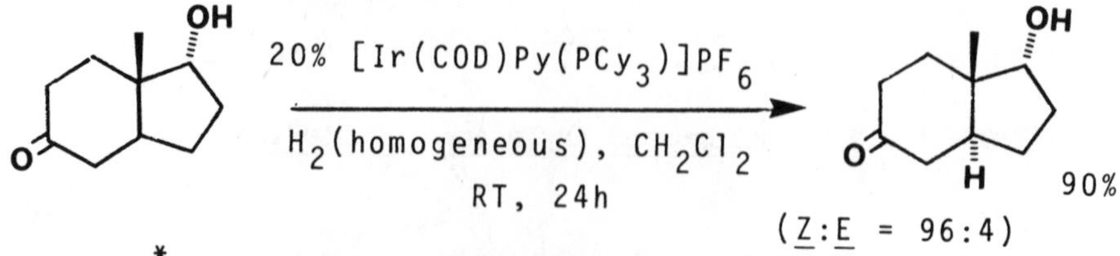

$$20\% \ [Ir(COD)Py(PCy_3)]PF_6$$

$$H_2(homogeneous), \ CH_2Cl_2$$

RT, 24h

90%

$(\underline{Z}:\underline{E} = 96:4)$

Stork, G.*; Kahne, D.E.   **J Am Chem Soc**, (1983), **105**, 1072

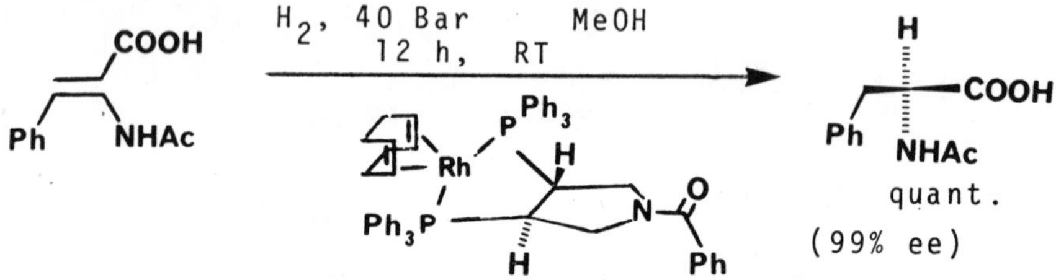

H_2, 40 Bar   MeOH

12 h,   RT

quant.

(99% ee)

Nagel, U.*   **Angew Chem Int Ed Engl**, (1984), **23**, 435

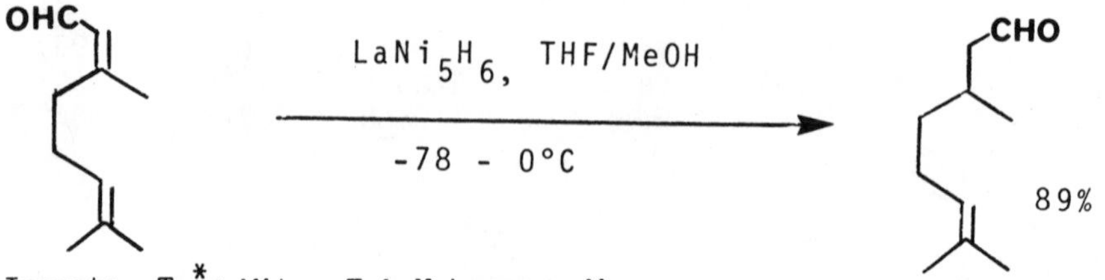

$LaNi_5H_6$, THF/MeOH

-78 - 0°C

89%

Inamoto, T.*; Mita, T.; Yokoyama, M.
JCS Chem Comm, (1984), 163

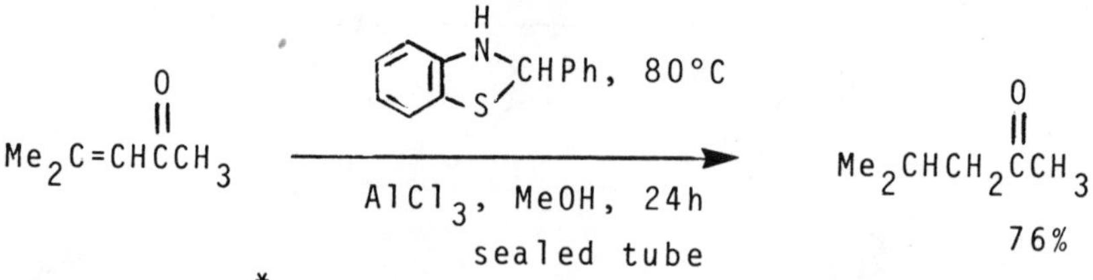

CHPh, 80°C

$Me_2C=CHCCH_3$

$AlCl_3$, MeOH, 24h

sealed tube

$Me_2CHCH_2CCH_3$

76%

Chikashita, H.*; Miyazaki, M.; Itoh, K.   **Synthesis**, (1984), 308

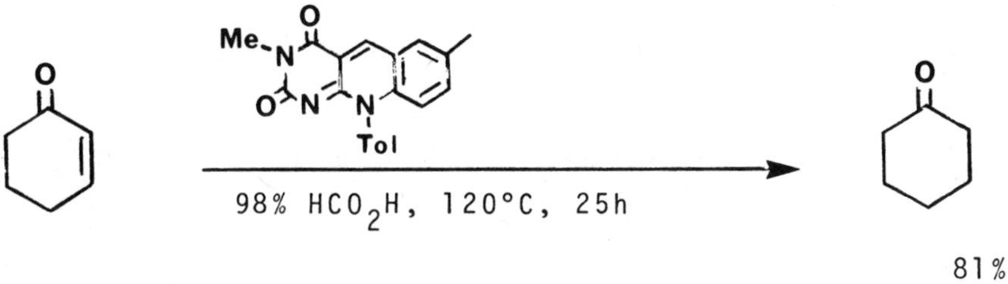
Me
N
O
O
N
N
Tol
O
O
98% HCO2H, 120°C, 25h
81%

Yoneda, F.*; Kuroda, K.; Tanaka, K.
JCS Chem Comm, (1984), 1194

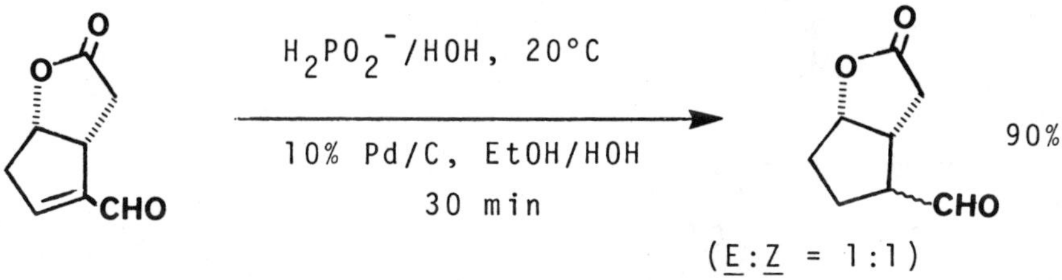
O
O
CHO
H2PO2-/HOH, 20°C
10% Pd/C, EtOH/HOH
30 min
O
O
CHO
90%
(E:Z = 1:1)

Sala, R.*; Doria, G.; Passarotti, C.
Tetrahedron Lett, (1984), **25**, 4565

$PhCH{=}CHCO_2Et$ (E) $\xrightarrow[Et_3SiH,\ PhH,\ 8h]{(PPh_3)_3RhCl,\ RT}$ $PhCH_2CH_2CO_2Et$ 98%

Liu, H.-J.*; Ramani, B. **Syn Commun**, (1985), **15**, 965

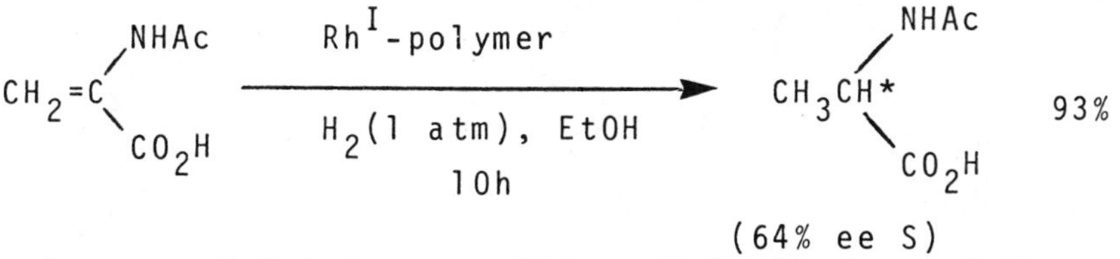
NHAc
CH2=C
CO2H
RhI-polymer
H2(1 atm), EtOH
10h
NHAc
CH3CH*
CO2H
93%
(64% ee S)

polymer - 2,2-O-isopropylidene-2,3-dihydroxy-1,4-bis(diphenylphosphino)butane

Deschenaux, R.; Stille, J.K.* **J Org Chem**, (1985), **50**, 2299

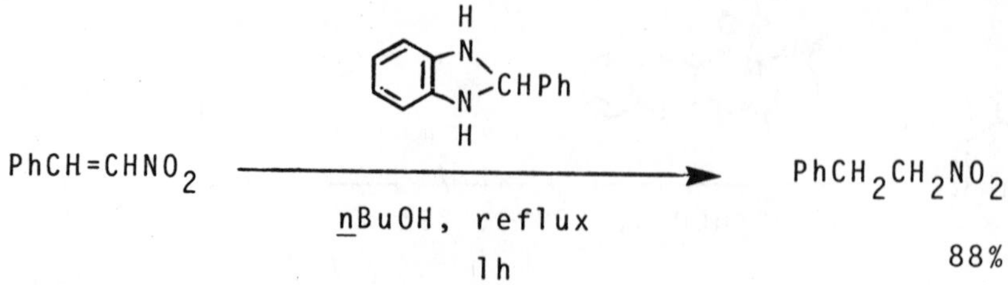

H
N
CHPh
N
H
PhCH=CHNO2
nBuOH, reflux
1h
PhCH2CH2NO2
88%

Chikashita, H.*; Morita, Y.; Itoh, K.
Syn Commun, (1985), **15**, 527

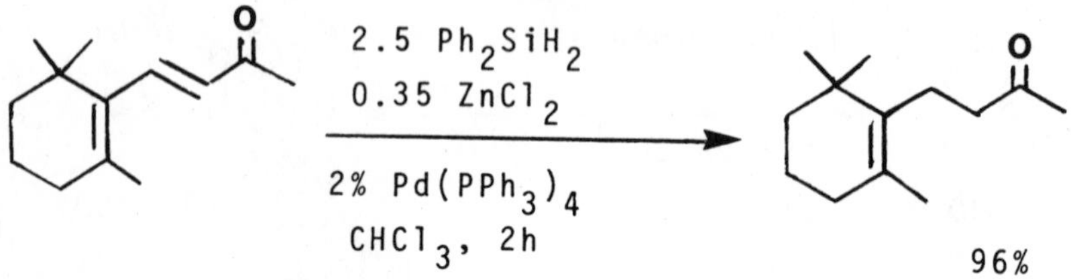

O
2.5 Ph2SiH2
0.35 ZnCl2
2% Pd(PPh3)4
CHCl3, 2h
O
96%

Keinan, E.*; Greenspoon, N. **Tetrahedron Lett**, (1985), **26**, 1353

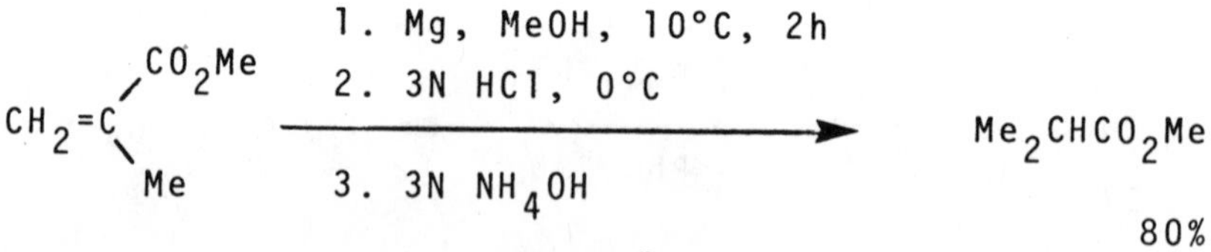

CO2Me
CH2=C
Me
1. Mg, MeOH, 10°C, 2h
2. 3N HCl, 0°C
3. 3N NH4OH
Me2CHCO2Me
80%

Youn, I.K.; Yon, G.H.; Pak, C.S.*
Tetrahedron Lett, (1986), **27**, 2409

SECTION 74E: Conjugate Alkylations

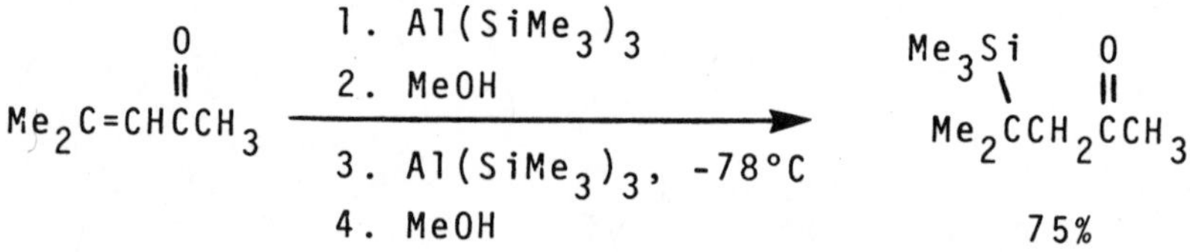

O
Me2C=CHCCH3
1. Al(SiMe3)3
2. MeOH
3. Al(SiMe3)3, -78°C
4. MeOH
Me3Si
O
Me2CCH2CCH3
75%

Altnau, G.; Rösch, L.* **Tetrahedron Lett**, (1983), **24**, 45

$CH_2=CHCH_2SiMe_3$

$PhCH=CHCO_2Me$ $\xrightarrow[\text{HMPA/DMF}]{\underline{n}Bu_4NF}$ $PhCHCH_2CH=CH_2$ with CH_2CO_2Me substituent

(E)

90%

Majetich, G.[*]; Casares, A.M.; Chapman, D.; Behnke, M.
Tetrahedron Lett, (1983), **24**, 1909

$\xrightarrow[\text{DMF/HMPA}]{\underline{n}Bu_4NF, \ RT, \ 1h}$

63%

Majetich, G.[*]; Desmond, R.; Casares, A.M.
Tetrahedron Lett, (1983), **24**, 1913

$\xrightarrow[\substack{DBN, \ 15 \ KBar, \ CH_3CN \\ 48h}]{EtO_2CCH_2CO_2Et, \ 25°C}$

60%   with CO_2Et, CO_2Et

Dauben, W.G.[*]; Gerdes, J.M.   **Tetrahedron Lett**, (1983), **24**, 3841

$I(CH_2)_4$ ... $C=C$ with Me and CO_2tBu $\xrightarrow[\text{2. EtOH}]{\substack{1. \ \underline{n}BuLi, \ -100°C \\ THF}}$ cyclopentane with CO_2tBu

86%

Cooke Jr., M.P.[*]   **J Org Chem**, (1984), **49**, 1144

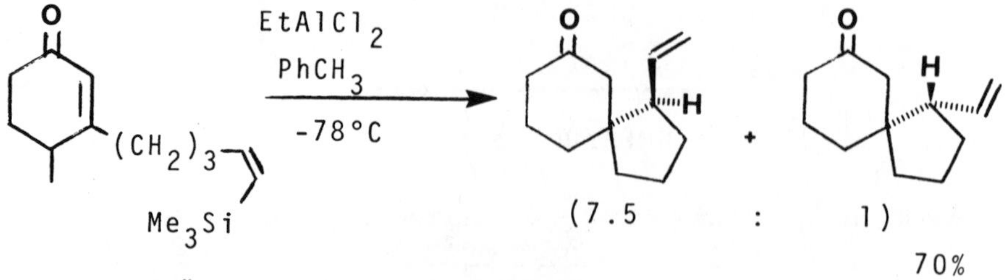

EtAlCl$_2$
PhCH$_3$
-78°C

(7.5   :   1)

70%

Schinzer, D.[*]   **Angew Chem Int Ed Engl**, (1984), **23**, 308

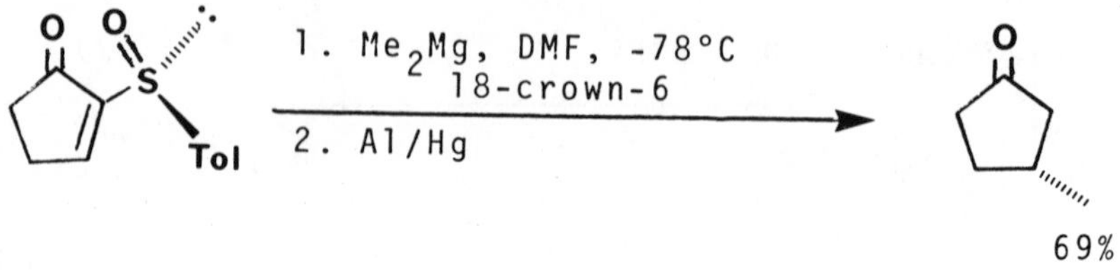

1. Me$_2$Mg, DMF, -78°C
   18-crown-6
2. Al/Hg

69%
(97% ee)

Posner, G.H.[*]; Hulce, M.
Tetrahedron Lett, (1984), **25**, 379, 383

CH$_2$=CHCN

PhCH$_2$I, $\underline{n}$Bu$_3$GeH
─────────────────────
AIBN, PhH, 12h
reflux

Ph(CH$_2$)$_3$CN
76%

Pike, P.; Hershberger, S.; Hershberger, J.[*]
Tetrahedron Lett, (1985), **26**, 6289

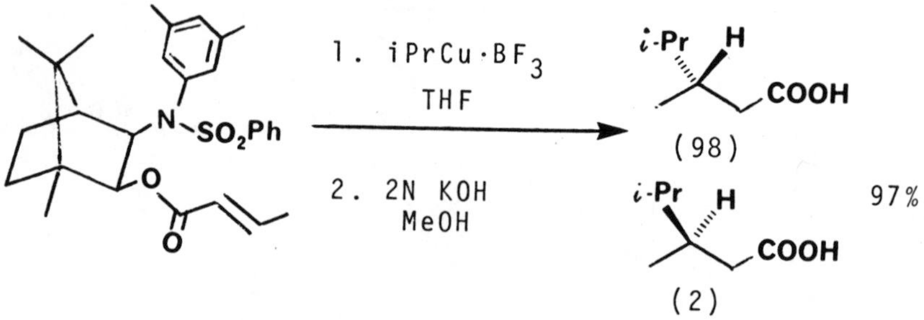

1. iPrCu·BF$_3$
   THF
2. 2N KOH
   MeOH

i-Pr H
╲╱
COOH
(98)

i-Pr H   97%
╲╱
COOH
(2)

Helmchen, G.[*]; Wegner, G.   **Tetrahedron Lett**, (1985), **26**, 6051

1. $2\begin{bmatrix} Bu_2Cu \cdot BF_3, & Et_2O \\ BuLi-CuI, & -78°C \\ PBu_3/BF_3 \cdot OEt_2 \end{bmatrix}$

$Et_2O/THF, -78°C$

2. NaOH, aq. EtOH

reflux   (95% ee)

89%

Oppolzer, W.[*]; Dudfield, P.; Stevenson, T.; Godel, T.
Helv Chim Acta, (1985), **68**, 212, 216

1. $PhCu \cdot BF_3, Et_2O$

-78 - -20°C

2. Ac_2O, DMAP

3. TsOH, aq. acetone

$CH_3\overset{*}{C}HCH_2CHO$
|
Ph   70%

(73% ee S)

Mangeney, P.; Alexakis, A.; Normant, J.F.[*]
Tetrahedron Lett, (1986), **27**, 3143

iPrBr, Zn/CuI
aq. EtOH

18°C, 45 min

))))

94%

Petrier, C.; Dupuy, C.; Luche, J.L.[*]
Tetrahedron Lett, (1986), **27**, 3149
de Souza Barboza, J.C.; Petrier, C.; Luche, J.-L.
Tetrahedron Lett, (1985), **26**, 829

$CH_2=C\begin{array}{c} CO_2Me \\ \\ NHAc \end{array}$

1. EtMgCl, Et_2O/PhH

5% CuI, 0°C

2. H_3O^+

$CH_3CH_2CH_2CH\begin{array}{c} CO_2Me \\ \\ NHAc \end{array}$

80%

Cardellicchio, C.; Fiandanese, V.; Marchese, G.; Naso, F.[*];
Ronzini, L.
Tetrahedron Lett, (1985), **26**, 4387

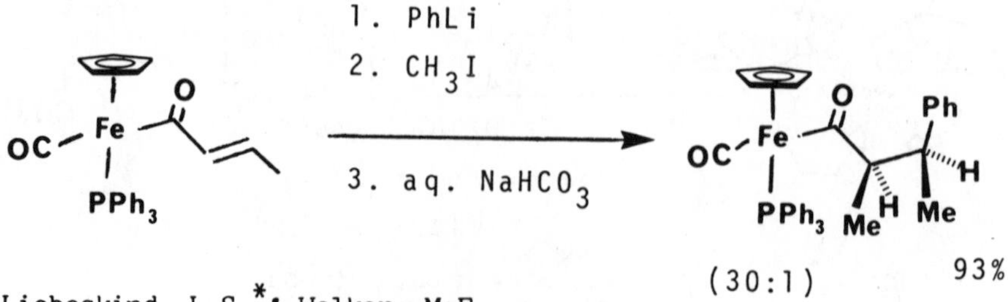
OC
Fe
PPh₃
O
1. PhLi
2. CH₃I
3. aq. NaHCO₃
Ph
H
Me
(30:1)
93%

Liebeskind, L.S.*; Welker, M.E.
Tetrahedron Lett, (1985), **26**, 3079

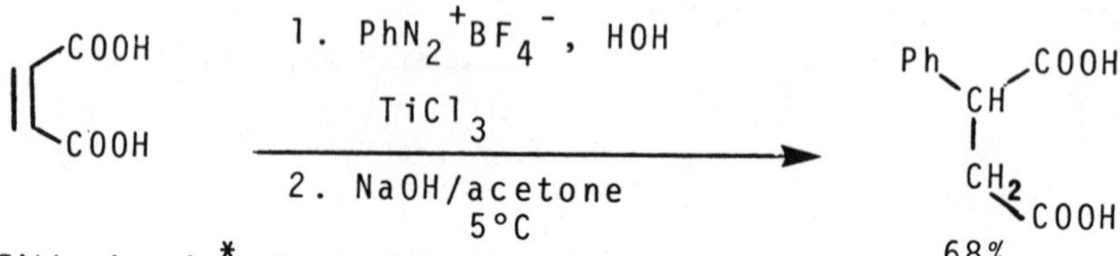
COOH
COOH
1. PhN₂⁺BF₄⁻, HOH
TiCl₃
2. NaOH/acetone
5°C
Ph
CH
CH₂
COOH
68%

Citterio, A.*; Cominelli, A.; Bonavoglia, F.
Synthesis, (1986), 308

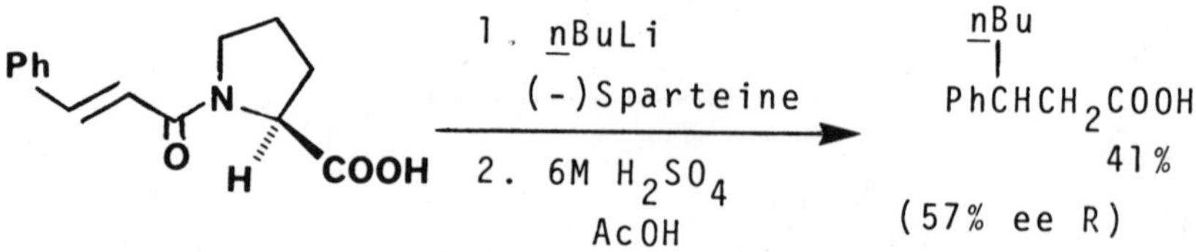
Ph
N
O
H
COOH
1. nBuLi
(-)Sparteine
2. 6M H₂SO₄
AcOH
nBu
PhCHCH₂COOH
41%
(57% ee R)

Soai, K.*; Ookawa, A. **JCS Perkin I**, (1986), 759

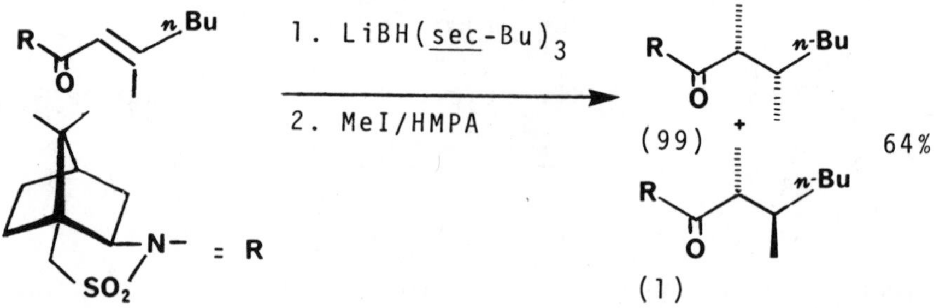
R
O
n-Bu
N
SO₂
= R
1. LiBH(sec-Bu)₃
2. MeI/HMPA
(99)
+
(1)
64%

Oppolzer, W.*; Poli, G. **Tetrahedron Lett**, (1986), **27**, 4717

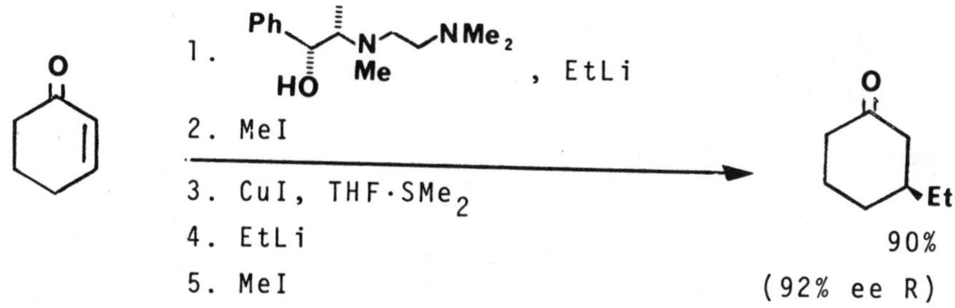
O
1.
Ph
N
NMe2
HO
Me
, EtLi
2. MeI
3. CuI, THF·SMe2
4. EtLi
5. MeI
O
Et
90%
(92% ee R)

Corey, E.J.*; Naef, R.; Hannon, F.J.
J Am Chem Soc, (1986), **108**, 7114

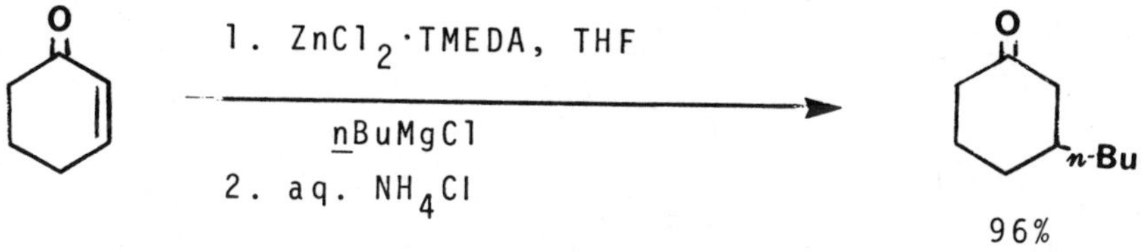
O
1. ZnCl2·TMEDA, THF
nBuMgCl
2. aq. NH4Cl
O
n-Bu
96%

Kjonaas, R.A.*; Vawter, E.J. **J Org Chem**, (1986), **51**, 3993

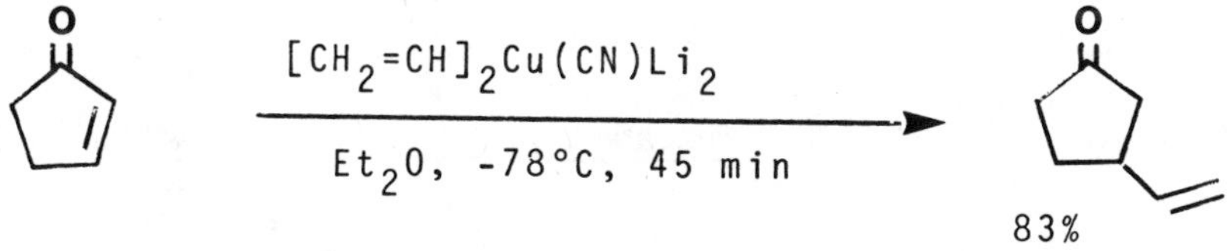
O
[CH2=CH]2Cu(CN)Li2
Et2O, -78°C, 45 min
O
83%

Lipshutz, B.H.*; Wilhelm, R.S.; Kozlowski, J.A.
J Org Chem, (1984), **49**, 3938

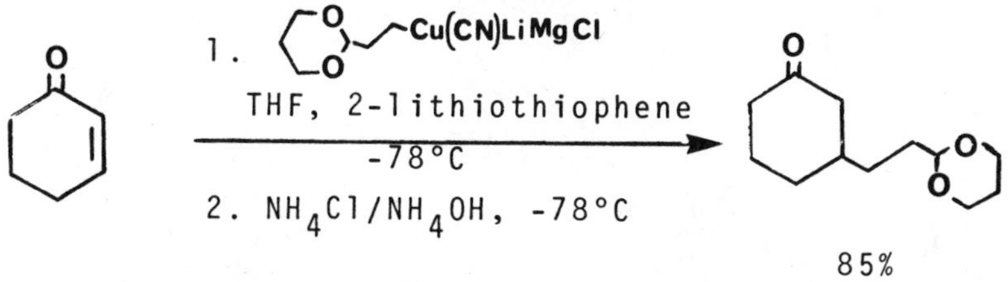
O
1.
O
O
Cu(CN)LiMgCl
THF, 2-lithiothiophene
-78°C
2. NH4Cl/NH4OH, -78°C
O
O
O
85%

Lipshutz, B.H.*; Parker, D.A.; Nguyen, S.L.; McCarthy, K.E.; Barton, J.C.; Whitney, S.E.; Kotsuki, H.
Tetrahedron, (1986), **42**, 2873

Reviews:

"The Chemistry of Higher Order Cuprates"

Lipshutz, B.H.[*]; Wilhelm, R.S.; Kozlowski, J.A.
                    Tetrahedron, (1984), **40**, 5005

"Selective Synthesis by Use of Lewis Acids in the
Presence of Organocopper and Related Reagents"

Yamamoto, Y.[*]   **Angew Chem Int Ed Engl**, (1986), **25**, 947

SECTION 74F: Cyclopropanations

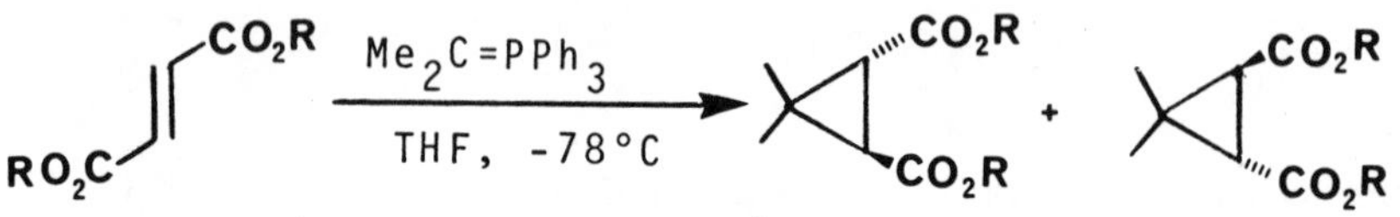

R = menthyl   (74% ee)   85%

de Vos, M.J.; Krief, A.[*]   **Tetrahedron Lett**, (1983), **24**, 103

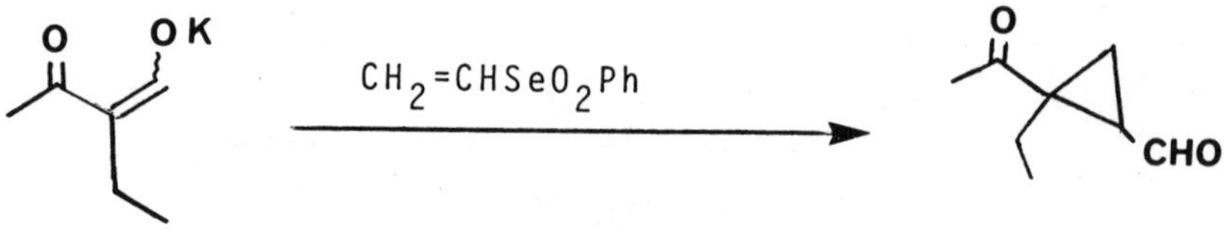

64%

Ando, R.; Sugawara, T.; Kuwajima, I.[*]
                    JCS Chem Comm, (1983), 1514

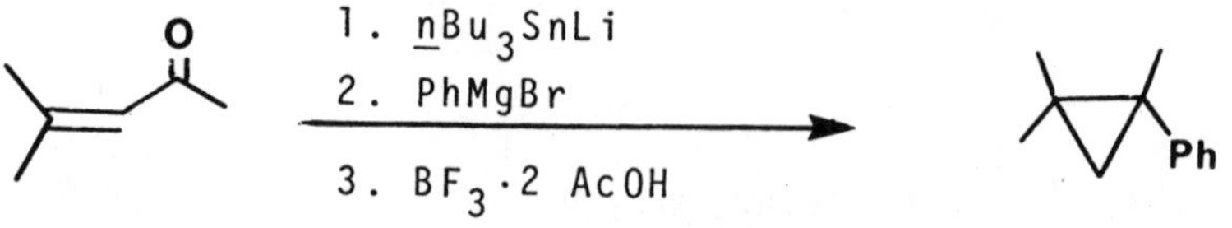

48%

Fleming, I.[*]; Urch, C.J.   **Tetrahedron Lett**, (1983), **24**, 4591

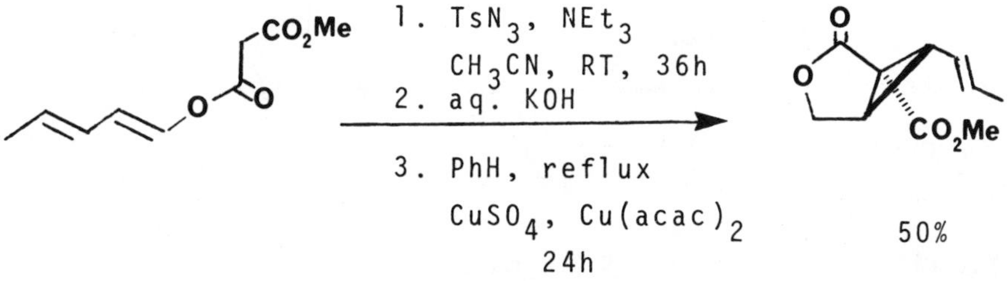
CO2Me
O
O
1. TsN3, NEt3
CH3CN, RT, 36h
2. aq. KOH
3. PhH, reflux
CuSO4, Cu(acac)2
24h
O
O
CO2Me
50%

Hudlicky, T.*; Reddy, D.B.; Govindan, S.V.; Kulp, T.; Still, B.; Sheth, J.P.
J Org Chem, (1983), **48**, 3423

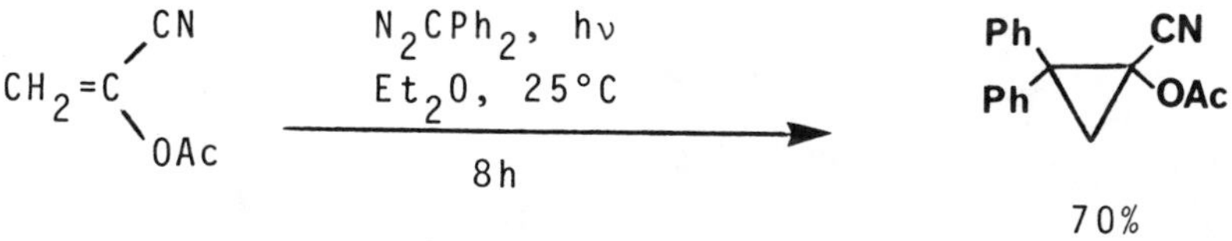
CN
CH2=C
OAc
N2CPh2, hν
Et2O, 25°C
8h
Ph
Ph
CN
OAc
70%

Oku, A.*; Yokoyama, T.; Harada, T.
J Org Chem, (1983), **48**, 5333

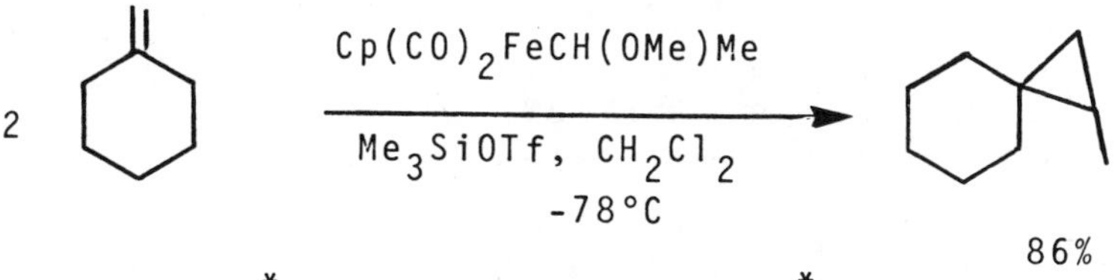
2
Cp(CO)2FeCH(OMe)Me
Me3SiOTf, CH2Cl2
-78°C
86%

Brookhart, M.*; Tucker, J.R.; Husk, G.R.*
J Am Chem Soc, (1983), **105**, 258

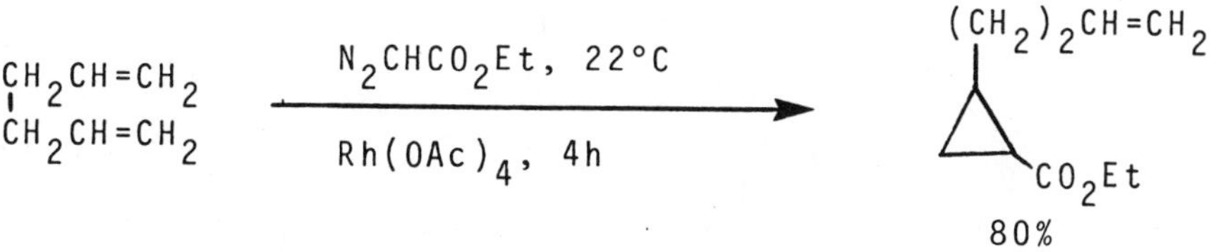
CH2CH=CH2
CH2CH=CH2
N2CHCO2Et, 22°C
Rh(OAc)4, 4h
(CH2)2CH=CH2
CO2Et
80%

Anciaux, A.J.; Demonceau, A.; Noels, A.F.; Warin, R.; Hubert, A.J.; Teyssie, P.
Tetrahedron, (1983), **39**, 2169

$$\left(Br\text{-}\bigcirc\text{-}\right)_3 \overset{+}{N}SbCl_6^-$$

N_2CHCO_2Et, CH_2Cl_2
0°C, 15 min

67%

Stufflebeme, G.; Lorenz, K.T.; Bauld, N.L.[*]
J Am Chem Soc, (1986), 108, 4234

$PhCH=CH_2$

N_2CHCO_2Et, 24°C
——————————————
3% phthalocyanine Co

quant.

(E:Z = 1.2:1)

Pazynina, G.V.; Luk'yanets, E.A.; Bolesov, I.G.
J Org Chem USSR, (1984), 20, 1731

$Pd(OAc)_2$ catalyst, PhH, 40°C   76%

(E:Z = 1:2)

Majchrzak, M.W.[*]; Kotelko, A.; Lambert, J.B.
Synthesis, (1983), 469

Et_2Zn, $CHBr_3$, 20°C
——————————————
pentane, RT, 2h

82%

Rousseau, G.[*]; Slougui, N.   J Am Chem Soc, (1984), 106, 7283

$$2\ CH_2=CH\overset{|}{\underset{CO_2Me}{}}$$

1. 2 , PhH, 80°C
———————————————
2. aq. THF, AcOH
25°C

69%

(Z:E = 95:1)

Boger, D.L.[*]; Brotherton, C.E.
Tetrahedron Lett, (1984), 25, 5611

1. LDA, THF, -30°C

2. Cl-⟨benzene⟩-Se(O)CH=CH$_2$

-30°C - RT   72%

Ando, R.; Sugawara, T.; Shimizu, M.; Kuwajima, I.*
Bull Chem Soc Jpn, (1984), **57**, 2897

N$_2$CHCO$_2$Et, 25°C

0.5% Rh$_2$(OAc)$_4$

8h   ($\underline{E}$:$\underline{Z}$ = 2.1:1)

96%

Doyle, M.P.*; Dorow, R.L.; Buhro, W.E.; Griffin, J.H.; Tamblyn,
W.H.; Trudell, M.L.
Organometallics, (1984), **3**, 44

EtCH=C(CO$_2$Me)(CO$_2$Me)

1. Me$_2$C(Li)SO$_2$Ph
   THF, 20°C, 30 min

2. MeOH, 70°C

3. H$_3$O$^+$

CO$_2$Me
CO$_2$Me

64%

Krief, A.*; de Vos, M.J.   **Tetrahedron Lett**, (1985), **26**, 6115

Ph$_2$P(O)-CH(CPh=O)-CH$_2$CH$_2$OH

tBuOK/tBuOH

30°C, 2h

quant.

Wallace, P.; Warren, S.*   **Tetrahedron Lett**, (1985), **26**, 5713

iBu_3Al, CH_2I_2

hexane

73%

Maruoka, K.; Fukutani, Y.; Yamamoto, H.
J Org Chem, (1985), 50, 4412

2 CH_2Br_2, 4 Zn

cat. CuCl, Et_2O, 45°C

60%

Friedrich, E.C.*; Domek, J.M.; Pong, R.Y.
J Org Chem, (1985), 50, 4640

$$PhCH=CHCPh$$
(with O above C)

$Ph_3\overset{+}{P}-CH_2Ph$

$KF-Al_2O_3$

70°C, 24h

(E:Z = 1:1)   72%

Texier-Boullet, F.*; Villemin, D.; Ricard, M.; Moison, H.;
Foucaud, A.
Tetrahedron, (1985), 41, 1259

1. $HC(OEt)_3$, NH_4NO_3

2. $HO-\!\!-CO_2Et$ $HO-\!\!-\!\!-CO_2Et$, TsOH·Py

3. Et_2Zn, CH_2I_2

4. TsOH, aq. MeOH

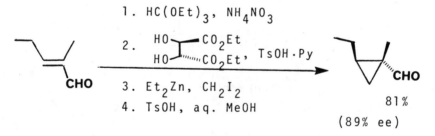

81%
(89% ee)

Arai, I.; Mori, A.; Yamamoto, H.*
J Am Chem Soc, (1985), 107, 8254
Tetrahedron, (1986), 42, 6447

Review: "Siloxy Cyclopropanes: Useful Synthetic Intermediates"

Murai, S.[*]; Ryu, I.; Sonoda, N.
J Organomet Chem, (1983), **250**, 121

SECTION 75: Alkyls, Methylenes, and Aryls from Miscellaneous Compounds

79%

Okuda, Y.; Sato, M.; Oshima, K.; Nozaki, H.
Tetrahedron Lett, (1983), **24**, 2015

$PhSO_2Ph$ $\xrightarrow[\text{reflux, 2h}]{\text{2 LinBu}_3\text{BH, THF}}$ $Ph(CH_2)_3CH_3$

79%

Brown, H.C.[*]; Kim, S.-C.; Krishnamurthy, S.
Organometallics, (1983), **2**, 779

71%

Baldwin, J.C.[*]; Adlington, R.M.; Newington, I.M.
JCS Chem Comm, (1986), 176

Reviews:

"Copper(I) Catalyzed Reactions of Organolithiums and Grigard Reagents"

Erdik, E.[*]   **Tetrahedron**, (1984), **40**, 641

"Syntheses with Radicals: C-C Bond Formation via Organotin and Organomercury Compounds"

Giese, B.[*]   **Angew Chem Int Ed Engl**, (1985), **24**, 553

CHAPTER 6
PREPARATION OF AMIDES

SECTION 76: Amides from Acetylenes

No Additional Examples

SECTION 77: Amides from Acid Derivatives

$$Ph(CH_2)_3COOH \xrightarrow[\substack{reflux \\ PhCH_2CH_2CH_2NH_2}]{pxylene, 3h} Ph(CH_2)_3\overset{\overset{\displaystyle O}{\|}}{C}NHCH_2CH_2Ph$$

86%

Mukaiyama, T.[*]; Ichikawa, J.; Asami, M. **Chem Lett**, (1983), 683

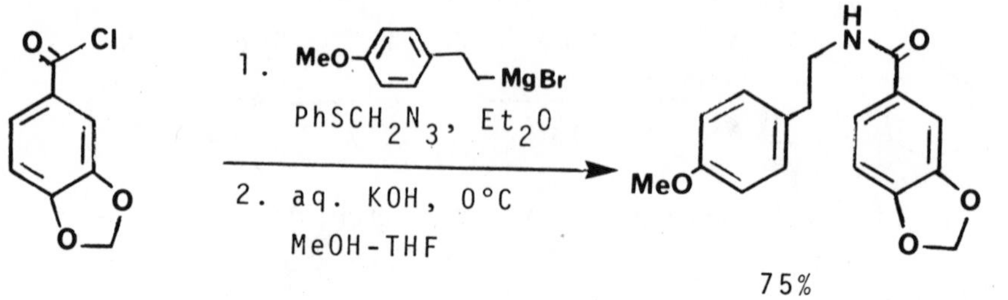

1. MeO—⟨⟩—CH$_2$—MgBr
   PhSCH$_2$N$_3$, Et$_2$O

2. aq. KOH, 0°C
   MeOH-THF

75%

Trost, B.M.[*]; Pearson, W.W. **J Am Chem Soc**, (1983), **105**, 1054

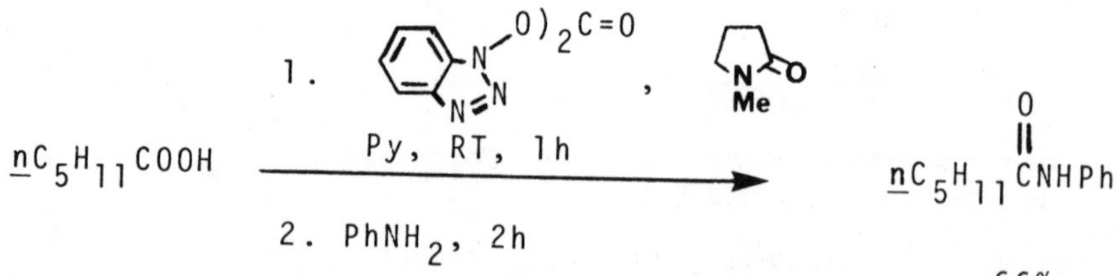

$$\underline{n}C_5H_{11}COOH \xrightarrow[\substack{Py, RT, 1h \\ 2. PhNH_2, 2h}]{1.} \underline{n}C_5H_{11}\overset{\overset{\displaystyle O}{\|}}{C}NHPh$$

66%

Ueda, M.[*]; Oikawa, H.; Teshirogi, J. **Synthesis**, (1983), 908

Cabré, J.; Palomo, A.L.*   **Synthesis**, (1984), 413

$$CH_3CH_2COOH \xrightarrow[\text{PPh}_3, \text{ PhH, reflux}]{N_3CH_2CH_2iPr, 24h} CH_3CH_2\overset{O}{\overset{\|}{C}}NHCH_2CH_2iPr$$

88%

Garcia, J.; Urpí, F.; Vilarrasa, J.*
                        Tetrahedron Lett, (1984), **25**, 4841

$$PhCOOH \xrightarrow{\text{PhCH}_2\text{NH}_2, \text{ Py}} PhC\overset{O}{\overset{\|}{}}NHPh$$

95%

Ueda, M.*; Kawaharasaki, N.; Imai, Y.
                        Bull Chem Soc Jpn, (1984), **57**, 85

$$PhCH_2COOH \xrightarrow[\substack{\text{PhCH}_2\text{NH}_2, \text{ CH}_2\text{Cl}_2 \\ RT, 30 min}]{} PhCH_2\overset{O}{\overset{\|}{C}}NHCH_2Ph$$

85%

Kim, S.*; Kim, S.S.   **JCS Chem Comm**, (1986), 719

Related Methods: Amides from Amines (Section 82)

SECTION 78: Amides from Alcohols and Thiols

No Additional Examples

SECTION 79: Amides from Aldehydes

$$\underline{n}C_7H_{15}CHO \xrightarrow[\begin{array}{c}PPh_3, \ K_2CO_3, \ DMP \\ PhBr, \ reflux, \ 24h\end{array}]{HN\bigcirc O, \ Pd(OAc)_2} \underline{n}C_7H_{15}\overset{O}{\overset{\|}{C}}-N\bigcirc O$$

59%

Tamaru, Y.; Yamada, Y.; Yoshida, Z.[*]   **Synthesis**, (1983), 474

PhCHO
$$\xrightarrow[\begin{array}{l}3. \ Me_2C=C\overset{OMe}{\underset{OTMS}{}}, \ ZnI_2 \\ Et_2O \\ 4. \ tBuOH \quad 5. \ 3 \ MeMgBr\end{array}]{\begin{array}{l}1. \ LiN(SiMe_3)_2, \ THF, \ 0°C \\ 2. \ Me_3SiCl, \ 30 \ min\end{array}}$$

60%

Colvin, E.W.[*]; McGarry, D.G.   **JCS Chem Comm**, (1985), 539

SECTION 80: Amides from Alkyl, Methylenes, and Aryls

No Additional Examples

SECTION 81: Amides from Amides

Conjugate reductions of unsaturated amides are listed in Section 74d (Alkyls from Olefins).

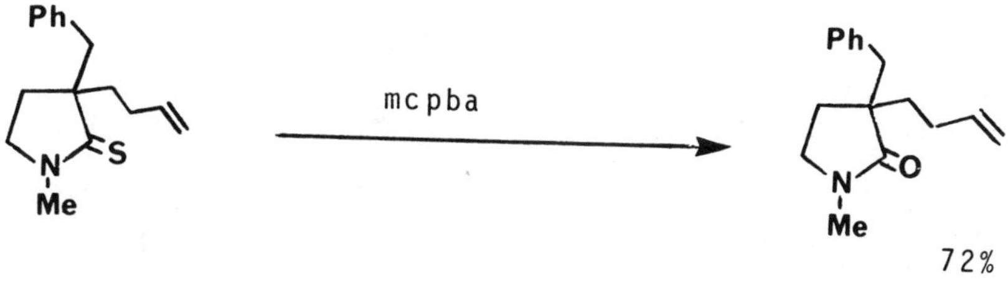

mcpba

72%

Kochhar, K.S.; Cottrell, D.A.; Pinnick, H.W.[*]
Tetrahedron Lett, (1983), **24**, 1323

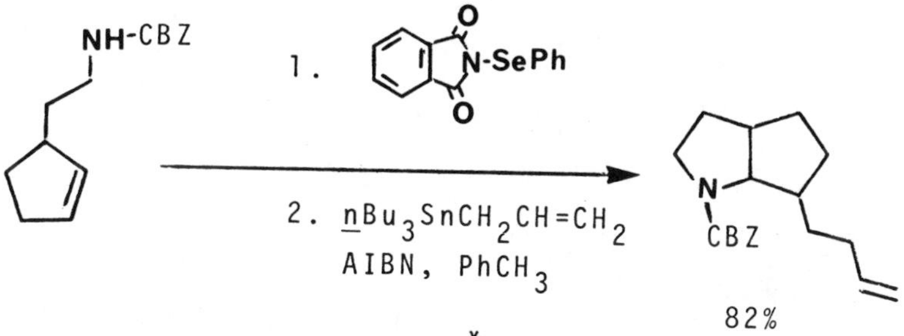

1. [image of N-SePh phthalimide]

2. $\underline{n}Bu_3SnCH_2CH=CH_2$
   AIBN, $PhCH_3$

82%

Webb II, R.R.; Danishefsky, S.[*]
Tetrahedron Lett, (1983), **24**, 1357

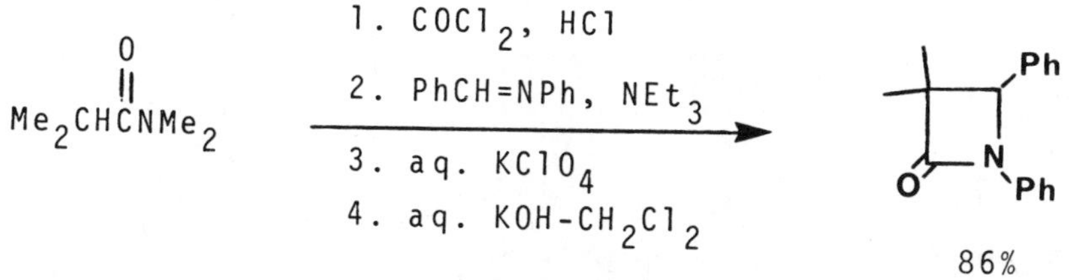

$Me_2CHCNMe_2$

1. $COCl_2$, HCl
2. $PhCH=NPh$, NEt_3
3. aq. $KClO_4$
4. aq. $KOH-CH_2Cl_2$

86%

Marchand-Brynaert, J.; Moya-Portuguez, M.; Huber, I.; Ghosez, L.[*]
JCS Chem Comm, (1983), 818

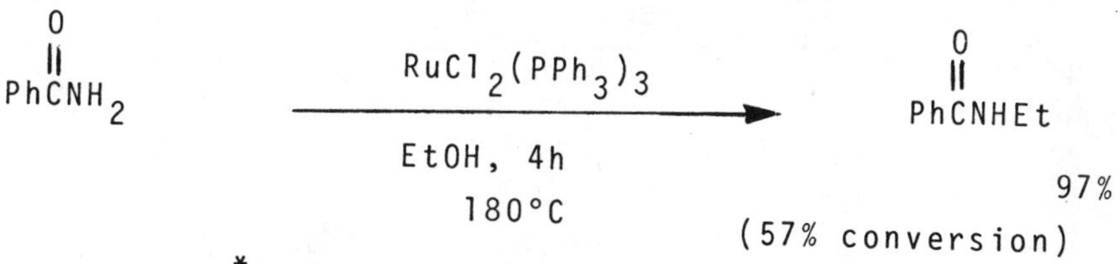

$PhCNH_2$

$RuCl_2(PPh_3)_3$

EtOH, 4h
180°C

$PhCNHEt$

97%
(57% conversion)

Watanabe, Y.[*]; Ohta, T.; Tsuji, Y.
Bull Chem Soc Jpn, (1983), **56**, 2647

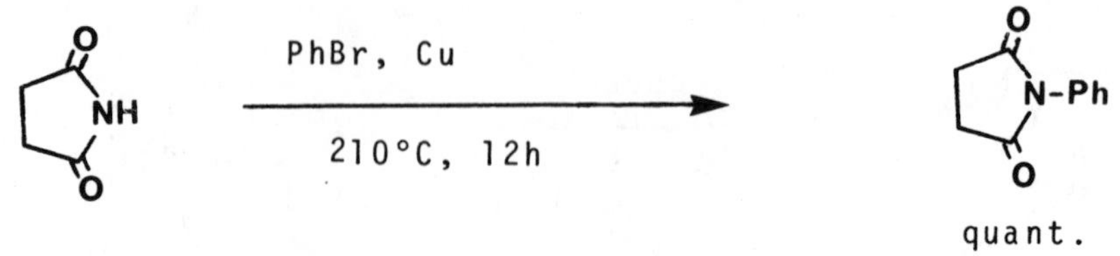

PhBr, Cu

210°C, 12h

quant.

Yamamoto, T.[*]; Kurata, Y. **Can J Chem**, (1983), **61**, 86

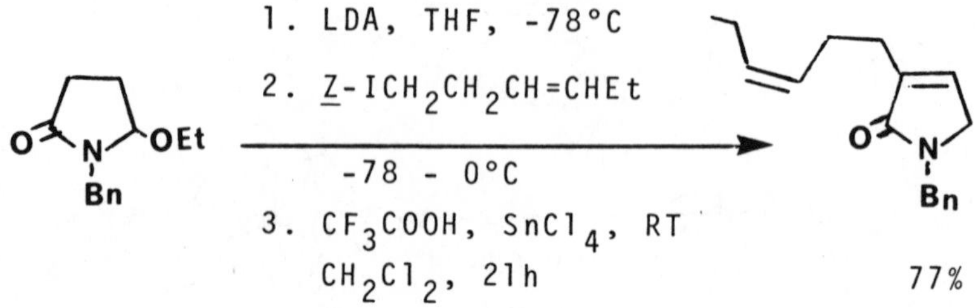

1. LDA, THF, -78°C

2. $\underline{Z}$-ICH$_2$CH$_2$CH=CHEt

-78 - 0°C

3. CF$_3$COOH, SnCl$_4$, RT
   CH$_2$Cl$_2$, 21h

77%

Hiemstra, H.[*]; Klaver, W.J.; Speckamp, W.N.
      J Org Chem, (1984), **49**, 1149

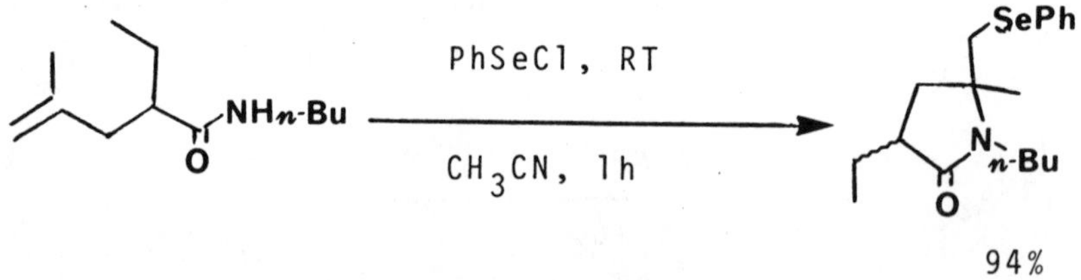

PhSeCl, RT

CH$_3$CN, 1h

94%

Toshimitsu, A.[*]; Terao, K.; Uemura, S.
      Tetrahedron Lett, (1984), **25**, 5917

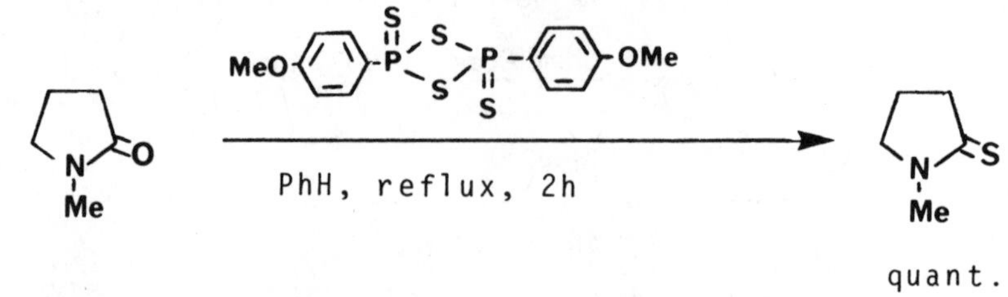

PhH, reflux, 2h

quant.

Thomsen, I.; Clausen, K.; Scheibye, S.; Lawesson, S.-O.[*]
      Org Syn, (1984), **62**, 158

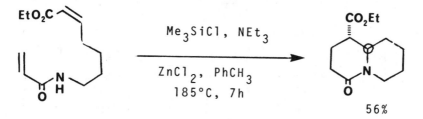

1. nBuLi
2. B(OMe)$_3$

3. H$^+$
4. Pd(PPh$_3$)$_4$, PhBr   82%

Sharp, M.J.; Snieckus, V.*   **Tetrahedron Lett**, (1985), **26**, 5997

nBu$_3$SnH, AIBN
PhH, reflux, 10h

84%

Padwa, A.*; Nimmesgern, H.; Wong, G.S.K.
                J Org Chem, (1985), **50**, 5620
                Tetrahedron Lett, (1985), **26**, 957

1. e$^-$, MeOH
2. ClCH$_2$OMe, NaH

3. CH$_2$=CHCH$_2$SiMe$_3$
     TiCl$_4$
4. H$_2$, Ni(R), KOH   31%

Shono, T.*; Matsumura, Y.; Uchida, K.; Kobayashi, H.
                J Org Chem, (1985), **50**, 3243

Me$_3$SiCl, NEt$_3$
ZnCl$_2$, PhCH$_3$
185°C, 7h

56%

Ihara, M.; Kirihara, T.; Fukumoto, K.*; Kametani, T.
                Heterocycles, (1985), **23**, 1097

$$\underset{\text{PhCNH}\underline{n}\text{Pr}}{\overset{\overset{\text{S}}{\parallel}}{}} \xrightarrow[\substack{\text{CH}_3\text{CN, } -35°\text{C} \\ 5\text{h}}]{\text{[2-NO}_2\text{-C}_6\text{H}_4\text{-SO}_2\text{Cl], KO}_2} \underset{\text{PhCNH}\underline{n}\text{Pr}}{\overset{\overset{\text{O}}{\parallel}}{}}$$

92%

Kim, Y.H.*; Chung, B.C.; Chang, H.S.
Tetrahedron <u>Lett</u>, (1985), <u>26</u>, 1079

$$\text{[pyrrolidinone]} \xrightarrow[\substack{e^-, \text{ Pt electrode} \\ \text{Et}_4\text{NOTs, RT}}]{\substack{\text{CH}_3\text{CHCH}_2\text{CH}_3, \text{ DMF} \\ \text{Cl}}} \text{[N-sec-butyl pyrrolidinone]}$$

68%

Shono, T.*; Kashimura, S.; Nagusa, H.   **Chem** <u>Lett</u>, (1986), <u>425</u>

SECTION 82: Amides from Amines

$$\text{[Ph-aziridine-}t\text{-Bu]} \xrightarrow[\substack{20 \text{ atm CO, PhH}}]{\text{[Rh(CO}_2)_2\text{Cl]}_2, \text{ 90°C}} \text{[β-lactam, Ph, }t\text{-Bu]}$$

quant.

Alper, H.*; Urso, F.; Smith, D.J.H.
J <u>Am</u> **Chem** <u>Soc</u>, (1983), <u>105</u>, 6737

$$\text{[N-Me piperidine N-oxide]} \xrightarrow[\substack{2. \text{ MeLi, THF, PhCCl} \\ 0°\text{C - RT}}]{\substack{1. \underline{t}\text{BuMe}_2\text{SiOSO}_2\text{CF}_3 \\ \text{CH}_2\text{Cl}_2, 0°\text{C}}} \text{[N-benzoyl piperidine]}$$

76%

Okazaki, R.*; Takitoh, N.   **JCS** **Chem** **Comm**, (1984), 192

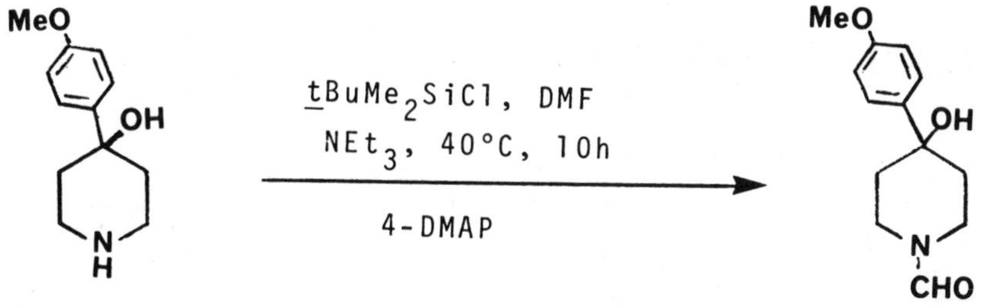

78%

Djurii, S.W.*   J Org Chem, (1984), 49, 1311

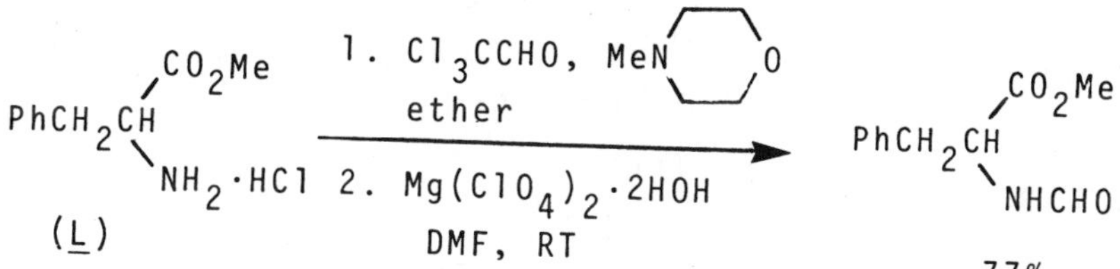

77%

Giesemann, G.; Ugi, I.   Synthesis, (1983), 788

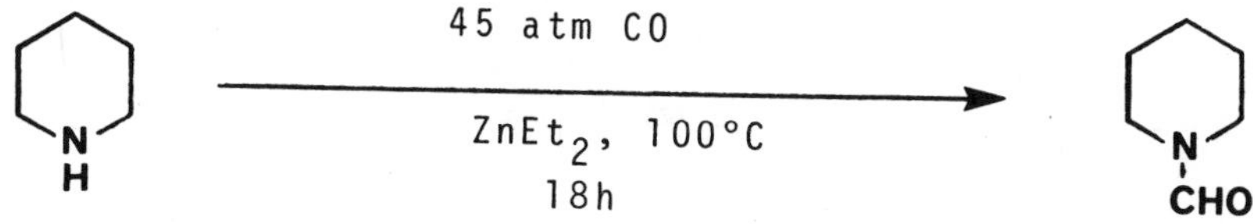

80%

Yoshida, Y.*; Asano, S.; Inoue, S.   Chem Lett, (1984), 1073

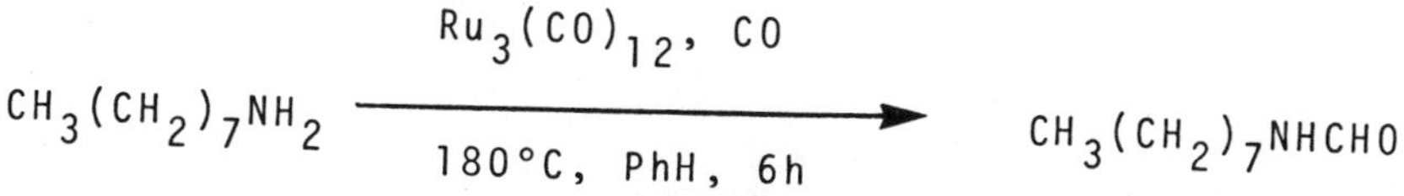

$$CH_3(CH_2)_7NH_2 \xrightarrow[\text{180°C, PhH, 6h}]{Ru_3(CO)_{12}, \ CO} CH_3(CH_2)_7NHCHO$$

91%

Tsuji, Y.; Ohsumi, T.; Kondo, T.; Watanabe, Y.*
            J Organomet Chem, (1986), 309, 333

Related Methods: Amides from Carboxylic Acids (Section 77)
                 Protection of Amines (Section 105A)

SECTION 83: Amides from Esters

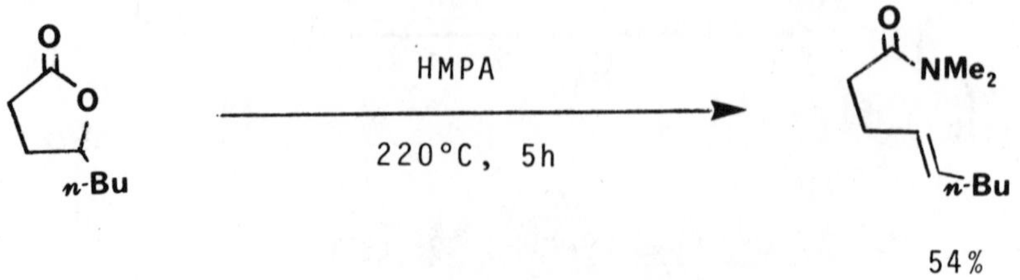
O
O
n-Bu
HMPA
220°C, 5h
O
NMe2
n-Bu
54%

Gupton, J.T.; Baran, D.; Bennett, R.; Hertel, G.R.; Idoux, J.P.*
Syn Commun, (1984), 14, 1001

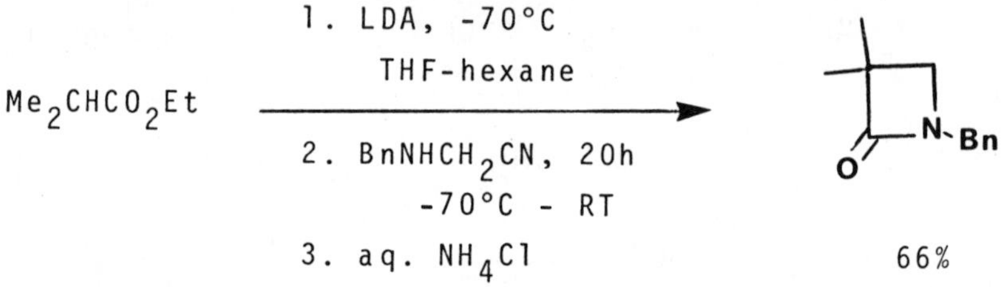
1. LDA, -70°C
THF-hexane
Me2CHCO2Et
2. BnNHCH2CN, 20h
-70°C - RT
3. aq. NH4Cl
N
Bn
O
66%

Overman, L.E.*; Osawa, T. J Am Chem Soc, (1985), 107, 1698

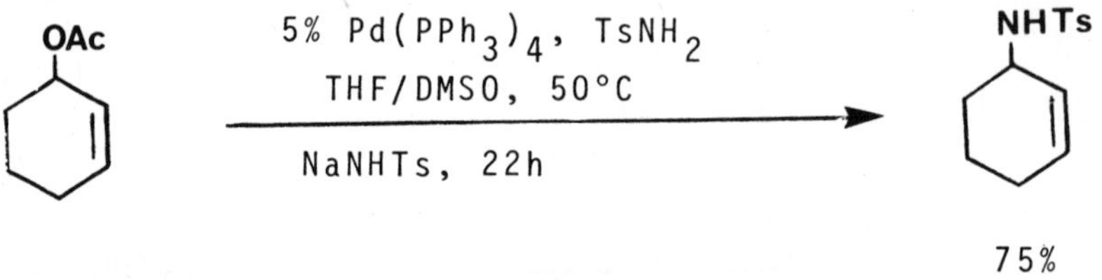
OAc
5% Pd(PPh3)4, TsNH2
THF/DMSO, 50°C
NaNHTs, 22h
NHTs
75%

Byström, S.E.; Aslanian, R.; Bäckvall, J.-E.*
Tetrahedron Lett, (1985), 26, 1749

SECTION 84: Amides from Ethers, Epoxides, and Thioethers

No Additional Examples

SECTION 85: Amides from Halides and Sulfonates

$CH_2=C$ (with Br and CH_2Br substituents)

1. $BnNH_2$, K_2CO_3
2. CO, NBu_3, 100°C

8% $Pd(OAc)_2$-PPh_3
3. H_2, PtO_2

58%

Mori, M.[*]; Chiba, K.; Okita, M.; Kayo, I.; Ban, Y.
Tetrahedron, (1985), **41**, 375

1. MeSiNEt$_2$, DMF

3 $Ni(CO)_4$, 70°C

2. HOH

85%

Hirao, T.[*]; Nagata, S.; Yamana, Y.; Agawa, T.
Tetrahedron Lett, (1985), **26**, 5061

$PhNH_2$, DMF

$Ni(CO)_4$, 70°C

63%

Hirao, T.[*]; Harano, Y.; Yamana, Y.; Hamada, Y.; Nagata, S.;
Agawa, T.
Bull Chem Soc Jpn, (1986), **59**, 1341

$tBuN=C=O$

1. PhBr,)))))
   RT, 45 min

2. Na, THF-HMPA
3. H_3O^+

$PhCNHtBu$ (amide, with C=O)

78%

Einhorn, J.; Luche, J.L.[*]   **Tetrahedron Lett**, (1986), **27**, 501

SECTION 86: Amides from Hydrides

No Additional Examples

SECTION 87: Amides from Ketones

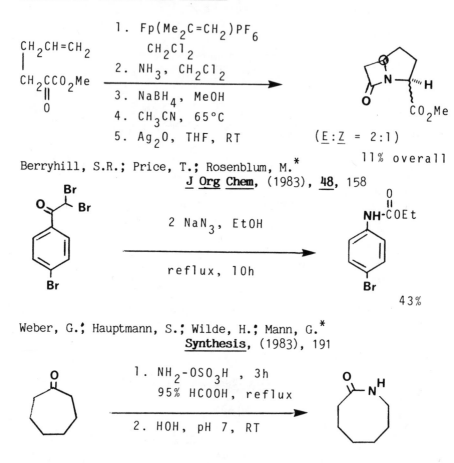

Berryhill, S.R.; Price, T.; Rosenblum, M.[*]
J Org Chem, (1983), **48**, 158

Weber, G.; Hauptmann, S.; Wilde, H.; Mann, G.[*]
Synthesis, (1983), 191

Olah, G.A.[*]; Fung, A.P.   **Org Syn**, (1984), **63**, 188

43%

Wälchli, R.; Bienz, S.; Hesse, M.[*]
　　　　　　　Helv Chim Acta, (1985), **68**, 484

SECTION 88: Amides from Nitriles

EtCN
$\xrightarrow[\text{2. HOH}]{\begin{array}{c}\text{1. \underline{t}BuBr, K-10 clay}\\ \text{FeNO}_3, \text{CH}_3\text{CN, RT}\end{array}}$
$\overset{\text{O}}{\overset{\|}{\text{EtCNH\underline{t}Bu}}}$

with 3° halides only　　　　　　　　　35%
Polla, E.[*]　　**Bull Chem Soc Belg**, (1985), **94**, 81

PhCN
$\xrightarrow{\begin{array}{l}\text{1. LiAl(OEt)}_3\text{H, Et}_2\text{O, 0°C}\\ \text{2. Me}_3\text{SiCl, 0°C - RT}\\ \text{3. Me}_2\text{C=C-OLi , -78 - -30°C}\\ \text{　　　　OEt}\\ \text{4. 0°C, aq NH}_4\text{Cl}\end{array}}$

57%

Andreoli, P.; Cainelli, G.[*]; Contento, M.; Giacomini, D.;
Martelli, G.; Panunzio, M.[*]
　　　　　　　Tetrahedron Lett, (1986), **27**, 1695

CH$_3$CN
$\xrightarrow[\text{3\% RuH}_2(\text{PPh}_3)_4, \text{ 24h}]{\begin{array}{c}\text{Pip, 2 HOH, DMF, 160°C}\\ \text{sealed tube}\end{array}}$

97%

Murahashi, S.-I.[*]; Naota, T.; Saito, E.
　　　　　　　J Am Chem Soc, (1986), **108**, 7846

SECTION 89: Amides from Olefins

1. $AcNH_2$, $Hg(NO_3)_2$
   reflux, 24h, CH_2Cl_2
2. $NaBH_4$, $BuNH_2$, 0°C
   aq. NaOH

NHAc

92%

Barluenga, J.[*]; Jiménez, C.; Nájera, C.; Yus, M.
                JCS **Perkin** I, (1983), 591

1. $R-NH_2$, CH_2Cl_2, 24h
   $Hg(NO_3)_2$, reflux
2. $NaBH_4$, $BuNH_2$, 0°C
   aq. NaOH

R = CO_2Me 96%

Barluenga, J.[*]; Jimenez, C.; Nájera, C.; Yus, M.
                J Heterocyclic Chem, (1983), **20**, 1733

R = Ts   63%

Barluenga, J.[*]; Jiménez, C.; Nájera, C.; Yus, M.
                JCS **Perkin** I, (1984), 721

$CH_2=CHPh$

1. NO_2^+ BF_4^-, -70°C
   CH_2Cl_2, CH_3CN
2. aq. $NaHCO_3$

Ph~CHNHAc
    |
   CH_2NO_2

84%

Bloom, A.J.; Fleischmann, M.; Mellor, J.M.[*]
                JCS **Perkin** I, (1984), 2357

SECTION 90: Amides from Miscellaneous Compounds

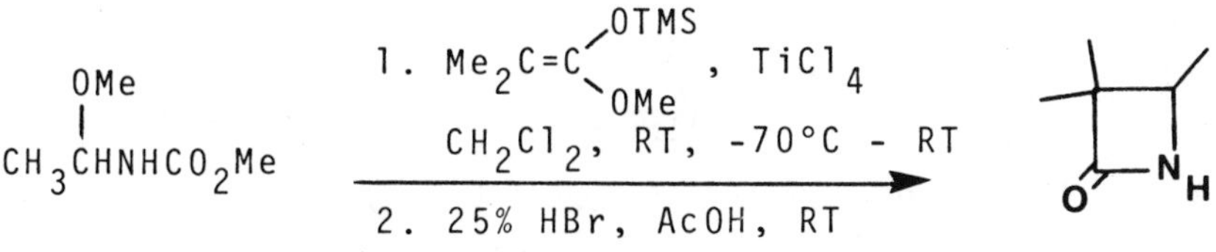
OMe
$CH_3CHNHCO_2Me$
1. $Me_2C=C(OTMS)(OMe)$, $TiCl_4$
CH_2Cl_2, RT, -70°C - RT
2. 25% HBr, AcOH, RT
3. RMgBr, THF, 0°C
O
N
H

65%

Shono, T.*; Tsubata, K.; Okinaga, N.
J Org Chem, (1984), **49**, 1056

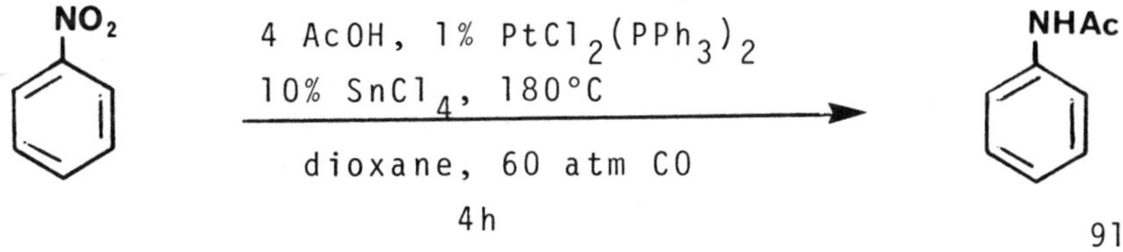
NO_2
4 AcOH, 1% $PtCl_2(PPh_3)_2$
10% $SnCl_4$, 180°C
dioxane, 60 atm CO
4h
NHAc

91%

Watanabe, Y.*; Tsuji, Y.; Kondo, T.; Takeuchi, R.
J Org Chem, (1984), **49**, 4451

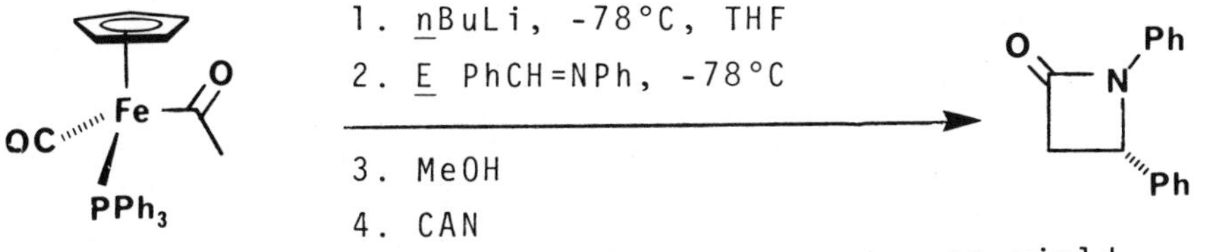
O
OC
Fe
PPh_3
1. nBuLi, -78°C, THF
2. E PhCH=NPh, -78°C
3. MeOH
4. CAN
O
N
Ph
Ph

no yield reported

Broadley, K.; Davies, S.G.* **Tetrahedron Lett**, (1984), **25**, 1743

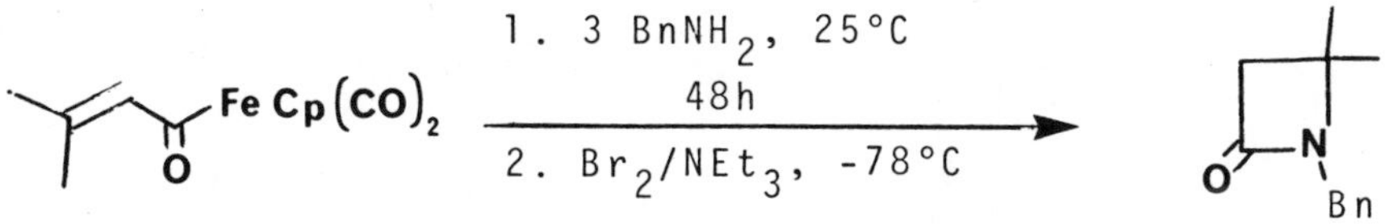
Fe Cp(CO)$_2$
O
1. 3 $BnNH_2$, 25°C
48h
2. Br_2/NEt_3, -78°C
O
N
Bn

56%

Ojima, I.*; Kwon, H.B. **Chem Lett**, (1985), **1327**

Reviews:

"Synthesis and Reactions of Sulfamides"

McDermott, S.D.; Spillane, W.J.

<u>Org</u> <u>Prep</u> <u>Proc</u> <u>Int</u>, (1984), **16**, 49

"α-Amide Alkylation at Carbon: Recent Advances"

Zaugg, H.E.*   <u>**Synthesis**</u>, (1984), 85, 181

CHAPTER 7
PREPARATION OF AMINES

SECTION 91: Amines from Acetylenes

No Additional Examples

SECTION 92: Amines from Acid Derivatives

1. NaN_3, $\underline{n}Bu_4NBr$, 0°C
   HOH/CH_2Cl_2, 2h
2. CF_3COOH, CH_2Cl_2
   6h
3. K_2CO_3/aq. MeOH

91%

Pfister, J.R.[*]; Wymann, W.E.   **Synthesis,** (1983), 38

CH_3CHCCl (O, Cl)

1. Pip, PhH, NEt_3
   0°C, 2h

2. $LiAlH_4$, Et_2O
   reflux, 20h

87%

Suzuki, K.; Okano, K.; Nakai, K.; Terao, Y.; Sekiya, M.[*]
Synthesis, (1983), 723

SECTION 93: Amines from Alcohols and Thiols

$\underline{n}BuOH$

1. $(EtO)_2\overset{O}{\overset{\|}{P}}NHCO_2\underline{t}Bu$, DEAD
   PPh_3, 0 - 25°C, 2h

2. HCl/PhH, RT, 12h

$\underline{n}BuNH_3^+Cl^-$

85%

Slusarska, E.; Zwierzak, A.[*]   **Liebigs Ann Chem,** (1986), 402

SECTION 94: Amines from Aldehydes

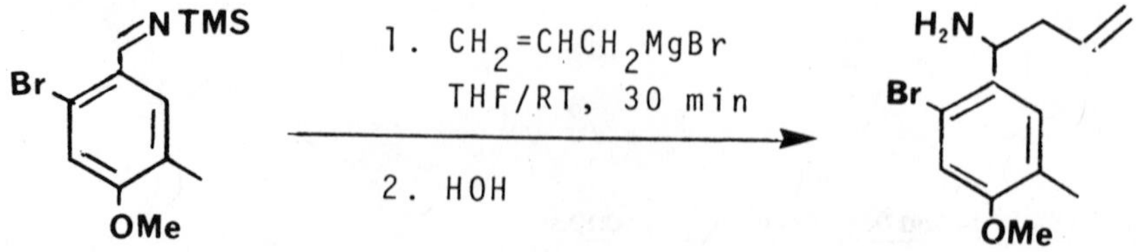

1. $CH_2=CHCH_2MgBr$
   THF/RT, 30 min

2. HOH

97%

Hart, D.J.[*]; Kanai, K.; Thomas, D.G.; Yang, T.-K.
J Org Chem, (1983), 48, 289

PhCHO
$\xrightarrow[\substack{PhNH_2, AcOH, RT \\ 2h}]{BH_3 \cdot Py, CH_2Cl_2}$
$PhCH_2NHPh$

93%

Pelter, A.; Rosser, R.M.; Mills, S.
JCS Perkin I, (1984), 717

PhCHO
$\xrightarrow[EtOH, RT, 20h]{2 PhNH_2, NaTeH}$
$PhCH_2NHPh$

73%

Yamashita, M.[*]; Kadokura, M.; Suemitsu, R.
Bull Chem Soc Jpn, (1984), 57, 3359

PhCHO
1. $NH_2SO_2NH_2$, H^+
2. MeLi, Et_2O

3. HOH-Py
4. NaOH

$\underset{\underset{NH_2}{|}}{PhCHCH_3}$

95%

Davies, F.A.[*]; Giangiordano, M.A.; Starner, W.E.
Tetrahedron Lett, (1986), 27, 3957

Related Methods: Amines from Ketones (Section 102)

SECTION 95: Amines from Alkyl, Methylenes, and Aryls

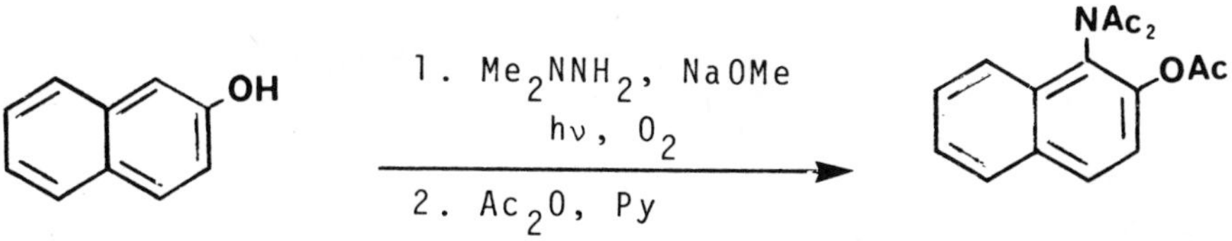

1. Me_2NNH_2, NaOMe
   $h\nu$, O_2
2. Ac_2O, Py

90%

Barton, D.H.R.; LeGreneur, S.; Motherwell, W.B.[*]
       Tetrahedron Lett, (1983), <u>24</u>, 1601

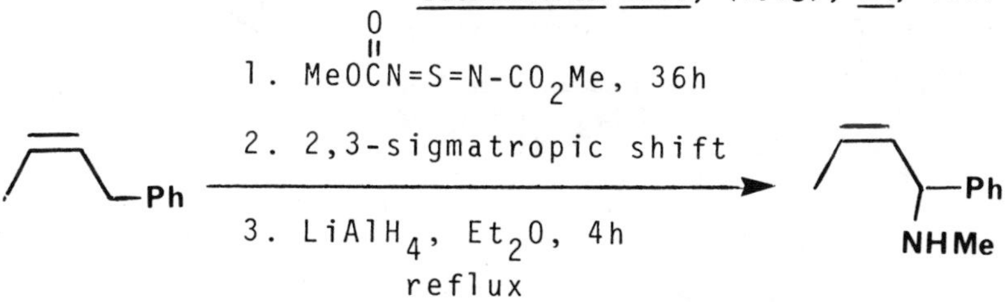

1. $MeO\overset{O}{\overset{||}{C}}N=S=N-CO_2Me$, 36h
2. 2,3-sigmatropic shift
3. $LiAlH_4$, Et_2O, 4h
   reflux

43%

Kresze, G.[*]; Münsterer, H.   **J Org Chem**, (1983), <u>48</u>, 3561

SECTION 96: Amines from Amides

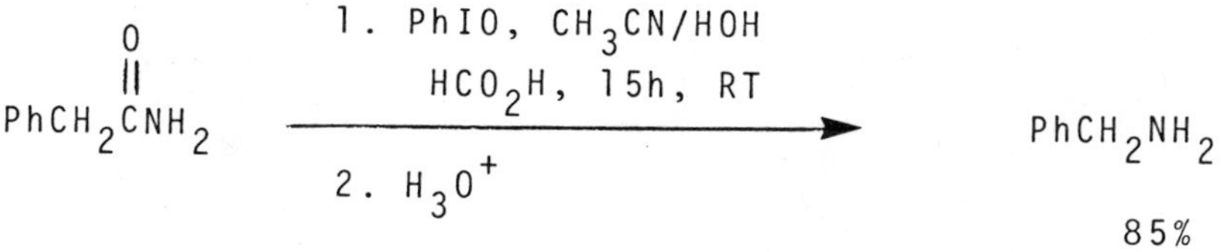

$PhCH_2\overset{O}{\overset{||}{C}}NH_2$

1. PhIO, CH_3CN/HOH
   HCO_2H, 15h, RT
2. H_3O^+

$PhCH_2NH_2$

85%

Radhakrishna, A.S.; Rao, C.G.[*]; Varma, R.K.[*]; Singh, B.B.[*];
Bhatnagar, S.P.[*]
       Synthesis, (1983), 538

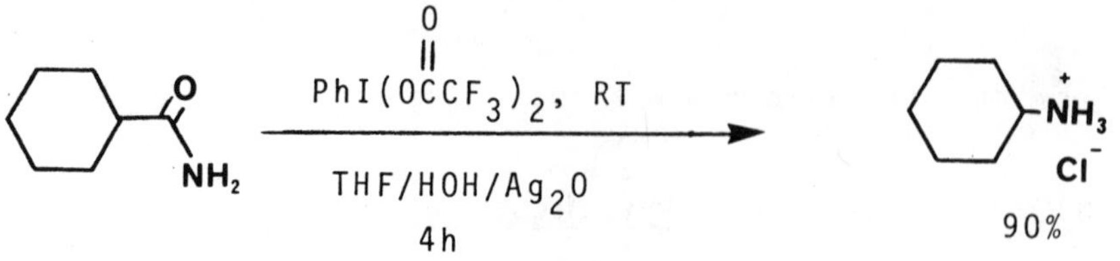

$$PhI(OCCF_3)_2, \ RT$$

THF/HOH/Ag$_2$O

4h

90%

Loudon, G.M.[*]; Radhakrishna, A.S.; Almond, M.R.; Blodgett, J.K.; Boutin, R.H.

J Org Chem, (1984), **49**, 4272

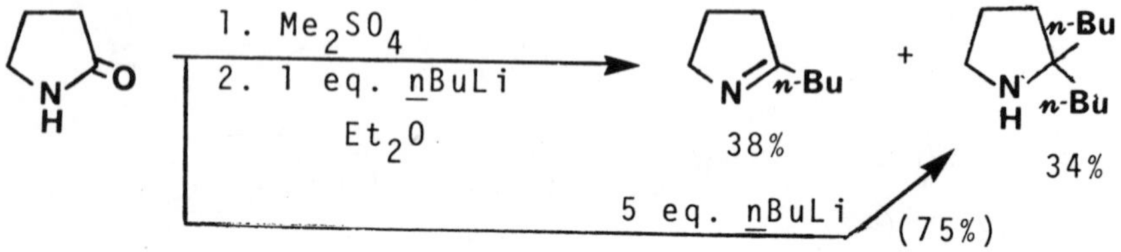

1. Me$_2$SO$_4$

2. 1 eq. nBuLi

Et$_2$O

38%

+

34%

5 eq. nBuLi (75%)

Zezza, C.A.; Smith, M.B.[*]; Ross, B.A.; Arhin, A.; Cronin, P.L.E.
J Org Chem, (1984), **49**, 4397

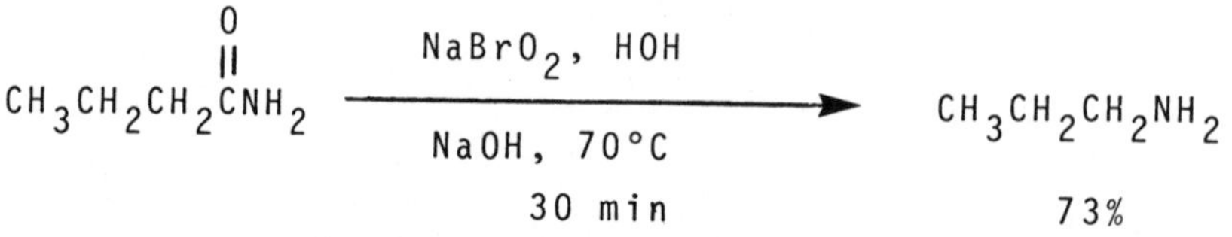

$$CH_3CH_2CH_2CNH_2 \xrightarrow[\substack{NaOH, \ 70°C \\ 30 \ min}]{NaBrO_2, \ HOH} CH_3CH_2CH_2NH_2$$

73%

Kajigaeshi, S.[*]; Nakagawa, T.; Fujisaki, S.
Chem Lett, (1984), 713

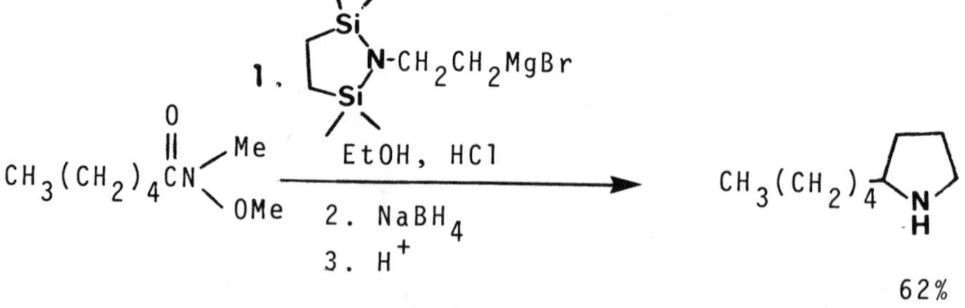

1.

$$CH_3(CH_2)_4CN \begin{matrix} Me \\ OMe \end{matrix}$$

EtOH, HCl

2. NaBH$_4$

3. H$^+$

$$CH_3(CH_2)_4 \text{—N}$$

62%

Basha, F.Z.; DeBernardis, J.F.[*]
Tetrahedron Lett, (1984), **25**, 5271

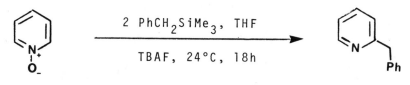

1. ⟨O⟩-Li , Et_2O, 0°C

2. $LiAlH_4$, Et_2O

3. 20% NaOH

82%

Hwang, Y.C.; Chu, M.; Fowler, F.W.[*]
J Org Chem, (1985), 50, 3885

1. $Et_3O^+BF_4^-$, CH_2Cl_2
   0°C

2. $CH_2=CHPh$, CH_3NO_2, 2d

3. $LiAlH_4$, THF, 2d

31%

Smith, M.B.[*]; Shroff, H.N.   Heterocycles, (1985), 23, 2229

NH $CO_2CH_2Ch=CH_2$

PPh_3, $CHCl_3$, THF

$Pd_2(dba)_3$, HCOOH, 30°C

NH_2

quant.

Minami, I.; Ohashi, Y.; Shimizu, I.; Tsuji, J.[*]
Tetrahedron Lett, (1985), 26, 2449

Related Methods: Protection of Amines (Section 105A)

SECTION 97: Amines from Amines

2 $PhCH_2SiMe_3$, THF

TBAF, 24°C, 18h

70%

Vorbrüggen, H.[*]; Krolikiewicz, K.
Tetrahedron Lett, (1983), 24, 889

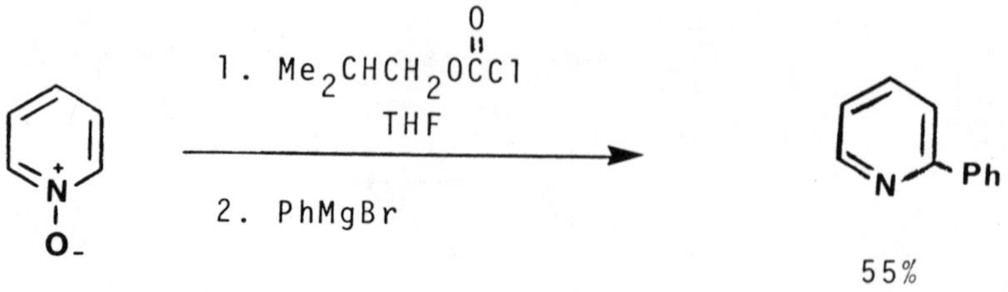

1. $Me_2CHCH_2O\overset{O}{\overset{\|}{C}}Cl$
   THF

2. PhMgBr

55%

Webb, T.R.[*]   **Tetrahedron Lett**, (1985), **26**, 3191

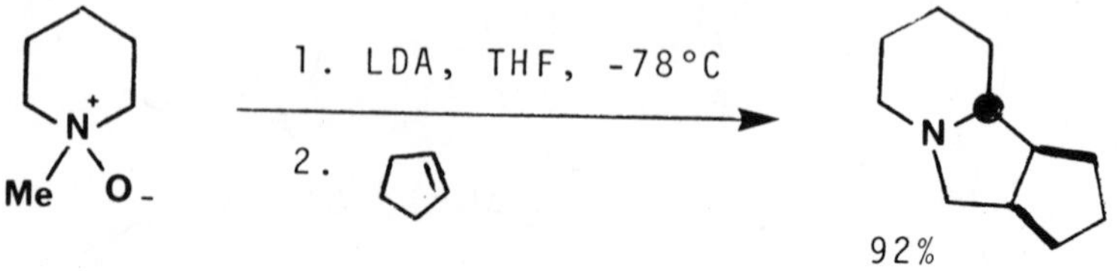

1. LDA, THF, -78°C

2.

92%

Chastanet, J.; Roussi, G.[*]   **J Org Chem**, (1985), **50**, 2910

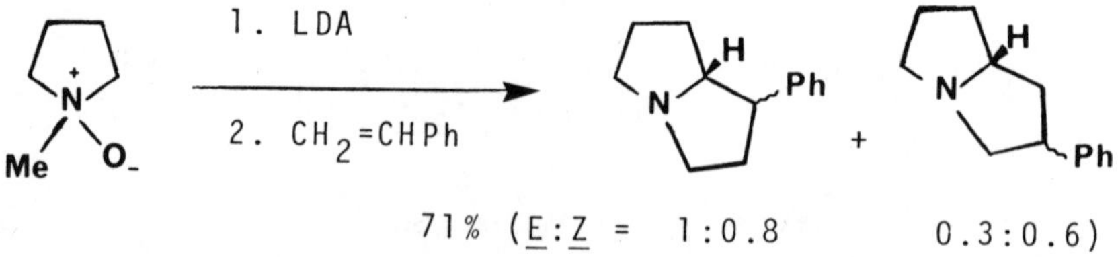

1. LDA

2. CH_2=CHPh

71% (E:Z = 1:0.8   0.3:0.6)

Chastanet, J.; Roussi, G.   **Heterocycles**, (1985), **23**, 653

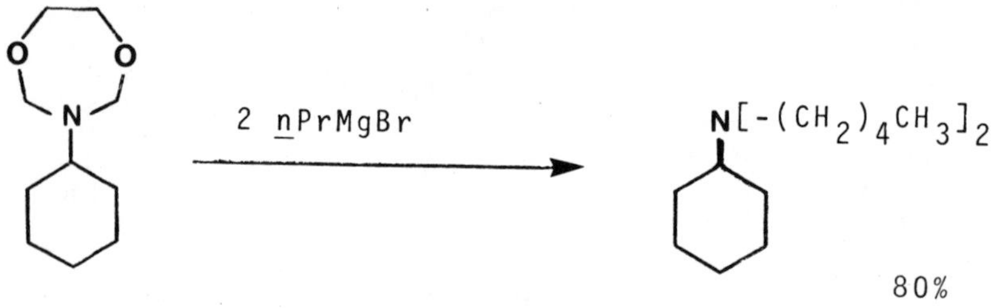

2 nPrMgBr

$N[-(CH_2)_4CH_3]_2$

80%

Kapnang, H.; Charles, G.[*]   **Tetrahedron Lett**, (1983), **24**, 1597

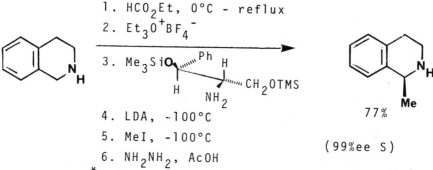

CH$_2$CH=CH$_2$
|
CH$_2$NHPh

1. HCl/MeOH, Et$_2$O
2. PhSCl, CH$_2$Cl$_2$
3. K$_2$CO$_3$, NaI, CH$_3$CN
4. Ni(R), EtOH

51%

Ohsawa, T.; Ihara, M.; Fukumoto, K.; Kametani, T.
J Org Chem, (1983), 48, 3644

1. HCO$_2$Et, 0°C - reflux
2. Et$_3$O$^+$BF$_4^-$
3. Me$_3$SiO...Ph H ...CH$_2$OTMS H NH$_2$
4. LDA, -100°C
5. MeI, -100°C
6. NH$_2$NH$_2$, AcOH

77%

(99%ee S)

Meyers, A.I.[*]; Fuentes, L.M.   J Am Chem Soc, (1983), 105, 117

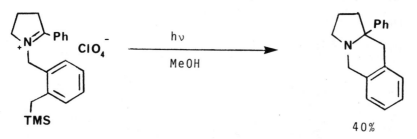

hν
MeOH

40%

Lan, A.J.Y.; Quillen, S.L.; Heuckeroth, R.O.; Mariano, P.S.[*]
J Am Chem Soc, (1984), 106, 6439
Yoon, U.C.; Quillen, S.L.; Mariano, P.S.[*]; Swanson, R.;
Stavinoha, J.L.; Bay, E.
J Am Chem Soc, (1983), 105, 1204

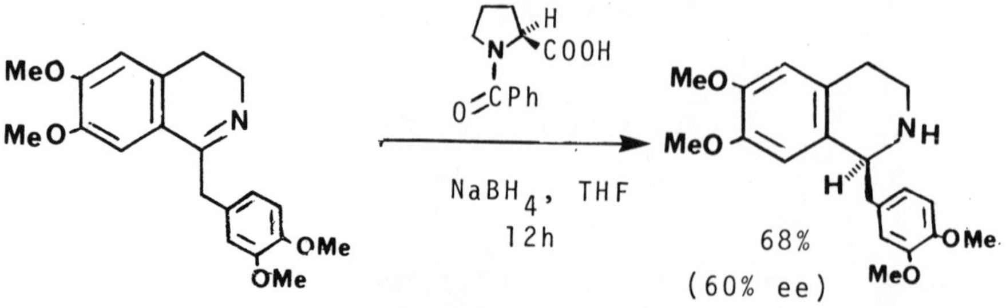
MeO
MeO
N
OMe
OMe
H
N
COOH
O=CPh
NaBH4, THF
12h
MeO
MeO
NH
H
OMe
MeO
68%
(60% ee)

Yamada, K.; Takeda, M.; Iwakuma, T.* __JCS Perkin I__, (1983), 265

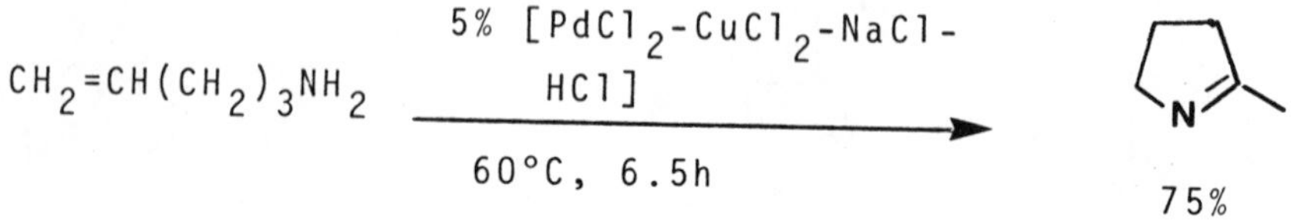
CH2=CH(CH2)3NH2
5% [PdCl2-CuCl2-NaCl-HCl]
60°C, 6.5h
N
75%

Pugin, B.; Venanzi, L.M.* __J Am Chem Soc__, (1983), __105__, 6877

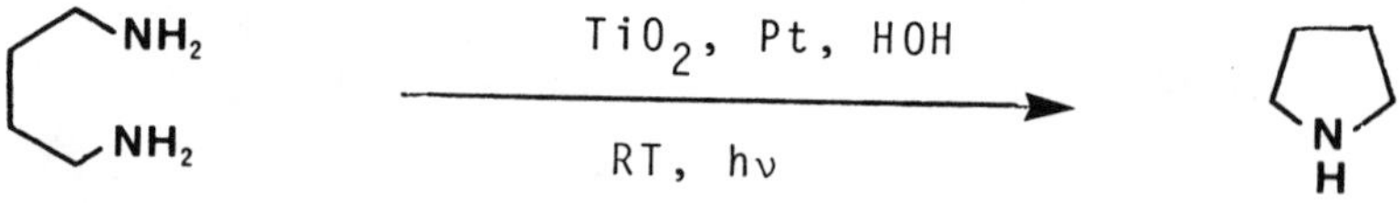
NH2
NH2
TiO2, Pt, HOH
RT, hν
N
H

Nishimoto, S.; Ohtani, B.; Yoshikawa, T.; Kagiya, T.*
__J Am Chem Soc__, (1983), __105__, 7180

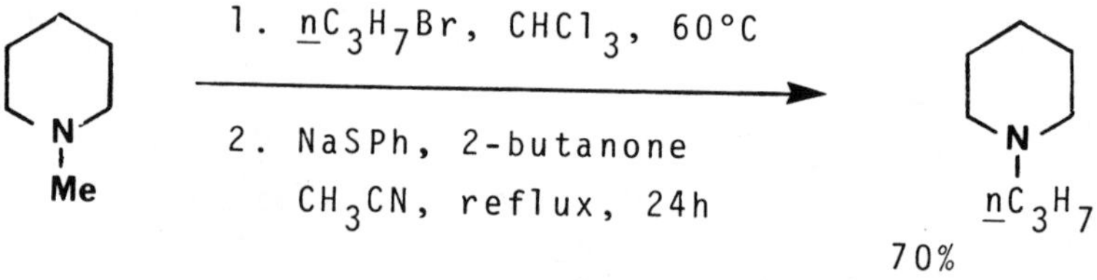
N
Me
1. nC3H7Br, CHCl3, 60°C
2. NaSPh, 2-butanone
CH3CN, reflux, 24h
N
nC3H7
70%

Manoharan, T.S.; Madyastha, K.M.*; Singh, B.B.; Bhatnagar, S.P.; Weiss, U.
__Synthesis__, (1983), 809
__Ind J Chem B__, (1984), __23__, 5

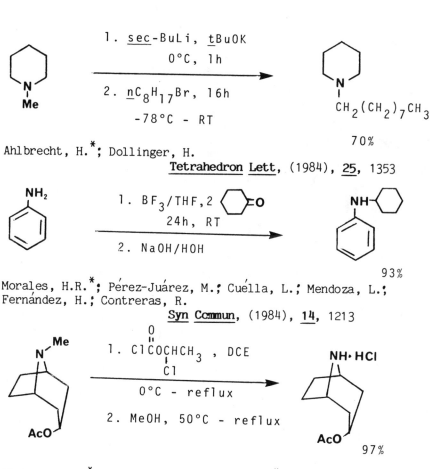

Ahlbrecht, H.[*]; Dollinger, H.

Tetrahedron Lett, (1984), **25**, 1353

Morales, H.R.[*]; Pérez-Juárez, M.; Cuélla, L.; Mendoza, L.;
Fernández, H.; Contreras, R.

Syn Commun, (1984), **14**, 1213

Olofson, R.A.[*]; Martz, J.T.; Senet, J.-P.[*]; Piteau, M.;
Malfroot, T.

J Org Chem, (1984), **49**, 2081

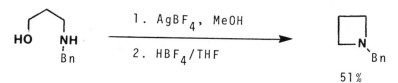

Sammes, P.G.[*]; Smith, S.   **JCS Perkin I**, (1984), 2415

1. $\underline{t}$BuLi, THF, -25°C

2. HMPA, $\underline{n}$BuBr

3. NH_2NH_2

53%

Meyers, A.I.[*]; Edwards, P.D.; Rieker, W.F.; Bailey, T.R.
J Am Chem Soc, (1984), 106, 3270
Meyers, A.I.[*]; Edwards, P.D.; Bailey, T.R.; Jagdmann Jr., G.E.
J Org Chem, (1985), 50, 1019

4 Me_3Al

CH_2Cl_2
0°C, 10h

Bn OCO_2Et

74%

Fujiwara, J.; Sano, H.; Maruoka, K.; Yamamoto, H.[*]
Tetrahedron Lett, (1984), 25, 2367

1. $\underline{t}$BuSiMe$_2$OTf, CH_2Cl_2
2. MeLi, THF

3. BnCl, 10h
4. TBAF, THF

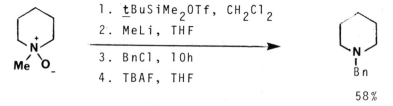

Me O$_-$

Bn

58%

Tokitoh, N.; Okazaki, R.[*]; Suzuki, H.[*]; Manabe, H.
Chem Lett, (1984), 1937

1. $\underline{t}$BuSiMe$_2$OTf, CH_2Cl_2
2. MeLi, THF

3. PhMgBr

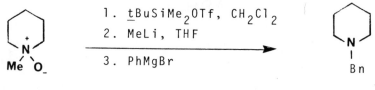

Me O$_-$

Bn

74%

Tokitoh, N.; Okazaki, R.[*]   Tetrahedron Lett, (1984), 25, 4677

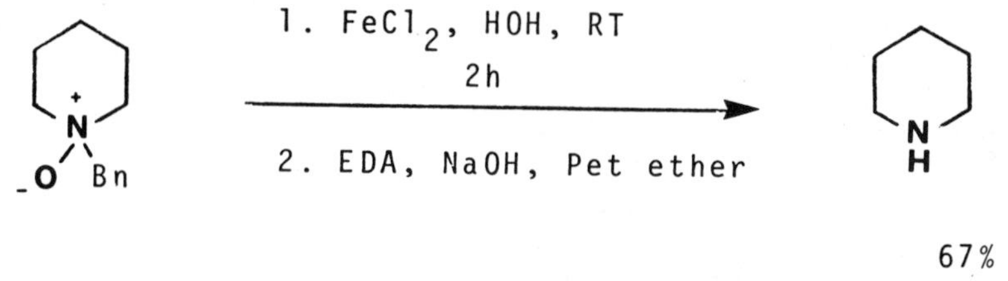

1. $FeCl_2$, HOH, RT
   2h

2. EDA, NaOH, Pet ether

67%

Monkovic, I.; Wong, H.; Bachand, C. **Synthesis**, (1985), 770

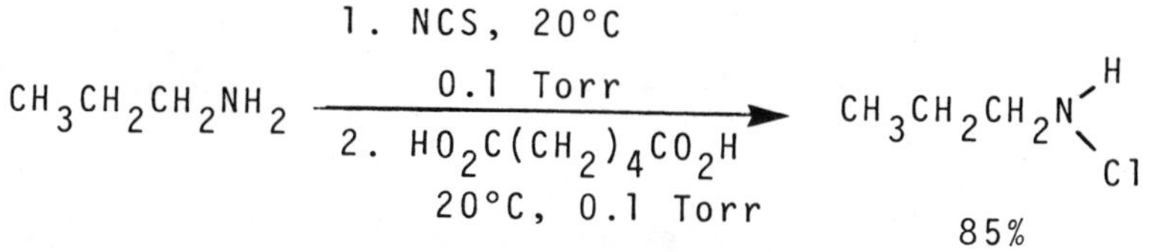

$CH_3CH_2CH_2NH_2$

1. NCS, 20°C
   0.1 Torr

2. $HO_2C(CH_2)_4CO_2H$
   20°C, 0.1 Torr

$CH_3CH_2CH_2N\overset{H}{\underset{Cl}{}}$

85%

Guillemin, J.C.; Denis, J.M.[*] **Synthesis**, (1985), 1131

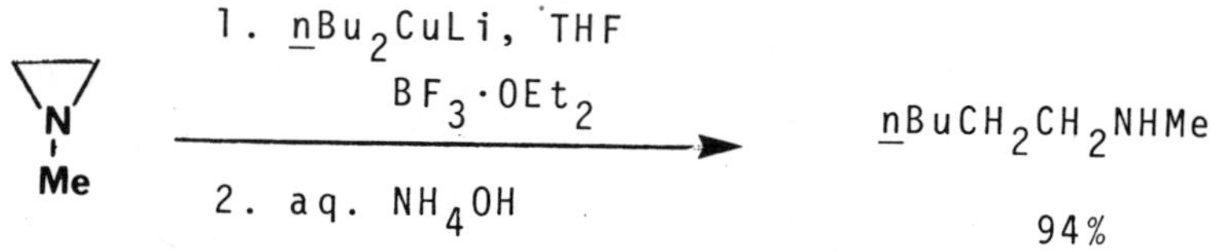

1. $\underline{n}Bu_2CuLi$, THF
   $BF_3 \cdot OEt_2$

2. aq. NH_4OH

$\underline{n}BuCH_2CH_2NHMe$

94%

Eis, M.J.; Ganem, B.[*] **Tetrahedron Lett**, (1985), **26**, 1153

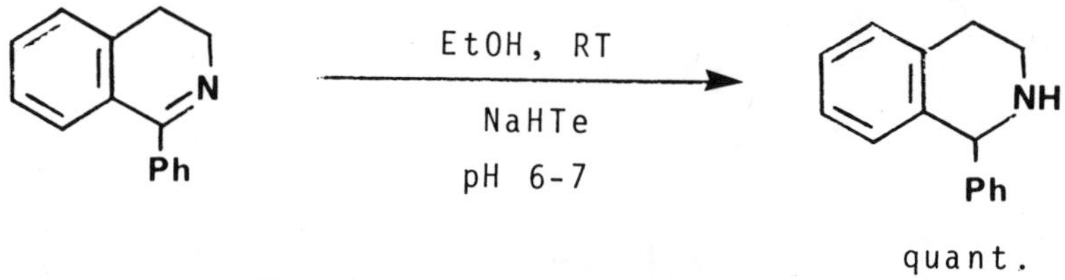

EtOH, RT

NaHTe

pH 6-7

quant.

Barton, D.H.R.[*]; Fekih, A.; Lusinchi, X.
                    Tetrahedron Lett, (1985), **26**, 3693

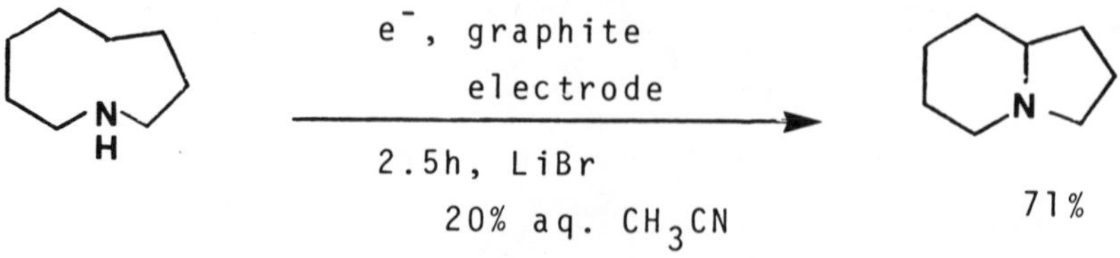

e⁻, graphite
electrode
$\xrightarrow{\hspace{3cm}}$
2.5h, LiBr
20% aq. CH_3CN

71%

Elofson, R.M.; Gadallah, F.F.; Laidler, J.K.
Can J Chem, (1985), **63**, 1170

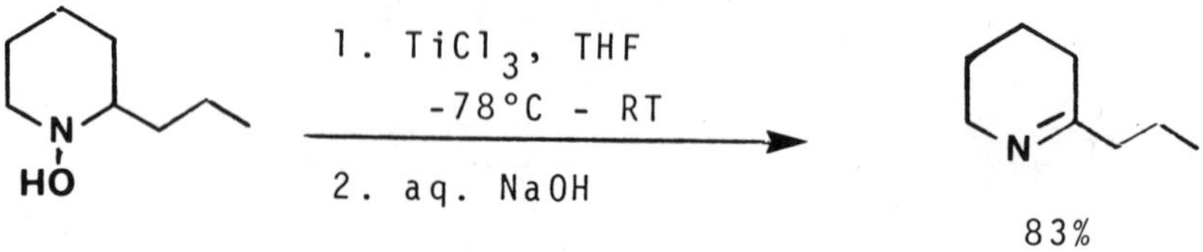

1. $TiCl_3$, THF
   -78°C - RT
$\xrightarrow{\hspace{3cm}}$
2. aq. NaOH

83%

Murahashi, S.[*]; Kodera, Y.   **Tetrahedron Lett**, (1985), **26**, 4633

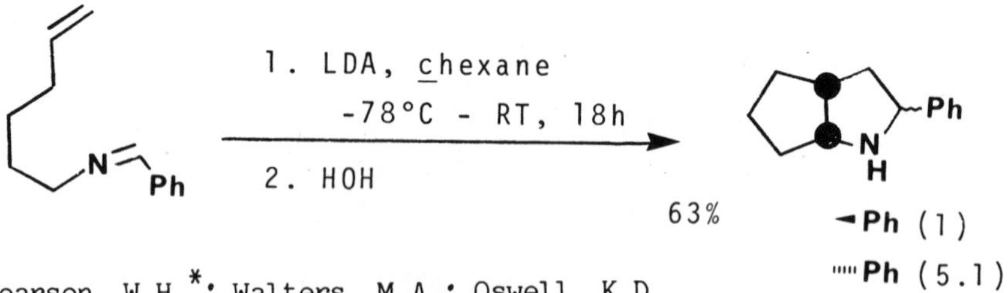

1. LDA, chexane
   -78°C - RT, 18h
$\xrightarrow{\hspace{3cm}}$
2. HOH

63%

-Ph (1)
''''Ph (5.1)

Pearson, W.H.[*]; Walters, M.A.; Oswell, K.D.
J Am Chem Soc, (1986), **108**, 2769

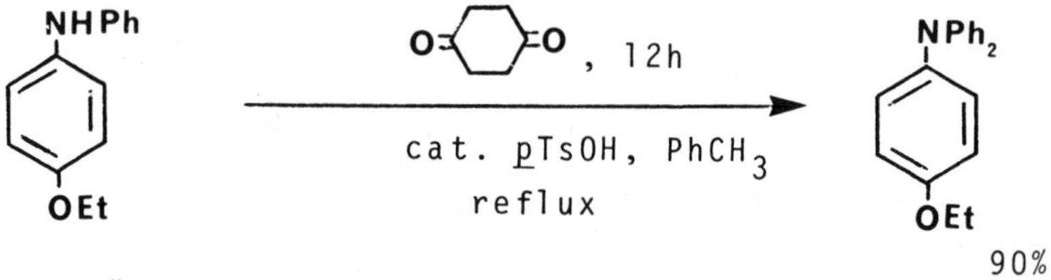

O=⟨ ⟩=O , 12h
$\xrightarrow{\hspace{3cm}}$
cat. pTsOH, $PhCH_3$
reflux

90%

Haga, K.[*]; Iwaya, K.; Kaneko, R.
Bull Chem Soc Jpn, (1986), **59**, 803
Haga, K.[*]; Oohashi, M.; Kaneko, R.
Bull Chem Soc Jpn, (1984), **57**, 1586

Review: "Nucleophilic Additions to Tetrahydropyridinium Salts: Applications to Alkaloid Syntheses"

Stevens, R.V.[*] **Accts Chem Res**, (1984), **17**, 289

SECTION 98: Amines from Esters

$$HOCH_2CH=CHCH_2OAc \xrightarrow[\substack{THF, 20°C \\ cat. Pd(PPh_3)_4 \\ 1h}]{PiP} HOCH_2CH=CHCH_2-N<$$

(E)   (E)   65%

Genêt, J.P.[*]; Balabane, M.; Bäckvall, J.E.[*]; Nyström, J.E.
Tetrahedron Lett, (1983), **24**, 2745

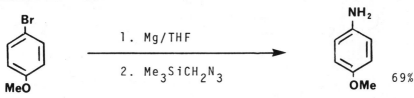

Bergbreiter, D.E.[*]; Chen, B.   **JCS Chem Comm**, (1983), 1238

SECTION 99: Amines from Ethers, Epoxides, and Thioethers

No Additional Examples

SECTION 100: Amines from Halides and Sulfonates

(p-bromoanisole) $\xrightarrow[\text{2. } Me_3SiCH_2N_3]{\text{1. Mg/THF}}$ (p-anisidine) 69%

Nishiyama, K.[*]; Tanaka, N.   **JCS Chem Comm**, (1983), 1322

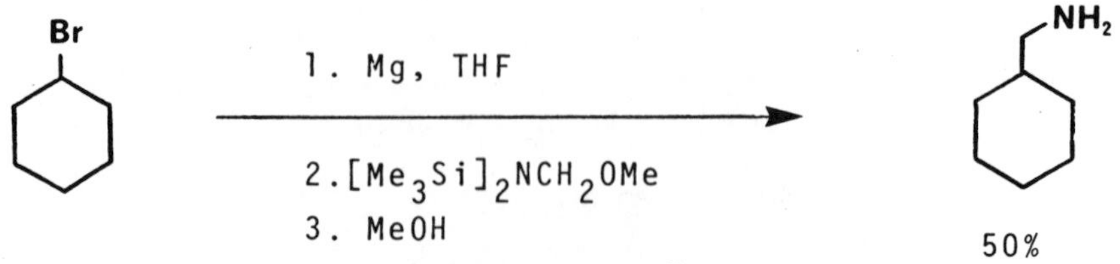
Br
1. Mg, THF
2. $[Me_3Si]_2NCH_2OMe$
3. MeOH
NH_2
50%

Morimoto, T.; Takahashi, T.; Sekiya, M.*
<u>JCS Chem Comm</u>, (1984), 794

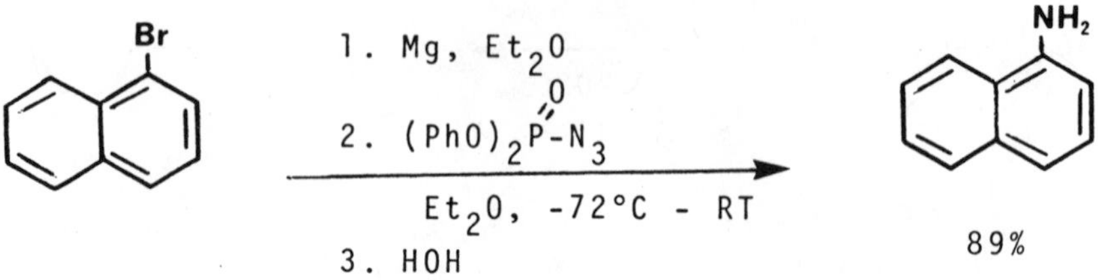
Br
1. Mg, Et_2O
2. $(PhO)_2P(O)-N_3$
Et_2O, -72°C - RT
3. HOH
NH_2
89%

Mori, S.; Aoyama, T.; Shioiri, T.*
<u>Tetrahedron Lett</u>, (1984), <u>25</u>, 429

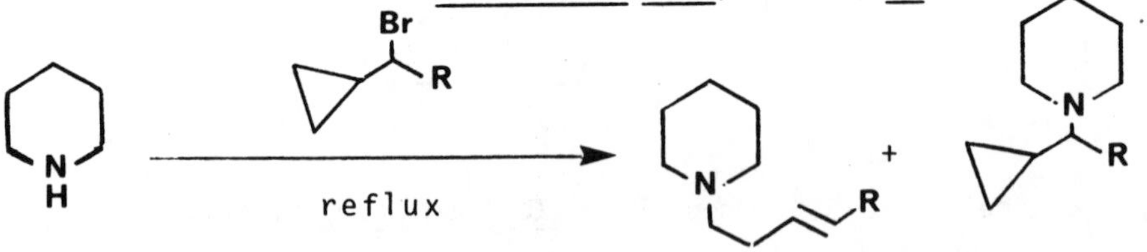
N
H
Br
R
reflux
N
R
+
N
R

R = Me 91% (8 : 92)
R = tBu 95% (92 : 8)

Hrubiec, R.T.; Smith, M.B.* <u>Tetrahedron Lett</u>, (1983), <u>24</u>, 5031

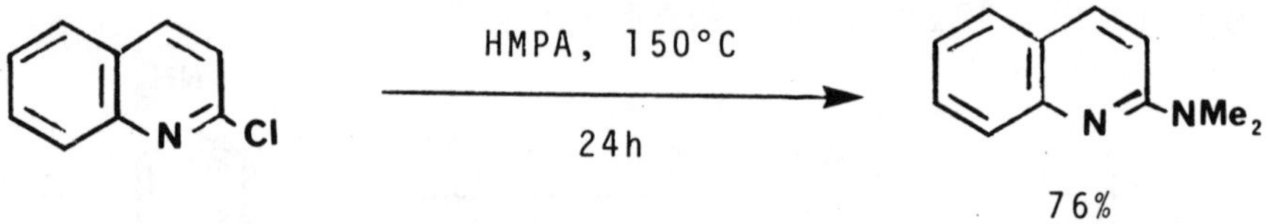
N
Cl
HMPA, 150°C
24h
N
NMe_2
76%

Gupton, J.T.*; Idoux, J.P.*; Baker, G.; Colon, C.; Crews, A.D.; Jurss, C.D.; Rampi, R.C.
<u>J Org Chem</u>, (1983), <u>48</u>, 2933

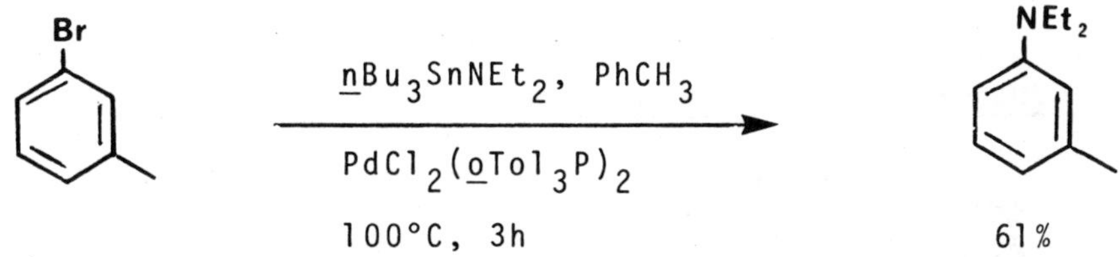

Br
nBu3SnNEt2, PhCH3
PdCl2(oTol3P)2
100°C, 3h
NEt2
61%

Kosugi, M.; Kameyama, M.; Migita, T.* **Chem Lett**, (1983), 927

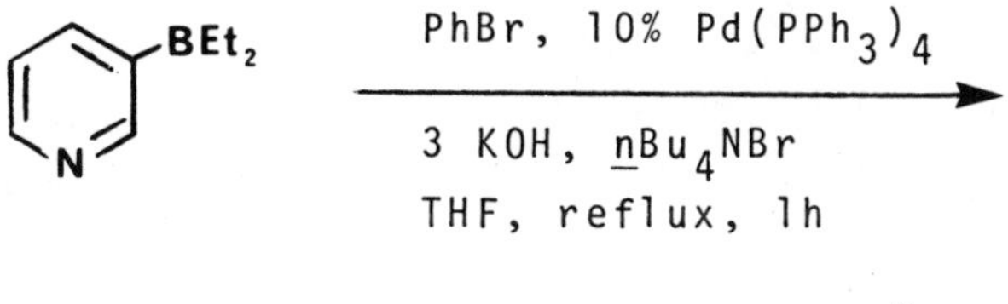

BEt2
N
PhBr, 10% Pd(PPh3)4
3 KOH, nBu4NBr
THF, reflux, 1h
Ph
N
85%

Ishikura, M.; Kamada, M.; Terashima, M.* **Heterocycles**, (1984), 22, 265

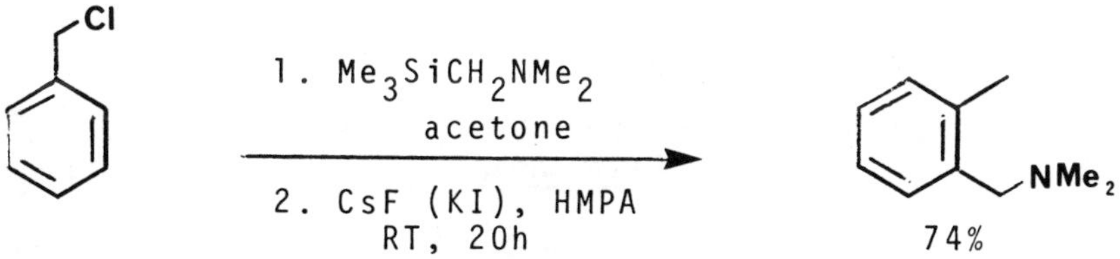

Cl
1. Me3SiCH2NMe2
acetone
2. CsF (KI), HMPA
RT, 20h
NMe2
74%

Nakano, M.; Sato, Y.* **JCS Chem Comm**, (1985), 1684

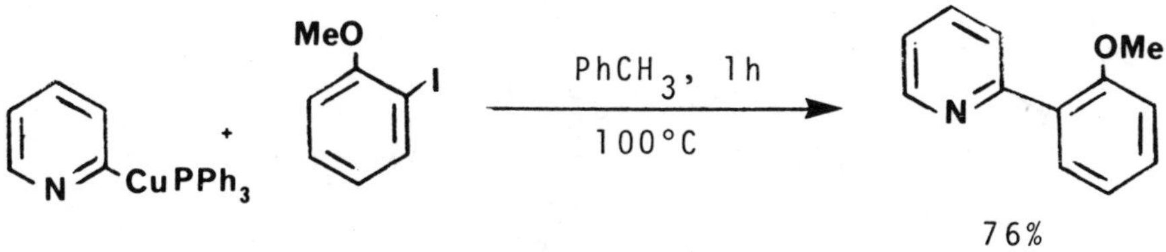

N
CuPPh3
+
MeO
I
PhCH3, 1h
100°C
N
OMe
76%

Malmberg, H.; Nilsson, M.* **Tetrahedron**, (1986), 42, 3981

Review: "Copper Assisted Nucleophilic Substitution of Aryl Halogen"

Lindley, J.* **Tetrahedron**, (1984), 40, 1433

SECTION 101: Amines from Hydrides

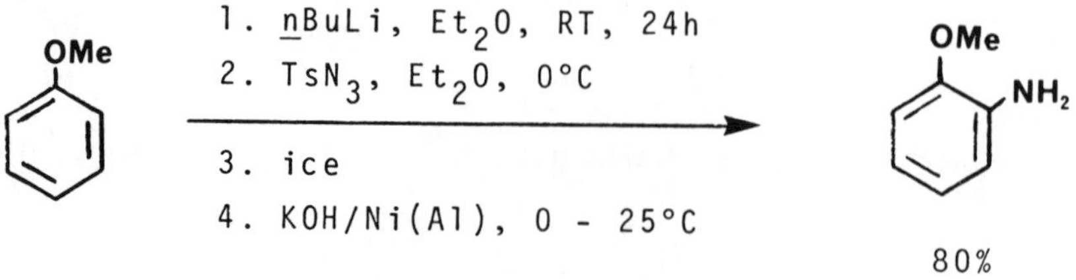
OMe
1. nBuLi, Et_2O, RT, 24h
2. TsN_3, Et_2O, 0°C
3. ice
4. KOH/Ni(Al), 0 - 25°C
OMe
NH_2
80%

Narasimhan, N.S.; Ammanamanchi, R.
Tetrahedron Lett, (1983), **24**, 4733

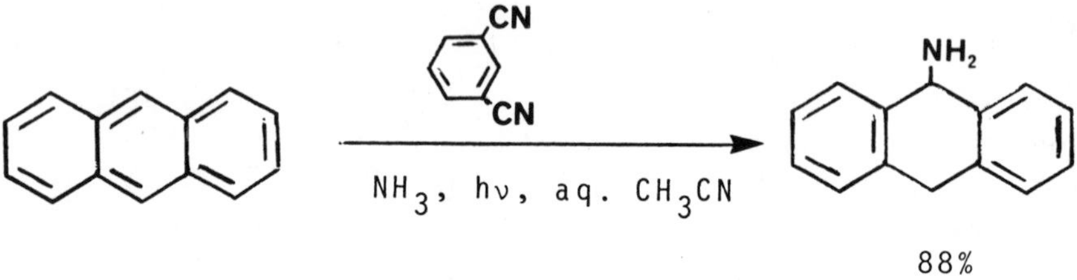
CN
CN
NH_3, hν, aq. CH_3CN
NH_2
88%

Yasuda, M.*; Yamashita, T.; Matsumoto, T.; Shima, K.; Pac, C.*
J Org Chem, (1985), **50**, 3667

SECTION 102: Amines from Ketones

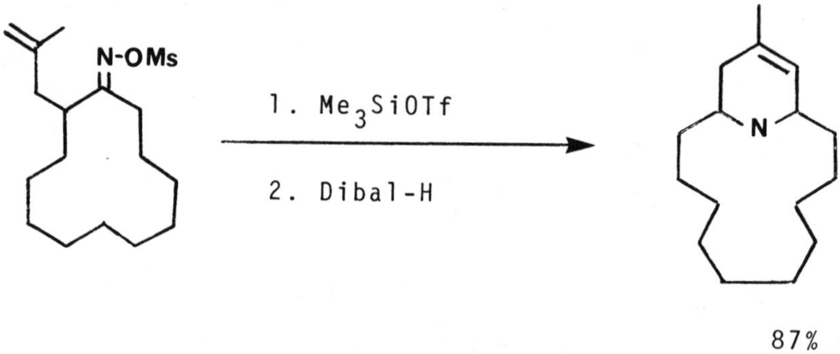
N-OMs
1. Me_3SiOTf
2. Dibal-H
N
87%

Sakane, S.; Maruoka, K.; Yamamoto, H.*
Tetrahedron Lett, (1983), **24**, 943

$$[CH_3(CH_2)_4]_2C=NOH \xrightarrow[\text{0-20°C}]{\text{5 Dibal-H}} [CH_3(CH_2)_4]_2NH$$

85%

Sasatani, S.; Miyazaki, T.; Maruoka, K.; Yamamoto, H.[*]
Tetrahedron Lett, (1983), **24**, 4711

1. Et_2AlCl

2. Dibal-H

53%

Sakane, S.; Matsumura, Y.; Yamamura, Y.; Ishida, Y.; Muruoka, K.; Yamamoto, H.[*]
J Am Chem Soc, (1983), **105**, 672

$\xrightarrow[\text{AcOH}]{\text{NaBH}_3\text{CN}}$ 94%

$\xrightarrow[\text{PhSSPh}]{\underline{n}Bu_3P}$ THF

$\xrightarrow[\text{NaCN}]{\text{AcOH}}$ 'good yield'

Barton, D.H.R.; Motherwell, W.B.; Simon, E.S.; Zard, S.Z.[*]
JCS Chem Comm, (1984), 337

1. $NaBH_3CN$, NaBr
   MeOH, RT, 4d

2. pH 3, RT, 3h

3. pH 11

80%

Abe, K.; Okumura, H.; Tsugoshi, T.; Nakamura, N.
Synthesis, (1984), 597

1. $\underline{n}$BuLi, THF, -78°C
2. PhSCu, THF, -78°C

$$\xrightarrow{\qquad\qquad}$$

3. $CH_2=CH\overset{\overset{O}{\|}}{C}CH_3$, -70°C

4. H_3O^+

82%

Kelly, T.R.[*]; Liu, H.   **J Am Chem Soc**, (1985), **107**, 4998

$$BrCH_2CH_2\overset{\overset{CH_3}{|}}{C}H\overset{\overset{}{C}}{\underset{\overset{\|}{O}}{}}CH_3$$

1. NaN_3, DMSO
   55°C, 18h

$$\xrightarrow{\qquad\qquad}$$

2. PPh_3, pentane
   RT, 12h

65%

Vaultier, M.; Lambert, P.H.; Carrié, R.[*]
                        Bull Chem Soc Fr, (1986), II083

1. $LiAlH_4$, THF

$$\xrightarrow{\qquad\qquad}$$

2. HOH

73%

(44% ee R)

Itsuno, S.[*]; Tanaka, K.; Ito, K.   **Chem Lett**, (1986), 1133

1. 2 $\underline{t}$BuOK, THF
   reflux, 2h

$$\xrightarrow{\qquad\qquad}$$

2. H_3O^+

95%

Sulmon, P.; De Kimpe, N.[*]; Schamp, N.
                        JCS Chem Comm, (1986), 1677

Related Methods: Amines from Aldehydes (Section 94)

SECTION 103: Amines from Nitriles

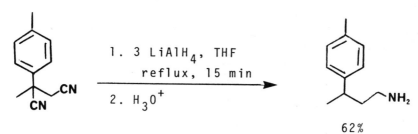

New, J.S.[*]; Yevich, J.P.   **Synthesis**, (1983), 388

1. Dibal-H, PhCH$_3$
   -70°C - RT

2. NaF, 0°C, 30 min

3. HOH, 0°C - RT   85%

Overman, L.E.[*]; Burk, R.M.   **Tetrahedron Lett**, (1984), **25**, 5737

HSiMe$_3$, CO

cat. Co$_2$(CO)$_8$
PhCH$_3$, -20 - -60°C

61%

Murai, T.[*]; Sakane, T.; Kato, S.
                    Tetrahedron Lett, (1985), **26**, 5145

1. iPrMgCl, THF
   reflux, 2h

2. LiNH$_2$, PhCO$_2$Na

60%

Weiberth, F.J.; Hall, S.S.[*]   **J Org Chem**, (1986), **51**, 5338

SECTION 104: Amines from Olefins

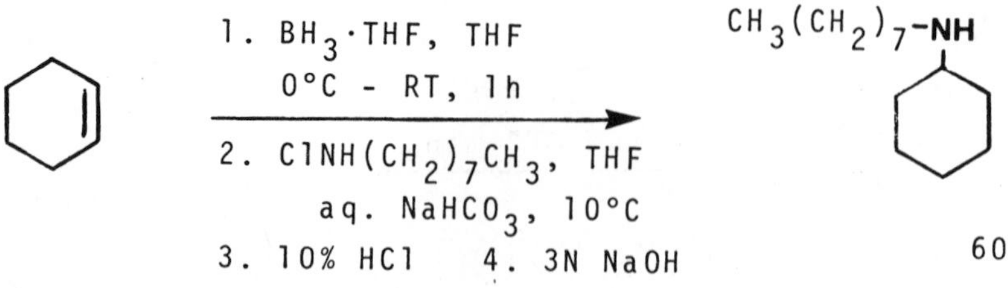
1. $BH_3 \cdot THF$, THF
0°C - RT, 1h
2. $ClNH(CH_2)_7CH_3$, THF
aq. $NaHCO_3$, 10°C
3. 10% HCl 4. 3N NaOH
$CH_3(CH_2)_7$-NH

60%

Kabalka, G.W.*; McCollum, G.W.; Kunda, S.A.
J Org Chem, (1984), **49**, 1656

Review: "Amination of Alkenes"

Gasc, M.B.; Lattes, A.; Perie, J.J.
Tetrahedron, (1983), **39**, 703

SECTION 105: Amines from Miscellaneous Compounds

$CH_3Li \cdot NHMeOMe$ —(1. nBuLi; 2. HOH)→ CH_3NHnBu

72%

Kokko, B.J.; Beak, P.* **Tetrahedron Lett**, (1983), **24**, 561

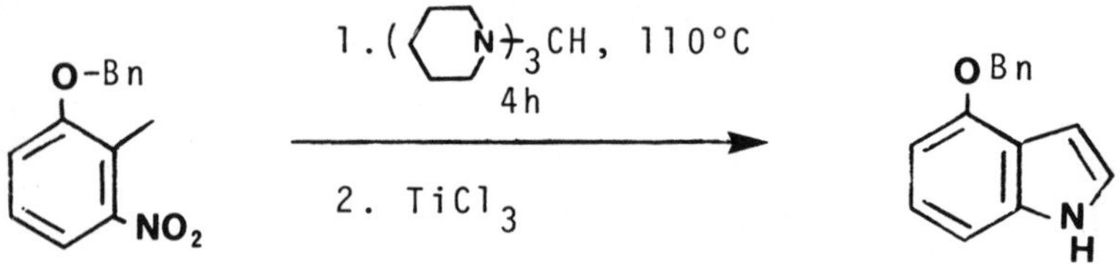
O-Bn
NO_2
1. (piperidino)$_3$CH, 110°C
4h
2. $TiCl_3$
OBn
N
H

62%

Lloyd, D.H.; Nichols, D.E.* **Tetrahedron Lett**, (1983), **24**, 4561

1. LDA, THF, -78°C
   $\underline{n}C_3H_7Br$, 3h

2. $NaBH_4$, EtOH,
   80°C, 35h

3. 70% H_2SO_4, 18h

94%

(98% ee)

Guerrier, L.; Royer, J.; Grierson, D.S.; Husson, H.-P.[*]
J Am Chem Soc, (1983), **105**, 7754

1. $SnCl_2$, 2HOH, EtOH
   70°C, 30 min

2. 5% aq. $NaHCO_3$

96%

Bellamy, F.D.[*]; Ou, K.   **Tetrahedron Lett**, (1984), **25**, 839

$$RN_3 \longrightarrow RNH_2$$

R =   1. PPh_3, THF, 12h ⟶   95%
       2. HOH, 12h

Vaultier, M.; Knouzi, N.; Carrie, R.
                    Tetrahedron Lett, (1983), **24**, 763

R = $\underline{n}C_6H_{13}$ 4 HCO_2NH_4, Pd/C, MeOH   93%

Gartiser, T.; Selve, C.[*]; Delpeuch, J.-J.
                    Tetrahedron Lett, (1983), **24**, 1609

R = Bn   NaTeH, EtOH/Et_2O, RT   81%

Suzuki, H.[*]; Takaoka, K.   **Chem Lett**, (1984), 1733

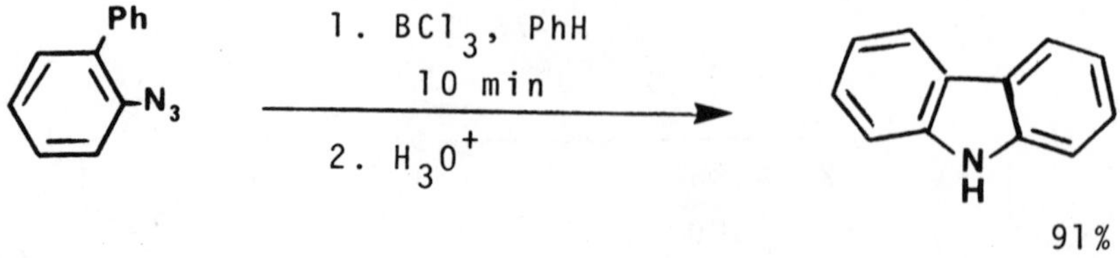

1. BCl_3, PhH
   10 min

2. H_3O^+

91%

Zanirato, P.[*]   **JCS Chem Comm**, (1983), 1065

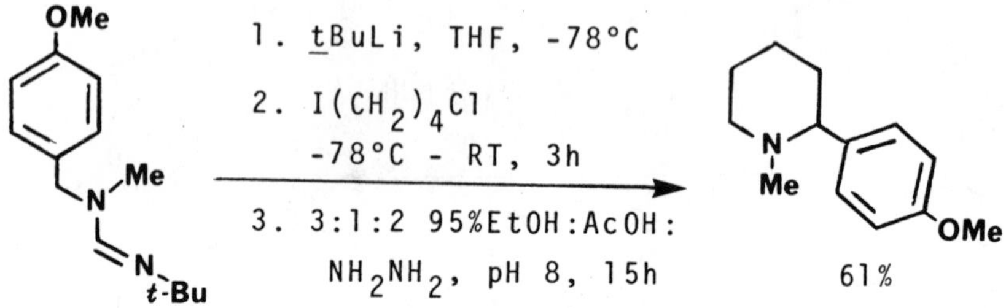

1. tBuLi, THF, -78°C

2. $I(CH_2)_4Cl$
   -78°C - RT, 3h

3. 3:1:2 95%EtOH:AcOH:
   NH_2NH_2, pH 8, 15h

61%

Meyers, A.I.[*]; Marra, J.M.   **Tetrahedron Lett**, (1985), **26**, 5863

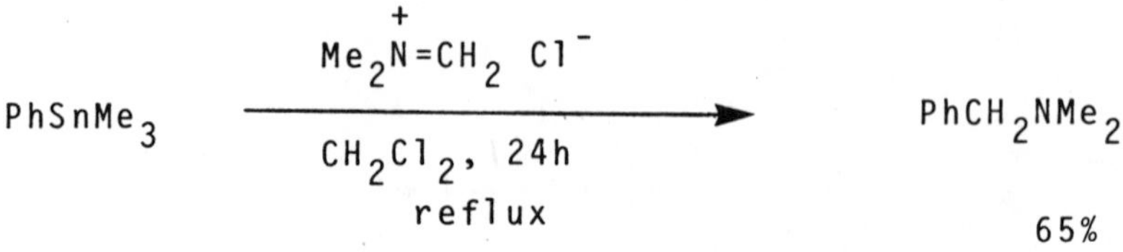

$PhSnMe_3$

$Me_2\overset{+}{N}=CH_2 \; Cl^-$
⟶
CH_2Cl_2, 24h
reflux

$PhCH_2NMe_2$

65%

Cooper, M.S.; Heaney, H.[*]   **Tetrahedron Lett**, (1986), **27**, 5011

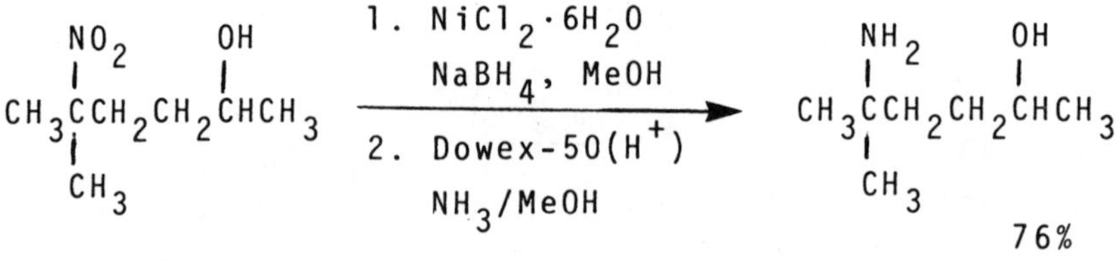

$CH_3CCH_2CH_2CHCH_3$ with NO_2, CH_3, OH substituents

1. $NiCl_2 \cdot 6H_2O$
   $NaBH_4$, MeOH

2. Dowex-50(H^+)
   NH_3/MeOH

$CH_3CCH_2CH_2CHCH_3$ with NH_2, CH_3, OH substituents

76%

Osby, J.O.; Ganem, B.[*]   **Tetrahedron Lett**, (1985), **26**, 6413

$$RNO_2 \longrightarrow RNH_2$$

R = [structure: trimethylphenyl group]
$$\xrightarrow{Na_2Te/aq.\ NaOH/dioxane/50°C}$$ 95%

Suzuki, H.[*]; Manabe, H.; Inouye, M.   **Chem Lett**, (1985), 1671

R = Ph $$\xrightarrow[2.\ TiCl_4/CH_2Cl_2 \quad 3.\ HOH]{1.\ iBu_2Te}$$ 74%

Suzuki, H.[*]; Hanazaki, Y.   **Chem Lett**, (1986), 549

R = [structure: phenyl]-OH $$\xrightarrow[2.\ aq.\ HCl \quad 3.\ Na_2CO_3]{1.\ \textcircled{P}\text{-}NMe_3Bn,\ HFe(CO)_4^-}$$ 80%

Boldrini, G.P.; Cainelli, G.; Umani-Ronchi, A.
                    J Organomet Chem, (1983), **243**, 195

R = [structure: phenyl with HO] $$\xrightarrow[2.\ TiCl_4,\ tBuOH,\ 0°C]{1.\ [HgCl_2/THF/Mg,\ RT]}$$ 94%

George, J.; Chandrasekaran, S.[*]   **Syn Commun**, (1983), **13**, 495

R = $-CH_2CH_2CHMe_2$ $$\xrightarrow[MeOH,\ RT,\ 40\ min]{HCO_2NH_4,\ 10\%Pd/C}$$ 82%

Ram, S.; Ehrenkaufer, R.E.[*]   **Tetrahedron Lett**, (1984), **25**, 3415

R = Ph $$\xrightarrow[EtOH/HOH,\ 80°C]{[PhTeTePh/NaBH_4],\ PhH}$$ 72%

Ohira, N.; Aso, Y.; Otsubo, T.; Ogura, F.[*]
                    Chem Lett, (1984), 853

R = [structure: phenyl]-OMe $$\xrightarrow[NEt_3,\ EtOH,\ 125°C,\ 5h]{RuCl_2(PPh_3)_3,\ HCOOH}$$ 97%

Watanabe, Y.[*]; Ohta, T.; Tsuji, Y.; Hiyoshi, T.; Tsuji, Y.
                    Bull Chem Soc Jpn, (1984), **57**, 2440

R = [structure: phenyl]-NO₂ $$\xrightarrow[\substack{NEt_3,\ aq.\ dioxane,\ CO \\ RT}]{PtCl_2(PPh_3)_2,\ SnCl_4}$$ quant.

Watanabe, Y.[*]; Tsuji, Y.; Ohsumi, T.; Takeuchi, R.
                    Bull Chem Soc Jpn, (1984), **57**, 2867

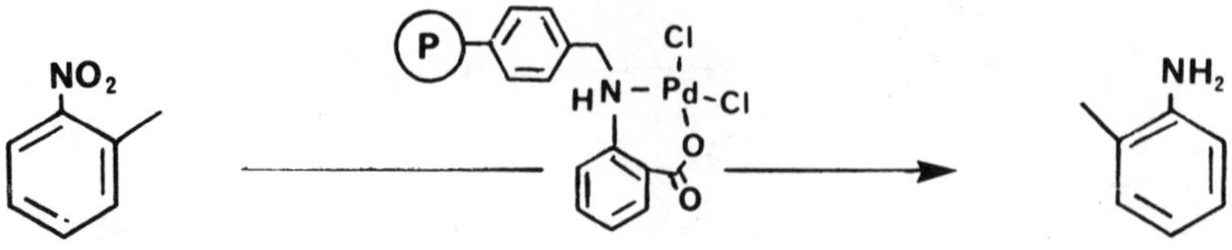

600 psi H_2, EtOAc, 100°C

$\boxed{P}$ = polystyrene

Baralt, E.; Holy, N.* **J Org Chem**, (1984), **49**, 2626

80%

SECTION 105A: Protection of Amines

Related Methods: Amides from Amines (Section 82)
Amines from Amides (Section 96)

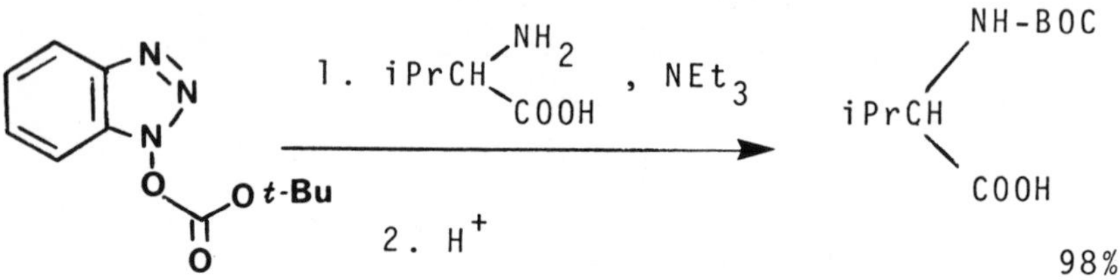

Kim, S.*; Chang, H. **JCS Chem Comm**, (1983), 1357

98%

$PhCH_2NH_2$ → $PhCH_2NHBOC$

BOC-OCHCCl$_3$, THF

20°C, satd. K_2CO_3

91%

Barcelo, G.; Senet, J.-P.; Sennyey, G.*
J Org Chem, (1985), **50**, 3951

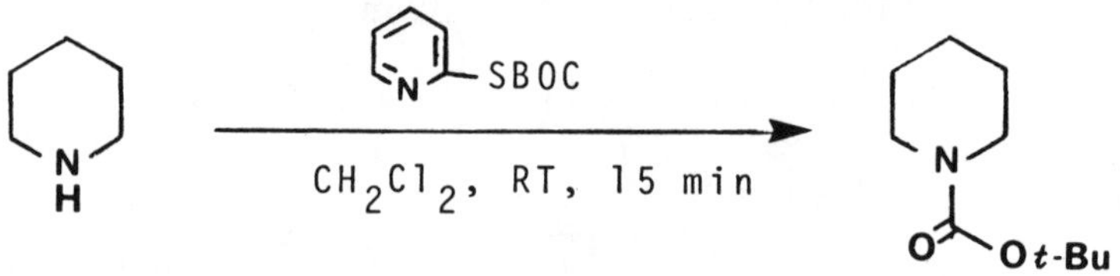

CH_2Cl_2, RT, 15 min

95%

Kim, S.*; Lee, J.I.; Yi, K.Y.
Bull Chem Soc Jpn, (1985), **58**, 3570

Me   Bn
   N
   (cyclohexane ring)

$$\text{ClCOEt, PhH}$$
reflux

Me   CO$_2$Et
   N
   (cyclohexane ring)   71%

Kapnang, H.; Charles, G.*   **Tetrahedron Lett**, (1983), **24**, 3233

NH$_2$
(cyclohexane ring)

$$\text{Ph}_3\text{CSCl, aq. THF, Na}_2\text{CO}_3$$
96%

NHSCPh$_3$
(cyclohexane ring)

1. TMS-I, CH$_2$Cl$_2$, -78°C

2. aq. EtOH
                          92%

TRS group
(trityl sulfenyl)

Branchaud, B.P.*   **J Org Chem**, (1983), **48**, 3538

NH$_2$
(meta-methyl benzene ring)

$$\text{Me}_2\text{Si} \overset{CH_2CH_2}{\underset{NMe_2 \quad Me_2N}{\diagup \diagdown}} \text{SiMe}_2, \text{ZnI}_2$$
140°C
                          92%

(bicyclic Si-N-Si ring on meta-methyl benzene)

2 MeOH, cat. pTsOH

quant.

Guggenheim, T.L.*   **Tetrahedron Lett**, (1984), **25**, 1253

PhCH$_2$NH$_2$ $\xrightarrow[\underset{O}{\text{ClCOCH=CH}_2}]{\text{(TMS)}_2\text{NAc}}$ PhCH$_2$NHCOCH=CH$_2$
                                                          97%

Raucher, S.*; Jones, D.S.   **Syn Commun**, (1985), **15**, 1025

$$PhCH_2CH_2NHAc \xrightarrow{\begin{array}{l}\text{1. } BOC_2O, \text{ DMAP, MeCN} \\ \text{2. } Et_2NCH_2CH_2NH_2 \\ \hline \text{MeOH, 3d} \\ \text{3. } NH_2NH_2/MeOH, \text{ 1h}\end{array}} PhCH_2CH_2NHBOC$$

94%

Grehn, L.; Gunnarsson, K.; Ragnarsson, U.
JCS Chem Comm, (1985), 1317

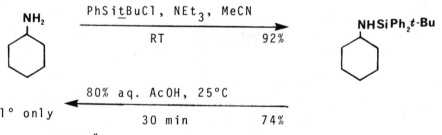

Kinoshita, H.*; Inomata, K.; Kameda, T.; Kotake, H.
Chem Lett, (1985), 515

$$PhCH_2CH_2NHCBZ \xrightarrow[CH_2Cl_2, \text{ RT, 2h}]{} PhCH_2CH_2NH_2$$

83%

King, P.F.*; Stroud, S.G.   **Tetrahedron Lett**, (1985), **26**, 1415

$$\xrightarrow[RT]{PhSi\underline{t}BuCl, NEt_3, MeCN} \quad 92\%$$

NH₂ ... **NHSiPh₂t-Bu**

1° only $\xleftarrow[\text{30 min} \quad 74\%]{80\% \text{ aq. AcOH, } 25°C}$

Overman, L.E.*; Okazaki, M.E.; Mishra, P.
Tetrahedron Lett, (1986), **27**, 4391

CHAPTER 8
PREPARATION OF ESTERS

SECTION 106: Esters from Acetylenes

$$\underline{n}BuC{\equiv}CCH_3 \quad \xrightarrow[\substack{O_2,\ 1\ atm\ CO,\ MeOH \\ RT,\ 2h}]{PdCl_2,\ CuCl_2,\ HCl} \quad \underset{H}{\overset{\underline{n}Bu}{\diagdown}}C=C\underset{CO_2Me}{\overset{CH_3}{\diagup}}$$

72%

Alper, H.[*]; Despeyroux, B.; Woell, J.B.
Tetrahedron Lett, (1983), **24**, 5691

$$\underline{n}BuC{\equiv}C-I \quad \xrightarrow[\substack{HMPT \\ 3.\ H_2O_2,\ NaOH}]{\substack{1.\ \quad\text{⊢⊣}_2 BH \\ 2.\ \underline{n}BuSMgBr,\ -50°C}} \quad \underset{}{\overset{O}{\underset{\parallel}{}}} \underline{n}BuCH_2CS\underline{n}Bu$$

83%

Hoshi, M.; Masuda, Y.; Arase, A.[*] **JCS Chem Comm**, (1985), 714

SECTION 107: Esters from Acid Derivatives

The following types of reactions are found in this section:

1. Esters from the reaction of Alcohols with Carboxylic acids, Acid Halides, and Anhydrides.

2. Lactones from Hydroxy Acids.

3. Esters from Carboxylic Acids and Halides, Sulfoxides, and Miscellaneous compounds.

1. 2 PhP(TMS)$_2$
   CH$_3$CN, 4h, -15°C

2. 4 MeOH, 24h
   reflux

62%

Appel, R.[*]; Hünerbein, J.; Knoch, F.
Angew Chem Int Ed Engl, (1983), **22**, 61

$CH_3(CH_2)_6COOH$ $\xrightarrow[\substack{\text{cat. DMAP, } CH_2Cl_2 \\ 30 \text{ min}}]{\substack{ClCOBn, NEt_3, 0°C}}$ $CH_3(CH_2)_6COBn$

97%

Kim, S.[*]; Lee, J.I.; Kim, Y.C.
J Org Chem, (1985), **50**, 560
Tetrahedron Lett, (1983), **24**, 3365

$PhCH_2SCl$ $\xrightarrow[\substack{2. \text{ aq. KF, } Et_2O}]{\substack{1. \underline{n}Bu_3SnS\underline{t}Bu, CHCl_3 \\ RT, 3 \text{ min}}}$ $PhCH_2SS\underline{t}Bu$

90%

Harpp, D.N.[*]; Aida, T.; Chan, T.H.
Tetrahedron Lett, (1983), **24**, 5173

$PhCH=CHCOOH$ $\xrightarrow[\substack{THF, DEAD, 25°C}]{\substack{(P) PPh_2OCH_2OH}}$ $PhCH=CHCO_2Bn$

(E)   (E)

86%

(P) = polystyryl

Amos, R.A.[*]; Emblidge, R.W.; Havens, N.
J Org Chem, (1983), **48**, 3598

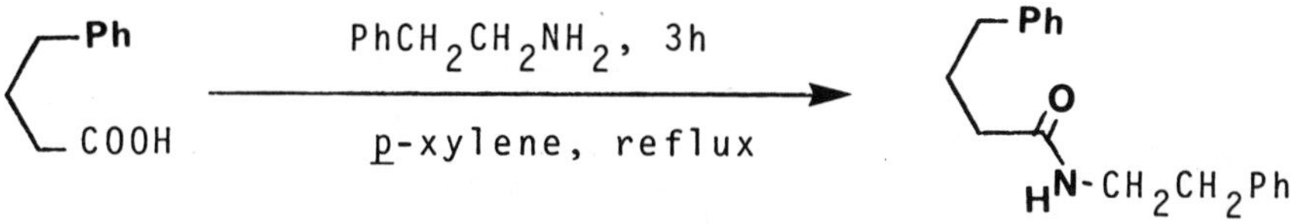
Ph
COOH
PhCH2CH2NH2, 3h
p-xylene, reflux
Ph
O
HN-CH2CH2Ph

86%

Mukaiyama, T.*; Ichikawa, J.; Asami, M. **Chem Lett**, (1983), 683

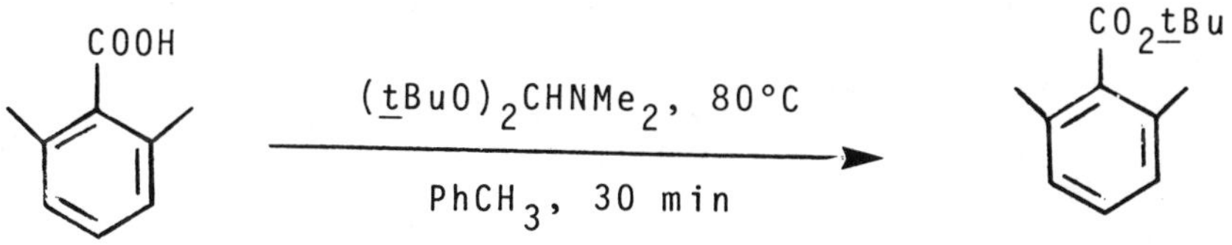
COOH
(tBuO)2CHNMe2, 80°C
PhCH3, 30 min
CO2tBu

82%

Widmer, U.* **Synthesis**, (1983), 135

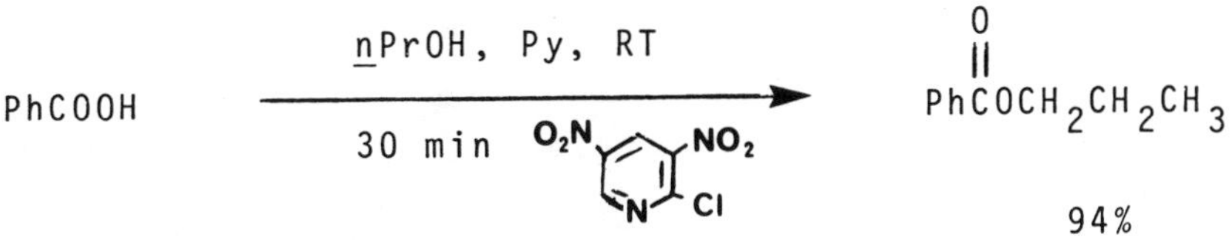
PhCOOH
nPrOH, Py, RT
30 min
O2N
NO2
N
Cl
O
PhCOCH2CH2CH3

94%

Takimoto, S.*; Abe, N.; Kodera, Y.; Ohta, H.
Bull Chem Soc Jpn, (1983), **56**, 639

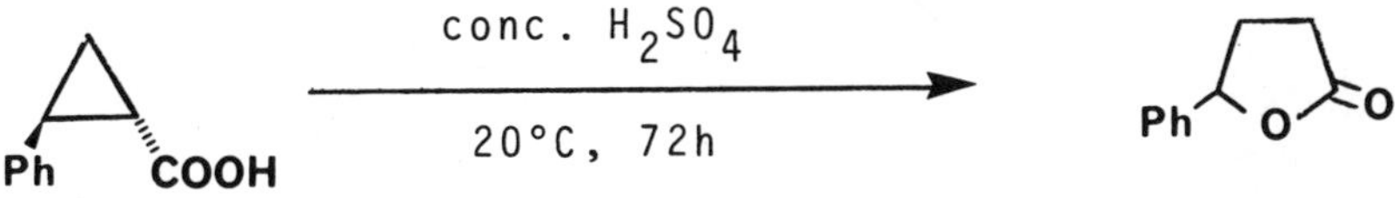
Ph
COOH
conc. H2SO4
20°C, 72h
Ph
O
O

84%

Sychkova, L.D.; Kharitonova, O.V.; Shabarov, Yu.S.
J Org Chem USSR, (1983), **19**, 1298

$$\underline{n}C_5H_{11}COOH \quad \xrightarrow[\substack{MeOH, \ reflux \\ 1h}]{Me_3SiCl, \ Et_2O} \quad \underline{n}C_5H_{11}CO_2Me$$

88%

Mandal, A.K.[*]   **Ind J Chem B**, (1983), **22**, 505

$$
\begin{array}{c}
HO \\
| \\
CHCH_2CH=CH(CH_2)_7COOH \\
| \\
\underline{n}C_6H_{13}
\end{array}
\xrightarrow[\substack{cat. \ camphor \\ sulfonic \ acid}]{\substack{1. \ Me_2NC{\equiv}CCO_2Me \\ 2. \ CH_2Cl_2, \ 40°C}}
$$

(E)

82%

Gais, H.-J.[*]   **Tetrahedron Lett**, (1984), **25**, 273

$$TMSO(CH_2)_{14}CO_2TMS \quad \xrightarrow[\substack{PhCH_3, \ reflux \\ 8h}]{\underline{n}Pr_2BOTf}$$

94%

Taniguchi, N.; Kinoshita, H.; Inomata, K.[*]; Kotake, H.
                Chem Lett, (1984), 1347

$$CH_3(CH_2)_6COOH \quad \xrightarrow[\substack{BnOH, \ CH_2Cl_2 \\ RT, \ 2h}]{\left(\substack{N} O\right)_2C=O, \ DMAP} \quad CH_3(CH_2)_6CO_2Bn$$

94%

Kim, S.[*]; Lee, J.I.; Ko, Y.K.
                Tetrahedron Lett, (1984), **25**, 4943

$$CH_3CH_2CH_2COOH \xrightarrow[\substack{EtI,\ Et_4NOTs,\ DMF \\ RT}]{e^-,\ Pt\ electrodes} CH_3CH_2CH_2CO_2Et$$

96%

Awata, T.; Baizer, M.M.; Nonaka, T.[*]; Fuchigami, T.
Chem Lett, (1985), 371

Further examples of the reaction RCOOH + ROH → RCOOR are
included in Section 108 (Esters from Alcohols and Phenols) and
Section 10A (Protection of Carboxylic Acids).

SECTION 108: Esters from Alcohols and Thiols

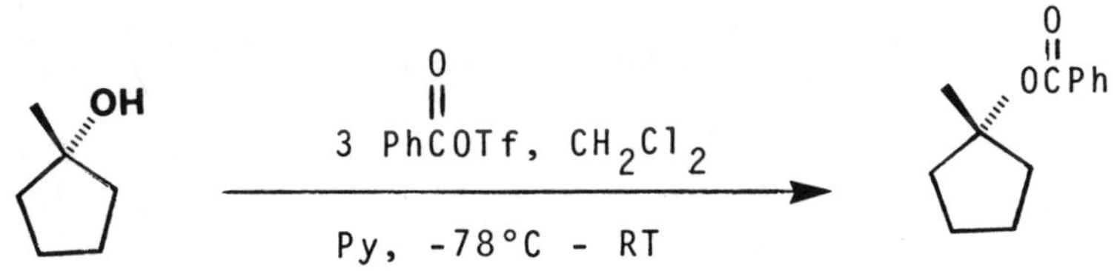

81%

Brown, L.; Koreeda, M.[*]   **J Org Chem**, (1984), **49**, 3875
                    JCS Chem Comm, (1983), 1113

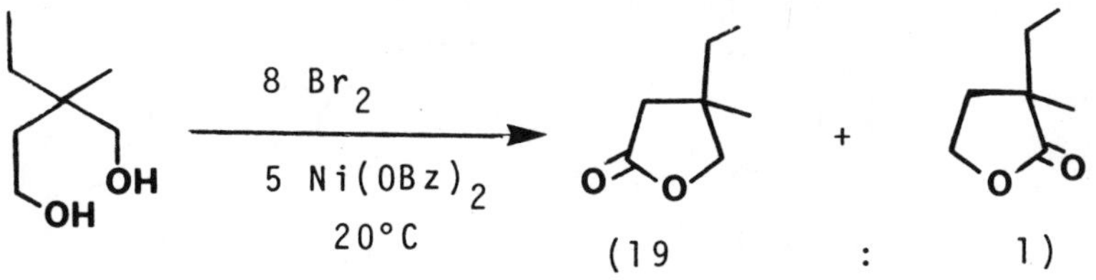

(19   :   1)

99%

Doyle, M.P.[*]; Dow, R.L.; Bagheri, V.; Patrie, W.J.
          J Org Chem, (1983), **48**, 476

$HO(CH_2)_6OH$ →[NaBrO$_2$, gl. AcOH / RT, 10h]

84%

Kageyama, T.[*]; Kawahara, S.; Kitamura, K.; Ueno, Y.; Okawara, M.
Chem Lett, (1983), 1097

$HO(CH_2)_4OH$ →[NaBrO$_3$-HBr / CCl$_4$, 5h, 37°C]

78%

Kajigaeshi, S.[*]; Nakagawa, T.; Nazasaki, N.; Yamasaki, H.;
Fujisaki, S.
Bull Chem Soc Jpn, (1986), **59**, 747

→[RhH(PPh$_3$)$_4$ / PhH, 50°C, 10h]

82%   (86   :   14)

Ishii, Y.; Suzuki, K.; Ikariya, T.; Saburi, M.[*]; Yoshikawa, S.[*]
J Org Chem, (1986), **51**, 2822
Ishii, Y.; Osakada, K.; Ikariya, T.; Saburi, M.[*]; Yoshikawa, S.[*]
J Org Chem, (1986), **51**, 2034

$(CH_2)_8$ with CH$_2$OH →[(bipy)H$_2$CrOCl$_5$ / CH$_2$Cl$_2$, RT]

90%

Chakraborty, T.K.; Bhushan, V.; Chandrasekaran, S.[*]
Ind J Chem B, (1983), **22**, 9

$$CH_3CH_2\overset{\overset{\displaystyle CH_3}{|}}{C}HCH_2OH \quad \xrightarrow[\text{2. dil. aq. } H_2SO_4]{\begin{array}{c}\text{1. } Me_2\overset{+}{N}=CHO\overset{\overset{\displaystyle O}{||}}{C}Ph, \ Et_2O \\ 20°C, \ \text{overnight}\end{array}} \quad CH_3CH_2\overset{\overset{\displaystyle CH_3}{|}}{C}HCH_2O\overset{\overset{\displaystyle O}{||}}{C}H$$

65%

Barluenga, J.[*]; Campos, P.J.; Gonzalez-Nuñez, C.; Asensio, G.
Synthesis, (1985), 426

$$CH_2=CHCH_2CH_2OH \quad \xrightarrow[\begin{array}{c}CuCl_2, \ HCl, \ RT \\ \text{overnight}\end{array}]{CO, \ THF, \ PdCl_2}$$

60%

Alper, H.[*]; Leonard, D.   **JCS Chem Comm**, (1985), 511

$$\xrightarrow[\begin{array}{c}900 \text{ psi}, \ 190°C \\ HOH, \ 24h\end{array}]{CO, \ Co_2(CO)_8, \ EtOH}$$

68%

Shim, S.C.; Antebi, S.; Alper, H.[*]   **J Org Chem**, (1985), **50**, 147

$$\xrightarrow[CH_2Cl_2, \ 30 \text{ min}]{CH_3\overset{\overset{\displaystyle O}{||}}{C}SH, \ ZnI_2}$$

85%

Gauthier, J.Y.[*]; Bourdon, F.; Young, R.N.
Tetrahedron Lett, (1986), **27**, 15

HLADH, pH 9, 20°C

NAD, FMN, 6d

86%

HLADH = Horse Liver Alcohol Dehydrogenase   (97% ee)

Lok, K.P.; Jakovac, I.J.; Jones, J.B.[*]

$\underline{J}$ $\underline{Am}$ $\underline{Chem}$ $\underline{Soc}$, (1985), **107**, 2521

Further examples of the reaction ROH → R'COOR are included in Section 107 (Esters from Carboxylic Acids and Acid Halides) and Section 45A (Protection of Alcohols and Phenols).

SECTION 109: Esters from Aldehydes

20 PhCHO

$\begin{array}{c} LiWO_2, THF \\ \hline 75°C, 72h \end{array}$

$$\overset{O}{\overset{\|}{PhCOBn}}$$

89%

Villacorta, G.M.; Filippo Jr., J.S.[*]

$\underline{J}$ $\underline{Org}$ $\underline{Chem}$, (1983), **48**, 1151

1. $\underline{n}$BuLi, THF, -78°C
2. $(NEt_2)_3 TiCl$, THF, -20°C

3. $\underline{n}C_8H_{17}CHO$, THF, reflux
4. MeOH, H[+]
5. $BF_3 \cdot OEt_2$, mcpba

$\underline{n}C_8H_{17}$

96%

Roder, H.; Helmchen, G.[*]; Peters, E.-M.; Peters, K.; van Schnering, H.-G.

Angew **Chem** **Int** **Ed** **Engl**, (1984), **23**, 898 iPrCHO

1. PhS-C-OMe (with Li above and SiMe$_3$ below), THF, -78-25°C

2. Me$_3$SiCl, NaI, MeCN

3. Al$_2$O$_3$

$$iPrCH_2\overset{O}{\overset{\|}{C}}SPh$$

86%

Hackett, S.; Livinghouse, T.*
                    Tetrahedron Lett, (1984), **25**, 3539

Ph—CHO

[cyclopropane with OTMS and OMe]

TiCl$_4$, CH$_2$Cl$_2$
0°C, 1h

[lactone products]   (61   :   39)

95%

Nakamura, E.*; Oshino, H.; Kuwajima, I.*
                    J Am Chem Soc, (1986), **108**, 3745

CHO (on cyclohexane)

CH$_3$CH=CHCO$_2$Me, SmI$_2$

HMPA, 0°C. 1 min

[lactone product]   96%

(E:Z = 1:1)

Otsubo, K.; Inanaga, J.*; Yamaguchi, M.
                    Tetrahedron Lett, (1986), **27**, 5763

Related Methods: Esters from Ketones (Section 117)

SECTION 110: Esters from Alkyl, Methylenes, and Aryls

No examples of the reaction RR → RCOOR' or R'COOR (R,R' = alkyl
aryl, etc.) occur in the literature. For the reaction RH →
RCOOR' or R'COOR see Section 116 (Esters from Hydrides).

$$CH_3(CH_2)_4CH_3 \xrightarrow[\substack{sunlight \\ 6d}]{Pb(OAc)_4, \ AcOH}$$

$$\overset{OAc}{\underset{|}{CH_3CHCH_2CH_2CH_2CH_3}}$$

+   50%

$$CH_3CH_2\underset{\underset{OAc}{|}}{CH}CH_2CH_2CH_3 \qquad 45\%$$

Bestre, R.D.; Cole, E.R.; Crank, G.[*]

Tetrahedron Lett, (1983), **24**, 3891

$$\xrightarrow[\substack{NaIO_4, \ HOH, \ CCl_4 \\ RT, \ 16h}]{RuCl_2 \cdot 3H_2O, \ DMF}$$

88%

Chakraborti, A.K.; Ghatak, U.R.   **Synthesis**, (1983), 746

SECTION 111: Esters from Amides

$$\overset{O}{\underset{\|}{CH_2}}=CHCNEt_2$$

1. $PhSO_2Na$, AcOH/HOH
   reflux
2. P_4S_{10}, $NaHCO_3$, DME
3. Me_2SO_4, MeCN, 12h
4. KCN, 15h
5. $\underline{n}C_3H_7I$, KOH, aq. THF
   tricaprylNH[+]Cl[-]
6. 3N HCl, MeOH, 15h

$$\overset{\underline{n}C_3H_7}{\underset{|}{PhSO_2CHCH_2CO_2Me}}$$

44%

De Lombaert, S.; Ghosez, L.[*]

Tetrahedron Lett, (1984), **25**, 3475

SECTION 112: Esters from Amines

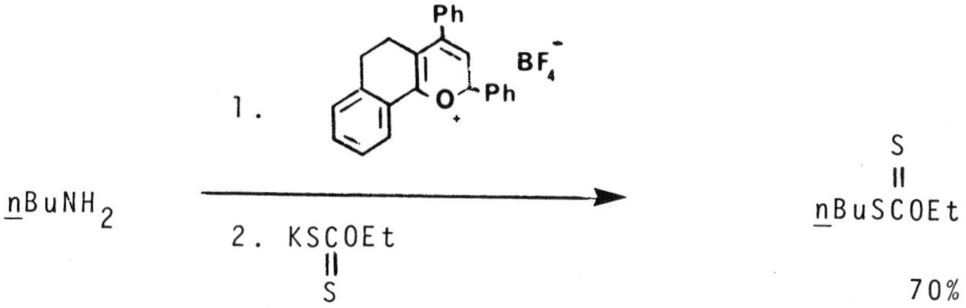

Ph
BF4
1.
O
Ph
nBuNH2
2. KSCOEt
S
S
nBuSCOEt
70%

Elshafie, S.M.M.* Org Prep Proc Int, (1983), 15, 225

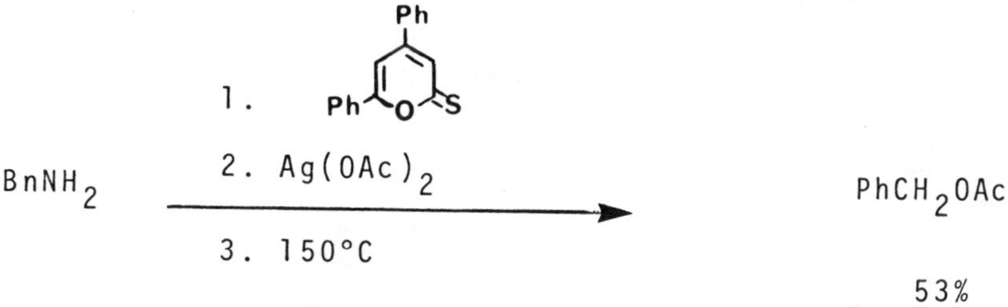

Ph
1.
Ph
O
S
2. Ag(OAc)2
BnNH2
PhCH2OAc
3. 150°C
53%

Molina, P.*; Alajarin, M.; De Vega, M.J.P.
Syn Commun, (1983), 13, 501

SECTION 113: Esters from Esters

Conjugate reductions and conjugate alkylations of unsaturated esters are found in Section 74 (Alkyls from Olefins).

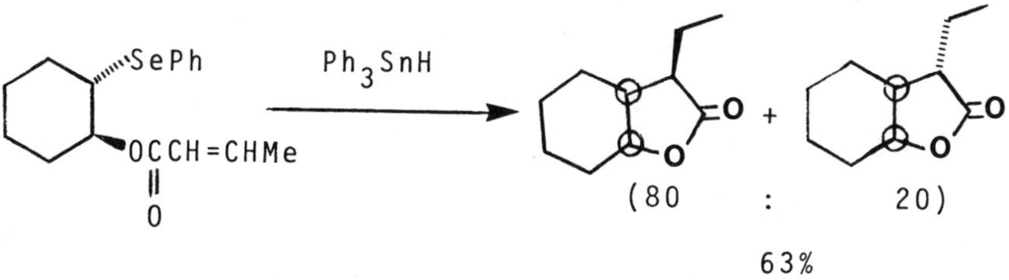

SePh
Ph3SnH
OCCH=CHMe
O
O
O
+
O
O
(80 : 20)
63%

Clive, D.L.J.*; Beaulieu, P.L. JCS Chem Comm, (1983), 307

$$\underset{\substack{R^*OCCH_2CH_3 \\ \parallel \\ O}}{} \xrightarrow[\text{2. BnBr, HMPT, -63°C}]{\text{1. } \underline{X}, \text{ -80°C}} \underset{\substack{R^*OC \\ \parallel \\ O}}{}\text{—Ph}$$

$\underline{X}$ = LICA/THF/2HMPT   89% (2$\underline{S}$:2$\underline{R}$ = 3:97)

$\underline{X}$ = LICA/THF/HMPT   94% (2$\underline{S}$:2$\underline{R}$ = 95:5)

R^* =

Helmchen, G.*; Selim, A.; Dorsch, D.; Taufer, I.
Tetrahedron Lett, (1983), **24**, 3213

1. LICA, THF, 65°C
   $CF_3SO_2OSiEt_3$
   THF

2. H_2, Pt, EtOH

3. CH_2N_2, Et_2O

84% CO_2Me

Brunner, R.K.; Borschberg, H.J.
Helv Chim Acta, (1983), **66**, 2608

$$\underset{\substack{\parallel \\ O}}{CS\underline{t}Bu} \xrightarrow[\text{CH}_3\text{CN, 25°C, 45 min}]{\underline{t}BuOH, \; Hg(O\overset{O}{\overset{\parallel}{C}}CF_3)_2} \underset{\substack{\parallel \\ O}}{CO\underline{t}Bu}$$

90%

Chan, W.K.; Masamune, S.*; Spessard, G.O.
Org Syn, (1983), **61**, 48

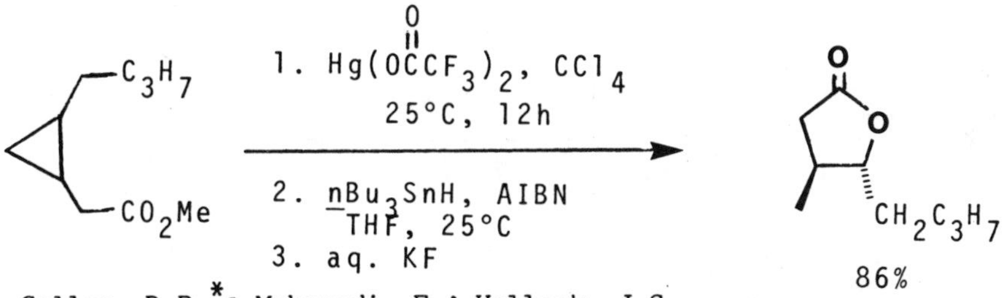

1. $Hg(OCCF_3)_2$, CCl_4
   25°C, 12h

2. nBu_3SnH, AIBN
   THF, 25°C
3. aq. KF

86%

Collum, D.B.[*]; Mohamadi, F.; Hallock, J.S.
J Am Chem Soc, (1983), 105, 6882

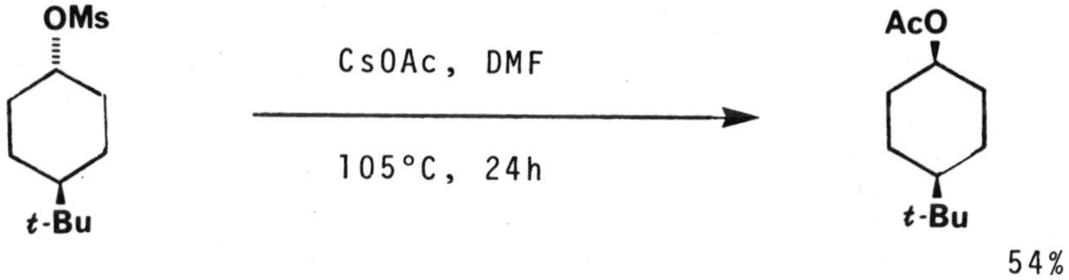

CsOAc, DMF

105°C, 24h

54%

Huffman, J.W.[*]; Desai, R.C.   Syn Commun, (1983), 13, 553

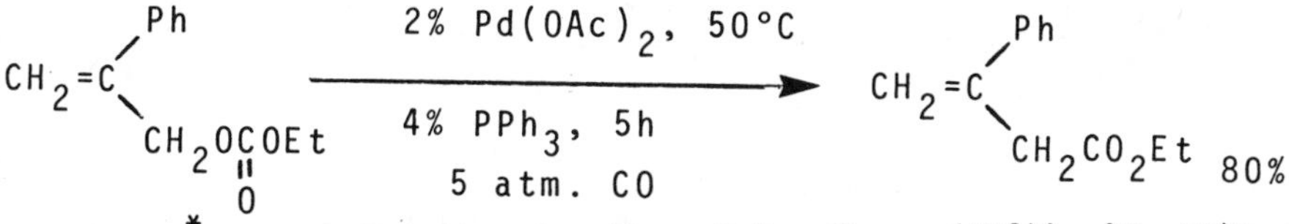

$CH_2=C$
    Ph
    CH_2OCOEt
        O

2% $Pd(OAc)_2$, 50°C

4% PPh_3, 5h

5 atm. CO

$CH_2=C$
    Ph
    CH_2CO_2Et 80%

Tsuji, J.[*]; Sato, K.; Okumoto, H.   J Org Chem, (1984), 49, 1341

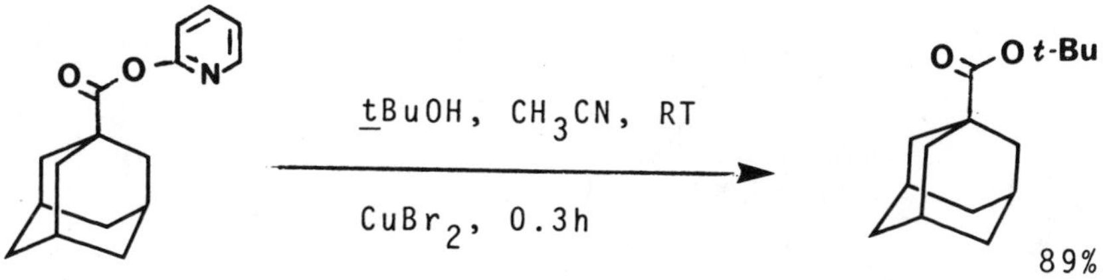

tBuOH, CH_3CN, RT

$CuBr_2$, 0.3h

89%

for the preparation of hindered esters

Kim, S.[*]; Lee, J.I.   J Org Chem, (1984), 49, 1712

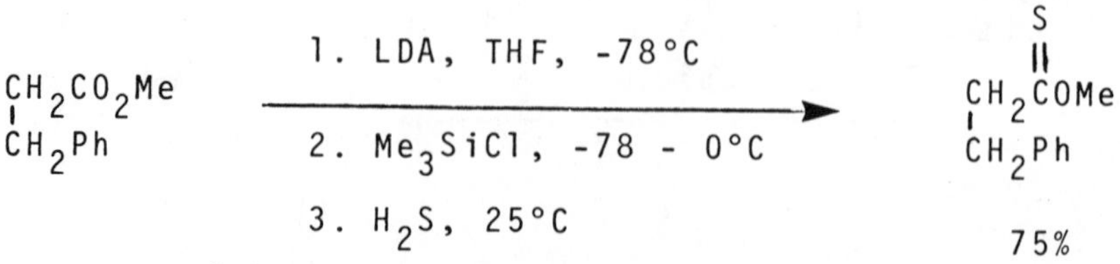

CH2CO2Me
CH2Ph
1. LDA, THF, -78°C
2. Me3SiCl, -78 - 0°C
3. H2S, 25°C
S
CH2COMe
CH2Ph
75%

Corey, E.J.*; Wright, S.W. **Tetrahedron Lett**, (1984), **25**, 2639

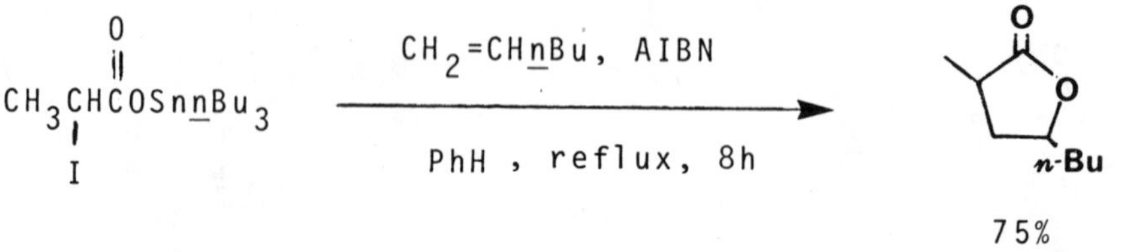

O
CH3CHCOSnnBu3
I
CH2=CHnBu, AIBN
PhH , reflux, 8h
O
O
n-Bu
75%

Kraus, G.A.*; Landgrebe, K. **Tetrahedron Lett**, (1984), **25**, 3939

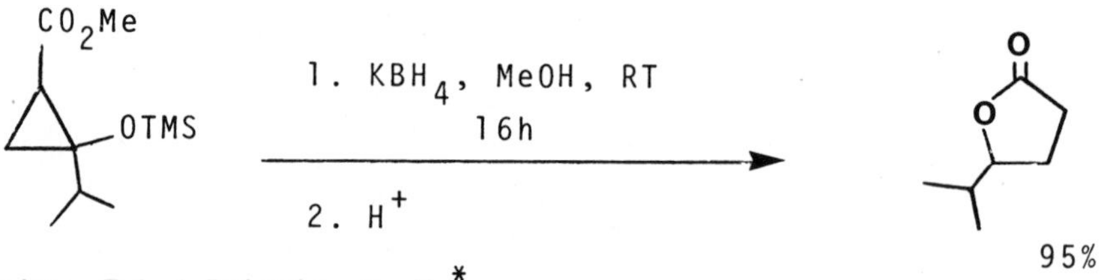

CO2Me
OTMS
1. KBH4, MeOH, RT
16h
2. H+
O
O
95%

Grimm, E.L.; Reissig, H.-U.*
J Org Chem, (1985), **50**, 242

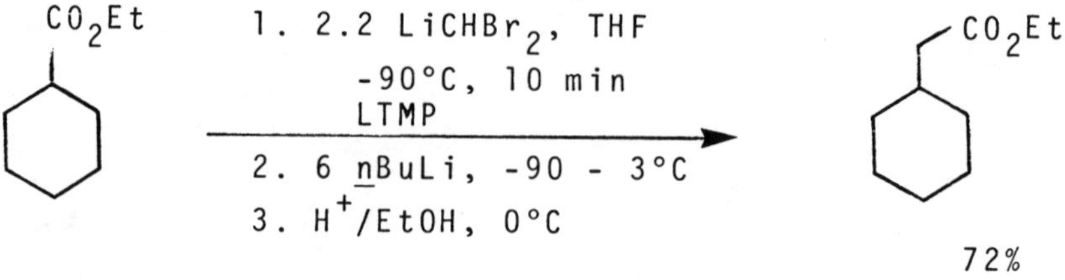

CO2Et
1. 2.2 LiCHBr2, THF
-90°C, 10 min
LTMP
2. 6 nBuLi, -90 - 3°C
3. H+/EtOH, 0°C
CO2Et
72%

Kowalski, C.J.*; Haque, M.S.; Fields, K.W.
J Am Chem Soc, (1985), **107**, 1429

$$PhCH_2Br \xrightarrow[\substack{HCO_2Me,\ KI \\ 100\ psi\ CO}]{[1,5-HDRhCl]_2,\ RT} PhCH_2CO_2Me$$

quant.

Buchan, C.; Hamel, N.; Woell, J.B.; Alper, H.[*]
JCS **Chem Comm**, (1986), 167

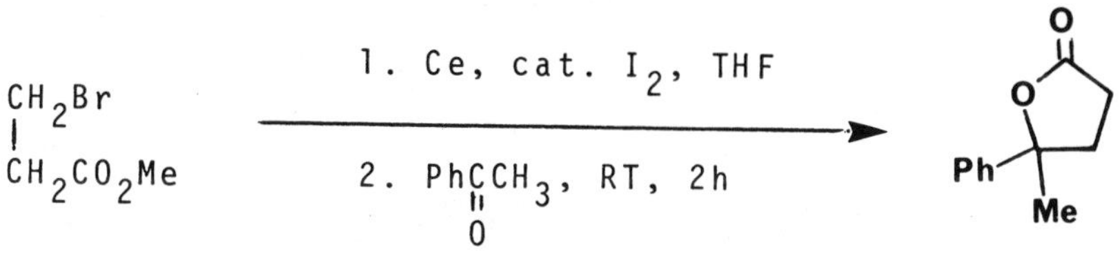

$$\substack{CH_2Br \\ | \\ CH_2CO_2Me} \xrightarrow[\substack{2.\ PhCCH_3,\ RT,\ 2h \\ \quad \| \\ \quad O}]{1.\ Ce,\ cat.\ I_2,\ THF}$$

70%

Fukuzawa, S.[*]; Fujinami, T.; Sakai, S.
JCS **Chem Comm**, (1986), 475

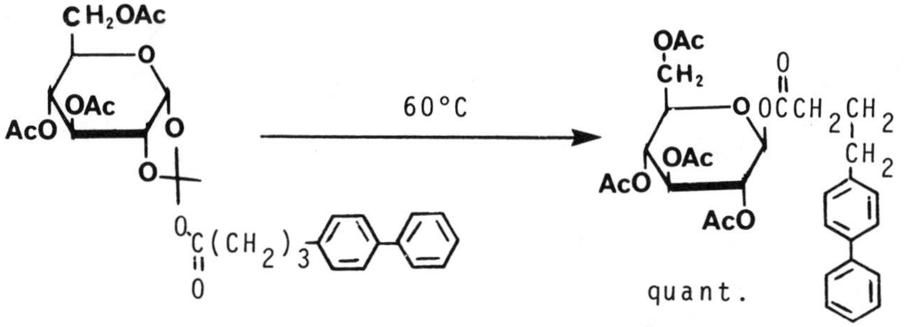

60°C

quant.

Wulff, G.; Patt, H.; Wichelhaus, J.
Nouv J Chem, (1986), **10**, 143

Review: "Sulfur Mediated Ring Expansions in Total
     Synthesis"

Vedejs, E.[*] **Accts Chem Res**, (1984), **17**, 358

SECTION 114: Esters from Ethers, Epoxides, and Thioethers

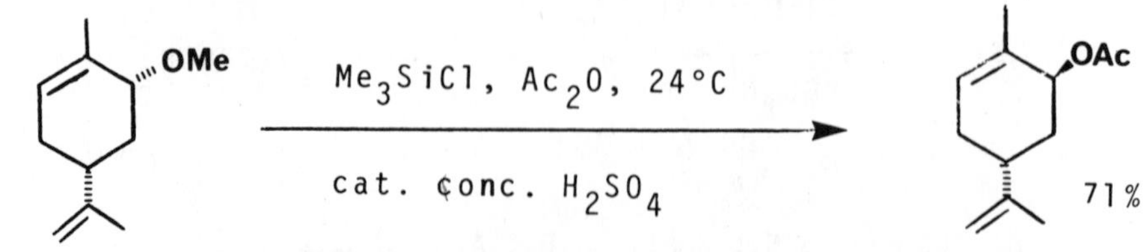
OMe
Me3SiCl, Ac2O, 24°C
cat. conc. H2SO4
OAc
71%

Sarma, J.C.; Barbaruah, M.; Sarma, D.N.; Barua, N.C.; Sharma, R.P.*
Tetrahedron, (1986), **42**, 3999

Barua, N.C.; Sharma, R.P.*; Baruah, J.N.
Tetrahedron Lett, (1983), **24**, 1189

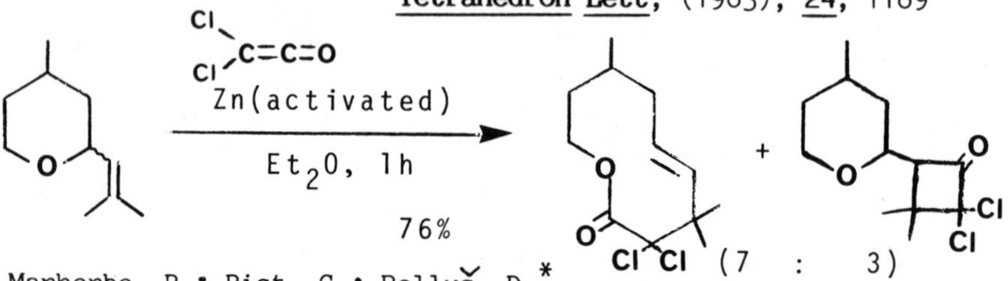
Cl
C=C=O
Cl
Zn(activated)
Et2O, 1h
76%
O
O
Cl Cl
+
O
O
Cl
Cl
(7 : 3)

Marherbe, R.; Rist, G.; Belluš, D.*
J Org Chem, (1983), **48**, 860

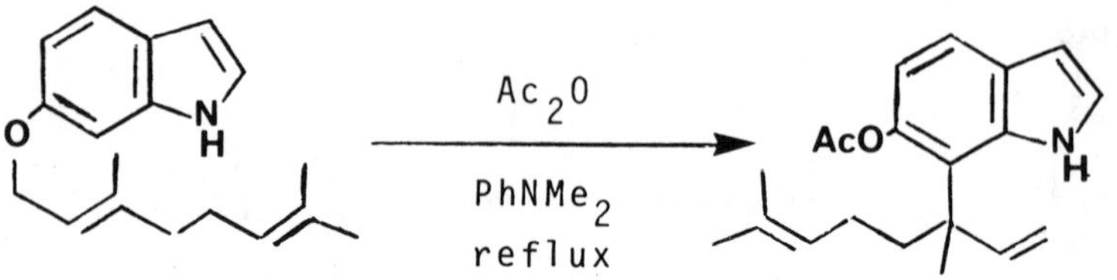
O
N
H
Ac2O
PhNMe2
reflux
AcO
N
H

Moody, C.J.* **JCS Chem Comm**, (1983), 1129 70%

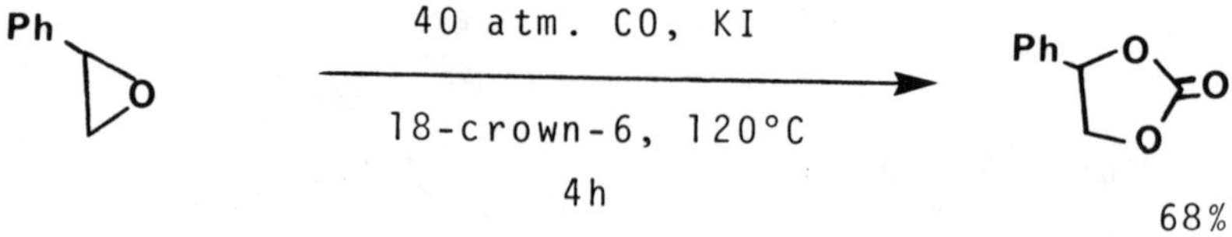
Ph
O
40 atm. CO, KI
18-crown-6, 120°C
4h
Ph
O
O
O
68%

Rokicki, G.; Kuran, W.*; Pogorzelska-Marciniak, B.
Monatsh Chem, (1983), **114**, 205

70%

Bonadies, F.[*]; DiFabio, R.; Bonini, C.[*]
            J Org Chem, (1984), 1491, 2647

$$Me_2C=CHCH_2SPh \xrightarrow[\substack{2AcOH/Ac_2O \\ 85-90°C, 10h}]{\substack{Cu, CuOAc_2 \\ LiOAc, HOH \ reflux}} Me_2C=CHCH_2OAc$$

45%

Uguen, D.[*]   **Tetrahedron Lett**, (1984), **25**, 541

SECTION 115: Esters from Halides and Sulfonates

$$R-X \xrightarrow{\hspace{4cm}} R-CO_2R'$$

R = Bn , X = Cl $\xrightarrow[\substack{Co(CO)_4^-/Co_2(CO)_8}]{\substack{CO/MeOH/MeOK/Al_2O_3}}$ 57%
R' = Me

Sawicki, R.A.[*]   **J Org Chem**, (1983), **48**, 5382

R = Bn , X = Br $\xrightarrow[\substack{Fe(CO)_5, \ RT}]{\substack{CO, \ MeOH, \ K_2CO_3}}$ 68%
R' = Me

Tustin, G.C.[*]; Hembre, R.T.
            J Org Chem, (1984), **49**, 1761

R = Bn , X = Br $\xrightarrow[\substack{cat. \ [1,5-HDRhCl]_2 \\ 2. \ aq. \ NaOH}]{\substack{1. \ Ti(OiPr)_4, \ CO, \ 75°C}}$ 98%
R' = iPr

Woell, J.B.; Fergusson, S.B.; Alper, H.[*]
            J Org Chem, (1985), **50**, 2134

R = Bn , X = Cl
R' = Et

$$\xrightarrow[\text{cat. } [1,5-HDRhCl]_2]{(EtO)_3B, \; CO, \; KI}$$

71%

Alper, H.; Hamel, N.; Smith, D.J.H.; Woell, J.B.
Tetrahedron Lett, (1985), **26**, 2273

R = Ph , X = Br
R' = Me

$$\xrightarrow[\text{Co(CO)}_4^-, \; \text{NaOH}, \; 25°C]{CO, \; MeOH, \; CH_3CO_2Et}$$

86%

Foà, M.; Francalanci, F.; Bencini, E.; Gardano, A.
J Organomet Chem, (1985), **285**, 293

R = α-naphthyl, X = I
R' = Me

$$\xrightarrow[\text{PhPdI(PPh}_3)_2, \; 20°C]{NaOMe, \; CO, \; HMPA}$$

90%

Bumagin, N.A.; Gulevich, Yu.V.; Beletskaya, I.P.[*]
J Organomet Chem, (1985), **285**, 415

R = Bn , X = Br
R' = nBu

$$\xrightarrow[\text{BuOBu}, \; KI, \; 75°C]{\text{cat. } [1,5-HDRhCl]_2, \; CO}$$

81%

Buchan, C.; Hamel, N.; Woell, J.B.; Alper, H.[*]
Tetrahedron Lett, (1985), **26**, 5743

$$\xrightarrow[\substack{h\nu, \; NaOMe/MeOH, \; 36h \\ 65°C}]{Co_2(CO)_8, \; CO}$$

95%

Kashimura, T.[*]; Kudo, K.; Mori, S.; Sugita, N.
Chem Lett, (1986), 851

$$CH_3(CH_2)_6I \xrightarrow[\substack{CHCl_3, \; RT, \; dark \\ 8h}]{PhI(O_2CCF_3)_2} CH_3(CH_2)_6CO_2CF_3$$

43%

Gallos, J.; Varvoglis, A.[*]   **JCS Perkin I**, (1983), 1999

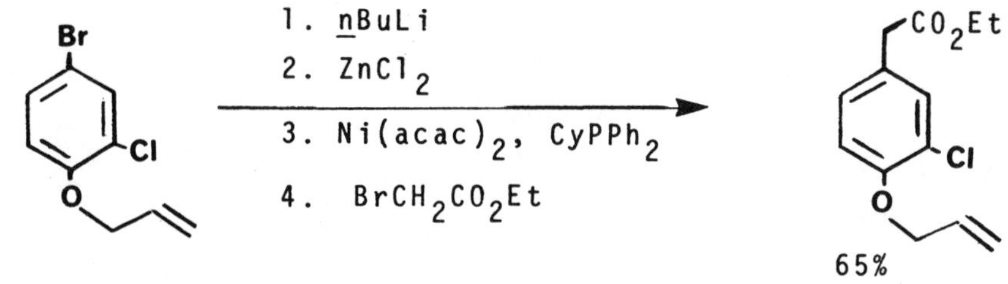

Klingstedt, T.; Frejd, T.*   **Organometallics**, (1983), **2**, 598

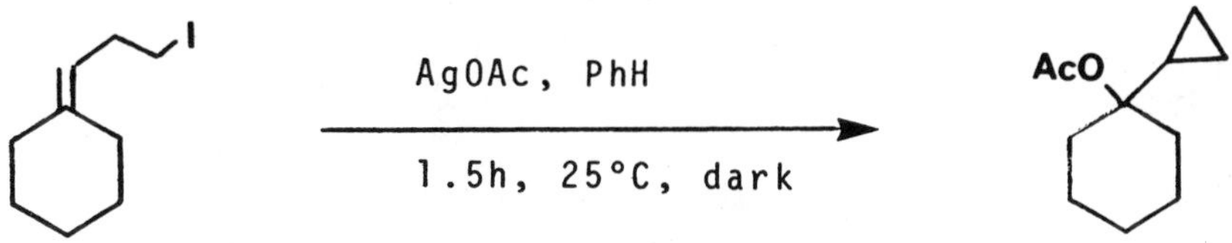

98%

Previtera, L.; Monaco, P.; Mangoni, L.
                    Tetrahedron Lett, (1984), **25**, 1293

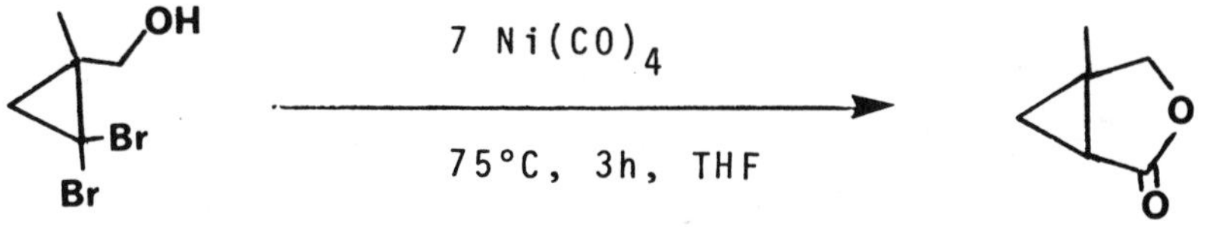

82%

Hirao, T.*; Harano, Y.; Yamana, Y.; Hamada, Y.; Nagata, S.;
Agawa, T.
                    Bull Chem Soc Jpn, (1986), **59**, 1341

Related Methods: Carboxylic Acids from Halides (Section 25)

SECTION 116: Esters from Hydrides

This section contains examples of the reaction RH → RCOOR' or
R'COOR (R = alkyl, aryl, etc.).

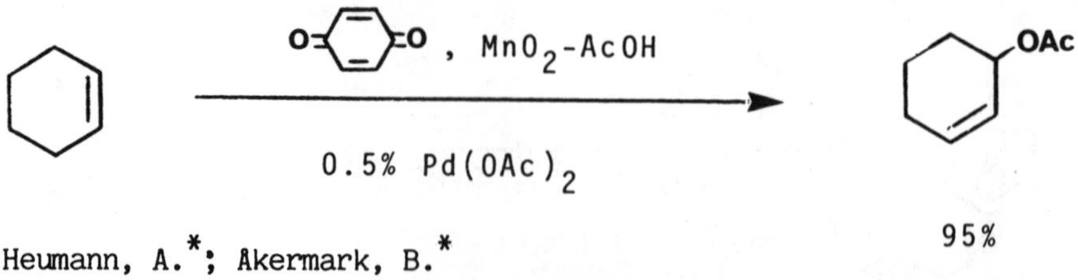

95%

Heumann, A.[*]; Akermark, B.[*]
Angew Chem Int Ed Engl, (1984), **23**, 453

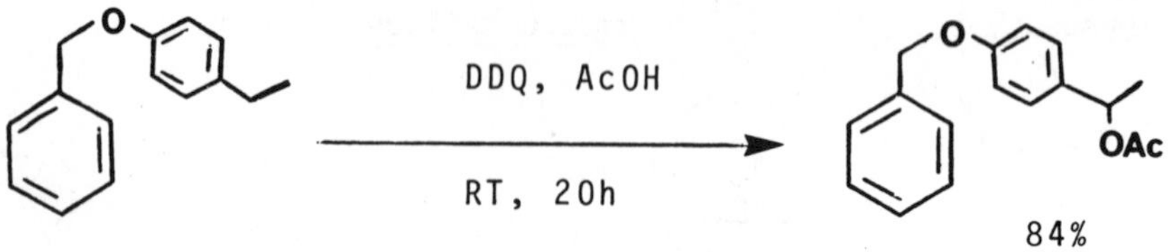

84%

Bouquet, M.; Guy, A.; Lemaire, M.; Guetté, J.P.
Syn Commun, (1985), **15**, 1153

Also via: Carboxylic Acids (Section 26); Alcohols (Section 41)

SECTION 117: Esters from Ketones

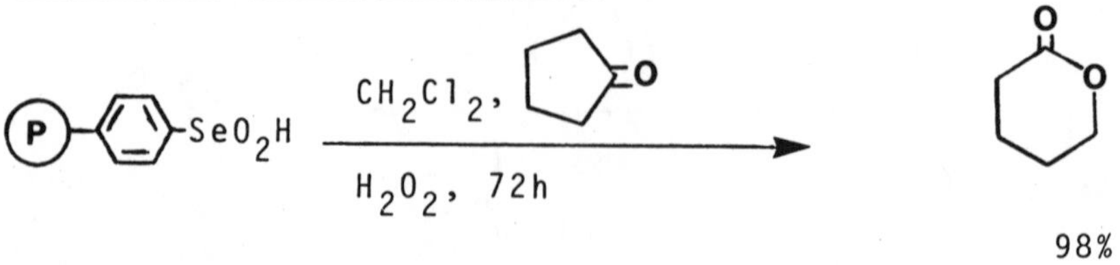

98%

Taylor, R.T.[*]; Flood, L.A.   **J Org Chem**, (1983), **48**, 5160

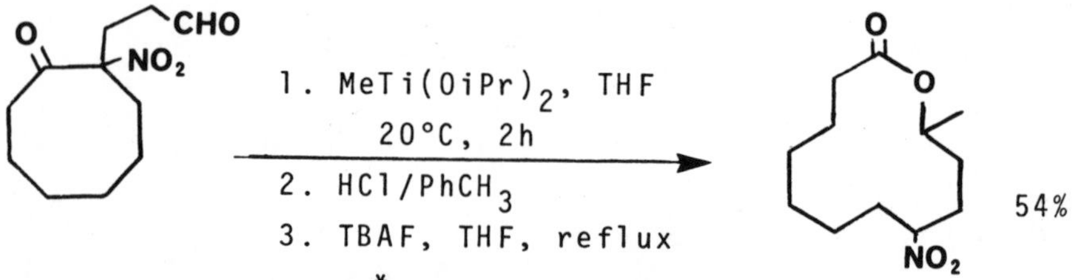

54%

Kostova, K.; Hesse, M.[*]   **Helv Chim Acta**, (1983), **66**, 741
Aono, T.; Hesse, M.[*]   **Helv Chim Acta**, (1984), **67**, 1448

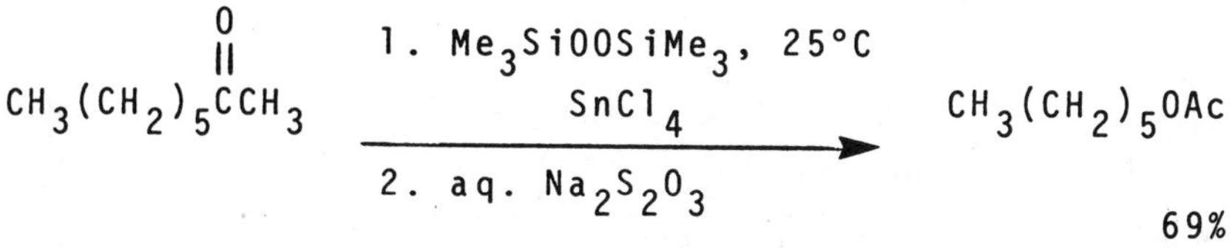

$$CH_3(CH_2)_5\overset{\overset{\displaystyle O}{\|}}{C}CH_3 \xrightarrow[\substack{\text{2. aq. } Na_2S_2O_3}]{\substack{\text{1. } Me_3SiOOSiMe_3, \ 25°C \\ SnCl_4}} CH_3(CH_2)_5OAc$$

69%

Matsubara, S.; Takai, K.[*]; Nozaki, H.
Bull Chem Soc Jpn, (1983), **56**, 2029

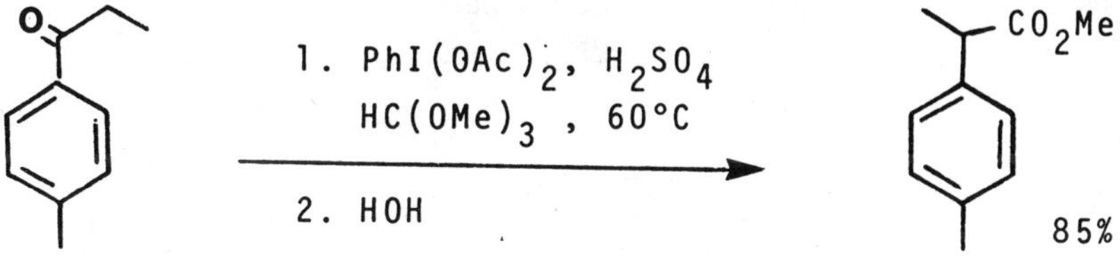

1. $PhI(OAc)_2$, H_2SO_4
   $HC(OMe)_3$, 60°C

2. HOH

85%

Tamura, Y.[*]; Shirouchi, Y.; Haruta, J.   **Synthesis**, (1984), 231

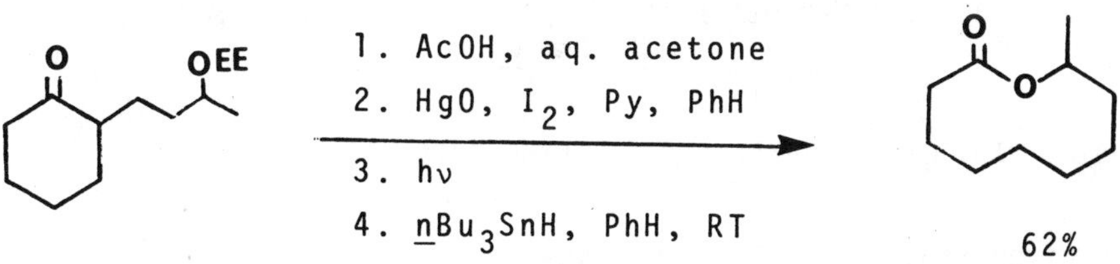

1. AcOH, aq. acetone
2. HgO, I_2, Py, PhH
3. $h\nu$
4. $\underline{n}Bu_3SnH$, PhH, RT

62%

Suginome, H.[*]; Yamada, S.   **Tetrahedron Lett**, (1985), **26**, 3715

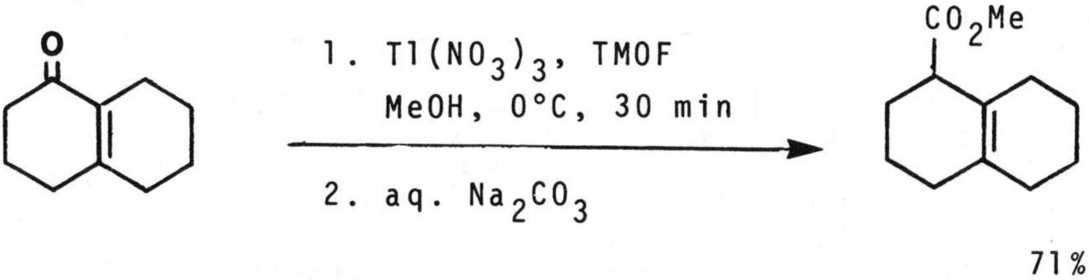

1. $Tl(NO_3)_3$, TMOF
   MeOH, 0°C, 30 min

2. aq. Na_2CO_3

71%

Mincione, E.[*]; Bovicelli, P.; Gil, J.B.; Forcellese, M.L.
Gazz Chim Ital, (1985), **115**, 37

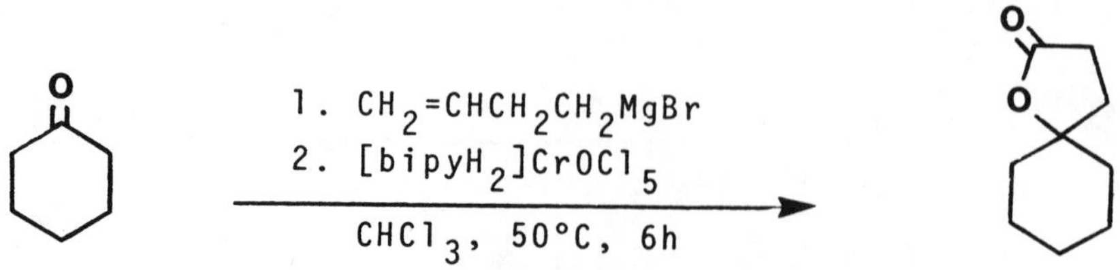
1. CH2=CHCH2CH2MgBr
2. [bipyH2]CrOCl5
CHCl3, 50°C, 6h

76%

Rathore, R.; Vankar, P.S.; Chandrasekaran, S.*
Tetrahedron Lett, (1986), **27**, 4079

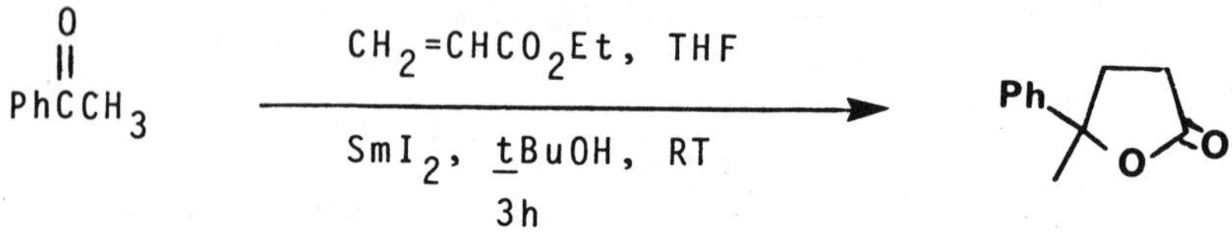
PhCCH3
CH2=CHCO2Et, THF
SmI2, tBuOH, RT
3h
Ph

70%

Fukuzawa, S.*; Nakanishi, A.; Fujinami, T.; Sakai, S.
JCS Chem Comm, (1986), 624

Also via: Carboxylic Acids (Section 27)

SECTION 118: Esters from Nitriles

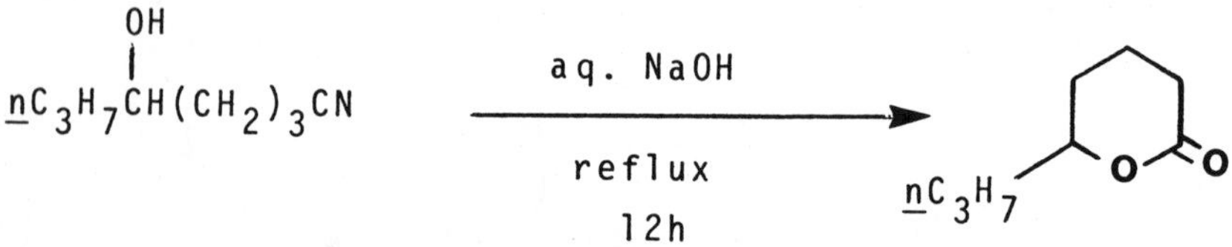
OH
nC3H7CH(CH2)3CN
aq. NaOH
reflux
12h
nC3H7

80%

Giese, B.*; Hasskerl, T.; Lüning, U.
Chem Ber, (1984), **117**, 859

SECTION 119: Esters from Olefins

$$\text{(alkene)} \xrightarrow[\text{AcOH, 90°C}]{25\% \text{ Mn(OAc)}_2} \text{(diacetate)}$$

49%

(meso:d,l = 32:68)

Fristad, W.E.[*]; Peterson, J.R.
                    Tetrahedron Lett, (1983), 24, 4547

$$\xrightarrow[\substack{CH_2Cl_2, \ 40°C \\ 8h}]{5 \ [bipyH_2]CrOCl_5}$$

80%

Chakraborty, T.K.; Chandrasekaran, S.[*]
                    Tetrahedron Lett, (1984), 25, 2895
                    Chem Lett, (1985), 551

1. PPh$_3$, Rh$_2$(OAc)$_4$
   EtOAc, 'bomb'

   CO/H$_2$, 350 psi
   100°C, 6h

2. PCC, CH$_2$Cl$_2$, RT   86%

Wuts, P.G.M.[*]; Obrzut, M.L.; Thompson, P.A.
                    Tetrahedron Lett, (1984), 25, 4051

1. CH$_3$Mn(CO)$_5$, THF
   3.9 KBar, 25°C

2. LiBEt$_3$H, -78°C

44%

DeShong, P.[*]; Slough, G.A.   Organometallics, (1984), 3, 636

OH
|
$CH_3CHCH=CHCH_3$ $\xrightarrow[\substack{THF,\ HCl,\ CO/O_2 \\ RT,\ 18h}]{PdCl_2,\ CuCl_2}$

65%
($\underline{E}:\underline{Z}$ = 1:1)

Alper, H.[*]; Leonard, D.   **Tetrahedron Lett**, (1985), **26**, 5639

$\xrightarrow[\substack{K_2CO_3,\ reflux \\ 2h}]{Mn(OAc)_3,\ AcOH}$

(3.3   :   1)
60%

Fristad, W.E.[*]; Peterson, J.R.   **J Org Chem**, (1985), **50**, 10

NO₂

1. $SnCl_4$, $PhCH_3$
   -29 - 0°C, 3h

2. $\underline{t}amylO^-K^+$, -70°C
3. HCl, CH_2O

40%

Denmark, S.E.[*]; Dappen, M.S.; Cramer, C.J.
J Am Chem Soc, (1986), **108**, 1306

Also via: Alcohols (Section 44)

SECTION 120: Esters from Miscellaneous Compounds

1. $\underline{n}Bu_3SnH$, AIBN
   PhH, reflux, 4h

2. aq. HCl, CrO_3
   RT

80%

Stork, G.[*]; Mook Jr., R.; Biller, S.A.; Rychnovsky, S.D.
J Am Chem Soc, (1983), **105**, 3741

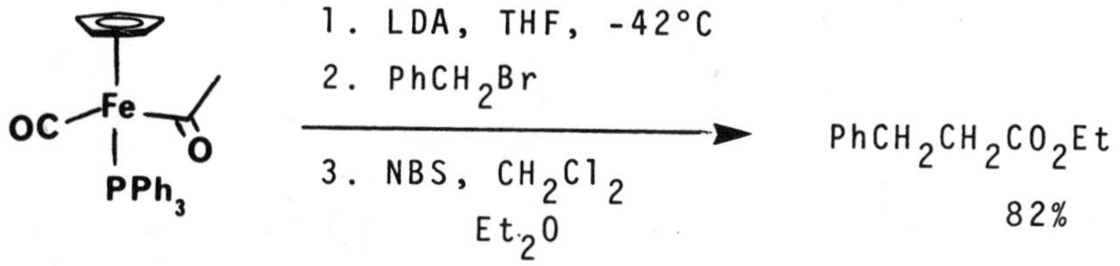
OC
Fe
O
PPh3
1. LDA, THF, -42°C
2. PhCH2Br
3. NBS, CH2Cl2
Et2O
PhCH2CH2CO2Et
82%

Liebeskind, L.S.*; Welker, M.E.; Fengl, R.W.
J Am Chem Soc, (1986), **108**, 6328

Liebeskind, L.S.*; Welker, M.E.
Organometallics, (1983), 2, 194

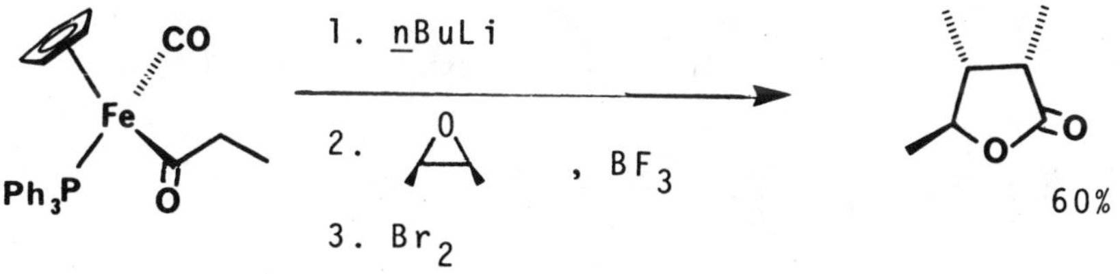
CO
Fe
Ph3P
O
1. nBuLi
2.
O
, BF3
3. Br2
O
O
60%

(RRS,SSR:RSR,SRS:RSS,SRR = 88:8:4)

Davies, S.G.*; Warner, P. **Tetrahedron Lett**, (1985), **26**, 4815

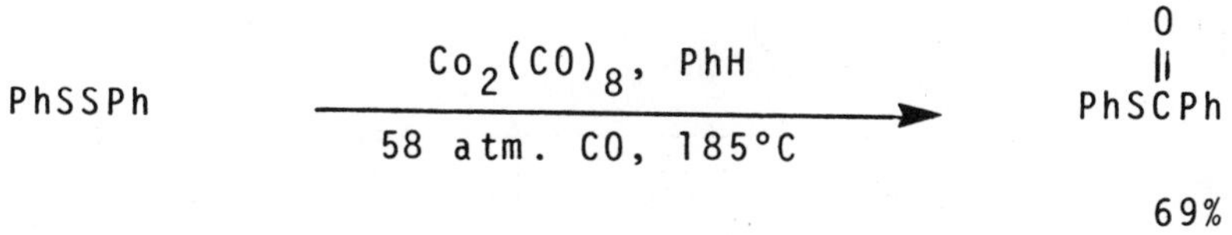
PhSSPh
Co2(CO)8, PhH
58 atm. CO, 185°C
O
PhSCPh
69%

Antebi, S.; Alper, H.* **Tetrahedron Lett**, (1985), **26**, 2609

CHAPTER 9
PREPARATION OF ETHERS, EPOXIDES, AND THIOETHERS

SECTION 121: Ethers, Epoxides, and Thioethers from Acetylenes

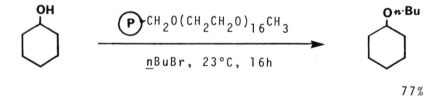

$HC{\equiv}CCH_2CH_2OAc$ $\xrightarrow{\quad 220°C, 48h \quad}$

70%

Liotta, D.[*]; Saindone, M.; Ott, W.
Tetrahedron Lett, (1983), **24**, 2473

SECTION 122: Ethers, Epoxides, and Thioethers from Acid Derivatives

No Additional Examples

SECTION 123: Ethers, Epoxides, and Thioethers from Alcohols and Thiols

OH
$\xrightarrow[\underline{n}BuBr, 23°C, 16h]{\quad \text{(P)}{-}CH_2O(CH_2CH_2O)_{16}CH_3 \quad}$
O*n*-Bu

77%

Kimura, Y.; Kirszensztejn, P.; Regen, S.L.[*]
J Org Chem, (1983), **48**, 385

Guindon, Y.[*]; Frenette, R.[*]; Fortin, R.; Rokach, J.
J Org Chem, (1983), **48**, 1357

Beebe, T.R.[*]; Adkins, R.L.; Bogardus, C.C.; Champney, B.; Hii,
P.S.; Reinking, P.; Shadday, J.; Weatherford III, W.D.; Webb,
M.W.; Yates, S.W.

J Org Chem, (1983), **48**, 3126

$CH_3(CH_2)_7OH$ $\xrightarrow[\text{hexane, reflux}]{\substack{\text{BnCl} \\ \text{KY-zeolite, 5h}}}$ $CH_3(CH_2)_7OBn$

69%

Onaka, M.[*]; Kawai, M.; Izumi, Y.   **Chem Lett**, (1983), 1101

Renga, J.M.[*]; Wang, P.-C.   **Syn Commun**, (1984), **14**, 69

$$HO(CH_2)_9CHCH_2OH \atop \quad\quad\quad\!|\atop \quad\quad\quad OH \xrightarrow[\text{2. NaOMe, MeOH}]{\text{1.} \quad\text{, AcOH}} HO(CH_2)_9CHCH_2$$

47%

Bhat, K.S.; Joshi, P.L.; Rao, A.S.*   **Synthesis**, (1984), 142

$$PhCH_2OH \xrightarrow[\text{DCE, 80°C, 6min}]{\text{PhSeH, ZnCl}_2} PhCH_2SePh$$

81%

Clarembeau, M.; Krief, A.*
                          Tetrahedron Lett, (1984), **25**, 3625

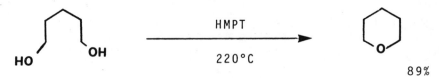

$$\text{P} = \text{polystyrene} \qquad 95\%$$

Kelly, J.W.; Robinson, P.L.; Evans Jr., S.A.*
                          J Org Chem, (1985), **50**, 5007
Robinson, P.L.; Barry, C.N.; Kelly, J.W.; Evans Jr., S.A.
                          J Am Chem Soc, (1985), **107**, 5210

$$\text{HO}\qquad\text{OH} \xrightarrow[\text{220°C}]{\text{HMPT}} \text{O}$$

89%

Diab, J.; Abou-Assali, M.; Gervais, C.; Anker, D.*
                          Tetrahedron Lett, (1985), **26**, 1501

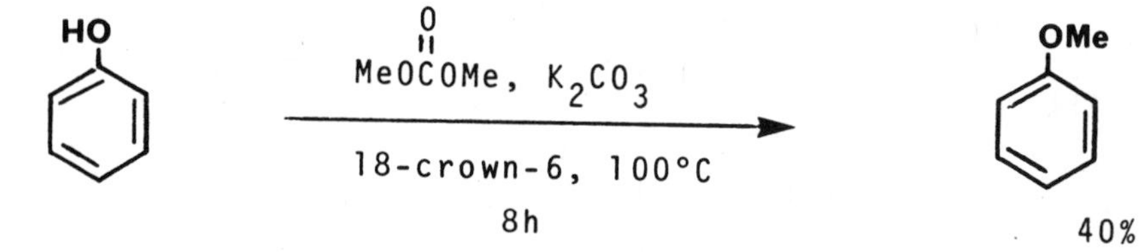

$$MeOCOMe, K_2CO_3$$
18-crown-6, 100°C
8h

40%

Lissel, M.*; Schmidt, S.; Neumann, B.   **Synthesis**, (1986), 382

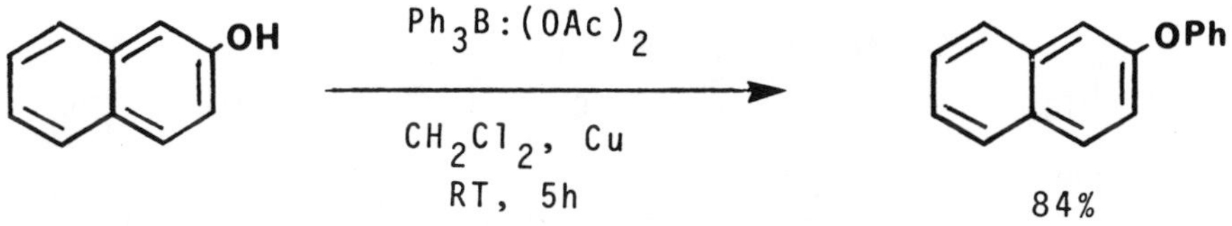

$$Ph_3B:(OAc)_2$$
$$CH_2Cl_2, Cu$$
RT, 5h

84%

Barton, D.H.R.*; Finet, J.-P.; Khamsi, J.; Pichon, C.
Tetrahedron Lett, (1986), **27**, 3619

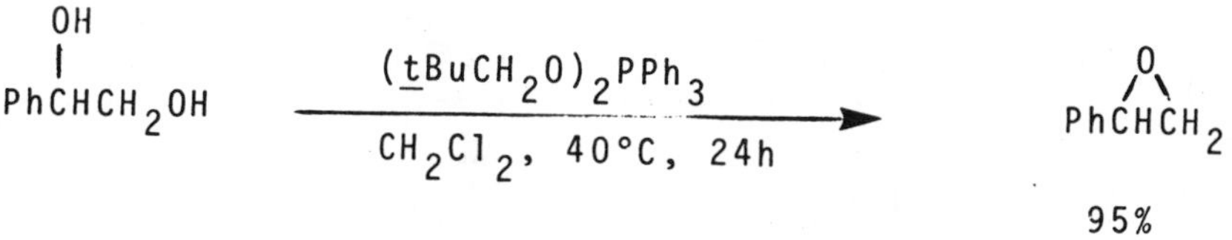

OH
|
PhCHCH$_2$OH

$$(\underline{t}BuCH_2O)_2PPh_3$$
$$CH_2Cl_2, 40°C, 24h$$

PhCHCH$_2$

95%

Kelly, J.W.; Evans Jr., S.A.   **J Org Chem**, (1986), **51**, 5490

SECTION 124: Ethers, Epoxides, and Thioethers from Aldehydes

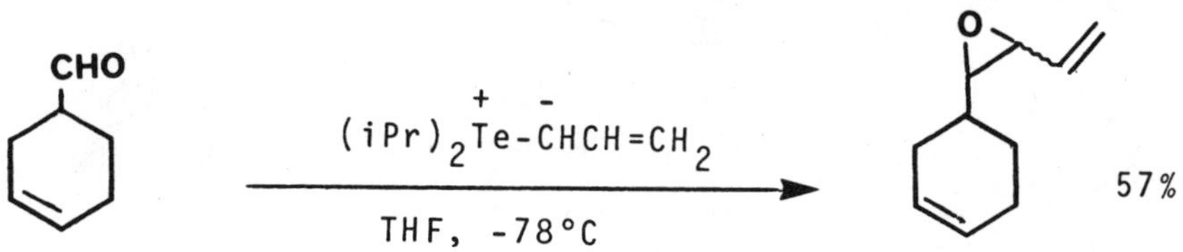

$$(iPr)_2\overset{+}{Te}-\overset{-}{C}HCH=CH_2$$
THF, -78°C

57%

($\underline{Z}$:$\underline{E}$ = 80:20)

Osuka, A.; Suzuki, H.   **Tetrahedron Lett**, (1983), **24**, 5109

CH_3CHBr_2

1. 5 $\underline{n}$BuLi, -115°C
2. PhCHO, -90°C
3. LiOEt
4. HOH

$\overset{O}{\overset{\triangle}{PhCHCHCH_3}}$

59%

($\underline{Z}$:$\underline{E}$ = 2:1)

Tarhouni, R.; Kirschleger, B.; Villieras, J.
J Organomet Chem, (1984), **272**, c1.

Tarhouni, R.; Kirschleger, B.; Rambaud, M.; Villieras, J.
Tetrahedron Lett, (1984), **25**, 835

PhCH=CHCHO
($\underline{E}$)

1. Ph$(CH_2)_3$OTMS
   5% $TrClO_4$, 0°C
   CH_2Cl_2, 5 min
2. Et_3SiH, 5 min

PhCH=CHCH$_2$O$(CH_2)_3$Ph
($\underline{E}$)

88%

Kato, J.; Iwasawa, N.; Mukaiyama, T. **Chem Lett**, (1985), 743

CHO / **NO₂** benzene

1. Me_3SiI, 50% NaOH
   CH_2Cl_2, Bu_4NI, reflux
2. HOH, 5°C

epoxide / **NO₂** benzene

75%

Rafizadeh, K.; Yates, K.
Org Prep Proc Int, (1985), **17**, 140

SECTION 125: Ethers, Epoxides, Thioethers from Alkyls,
Methylenes, and Aryls

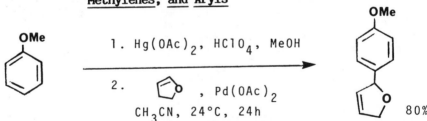

OMe benzene

1. Hg(OAc)$_2$, HClO$_4$, MeOH

2. (furan ring) , Pd(OAc)$_2$
   CH_3CN, 24°C, 24h

80%

Lee, T.D.; Daves Jr., G.D.*   **J Org Chem**, (1983), **48**, 399

SECTION 126: Ethers, Epoxides, and Thioethers from Amides

No Additional Examples

SECTION 127: Ethers, Epoxides, and Thioethers from Amines

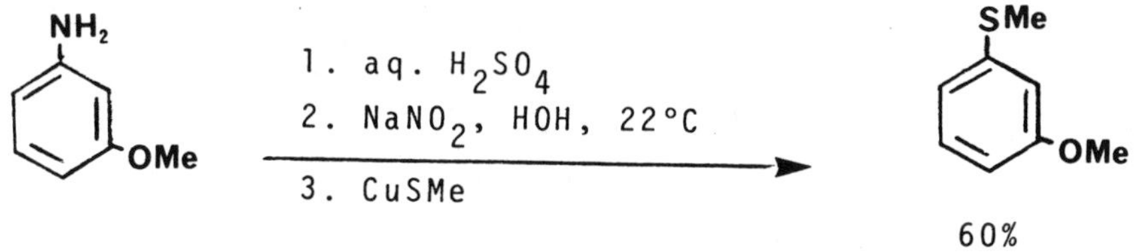

NH2
OMe
1. aq. H2SO4
2. NaNO2, HOH, 22°C
3. CuSMe
SMe
OMe
60%

Baleja, J.D.* **Syn Commun**, (1984), **14**, 215

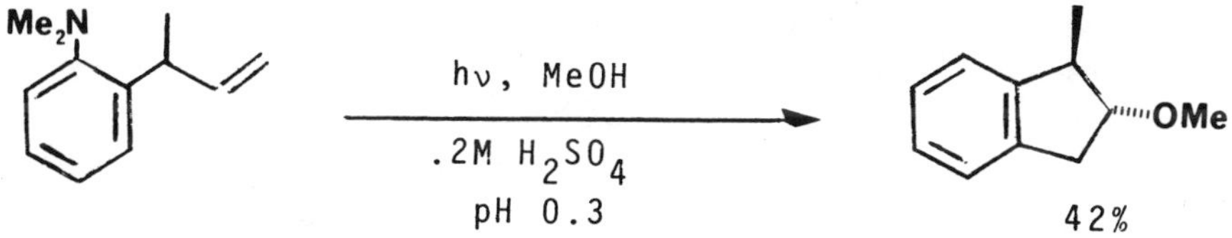

Me2N
hν, MeOH
.2M H2SO4
pH 0.3
OMe
42%

Scholl, B.; Jolidon, S.; Hansen, H.-J. **Helv Chim Acta**, (1986), **69**, 184

SECTION 128: Ethers, Epoxides, and Thioethers from Esters

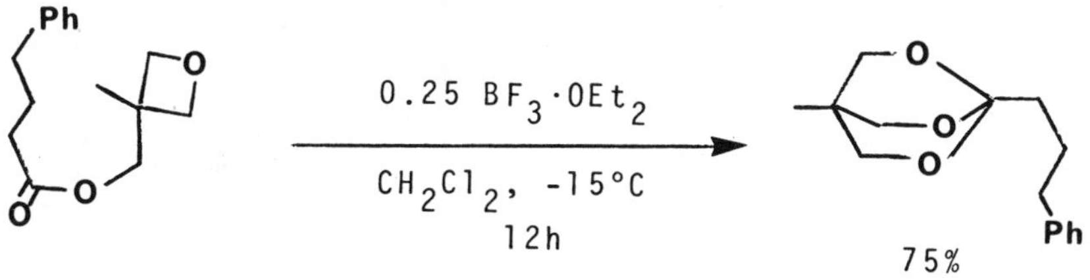

Ph
O
O
O
0.25 BF3·OEt2
CH2Cl2, -15°C
12h
O
O
O
Ph
75%

Corey, E.J.*; Raju, N. **Tetrahedron Lett**, (1983), **24**, 5571

SECTION 129: Ethers, Epoxides, and Thioethers from Ethers, Epoxides, and Thioethers

1. nBuLi, THF, -4°C
2. BnI
3. O_2, Et_2O

65-80%

Davies, S.G.[*]; Holman, N.J.; Laughton, C.A.; Mobbs, B.E.
JCS Chem Comm, (1983), 1316

$(CO)_5W=C$

-78°C - RT

90%

Fischer, H.[*]; Schmid, J.; Märkl, R.
JCS Chem Comm, (1985), 572

Et_3Al, 0°C

$PhCH_3$, 10 min

79%
($\alpha:\beta$ = 20:1)

Posner, G.H.[*]; Haines, S.R.   Tetrahedron Lett, (1985), 26, 1823

$HSiMeEt_2$

PhH, 25°C
20h

$Et_2MeSiO(CH_2)_4CHOSiMeEt_2$
(62) CH_3

+   77%

$Et_2MeSiO(CH_2)_3CHCH_2OSiMeEt_2$
(38) CH_3

Murai, T.; Hatayama, Y.; Murai, S.[*]; Sonoda, N.
Organometallics, (1983), 2, 1883

Murai, T.; Furuta, K.; Kato, S.
J Organomet Chem, (1986), 302, 249

1. Br_2, CCl_4, 20°C

2. $\underline{n}Bu_3SnH$, 24h
   Et_2O, 0°C - reflux

66%

Davies, S.G.[*]; Polywka, M.E.C.; Thomas, S.E.
JCS Perkin I, (1986), 1277

SECTION 130: Ethers, Epoxides, and Thioethers from Halides and Sulfonates

$$PhCH=CHBr$$
$$(\underline{E})$$

$iPrS^-Na^+$, RT
────────────────→
HMPA, 0.5h

$$PhCH=CHSiPr$$
95%
(98% $\underline{E}$)

Tiecco, M.[*]; Testaferri, L.[*]; Tingoli, M.; Chianelli, D.;
Montanucci, M.
J Org Chem, (1983), **48**, 4795

$$2 \ \underline{n}C_8H_{17}Br$$

$$\overset{O}{\overset{\|}{MeSCSMe}}, 45 \ min$$
────────────────→
Bu_4NBr, reflux
30% aq. KOH

$$2 \ CH_3S\underline{n}C_8H_{17}$$
quant.

Degani, I.; Fochi, R.; Regondi, V.   **Synthesis**, (1983), 630

$\underline{n}C_7H_{15}$, KOH, 3h
$PhCH_3$, 3h
────────────────→
$Cy_3\overset{+}{P}-C_{12}H_{25} \ Br^-$

86%

Brunelle, D.J.[*]   **J Org Chem**, (1984), **49**, 1288

$$\text{(2-Br-C}_6\text{H}_4\text{-OMe)} \xrightarrow[\substack{\text{PdCl}_2(\text{PPh}_3)_2, \ 180°\text{C} \\ 70\text{h}}]{\underline{n}\text{Bu}_3\text{SnCH}_2\text{OMe, HMPA}} \text{(2-MeOCH}_2\text{-C}_6\text{H}_4\text{-OMe)}$$

80%

Kosugi, M.; Sumiya, T.; Ogata, T.; Sano, H.; Migita, T.[*]
Chem Lett, (1984), 1225

$$\text{Br(CH}_2)_6\text{Br} \xrightarrow[\substack{2 \ \text{MeLi, THF} \\ \text{reflux, 6h}}]{(\text{Me}_3\text{Si})_2\text{S}}$$

61%

Steliou, K.[*]; Salama, P.; Corriveau, J.
J Org Chem, (1985), **50**, 4969

$$\text{EtI} \xrightarrow[\substack{2. \ \text{BnOH, DMF, NaH}}]{\substack{1. \ \text{Me}_2\text{N}\overset{\text{S}}{\overset{\|}{\text{C}}}\text{NMe}_2, \ \text{DMF}}} \text{EtSCH}_2\text{Ph}$$

76%

Fujisaki, S.; Fujiwara, I.; Norisue, Y.; Kajigaeshi, S.[*]
Bull Chem Soc Jpn, (1985), **58**, 2429

$$\text{(3-MeC}_6\text{H}_4\text{I)} \xrightarrow[\substack{(\text{bipy})_2\text{NiBr}_2, \ 120°\text{C} \\ 30\text{h}}]{(\text{3-MeC}_6\text{H}_4)\text{Se}^-\text{Na}^+, \ \text{EtOH}} \text{(3-MeC}_6\text{H}_4)_2\text{Se}$$

90%

Cristau, H.J.[*]; Chabaud, B.[*]; Labaudiniere, R.; Christol, H.
Organometallics, (1985), **4**, 657

Review: "Copper Assisted Nucleophilic Substitution of Aryl Halogen"

Lindley, J.[*] **Tetrahedron**, (1984), **40**, 1433

Related Methods: Ethers from Alcohols (Section 123)

SECTION 131: Ethers, Epoxides, and Thioethers from Hydrides

No Additional Examples

SECTION 132: Ethers, Epoxides, and Thioethers from Ketones

1. $Et_3O^+BF_4^-$

2. $\underline{t}BuOK$

73%

(74 : 26)

Garst, M.E.[*]; McBride, B.J.; Johnson, A.T.; Arrhenius, P.
J Org Chem, (1983), **48**, 8, 16

$$3 \ CH_3SCH_2Na$$
N-$\underline{p}$Ts

DMSO, 45°C

79%

(88 : 12)

Welch, S.C.[*]; Prakasa Rao, A.S.C.; Lyon, J.T.; Assercq, J.-M.
J Am Chem Soc, (1983), **105**, 252

$CH_3CH(OEt)_2$ → PhCu, LiCl / -30°C, 30 min

$$\underset{PhCHOEt}{\overset{CH_3}{|}}$$

89%

Ghribi, A.; Alexakis, A.[*]; Normant, J.F.
Tetrahedron Lett, (1984), **25**, 3075, 3079, 3083

1. mcpba
2. Dibal-H, -78°C
3. HgO-I$_2$, Py, PhH
4. hν, 3h
5. I$_2$
6. NaBH$_4$, THF, reflux   39%

Suginome, H.*; Yamada, S.   **Tetrahedron Lett**, (1984), **25**, 3995

EtO OEt

1. mcpba, CH$_2$Cl$_2$
   reflux, 8.5h

2. aq. NaOH
                                              44%

Bailey, W.F.*; Bischoff, J.J.   **J Org Chem**, (1985), **50**, 3009

S Me
S Me

1. CH$_2$=C$<$Me$/$CH$_2$MgCl

2. THF:HOH
   10% HBF$_4$

S Me
                                              80%

Dieter, R.K.*; Lin, Y.L.   **Tetrahedron Lett**, (1985), **26**, 39

1. Dibal-H, THF, -78°C
2. LiAlH$_4$
3. Me$_3$O$^+$BF$_4^-$
4. NaOH   95%

Salladie, G.*; Demailly, G.; Greck, C.
                    Tetrahedron Lett, (1985), **26**, 435

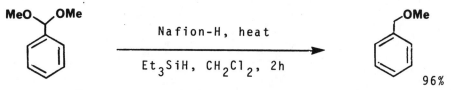

ICH_2Cl, nBuLi

0 - 20°C, overnight

91%

Sadhu, K.M.; Matteson, D.S.[*]   **Tetrahedron Lett**, (1986), **27**, 795

MeO   OMe

Nafion-H, heat

Et_3SiH, CH_2Cl_2, 2h

OMe

96%

Olah, G.A.[*]; Yamato, T.; Iyer, P.S.; Surya Prakash, G.K.
J Org Chem, (1986), **51**, 2826

Review: "Recent Advances in the Preparation and Synthetic
Applications of Oxiranes"

Rao, A.S.[*]; Paknikar, S.K.; Kirtane, J.G.
Tetrahedron, (1983), **39**, 2323

Related Methods: Epoxides from Aldehydes (Section 124)

SECTION 133: Ethers, Epoxides, and Thioethers from Nitriles

No Additional Examples

SECTION 134: Ethers, Epoxides, and Thioethers from Olefins

SPh

CH_2=$CHCH_2TMS$, PhSCl

$ZnBr_2$, CH_3NO_2, 20°C
16h

78%

Alexander, R.; Paterson, I.[*]
Tetrahedron Lett, (1983), **24**, 5911

Kim, Y.H.*; Chung, B.C.   **J Org Chem**, (1983), **48**, 1562

Venturello, C.*; Alneri, E.; Ricci, M.
   J Org Chem, (1983), **48**, 3831

Miyaura, N.; Kochi, J.K.*   **J Am Chem Soc**, (1983), **105**, 2368

Tezuka, T.*; Iwaki, M.   **JCS Perkin I**, (1984), 2507
                         Heterocycles, (1984), **22**, 725

TDCPP = tetra-2,6-dichlorophenyl porphyrin

Renaud, J.-P.; Battioni, P.; Bartoli, J.F.; Mansuy, D.
JCS Chem Comm, (1985), 888

Prandi, J.; Kagan, H.B.; Mimoun, H.
Tetrahedron Lett, (1986), **27**, 2617

TFPP = 5,10,15,20-tetraperfluorophenylporphydrinato^{-2}
De Poorter, B.; Meunier, B.*   **Nouv J Chem**, (1985), **9**, 393

$\underline{1}$-M = $(\underline{n}Bu_4N)_4HMPW_{11}O_{39}$ M = Co   82%
Hill, C.L.*; Brown Jr., R.B.   **J Am Chem Soc**, (1986), **108**, 536

          M = Fe (cyclooctene)   84%
Groves, J.T.*; Nemo, T.E.
          J Am Chem Soc, (1983), **105**, 5786, 5791

10

$$PhCl, 25°C$$

$$Ph-I-O-Mn-O-I-Ph$$
$$\quad \overset{|}{OAc} \qquad \overset{|}{OAc}$$

73%

Smegal, J.A.; Hill, C.L.[*]   **J Am Chem Soc**, (1983), **105**, 2920

$$CH_2Cl_2$$
2 TMSOTf, -78°C

91%

Edstrom, E.D.; Livinghouse, T.[*]
                Tetrahedron Lett, (1986), **27**, 3483

$$CHCl_3, 25°C, 48h$$

60-70%

Z = (-)-3-bromocamphor
Ar = 2-chloro-5-nitrophenyl   (53% ee)
Davis, F.A.[*]; Chattopadhyay, S.
                Tetrahedron Lett, (1986), **27**, 5079

$$N_2/F_2, -15°C$$

aq. CH_3CN

80%

Rozen, S.[*]; Brand, M.   **Angew Chem Int Ed Engl**, (1986), **25**, 554

Review: "Preparation and Synthetic Applications of Oxiranes"

Rao, A.S.[*]; Paknikar, S.K.; Kirtane, J.G.
                Tetrahedron, (1983), **39**, 2323

SECTION 135: Ethers, Epoxides, and Thioethers from Miscellaneous Compounds

1. $BF_3 \cdot OEt_2$

2. THF, -78°C, 2h

$CH_2 = C \overset{Ph}{\underset{OLi}{}}$

3. TFA:HOH

80%

Meltz, C.N.; Volkmann, R.A.[*]

Tetrahedron Lett, (1983), 24, 4507

$$PhCH_2 \overset{O}{\overset{\|}{S}} CH_3$$

B-Br, CH_2Cl_2

-23 - 30 - 0°C

$PhCH_2SCH_3$

90%

Guindon, Y.[*]; Atkinson, J.G.; Morton, H.E.

J Org Chem, (1984), 49, 4538

$$PhCH_2 \overset{O}{\overset{\|}{S}} Ph$$

BHCl, CH_2Cl_2

0°C, 10 min

$PhCH_2SPh$

93%

Cha, J.S.[*]; Kim, J.E.; Kim, J.D.

Tetrahedron Lett, (1985), 26, 6453

$$\underline{n}BuS\underline{n}Bu \overset{O}{\overset{\|}{}}$$

(P)-PPh_2Cl_2, reflux

THF/CCl_4, 1.5h

$\underline{n}BuS\underline{n}Bu$

(P) = polystyryl

98%

Amos, R.A.[*]   J Org Chem, (1985), 50, 1311

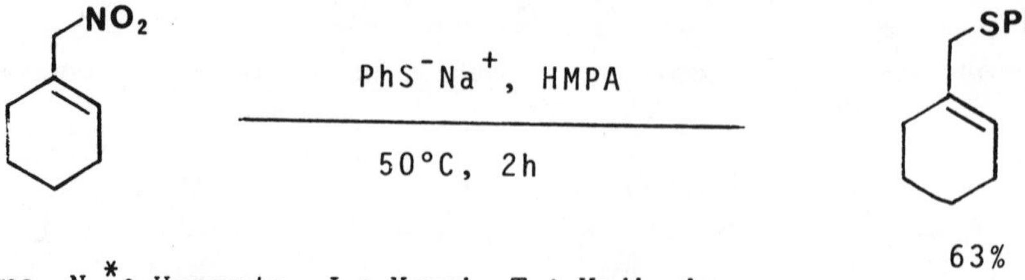

PhS^-Na^+, HMPA

50°C, 2h

63%

Ono, N.*; Hamamoto, J.; Yanai, T.; Kaji, A.
JCS Chem Comm, (1985), 523

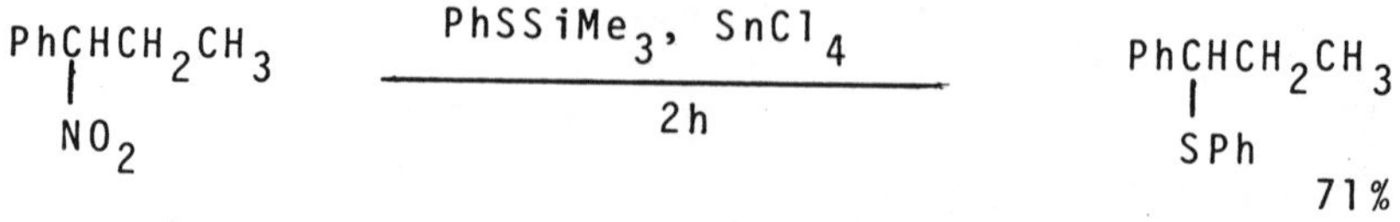

$PhCHCH_2CH_3$
|
NO_2

$PhSSiMe_3$, $SnCl_4$

2h

$PhCHCH_2CH_3$
|
SPh

71%

Ono, N.*; Yanai, T.; Kaji, A.   **JCS Chem Comm**, (1986), 1040

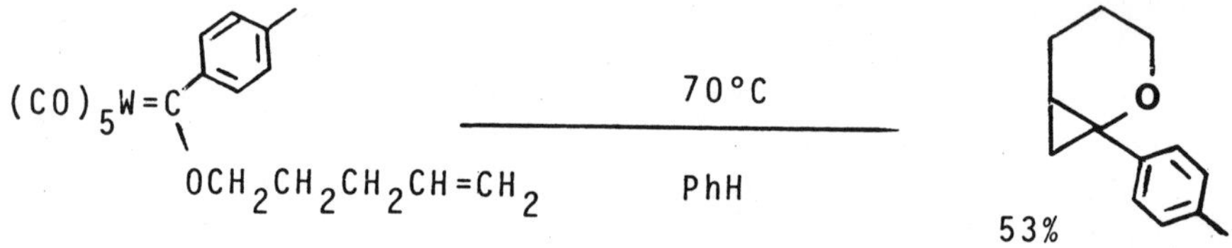

$(CO)_5W=C$

$OCH_2CH_2CH_2CH=CH_2$

70°C

PhH

53%

Casey, C.P.*; Shusterman, A.J.   **Organometallics**, (1985), **4**, 736

SECTION 135A: Protection of Ethers, Epoxides, and Thioethers

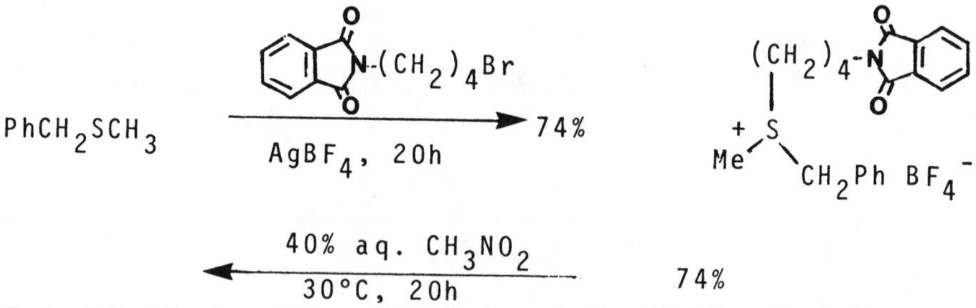

N-$(CH_2)_4Br$

$PhCH_2SCH_3$

$AgBF_4$, 20h

74%

$(CH_2)_4$-N

$Me\overset{+}{S}CH_2Ph$ BF_4^-

40% aq. CH_3NO_2

30°C, 20h

74%

Doi, J.T.; Luehr, G.W.   **Tetrahedron Lett**, (1985), **26**, 6143

CHAPTER 10
PREPARATION OF HALIDES AND SULFONATES

<u>SECTION 136</u>: <u>Halides and Sulfonates from Acetylenes</u>

No Additional Examples

<u>SECTION 137</u>: <u>Halides and Sulfonates from Acid Derivatives</u>

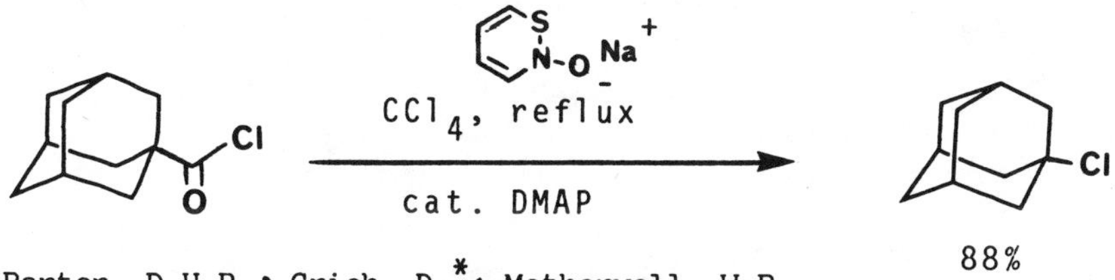

Barton, D.H.R.; Crich, D.*; Motherwell, W.B.
Tetrahedron Lett, (1983), <u>24</u>, 4979

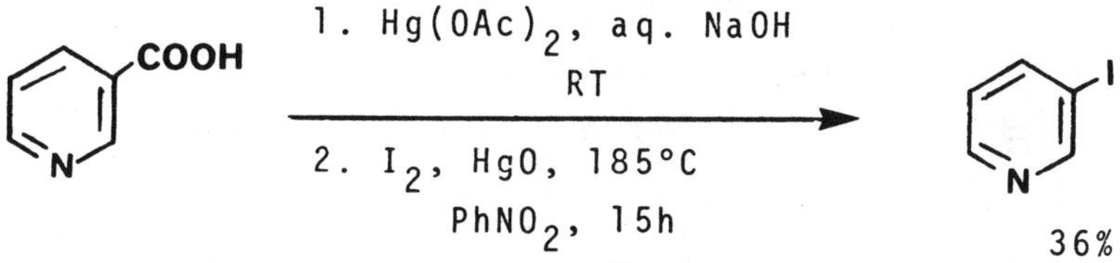

88%

Uemura, S.; Tanaka, S.; Okano, M.*
J Org Chem, (1983), <u>48</u>, 3297

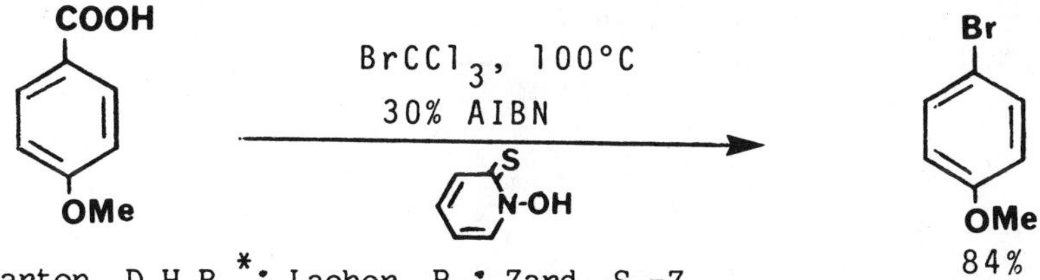

36%

Barton, D.H.R.*; Lacher, B.; Zard, S.-Z.
Tetrahedron Lett, (1985), <u>26</u>, 5939

84%

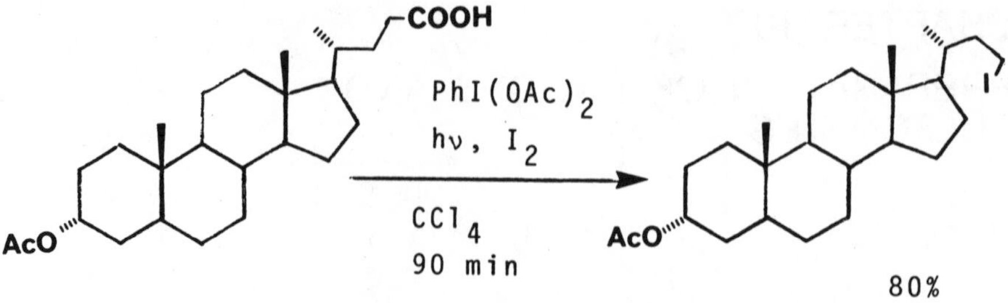
COOH
PhI(OAc)2
hν, I2
CCl4
90 min
AcO
AcO
I
80%

Concepción, J.I.; Francisco, C.G.; Freire, R.; Hernández, R.; Salazar, J.A.; Suárez, E.*
J Org Chem, (1986), 51, 402

SECTION 138: Halides and Sulfonates from Alcohols and Thiols

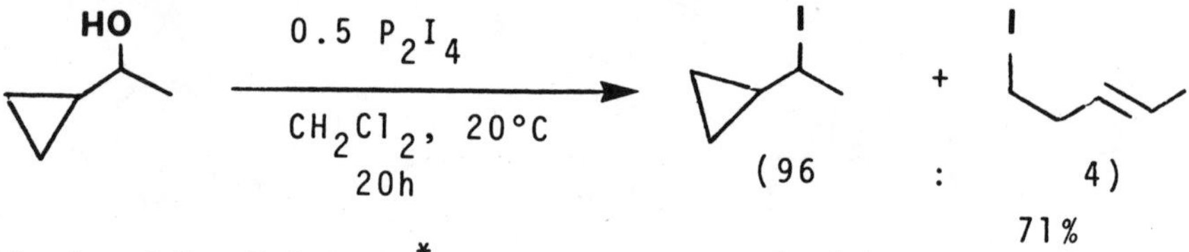
HO
0.5 P2I4
CH2Cl2, 20°C
20h
I
+
I
(96 : 4)
71%

Denis, J.N.; Krief, A.* JCS Chem Comm, (1983), 229

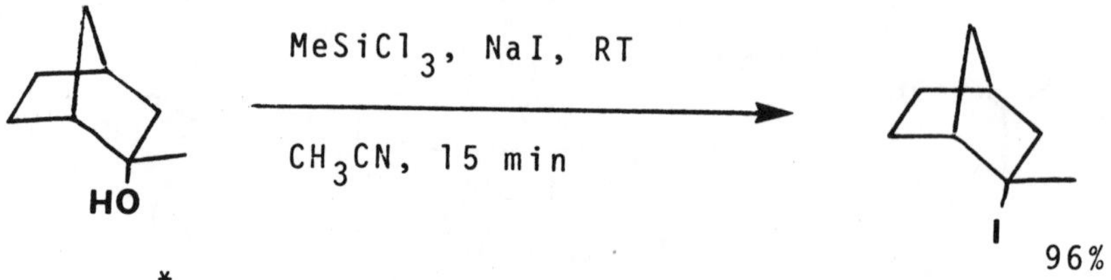
MeSiCl3, NaI, RT
CH3CN, 15 min
HO
I
96%

Olah, G.A.*; Husain, A.; Singh, B.P.; Mehrota, A.K.
J Org Chem, (1983), 48, 3667

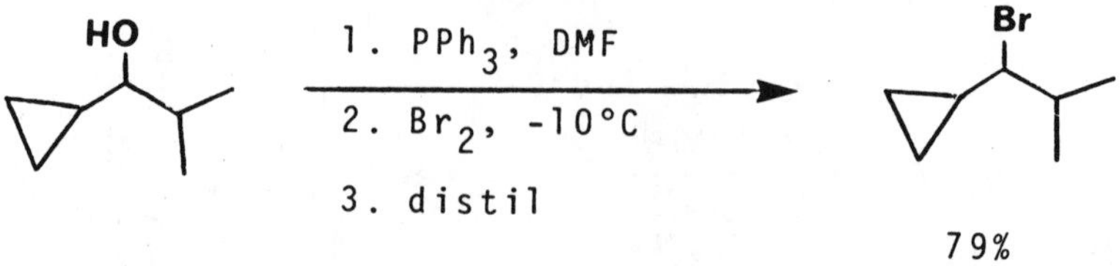
HO
1. PPh3, DMF
2. Br2, -10°C
3. distil
Br
79%

Hrubiec, R.T.; Smith, M.B.* J Org Chem, (1984), 49, 431

$$\underline{n}C_6H_{13}\overset{\displaystyle OH}{\underset{|}{C}}HCH_3 \quad \xrightarrow[\text{CBr}_4,\ \text{CHCl}_3,\ 40\ \text{min}]{\textcircled{P}\!\!-\!\!\langle\ \rangle\!\!-\!\!PPh_2,\ 20°C} \quad \underline{n}C_6H_{13}\overset{\displaystyle Br}{\underset{|}{C}}HCH_3$$

76%

Hodge, P.[*]; Khoshdel, E.   **JCS Perkin I**, (1984), 195

$$\underline{n}C_7H_{15}CH_2OH \quad \xrightarrow[\text{CCl}_4,\ \text{PhCH}_3,\ 90°C]{H(CH_2CH_2)_n\text{-}CH_2CH_2PPh_2} \quad \underline{n}C_7H_{15}CH_2Cl$$

96%

Bergbreiter, D.E.[*]; Blanton, J.R.   **JCS Chem Comm**, (1985), 337

$$PhCH_2CH_2OH \quad \xrightarrow[\text{Et}_2O,\ RT,\ 1h]{\textcircled{P}\text{-}CH_2NR_2/PBr_3} \quad PhCH_2CH_2Br$$

82%

$\textcircled{P}$ = Amberlite IRA93

Cainelli, G.; Contento, M.; Manescalchi, F.; Plessi, L.;
Panunzio, M.
                        Synthesis, (1983), 306

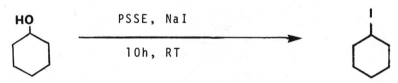

PSSE, NaI

10h, RT

PSSE = trimethylsilylpolyphosphate   99%
Imamoto, T.[*]; Matsumoto, T.; Kusumoto, T.; Yokoyama, M.
                        Synthesis, (1983), 460

$$TsCl, DMAP, NEt_3$$
$$CH_2Cl_2, 25°C, 24h$$

85%

Hwang, C.K.; Li, W.S.; Nicolaou, K.C.
                    Tetrahedron Lett, (1984), **25**, 2295

$$\underline{n}C_{11}H_{23}OH$$

1. $\begin{array}{c}Cl \\ Ph\end{array}C=NPh_2{}^+ \; Cl^-, \; NEt_3$

$$CH_2Cl_2, \; RT$$

2. HOH

$$\underline{n}C_{11}H_{23}Cl$$

96%

Fujisawa, T.[*]; Iida, S.; Sato, T.   **Chem Lett**, (1984), 1173

HO

1. $N=\!\!\!\!\!\!\!\!\!\!\begin{array}{c}O\\ \\ \end{array}\!\!N\text{-}Me$, PPh$_3$

2. CH$_3$I

I

74%

Oshikawa, T.[*]; Yamashita, M.
                    Bull Chem Soc Jpn, (1984), **57**, 2675

OH
|
$\underline{n}BuCH(CH_2)_9CH_2OH$

$\langle\!\!\!\bigcirc\!\!\!\rangle\!\!-\!SO_2F$

Bu$_4$NF, THF, RT
4Å sieve, 1d

OH
|
$\underline{n}BuCH(CH_2)_9CH_2F$

69%

Shimizu, M.[*]; Nakahara, Y.; Yoshioka, H.[*]
                    Tetrahedron Lett, (1985), **26**, 4207

$$CH_3CH_2CH_2CH_2OH \xrightarrow[\substack{2. \ PhCH_3, \ 75°C}]{\substack{1. \ ClCOCHClMe, \ NaI \\ acetone, \ reflux}} CH_3CH_2CH_2CH_2I$$

65%

Brunet, J.J.; Laurent, H.; Caubere, P.[*]
Tetrahedron **Lett**, (1985), **26**, 5445

$$ArCH_2CH_2OH \xrightarrow{2.45 \ DEAD, \ 2.5 \ PPh_3} ArCH_2CH_2X$$

Ar = Ph   5 LiBr/THF   X = Br   95%

Manna, S.; Falck, J.R.[*]; Mioskowski, C.
Syn **Commun**, (1985), **15**, 663

Ar = [structure]Cl   Py·HCl/CH₂Cl₂   X = Cl   96%

Alpegiani, M.; Bedeschi, A.; Perrone, E.
Gazz **Chim** **Ital**, (1985), **115**, 393

$$\xrightarrow[\substack{30°C, \ 90 \ min}]{2 \ TMS-Cl, \ CH_2Cl_2}$$

90%

Balme, G.[*]; Fournet, G.; Gore, J.
Tetrahedron **Lett**, (1986), **27**, 1907

SECTION **139**: **Halides** **and** **sulfonates** **from** **Aldehydes**

$$PhCHO \xrightarrow[\substack{TMS-Cl, \ CH_3CN, \ 30 \ min}]{\substack{(HSiMe_2)_2O, \ NaI, \ 0°C}} PhCH_2I$$

91%

Aizpurua, J.M.; Palomo, C.[*]   **Tetrahedron** **Lett**, (1984), **25**, 1103

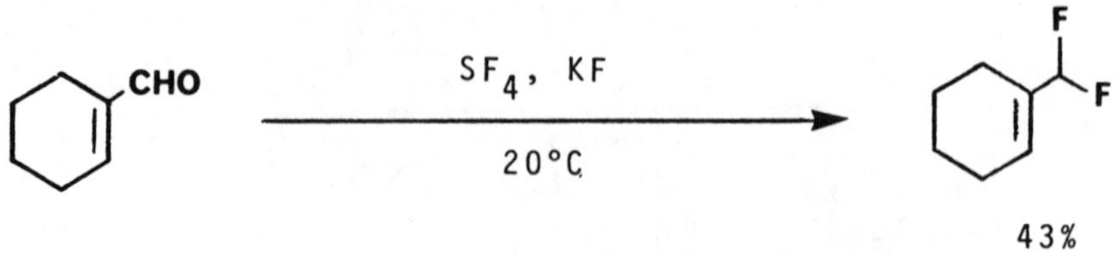

43%

Haas, A.*; Plümer, R.; Schiller, A.
Chem Ber, (1985), **118**, 3004

SECTION 140: Halides and Sulfonates from Alkyls, Methylenes, and Aryls

For the conversion RH → RHal see Section 146 (Halides from Hydrides).

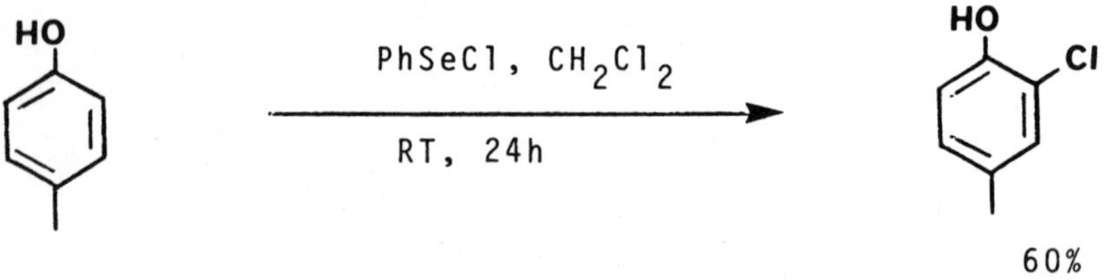

60%

Ayorinde, F.O.*   **Tetrahedron Lett**, (1983), **24**, 2077

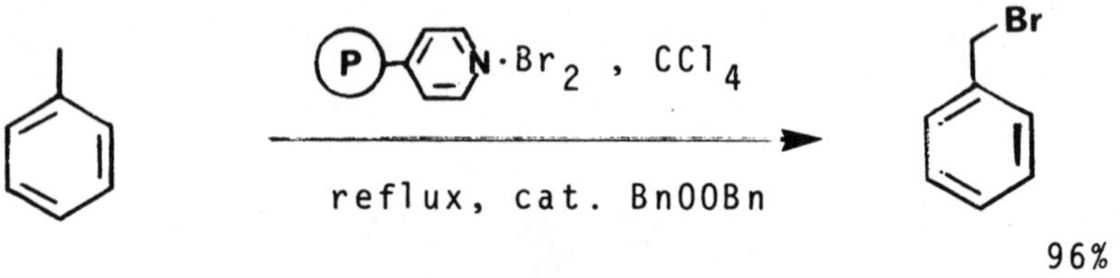

96%

Šket, B.; Zupan, M.*   **J Org Chem**, (1986), **51**, 929

SECTION 141: Halides and Sulfonates from Amides

No Additional Examples

SECTION 142: Halides and Sulfonates from Amines

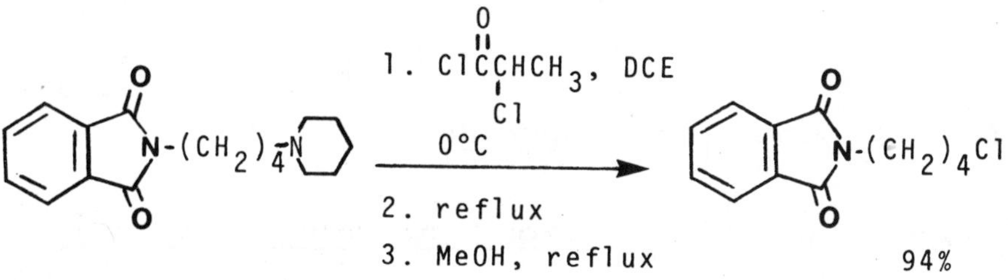

$$1. \quad ClCCHCH_3, \ DCE$$

(with structure: Cl–C(=O)–CH(Cl)–CH₃)

2. reflux

3. MeOH, reflux   94%

Olofson, R.A.[*]; Abbott, D.E.   **J Org Chem**, (1984), **49**, 2795

SECTION 143: Halides and Sulfonates from Esters

No Additional Examples

SECTION 144: Halides and Sulfonates from Ethers, Epoxides, and Thioethers

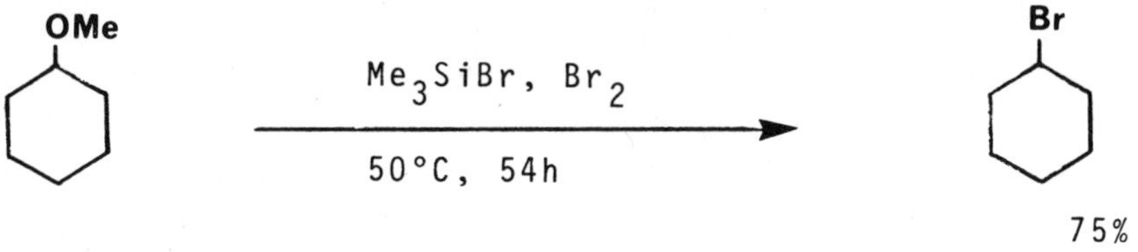

$$Me_3SiBr, \ Br_2$$

50°C, 54h

75%

Friedrich, E.C.[*]; DeLucca, G.   **J Org Chem**, (1983), **48**, 1678

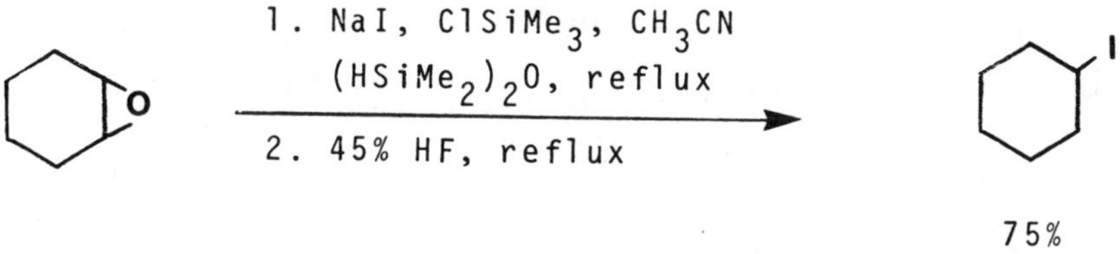

1. NaI, ClSiMe₃, CH₃CN
   $(HSiMe_2)_2O$, reflux

2. 45% HF, reflux

75%

Aizpurua, J.M.; Palomo, C.[*]   **Tetrahedron Lett**, (1984), **25**, 3123

$$\text{nBuOTMS} \xrightarrow[\text{CH}_2\text{Cl}_2,\ 15\ \text{min}]{(\text{CF}_3\text{SO}_2)_2\text{O},\ 0^\circ\text{C}} \text{nBuOSO}_2\text{CF}_3$$

72%

Aubert, C.; Bégué, J.-P.* **Synthesis**, (1985), 759

SECTION 145: Halides and Sulfonates from Halides and Sulfonates

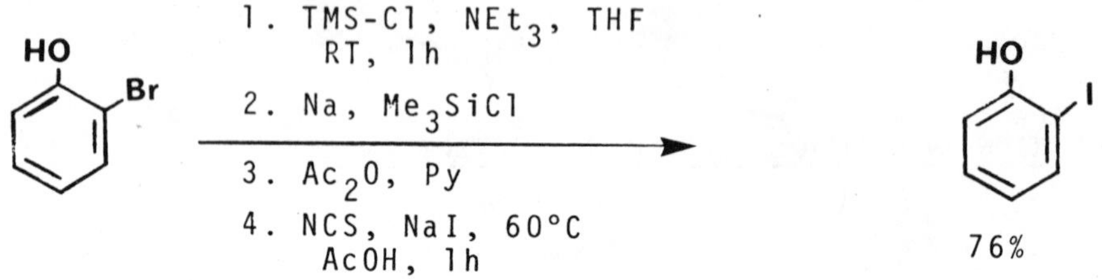
HO
Br
1. TMS-Cl, NEt3, THF
RT, 1h
2. Na, Me3SiCl
3. Ac2O, Py
4. NCS, NaI, 60°C
AcOH, 1h
HO
I

76%

Wilbur, D.S.*; Stone, W.E.; Anderson, K.W.
J Org Chem, (1983), **48**, 1542

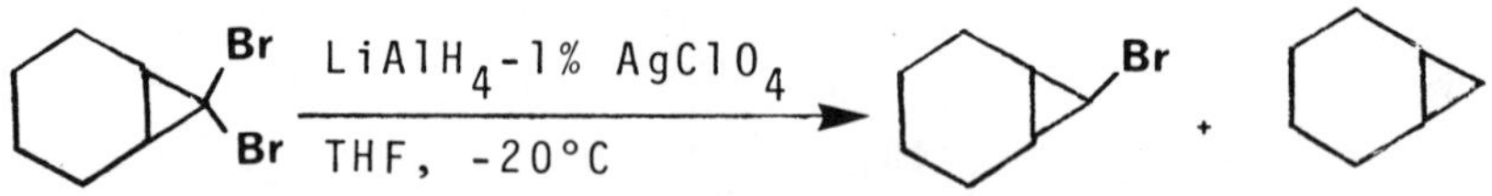
Br
Br
LiAlH4-1% AgClO4
THF, -20°C
Br
+

THF, 1h	59%(Z:E=8:1)	35%
THF, 24h	0%	77%

Shimizu, N.*; Watanabe, K.; Tsuno, Y. **Chem Lett**, (1983), 1877

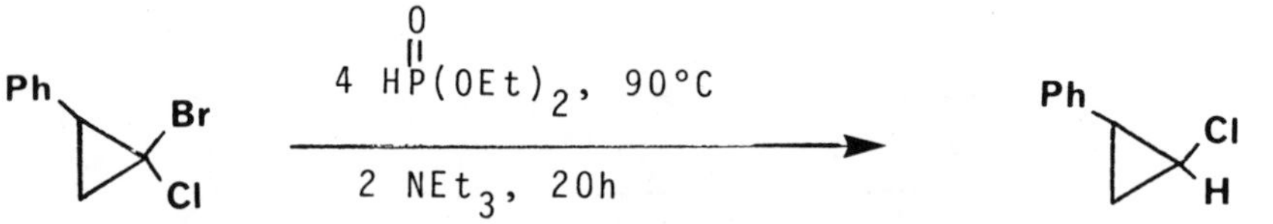
Ph
Br
Cl
4 HP(O)(OEt)2, 90°C
2 NEt3, 20h
Ph
Cl
H

76%

Hirao, T.; Kohno, S.; Ohshiro, Y.*; Agawa, T.
Bull Chem Soc Jpn, (1983), **56**, 1881

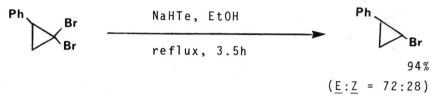

94%

(E:Z = 72:28)

Osuka, A.*; Takechi, K.*; Suzuki, H.*
Bull Chem Soc Jpn, (1984), **57**, 303

$$\underline{n}C_8H_{17}Cl \xrightarrow[Bu_4NBr]{KF} \underline{n}C_8H_{17}F$$

69%

Escoula, B.; Rico, I.; Lattes, A.
Tetrahedron Lett, (1986), **27**, 1499

$$\text{Cl} \xrightarrow[5h]{\substack{NaNO_3, (HF)_x \cdot Py \\ CH_2Cl_2, \ 0°C}} \text{F}$$

95%

Olah, G.A.*; Shih, J.G.; Singh, B.P.; Gupta, B.G.B.
Synthesis, (1983), 713

$$PhCH_2Br \xrightarrow[CH_3CN, \ reflux, \ 15h]{KF-CaF_2 \ (1:2)} PhCH_2F$$

81%

Ichihara, J.*; Matsuo, T.; Hanafusa, T.; Ando, T.
JCS Chem Comm, (1986), 793

$$CH_2=CHCH_2Br \xrightarrow[\text{vacuum, 48h}]{Bu_4NF \cdot 3H_2O} CH_2=CHCH_2F$$

85%

Cox, D.P.[*]; Terpinski, J.; Lawrynowicz, W.
                    J Org Chem, (1984), 49, 3216

KI, CuI

HMPT, 160°C
7h

80%

Suzuki, H.[*]; Kondo, A.; Inouye, M.; Ogawa, T.
                    Synthesis, (1986), 121
Suzuki, H.[*]; Kondo, A.; Ogawa, T.   Chem Lett, (1985), 411

Review: "Radical Brominations of Alkyl Bromides and the Nature
        of β-Bromoalkyl Radicals"

Skell, P.S.[*]; Traynham, J.E.[*]   Accts Chem Res, (1984), 17, 160

SECTION 146: Halides and Sulfonates from Hydrides

α-Halogenations of Aldehydes, Ketones, and Acids are found in
Sections 338 (Haloaldehydes), 369 (Haloketones), 359
(Halo-Esters), and 319 (Haloacids).

CF_3CO_2H

N-Cl, RT

99%   1%

Lindsay-Smith, J.R.[*]; McKeer, L.C.
                    Tetrahedron Lett, (1983), 24, 3117

$$NO^+BF_4^-, \text{ 6h}$$

PPHF, CH_2Cl_2, RT

95%

PPHF = pyridinepolyhydrogen fluoride

Olah, G.A.[*]; Shih, J.G.; Singh, B.P.; Gupta, B.G.B.
J Org Chem, (1983), **48**, 3356

OMe

1. CH_3COF, CH_2Cl_2, $CFCl_3$
   -75°C

2. HOH

OMe

F

(o:p = 9:1) 85%

Lerman, O.; Tor, Y.; Hebel, D.; Rozen, S.[*]
J Org Chem, (1984), **49**, 806

$\underline{t}BuOCl$, SiO_2, CCl_4

RT, 1h   quant.
[Zeolite, NaCl, 40°C]

Cl

(o:p = 65:35)
[o:p = 82:18]

Smith, K.[*]; Butlers, M.; Paget, W.E.
Synthesis, (1985), 1155, 1157

SECTION 147: Halides and Sulfonates from Ketones

$\underline{n}C_5H_{11}C\underline{n}C_5H_{11}$

1. $HS(CH_2)_3SH$, BF_3
   2 AcOH

2. DBH, 20 PPHF
   CH_2Cl_2, -78°C

$\underline{n}C_5H_{11}C\underline{n}C_5H_{11}$

F

F

97%

DBH = 1,3-dibromo-5,5-dimethylhydantoin
PPHF = pyridinium poly(hydrogen) fluoride
Sondej, S.C.; Katzenellenbogen, J.A.
J Org Chem, (1986), **51**, 3508

SECTION 148: Halides and Sulfonates from Nitriles

No Additional Examples

SECTION 149: Halides and Sulfonates from Olefins

For halocyclopropanations see Section 74E (Alkyls from Olefins).

$CH_2=CHPh$, PhH

$\xrightarrow{}$

4% $Ru_2ClO_4(diop)_3$
60°C, 1h

Cl
$PhCHCH_2SO_2$—⟨ ⟩—
92%(49%conversion)
(29% ee R)

Kameyama, M.; Kamigata, N.[*]; Kobayashi, M.
Chem Lett, (1986), 527

$\underline{n}C_{10}H_{21}CH=CH_2$ $\xrightarrow[\substack{2.\ I_2,\ Py,\ -78°C \\ 1h}]{\substack{1.\ BEt_3,\ Cl_2AlH \\ RT,\ 2h}}$ $\underline{n}C_{10}H_{21}CH_2CH_2I$
85%

Maruoka, K.; Sano, H.; Shinoda, K.; Nakai, S.; Yamamoto, H.[*]
J Am Chem Soc, (1986), **108**, 6036

SECTION 150: Halides and Sulfonates from Miscellaneous Compounds

TMS-I, CH_2Cl_2
$\xrightarrow{}$
25°C, 16h

98%

Olah, G.A.[*]; Narang, S.C.; Field, L.D.; Fung, A.P.
J Org Chem, (1983), **48**, 2766

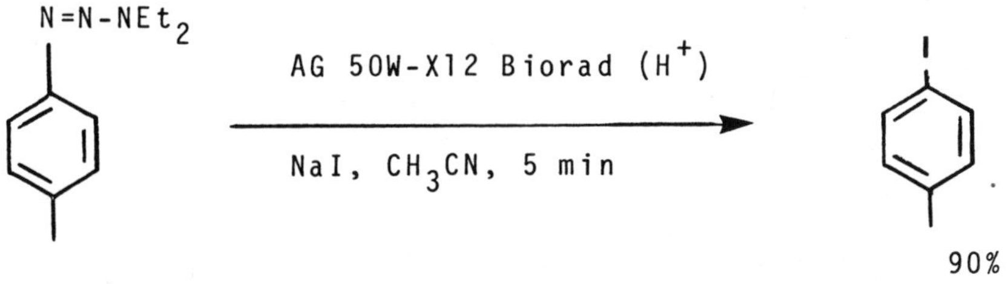

90%

Satyamurthy, N.; Barrio, J.R.[*]   **J Org Chem**, (1983), **48**, 4394

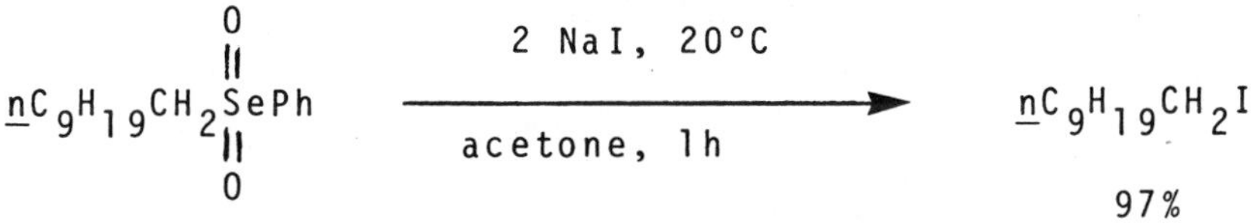

97%

Krief, A.[*]; Dumont, W.; Denis, J.-N.
                    JCS Chem Comm, (1985), 571

CHAPTER 11
PREPARATION OF HYDRIDES

This chapter lists hydrogenolysis and related reactions by which functional groups are replaced by hydrogens, e.g. $RCH_2X \rightarrow RCH_2$-H or R-H.

SECTION 151: Hydrides from Acetylenes

No Additional Examples

SECTION 152: Hydrides from Acid Derivatives

This section lists examples of decarboxylation (R-COOH → R-H) and related reactions.

$$CH_3(CH_2)_{16}COOH \xrightarrow[\text{2. } \underline{n}Bu_3SnH]{\text{1. } \quad , \text{ PhH, } 80°C} CH_3(CH_2)_{16}\text{-H}$$

70%

Barton, D.H.R.[*]; Crich, D.; Motherwell, W.B.
JCS Chem Comm, (1983), 939

Py/Cu	67%	($\underline{Z}:\underline{E}$ = 84:16)
Quinoline/Cu	75%	($\underline{Z}:\underline{E}$ = 16:84)

Schmidt, U.[*]; Lieberknecht, A.
Angew Chem Int Ed Engl, (1983), **22**, 550

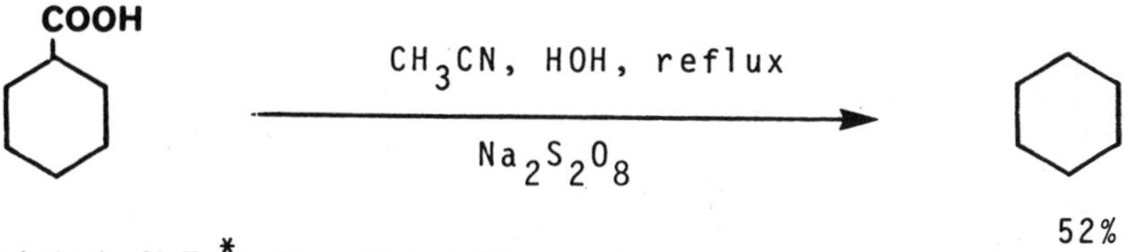

52%

Fristad, W.E.*; Fry, M.A.; Klang, J.A.
J Org Chem, (1983), **48**, 3575

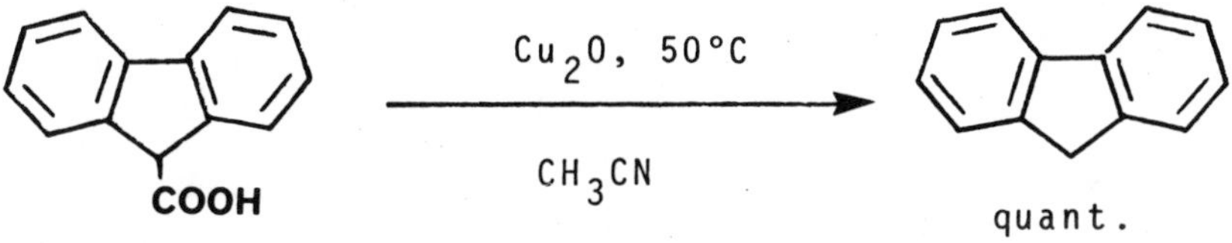

quant.

Toussaint, O.; Capdevielle, P.; Maumy, M.
Tetrahedron, (1984), **40**, 3229

SECTION 153: Hydrides from Alcohols and Thiols

This section lists examples of the hydrogenolysis of alcohols and phenols, ROH → RH.

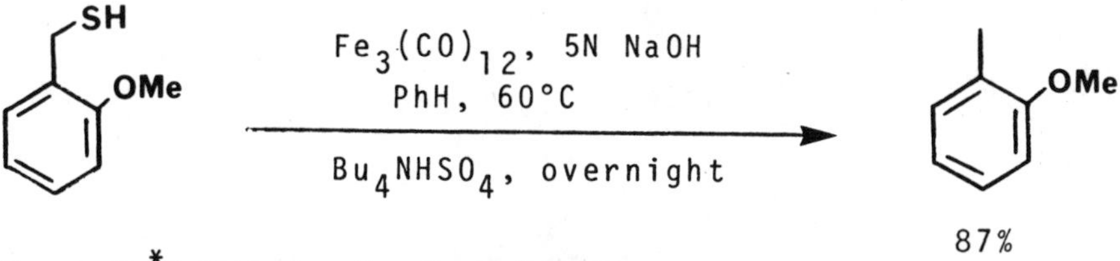

87%

Alper, H.*; Sibtain, F.; Heveling, J.
Tetrahedron Lett, (1983), **24**, 5329

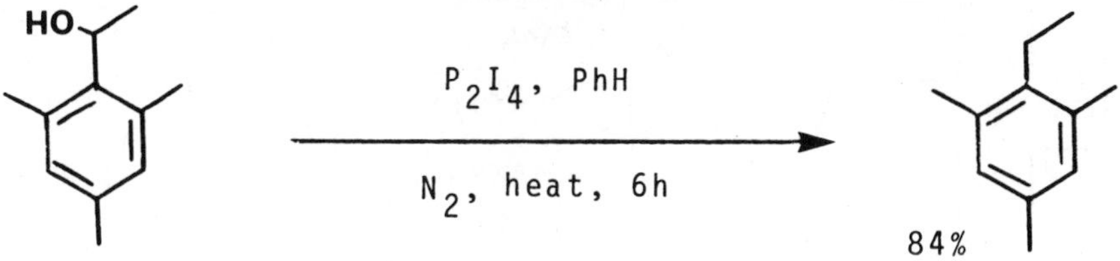

84%

Suzuki, H.; Tani, H.; Kubota, H.; Sato, N.; Tsuiji, J.; Osuka, A.

Chem Lett, (1983), 247

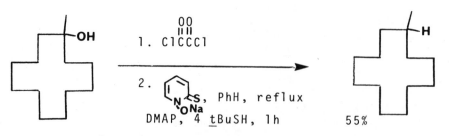

$$
\begin{array}{c}
\text{1. } \overset{\overset{O\ O}{||\ ||}}{ClCCCl}
\end{array}
$$

2. (structure), PhH, reflux
DMAP, 4 tBuSH, 1h   55%

Barton, D.H.R.[*]; Crich, D.   **JCS** **Chem** **Comm**, (1984), 774

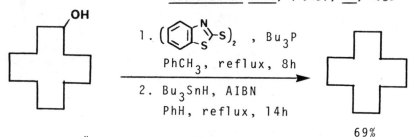

1. $\overset{\overset{O\ O}{||\ ||}}{ClCCOMe}$

2. Bu₃SnH

   AIBN, PhMe
   reflux
                                    65%

Dolan, S.C.; MacMillan, J.   **JCS** **Chem** **Comm**, (1985), 1588

(structure with SH)

$$\xrightarrow[\text{900 psi CO, 190°C}]{Co_2(CO)_8,\ HOH/PhH}$$

(structure)   84%

Shim, S.C.; Antebi, S.; Alper, H.[*]
                    Tetrahedron **Lett**, (1985), **26**, 1935

(structure with OH)

1. $\left(\text{(structure)} -S \right)_2$, Bu₃P

   PhCH₃, reflux, 8h

2. Bu₃SnH, AIBN
   PhH, reflux, 14h
                                    69%

Watanabe, Y.[*]; Araki, T.; Ueno, Y.; Endo, T.
                    Tetrahedron **Lett**, (1986), **27**, 5385

Review: "Radical Deoxygenation of Alcohols"

Hartwig, W.[*]   **Tetrahedron**, (1983), **39**, 2609

Also via: Halides and Sulfonates (Section 160)

SECTION 154: Hydrides from Aldehydes

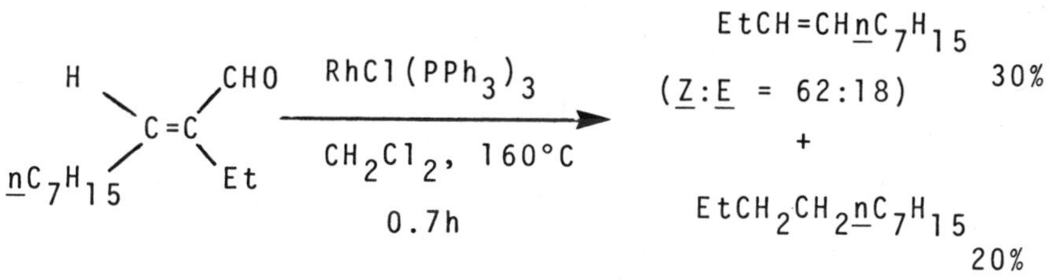

Grigor'eva, N.Ya.; Pinsker, O.A.; Semenovskii, A.V.[*]
Bull Acad Sci USSR, (1983), **32**, 593

For the conversion RCHO → RMe, etc. see Section 64 (Alkyls from Aldehydes).

SECTION 155: Hydrides from Alkyls, Methylenes, and Aryls

No Additional Examples

SECTION 156: Hydrides from Amides

No Additional Examples

SECTION 157: Hydrides from Amines

This section lists examples of the conversion RNH_2 → RH.

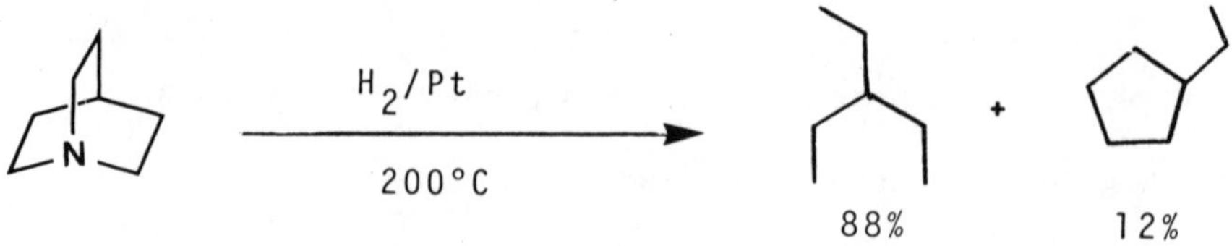
H_2/Pt
200°C
88%
12%

Guttieri, M.J.; Maier, W.F.* **J Org Chem**, (1984), **49**, 2875

SECTION 158: Hydrides from Esters

This section lists examples of the reactions RCOOR' → RH and RCOOR' → R'H.

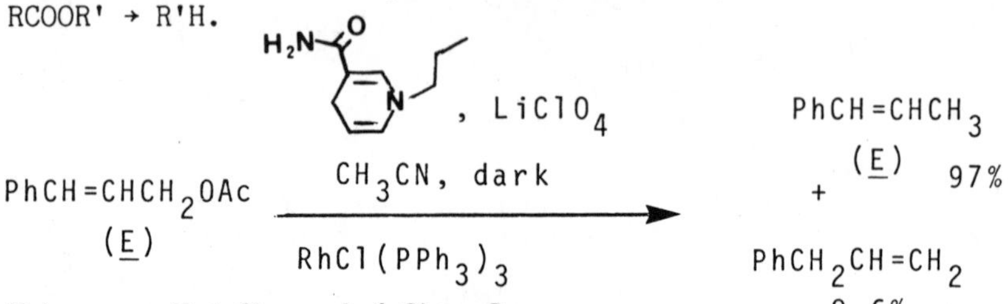
H_2N
O
N
, $LiClO_4$
CH_3CN, dark
$RhCl(PPh_3)_3$
PhCH=CHCH$_2$OAc
(E)
PhCH=CHCH$_3$
(E) 97%
+
PhCH$_2$CH=CH$_2$
0.6%

Nakamura, K.; Ohno, A.; Oka, S.
Tetrahedron Lett, (1983), **24**, 3335

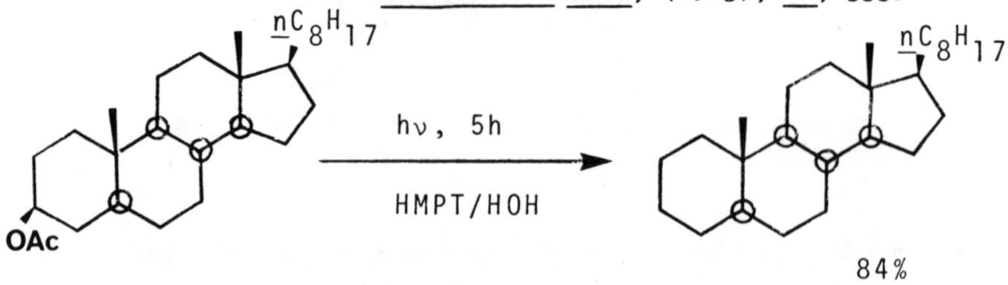
nC$_8$H$_{17}$
OAc
hν, 5h
HMPT/HOH
nC$_8$H$_{17}$
84%

Portella, C.; Deshayes, H.; Pete, J.P.; Scholler, D.
Tetrahedron, (1984), **40**, 3635

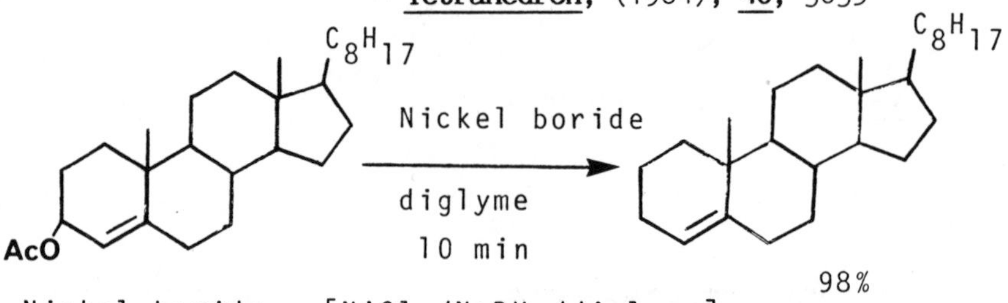
C$_8$H$_{17}$
AcO
Nickel boride
diglyme
10 min
C$_8$H$_{17}$
98%

Nickel boride = [$NiCl_2$/$NaBH_4$/diglyme]
Sarma, D.N.; Sharma, R.P.* **Tetrahedron Lett**, (1985), **26**, 2581

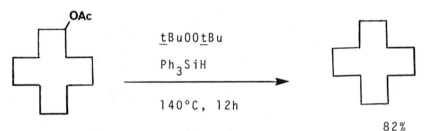

$$PhCH=CHCHOAc \xrightarrow[\substack{THF, RT, 1h \\ 1\% Pd(PPh_3)_4}]{SMI_2, \underline{t}BuOH} PhCH=CHCH_2CH_3 \ (\underline{E})$$

CH₃ over the CHOAc carbon.

PhCH=CHCH₂CH₃ (E)
(83)
+
                                        81%
PhCH₂CH=CHCH₃ (E+Z)
(17)

Tabuchi, T.; Inanaga, J.*; Yamaguchi, M.
                   Tetrahedron Lett, (1986), **27**, 601

OAc

$$\xrightarrow[140°C, 12h]{\substack{\underline{t}BuOO\underline{t}Bu \\ Ph_3SiH}}$$

                                        82%

Sano, H.*; Ogata, M.; Migita, T.*   **Chem Lett**, (1986), 77

SECTION 159: Hydrides from Ethers, Epoxides, and Thioethers

This section lists examples of the reaction of R-O-R' → R-H.

SePh

$$\xrightarrow[\substack{THF/MeOH, 0°C \\ 15 \ min}]{\substack{NiCl_2 \cdot 6H_2O \\ NaBH_4}}$$

                                        93%

Back, T.G.*   **JCS Chem Comm**, (1984), 1417

SECTION 160: Hydrides from Halides and Sulfonates

This section lists the reductions of halides and sulfonates,
RX → RH.

OMe
⟨benzene ring with I at para position⟩

1. LiAlH$_4$, THF, hν
   tBuOOtBu, 2.2h

2. H$_3$O$^+$

OMe
⟨benzene ring⟩

quant.

Beckwith, A.L.J.*; Goh, S.H.   **JCS Chem Comm**, (1983), 907

Et
|
nBuC-X   [2 NaBH$_3$CN, MCl$_2$], Et$_2$O
| ————————————————————————————→
Et   2.5h, RT

Et
|
nBuC-H
|
Et

X = Cl, M = Zn   96%

Kim, S.*; Kim, Y.J.; Ahn, K.H.
                          Tetrahedron Lett, (1983), **24**, 3369

X = Br, M = Sn   95%

Kim, S.*; Ko, J.S.   **Syn Commun**, (1985), **15**, 603

Br
|
PhCHPh   Zn(BH$_4$)$_2$
           ————————————————————→   PhCH$_2$Ph
              Et$_2$O, 0.2h
                                                    99%

Kim, S.*; Hong, C.Y.; Yang, S.
                          Angew Chem Int Ed Engl, (1983), **22**, 562

O
‖
PhCH$_2$SCH$_2$Br   Co$_2$(CO)$_8$, Al$_2$O$_3$
              ————————————————————————→
                   C$_6$H$_{14}$, RT, 2h

O
‖
PhCH$_2$SCH$_3$

quant.

Alper, H.*; Gopal, M.   **J Org Chem**, (1983), **48**, 4390

[Te, NaBH$_4$], EtOH

EtOAc, RT, 1h

78%

Osuka, A.; Suzuki, H. **Chem Lett**, (1983), 119

O
‖
PhCCHCH$_3$
|
Br

1. SnCl$_2$, 2 Dibal-H
   TMEDA, THF, PhCH$_3$

   -78°C, 10 min

2. pH 7

O
‖
PhCCH$_2$CH$_3$

81%

Oriyama, T.; Mukaiyama, T. **Chem Lett**, (1984), 2069

Cl

NaH$_2$PO$_2$, THF, 65°C

Na$_2$CO$_3$, cat. Pd-C
HOH

96%

Boyer, S.K.[*]; Bach, J.; McKenna, J.; Jagdmann Jr., E.
J Org Chem, (1985), **50**, 3408

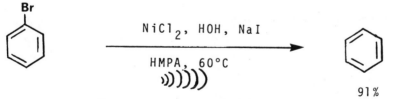

Br

NiCl$_2$, HOH, NaI

HMPA, 60°C

91%

Yamashita, J.[*]; Inoue, Y.; Kondo, T.; Hashimoto, H.
Bull Chem Soc Jpn, (1985), **58**, 2709

$$\text{Ph-C}_6\text{H}_4\text{-C}_6\text{H}_4\text{-Br} \xrightarrow[\substack{h\nu, \ DMF, \ 80°C \\ 6h}]{\underline{t}BuOO\underline{t}Bu, \ NaBH_4} \text{Ph-C}_6\text{H}_4\text{-C}_6\text{H}_5}$$

74%

Abeywickrema, A.N.; Beckwith, A.L.J.
Tetrahedron Lett, (1986), 27, 109

$$\underset{\substack{|| \\ PhCCH_2Br}}{\overset{O}{}} \xrightarrow[\substack{PMHS, \ CH_3CN/DMSO \\ \hline Bn_3N, \ 3h}]{cat. \ Pd(PPh_3)_4, \ 160°C} \underset{\substack{|| \\ PhCCH_3}}{\overset{O}{}}$$

80%

PMHS = polymethyl hydrosiloxane

Pri-Bar, I.[*]; Buchman, O.   J Org Chem, (1986), 51, 734

$$\text{naphthyl-OTf} \xrightarrow[\substack{4\% \ PPh_3, \ 60°C, \ 1h}]{2\% \ Pd(OAc)_2, \ NEt_3} \text{naphthalene}$$

91%

Cacchi, S.; Ciattini, P.G.; Morera, E.; Ortar, G.[*]
Tetrahedron Lett, (1986), 27, 5541

SECTION 161: Hydrides from Hydrides

No Additional Examples

SECTION 162: Hydrides from Ketones

This section lists examples of the reaction $R_2CCOR \rightarrow R_2C\text{-}H$.

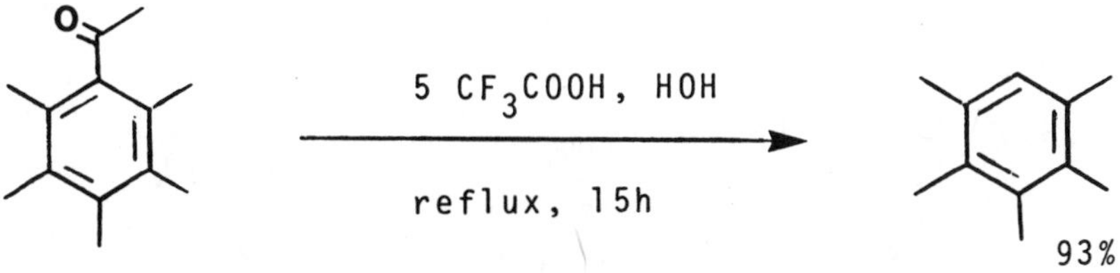

5 CF_3COOH, HOH

reflux, 15h

93%

Keumi, T.[*]; Morita, T.; Inui, Y.; Teshima, N.; Kitajima, H.
Synthesis, (1985), 979

SECTION 163: Hydrides from Nitriles

This section lists examples of the reaction RCN → RH.

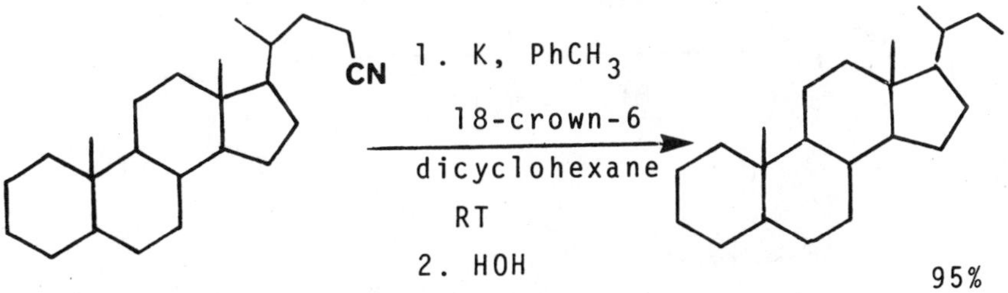

1. K, $PhCH_3$

18-crown-6

dicyclohexane

RT

2. HOH

95%

Ohsawa, T.[*]; Kobayashi, T.; Mizuguchi, Y.; Saitoh, T.; Oishi, T.[*]

Tetrahedron Lett, (1985), **26**, 6103

SECTION 164: Hydrides from Olefins

No Additional Examples

SECTION 165: Hydrides from Miscellaneous Compounds

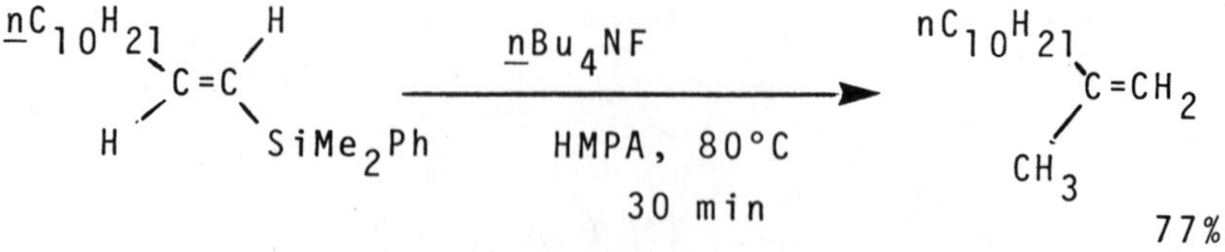

$$\underset{H}{\overset{\underline{n}C_{10}H_{21}}{}}C=C\underset{SiMe_2Ph}{\overset{H}{}} \quad \xrightarrow[\substack{HMPA, \ 80°C \\ 30 \ min}]{\underline{n}Bu_4NF} \quad \underset{CH_3}{\overset{nC_{10}H_{21}}{}}C=CH_2$$

77%

Oda, H.; Sato, M.; Morizawa, Y.; Oshima, K.*; Nozaki, H.
                    Tetrahedron Lett, (1983), **24**, 2877

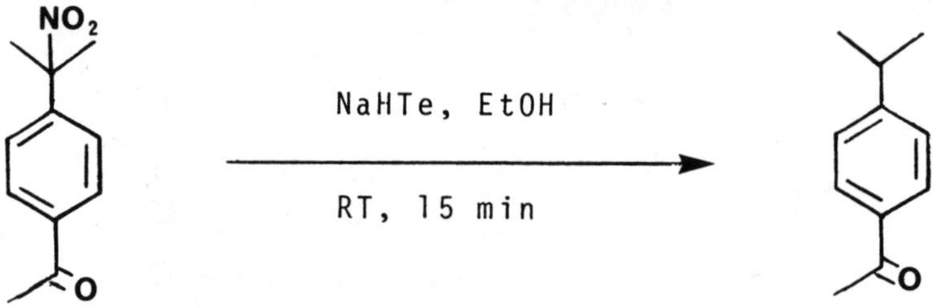

NaHTe, EtOH

RT, 15 min

85%

Suzuki, H.*; Takaoka, K.; Osuka, A.
                    Bull Chem Soc Jpn, (1985), **58**, 1067

CHAPTER 12
PREPARATION OF KETONES

SECTION 166: Ketones from Acetylenes

$$\underline{n}BuC\equiv C\underline{n}Bu \xrightarrow[\substack{3.\ 30\%\ H_2O_2,\ KHCO_3 \\ MeOH/THF,\ reflux}]{\substack{1.\ HSiMe(OEt)_2 \\ 2.\ H_2PtCl_6\cdot 6H_2O \\ RT,\ 30\ min}} \underset{82\%}{\underline{n}BuCH_2\overset{O}{\overset{\|}{C}}\underline{n}Bu}$$

Tamao, K.[*]; Ishida, N.; Tanaka, T.; Kumada, M.
Organometallics, (1983), **2**, 1694

$$BnOCH_2C\equiv CH \xrightarrow[\substack{2.\ H_3O^+}]{\substack{1.\ HgCl_2,\ K_2CO_3 \\ THF,\ RT}} \underset{60\%}{BnOCH_2\overset{O}{\overset{\|}{C}}CH_3}$$

Barluenga, J.[*]; Aznar, F.; Liz, R.   **Synthesis**, (1984), 304

$$\underline{n}BuC\equiv CH \xrightarrow[\substack{2.\ 95^{\circ}C,\ 15h}]{\substack{1.\ \ \ \ \ \ \ \ \ \ C=Cr(CO)_5 \\ DMF,\ 125^{\circ}C,\ 5h}}$$

95%

Yamashita, A.[*]   **Tetrahedron Lett**, (1986), **27**, 5915

SECTION 167: Ketones from Acid Derivatives

1. $Cp_2Ti \overset{CH_2}{\underset{Cl}{\diagdown}} AlMe_2$

2. HOH

35%

Chou, T.-S.[*]; Huang, S.-B.   **Tetrahedron Lett**, (1983), **24**, 2169

$CH_3CH_2CH_2COOH$ $\xrightarrow[\substack{Ni(dppe)Cl_2 \\ reflux}]{PhMgX, THF}$ $CH_3CH_2CH_2\overset{O}{\overset{\|}{C}}Ph$

65%

Fiandanese, V.; Marchese, G.[*]; Ronzini, L.
Tetrahedron Lett, (1983), **24**, 3677

$Ph\overset{O}{\overset{\|}{C}}Cl$ $\xrightarrow[\substack{Ni^\circ, 45 min}]{PhCH_2Cl, glyme, 85°C}$ $PhCH_2\overset{O}{\overset{\|}{C}}Ph$

73%

Inaba, S.; Rieke, R.D.[*]   **J Org Chem**, (1985), **50**, 1373

$Ph\overset{O}{\overset{\|}{C}}Cl$ $\xrightarrow[\substack{5\% Pd(PPh_3)_4}]{\substack{Cl-\bigcirc-ZnCl, THF \\ 25°C, 6h}}$

95%

Negishi, E.[*]; Bagheri, V.; Chatterjee, S.; Luo, F.-T.; Miller, J.A.; Stoll, A.T.
Tetrahedron Lett, (1983), **24**, 5181

$$\underline{n}C_3H_7\overset{\overset{\textstyle O}{\|}}{C}Cl \quad \xrightarrow[\text{2. 3N HCl, 16°C}]{\begin{array}{c}\text{1. iPr}_2\text{Zn, Et}_2\text{O, 0-23°C}\\ \text{BnPdCl(PPh}_3)_2\end{array}} \quad \underline{n}C_3H_7\overset{\overset{\textstyle O}{\|}}{C}iPr$$

98%

Grey, R.A. **J Org Chem**, (1984), **49**, 2288

$$Ph\overset{\overset{\textstyle O}{\|}}{C}Cl \quad \xrightarrow[\begin{array}{c}\text{cat. Pd(PPh}_3)_4\\ \text{PhH-DMF}\end{array}]{\text{CH}_3\text{CH}_2\text{CH}_2\text{I, Zn-Cu}} \quad Ph\overset{\overset{\textstyle O}{\|}}{C}CH_2CH_2CH_3$$

85%

Tamaru, Y.; Ochiai, H.; Sanda, F.; Yoshida, Z.[*]
Tetrahedron Lett, (1985), **26**, 5529

$$PhCH_2\overset{\overset{\textstyle O}{\|}}{C}Cl \quad \xrightarrow[\text{2. HCl}_{(g)}, \text{ -10°C}]{\text{1. Cp}_2\text{Ti=CH}_2, \text{ -20 - 3°C}} \quad PhCH_2\overset{\overset{\textstyle O}{\|}}{C}CH_3$$

97%

Stille, J.R.; Grubbs, R.H.[*]   **J Am Chem Soc**, (1983), **105**, 1664

$$CH_3(CH_2)_4COOH \quad \xrightarrow[\text{3. aq. HCl}]{\begin{array}{c}\text{1. ClCH}_2\text{C} \diagdown \text{N} \diagup \\ \text{CH}_2\\ \text{CH}_2\text{Cl}_2/\text{THF, 0°C}\\ \text{2. CuI, EtCH(Me)MgCl}\end{array}} \quad CH_3(CH_2)_4\overset{\overset{\textstyle O}{\|}}{C}\underset{\underset{\textstyle Me}{|}}{C}HEt$$

77%

Fujisawa, T.[*]; Mori, T.; Higuchi, K.; Sato, T.
Chem Lett, (1983), **1791**

$Ph_2CHCOOH$ $\xrightarrow[\text{CH}_3\text{CN, 50°C, 15h}]{\text{Cu}_2\text{O/O}_2}$

$$\underset{\text{PhCPh}}{\overset{\text{O}}{\|}}$$

57%

Toussaint, O.; Capdevielle, P.; Maumy, M.
Tetrahedron Lett, (1984), **25**, 3819

$\underset{CO_2SiMe_3}{\overset{CO_2SiMe_3}{\underset{|}{\overset{|}{CH_2}}}}$ 1. $MgCl_2$, NEt_3, Et_2O
2. $Ph\overset{O}{\overset{\|}{C}}Cl$, 0°C, 1h
$\xrightarrow{\hspace{3cm}}$
3. 5N HCl, reflux

$$\underset{\text{PhCCH}_3}{\overset{\text{O}}{\|}}$$

90%

Rathke, M.W.[*]; Nowak, M.A.   **Syn Commun**, (1985), **15**, 1039

$\underset{\overset{|}{\underset{\|}{C-Cl}}\\{O}}{\overset{CH_2OAc}{\underset{|}{(CH_2)_4}}}$ 1. nBuMnI, Et_2O
−50°C − RT
$\xrightarrow{\hspace{3cm}}$
2. aq. HCl, 0°C
2h

$\underset{O=\overset{C}{\underset{\text{nBu}}{\diagdown}}}{\overset{CH_2OAc}{\underset{|}{(CH_2)_4}}}$

95%

Friour, G.; Cahiez, G.[*]; Normant, J.F.   **Synthesis**, (1985), 50

$$\underset{\text{PhCCl}}{\overset{\text{O}}{\|}}$$ $\xrightarrow[\text{Pd(PPh}_3)_4]{\text{Et}_3\text{Al, THF, 25°C}}$

$$\underset{\text{PhCEt}}{\overset{\text{O}}{\|}}$$

70%

Wakamatsu, K.; Okuda, Y.; Oshima, K.[*]; Nozaki, H.
Bull Chem Soc Jpn, (1985), **58**, 2425

$$\underset{PhSCCl}{\overset{O}{\underset{\|}{}}} \quad \xrightarrow[\substack{2.\ \underline{n}BuMgBr,\ Fe(acac)_3 \\ THF,\ 0°C}]{\substack{1.\ \underline{n}C_7H_{15}MgBr,\ THF,\ RT \\ Ni(dppe)Cl_2}} \quad \underline{n}C_7H_{15}\overset{O}{\overset{\|}{C}}\underline{n}Bu$$

84%

Cardellicchio, C.; Fiandanese, V.; Marchese, G.[*]; Ronzini, L.
Tetrahedron Lett, (1985), **26**, 3595

SECTION 168: Ketones from Alcohols and Thiols

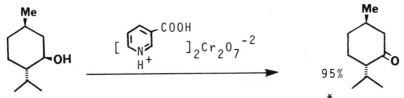

1. $Cr(CO)_6$
2. Ac_2O/Py

3. $CH_2CHMeOTMS$, 22°C
   $ZnCl_2$, CH_2Cl_2
4. I_2

67%

Reetz, M.T.[*]; Sauerwald, M.   **Tetrahedron Lett**, (1983), **24**, 2837

$$\left[\underset{H^+}{\text{pyridine-COOH}} \right]_2 Cr_2O_7^{-2}$$

95%

Lopez, C.; Gonzalez, A.; Cossio, F.P.; Palomo, C.[*]
Syn Commun, (1985), **15**, 1197

2%   (chromium complex)

CH_2Cl_2-CCl_4, 0°C

2 CH_3CO_3H, 30 min

96%

Corey, E.J.[*]; Barrette, E.P.; Magriotis, P.A.
Tetrahedron Lett, (1985), **26**, 5855

$$4 \ [\ HN \overset{\frown}{\underset{\smile}{\hspace{1.2em}}} NH]^+ CrO_3Cl$$

$$CH_2Cl_2, \ 7h$$

82%

Davis, H.B.; Sheets, R.M.; Brannfors, J.M.; Paudler, W.W.; Gard, G.L.
Heterocycles, (1983), **20**, 2029

5% Pd(OAc)$_2$
6% PPh$_3$, NaH, THF

PhBr, 50°C

80%

Tamaru, Y.; Yamada, Y.; Inoue, K.; Yamamoto, Y.; Yoshida, Z.[*]
J Org Chem, (1983), **48**, 1286

Pd(OAc)$_2$/NaHCO$_3$

PhI, Bu$_4$NCl, DMF
RT, 48h

95%

Choudary, B.M.[*]; Reddy, N.P.; Kantam, M.L.; Jamil, Z.
Tetrahedron Lett, (1985), **26**, 6257

$$CH_3CH(CH_2)_8CH_2OH \xrightarrow[\substack{30\% \ H_2O_2, \ K_2CO_3 \\ Bu_4NCl}]{\substack{(NH_4)_6Mo_7O_{24}\cdot 2H_2O \\ THF, \ 7d, \ RT}} CH_3C(CH_2)_8CH_2OH$$

with OH on first and O on the product carbonyl

79%

Trost, B.M.[*]; Masuyama, Y.   **Tetrahedron Lett**, (1984), **25**, 173

$$\underset{\overset{|}{OH}}{\underset{\overset{|}{HOCH_2CHCH\underline{n}Pr}}{\overset{\overset{Et}{|}}{}}} \xrightarrow[\text{80°C, 24h}]{\text{3 } \underline{t}\text{BuOOH, PhH}} \underset{\overset{||}{O}}{\overset{\overset{Et}{|}}{HOCH_2CHCnPr}}$$

97%

Kaneda, K.; Kawanishi, Y.; Jitsukawa, K.; Teranishi, S.
                        Tetrahedron Lett, (1983), **24**, 5009

Mo(CO)$_6$, cetylPy·HCl
MgSO$_4$, 4h   96%

Yamawaki, K.; Yoshida, T.; Suda, T.; Ishii, Y.[*]; Ogawa, M.
                        Synthesis, (1986), 59

PMo$_{12}$O$_{40}$]$^{-3}$cetylPy$_3$ 75°C
18h   quant.

Yamawaki, K.; Yoshida, T.; Nishihara, H.; Ishii, Y.[*]; Ogawa, M.[*]
                        Syn Commun, (1986), **16**, 537

$$\xrightarrow[\text{RT, 30 min}]{\text{NaBrO}_2, \text{ aq. AcOH}}$$

84%

Kageyama, T.; Ueno, Y.[*]; Okawara, M.   **Synthesis**, (1983), 815

$$\underset{}{\overset{\overset{OH}{|}}{CH_3(CH_2)_4CHCH_3}} \xrightarrow[\text{NaBr, 60°C, 1h}]{\text{AcOH, Co(OAc)}_2} \overset{\overset{O}{||}}{CH_3(CH_2)_4CCH_3}$$

67%

Morimoto, T.[*]; Hirano, M.; Wachi, M.; Murakami, T.
                        JCS Perkin II, (1983), 1949

$$\xrightarrow[\text{reflux, 15h}]{\text{BaMnO}_4, \text{ PhH}}$$

90%

Firouzabadi, H.[*]; Mostafavipoor, Z.
                        Bull Chem Soc Jpn, (1983), **56**, 914

$$CH_2=CHCHCH_3 \text{ (OH)} \xrightarrow[\text{PhH, reflux, 1h}]{5\% \underline{trans} \text{ Mo}(N_2)_2(dpe)_2} CH_3\overset{O}{\overset{\|}{C}}CH_2CH_3$$

95%

Tatsumi, T.; Hashimoto, K.; Tominaga, H.; Mizuta, Y.; Hata, K.;
Hidai, M.; Uchida, Y.
J Organomet Chem, (1983), <u>252</u>, 105

1. BF$_3$·OEt$_2$, DCC
   CH$_2$Cl$_2$, RT, 1h
2. PhIO, 0°C-RT
   4h

$$PhC\overset{O}{\overset{\|}{}}(CH_2)_3CH=CH_2$$

81%

Ochiai, M.; Ukita, T.; Nagao, Y.; Fujita, E.[*]
JCS Chem Comm, (1984), 1007

1. Cl$\overset{O}{\overset{\|}{C}}OCH_2$CH=CH$_2$
2. Pd(OAc)$_2$, CH$_3$CN

80°C, 2h

76%

Tsuji, J.[*]; Minami, I.; Shimizu, I.
Tetrahedron Lett, (1984), <u>25</u>, 2791

$$PhCHCH_3 \text{ (OH)} \xrightarrow[\text{h}\nu, \text{ 15h}]{Pt-TiO_2, \text{ PhH}} Ph\overset{O}{\overset{\|}{C}}CH_3$$

83%

Hussein, F.H.; Pattenden, G.[*]; Rudham, R.; Russell, J.J.
Tetrahedron Lett, (1984), <u>25</u>, 3363

$$BnNMe_3^+ \ ^-OMoBr_4$$

$$CH_3(CH_2)_6\overset{\overset{\displaystyle OH}{|}}{CH}(CH_2)_3OH \xrightarrow[\ 60°C, \ 24h\ ]{\underline{t}BuOOH \ , \ THF} CH_3(CH_2)_6\overset{\overset{\displaystyle O}{||}}{C}(CH_2)_3OH$$

60%

Masuyama, Y.[*]; Takahashi, M.; Kurusu, Y.
Tetrahedron Lett, (1984), **25**, 4417

$$\xrightarrow[\substack{CH_2Cl_2, \ 30°C \\ 1.5h}]{cetylNMe_3^+ \ ^-MnO_4}$$

90%

Rathore, R.; Bhushan, V.; Chandrasekaran, S.[*]
Chem Lett, (1984), 2131

$$CH_3(CH_2)_5\overset{\overset{\displaystyle OH}{|}}{CH}CH_3 \xrightarrow[\substack{Aliquat \ 336, \ RT \\ CHCl_3/HOH, \ 4.5h}]{RuCl_3(H_2O)_n/NaBrO_3} CH_3(CH_2)_5\overset{\overset{\displaystyle O}{||}}{C}CH_3$$

82%

Yamamoto, Y.; Suzuki, H.[*]; Moro-Oka, Y.[*]
Tetrahedron Lett, (1985), **26**, 2107

$$\overset{\overset{\displaystyle OH}{|}}{PhCHPh} \xrightarrow[\ reflux, \ 0.8h\ ]{3 \ Ag_2FeO_4, \ PhH} \overset{\overset{\displaystyle O}{||}}{PhCPh}$$

97%

Firouzabadi, H.; Mohajer, D.; Moghaddam, M.E.
Syn Commun, (1986), **16**, 211

$$CH_3(CH_2)_7\overset{\overset{\displaystyle OH}{|}}{C}HnPr \xrightarrow{\text{CAN}} CH_3(CH_2)_7\overset{\overset{\displaystyle O}{||}}{C}nPr$$

94%

Kanemoto, S.; Tomioka, H.; Oshima, K.[*]; Nozaki, H.
Bull Chem Soc Jpn, (1986), **59**, 105

$$\xrightarrow[\substack{\text{PhH, reflux} \\ 24h}]{\text{BaFeO}_4}$$

80%

Firouzabadi, H.[*]; Mohajer, D.; Enterzari-Moghaddam, M.
Syn Commun, (1986), **16**, 723

$$CH_3\overset{\overset{\displaystyle OH}{|}}{C}H(CH_2)_5CH_3 \xrightarrow[\substack{\text{Dean Stark trap} \\ 3h}]{\text{PhH, Ni(R), reflux}} CH_3\overset{\overset{\displaystyle O}{||}}{C}(CH_2)_5CH_3$$

93%

Krafft, M.E.[*]; Zorc, B.   **J Org Chem**, (1986), **51**, 5482

Review: "Chromium (VI) Based Oxidants"

Firouzabadi, H.[*]; Iranpoor, N.; Kiaeezadeh, F.; Toofan, J.
Tetrahedron, (1986), **42**, 719

Related Methods: Aldehydes from Alcohols and Phenols (Section 48)

SECTION 169: Ketones from Aldehydes

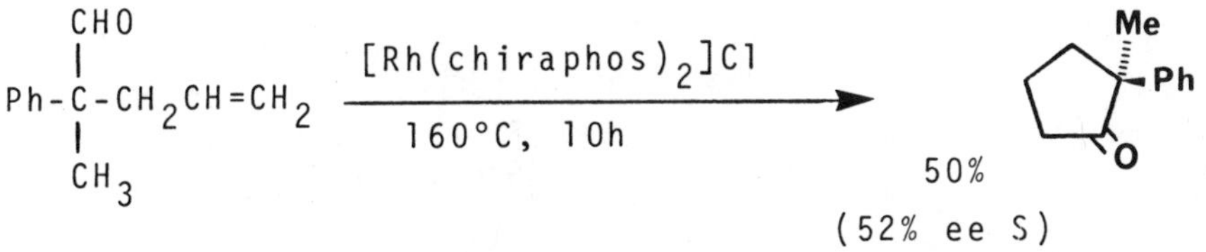
CHO
Ph-C-CH2CH=CH2
CH3
[Rh(chiraphos)2]Cl
160°C, 10h
Me
Ph
O
50%
(52% ee S)

chiraphos = 2S, 3S-bis-(diphenylphosphino) butane

James, B.R.*; Young, C.G. **JCS Chem Comm**, (1983), 1215

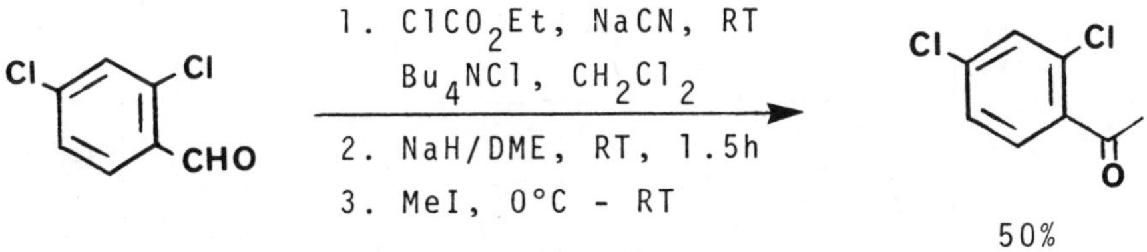
Cl
Cl
CHO
1. ClCO2Et, NaCN, RT
Bu4NCl, CH2Cl2
2. NaH/DME, RT, 1.5h
3. MeI, 0°C - RT
Cl
Cl
O
50%

Au, A.T.* **Syn Commun**, (1984), **14**, 743, 749

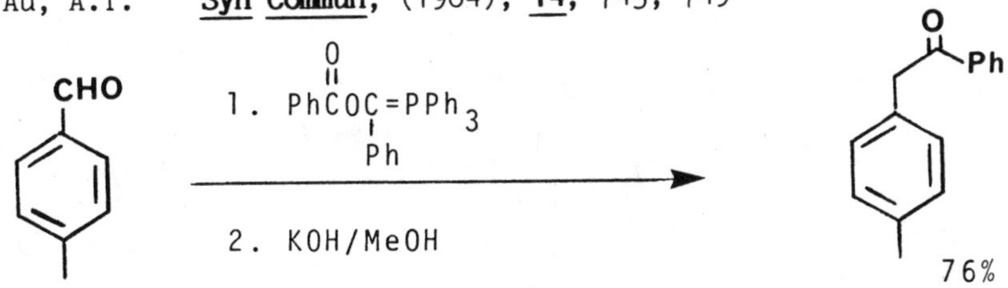
CHO
1. PhCOC(Ph)=PPh3
2. KOH/MeOH
O
Ph
76%

Anders, E.*; Gassner, T. **Chem Ber**, (1984), **117**, 1034

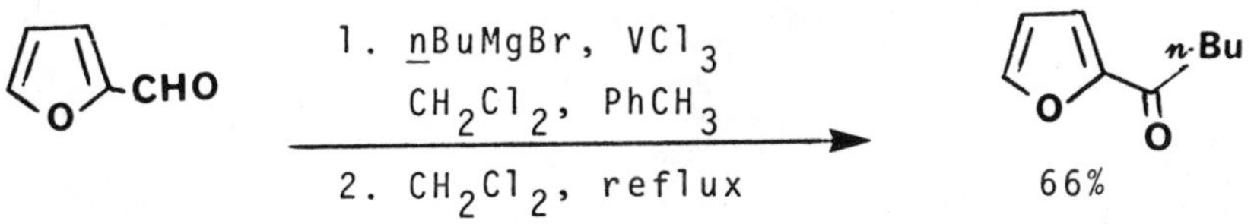
CHO
O
1. nBuMgBr, VCl3
CH2Cl2, PhCH3
2. CH2Cl2, reflux
n-Bu
O
O
66%

Hirao, T.*; Misu, D.; Agawa, T.
J Am Chem Soc, (1985), **107**, 7179

$$CH_3CHO \xrightarrow[\text{MeO}^- \text{K}^+, \ K_2CO_3]{}$$

72%

Kirmse, W.[*]; Hellwig, G.; Van Chiem, P.
Chem Ber, (1986), **119**, 1511

SECTION 170: Ketones from Alkyls, Methylenes, and Aryls

This section lists examples of the reaction R-CH$_2$-R' →
R-C(=O)-R'

1. nBuLi
2. CuBr·SMe$_2$
3. CH$_2$CHCCl, THF, 2h
   $\parallel$
   O   -78°C-RT

62%

Beswick, P.J.; Leach, S.J.; Masters, N.F.; Widdowson, D.A.[*]
JCS Chem Comm, (1984), 46

RuCl$_3$, NaIO$_4$, 21h

CH$_3$CN, RT, pH 7

62%

Hasegawa, T.; Niwa, H.; Yamada, K.[*]   **Chem Lett**, (1985), 1385

0.1   , tBuOOH

PhCH$_2$Ph   ─────────────────────────→

CH$_2$Cl$_2$/CCl$_4$, 0°C, 8h

O
$\parallel$
PhCPh
87%
(46% conversion)

Muzart, J.   **Tetrahedron Lett**, (1986), **27**, 3139

SECTION 171: Ketones from Amides

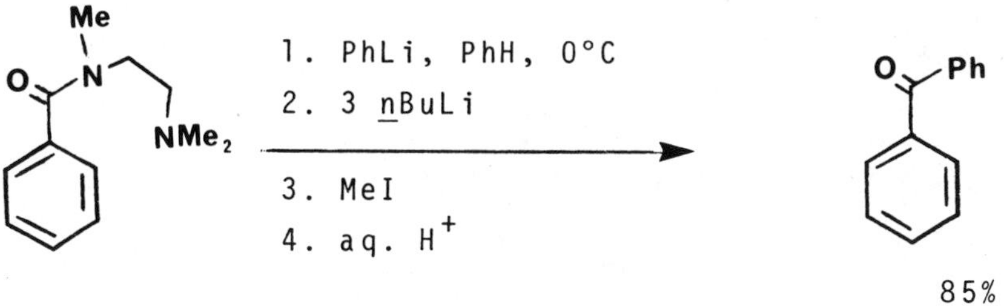

1. PhLi, PhH, 0°C
2. 3 nBuLi

3. MeI
4. aq. H$^+$

85%

Comins, D.L.*; Brown, J.D.   **Tetrahedron Lett**, (1983), **24**, 5465

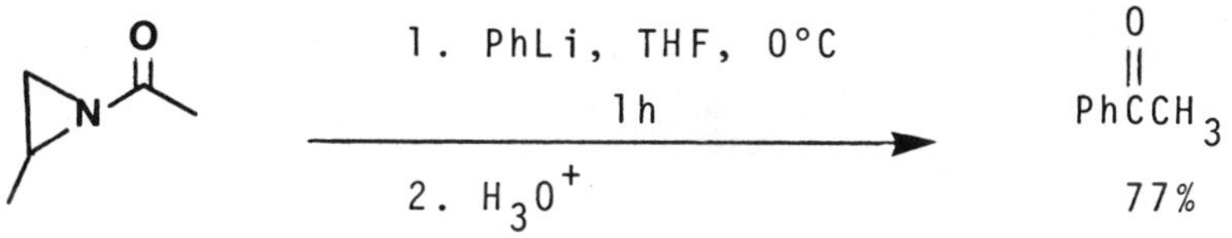

1. PhLi, THF, 0°C
      1h

2. H$_3$O$^+$

$$PhCCH_3$$

77%

Wattanasin, S.; Kathawala, F.G.
                    Tetrahedron Lett, (1984), **25**, 811

SECTION 172: Ketones from Amines

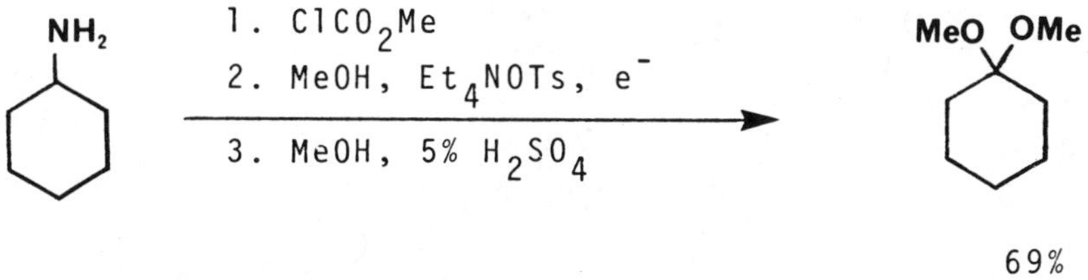

1. ClCO$_2$Me
2. MeOH, Et$_4$NOTs, e$^-$

3. MeOH, 5% H$_2$SO$_4$

69%

Shono, T.*; Matsumura, Y.; Kashimura, S.
                J Org Chem, (1983), **48**, 3338

SECTION 173: Ketones from Esters

$$CH_3(CH_2)_6\overset{\overset{O}{\|}}{C}SeMe \quad \xrightarrow[-78°C,\ 15\ min]{\underline{n}Bu_2CuLi,\ Et_2O} \quad CH_3(CH_2)_6\overset{\overset{O}{\|}}{C}\underline{n}Bu$$

96%

Sviridov, A.F.; Ermolenko, M.S.; Yashunsky, D.V.; Kochetkov, N.K.[*]

Tetrahedron Lett, (1983), **24**, 4355, 4359

$$\xrightarrow[30\ min]{\underline{n}Bu_2CuLi,\ Et_2O} \quad \underline{n}BuC\overset{\overset{O}{\|}}{(CH_2)_5}Br$$

90%

Kim, S.[*]; Lee, J.I.   **J Org Chem**, (1983), **48**, 2608

1. LiCH($\underline{n}C_5H_{11}$)$\overset{\overset{O}{\|}}{P}(OEt)_2$
   THF, -78°C

2. NH$_4$Cl, HOH, RT

3. (COCl)$_2$, DMSO, NEt$_3$
   CH$_2$Cl$_2$, -60°C

4. NaH, THF, RT

27%

Attenbach, H.-J.[*]; Holzapfel, W.; Smeret, G.; Finkler, S.H.
Tetrahedron Lett, (1985), **26**, 6329

1. CH$_3$CH$_2$CHMgBr (CH$_3$)
   CH$_2$Cl$_2$/THF, 0°C

2. HOH

$$PhC\overset{\overset{O}{\|}}{C}CHCH_2CH_3 \quad (CH_3)$$

72%

Miyasaka, T.[*]; Monobe, H.; Noguchi, S.   **Chem Lett**, (1986), 449

$$\underset{PhCOCH_2CH_3}{\overset{O}{\parallel}}$$

1. $Me_3SnCH_2^-$, THF
   $-78°C$, 3h
2. aq. NH_4Cl

$$\underset{PhCCH_3}{\overset{O}{\parallel}}$$

84%

Sato, T.*; Matsuoka, H.; Igarashi, T.; Murayama, E.
Tetrahedron Lett, (1986), **27**, 4339

SECTION 174: Ketones from Ethers, Epoxides, and Thioethers

$$Me_2\overset{O}{\overset{/\backslash}{C}}-CH(CH_2)_2\overset{\overset{CH_3}{|}}{\underset{BnOCH_2CH_2}{CH}}$$

5% $LiClO_4$, e^-
CH_2Cl_2/THF, RT
Pt electrodes   91%

$$Me_2\overset{O}{\overset{\parallel}{C}}HC(CH_2)_2\overset{\overset{CH_3}{|}}{\underset{BnOCH_2CH_2}{CH}}$$

Uneyama, K.; Isimura, A.; Fujii, K.; Torii, S.*
Tetrahedron Lett, (1983), **24**, 2857

$$\underset{PhCHPh}{\overset{OSiMe_3}{|}}$$

CrO_3, H_2SO_4
acetone

$$\underset{PhCPh}{\overset{O}{\parallel}}$$

82%

Baker, R.*; Rao, V.B.; Ravenscroft, P.D.; Swain, C.J.
Synthesis, (1983), 572

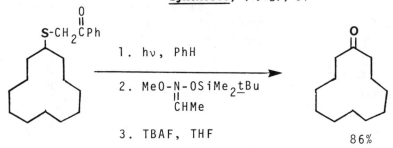

1. hν, PhH
2. MeO-N-OSiMe$_2$tBu
   $\overset{\parallel}{CHMe}$
3. TBAF, THF   86%

Vedejs, E.*; Perry, D.A.   **J Org Chem**, (1984), **49**, 573

$$PhS \overset{O}{\underset{Bn}{\diagdown}} \overset{O}{\diagup} C-CHMe \qquad \xrightarrow[\text{RT, 20 min}]{\text{3 NaSePh}} \qquad Bn\overset{O}{\overset{\|}{C}}CH_2CH_3$$

92%

Satoh, T.; Kaneko, Y.; Kumagawa, T.; Izawa, T.; Sakata, K.; Yamakawa, K.[*]

Chem Lett, (1984), 1957

Satoh, T.; Kaneko, Y.; Izawa, T.; Sakata, K.; Yamakawa, K.[*]
Bull Chem Soc Jpn, (1985), **58**, 1983

$$\underline{n}C_8H_{17}\overset{O}{\overset{\diagup \diagdown}{CH-CH_2}} \qquad \xrightarrow[\text{RT, 20h}]{\text{SmIO, THF}} \qquad \underline{n}C_8H_{17}\overset{O}{\overset{\|}{C}}CH_3$$

85%

SmIO = 2 SmI$_2$ + 1/2 O$_2$
Prandi, J.; Namy, J.L.; Menoret, G.; Kagan, H.B.[*]
J Organomet Chem, (1985), **285**, 449

SECTION 175: Ketones from Halides and Sulfonates

$$CH_3CH_2CH_2I \qquad \xrightarrow[\text{Zn-Cu, THF, 50°C, 1d}]{\text{PhI, CO, Pd(PPh}_3)_4} \qquad Ph\overset{O}{\overset{\|}{C}}CH_2CH_2CH_3$$

(75% conversion) 90%

Tamaru, Y.; Ochiai, H.; Yamada, Y.; Yoshida, Z.[*]
Tetrahedron Lett, (1983), **24**, 3869

$$CH_3(CH_2)_6Cl \qquad \xrightarrow[\substack{\text{3. Et}\overset{O}{\overset{\|}{C}})_2O, \text{ THF:PhH}}]{\substack{\text{1. Mg, THF} \\ \text{2. MnCl}_2, \text{ 2 LiCl, THF}}} \qquad CH_3(CH_2)_6\overset{O}{\overset{\|}{C}}Et$$

82%

Friour, G.; Alexakis, A.; Cahiez, G.[*]; Normant, J.[*]
Tetrahedron, (1984), **40**, 683

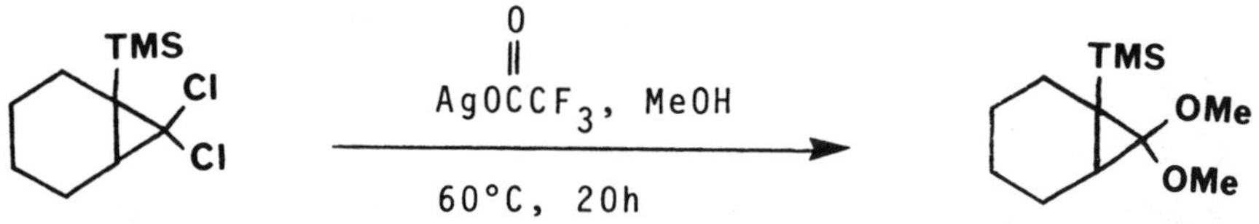
TMS
Cl
Cl
O
AgOCCF3, MeOH
60°C, 20h
TMS
OMe
OMe

66%

Ishihara, T.*; Kudaka, T.; Ando, T.
Tetrahedron Lett, (1984), 25, 4765

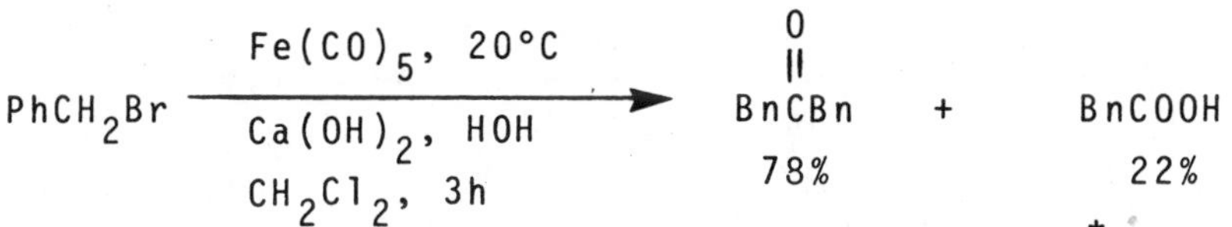
PhCH2Br
Fe(CO)5, 20°C
Ca(OH)2, HOH
CH2Cl2, 3h
O
BnCBn
78%
+
BnCOOH
22%

Tanguay, G.; Weinberger, B.; Des Abbayes, H.*
Tetrahedron Lett, (1984), 25, 5529

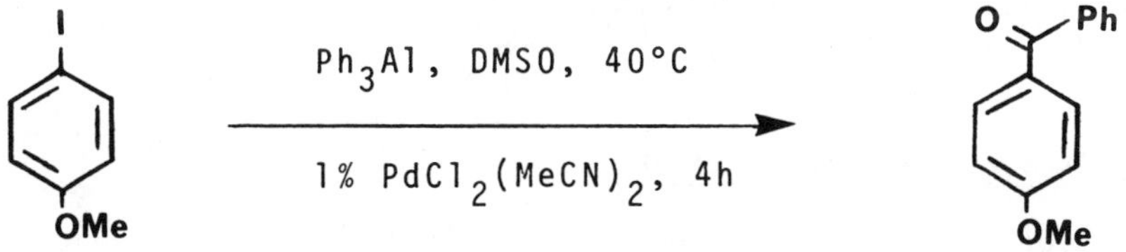
I
OMe
Ph3Al, DMSO, 40°C
1% PdCl2(MeCN)2, 4h
O
Ph
OMe

98%

Bumagin, N.A.; Ponomaryov, A.B.; Beletskaya, I.P.*
Tetrahedron Lett, (1985), 26, 4819

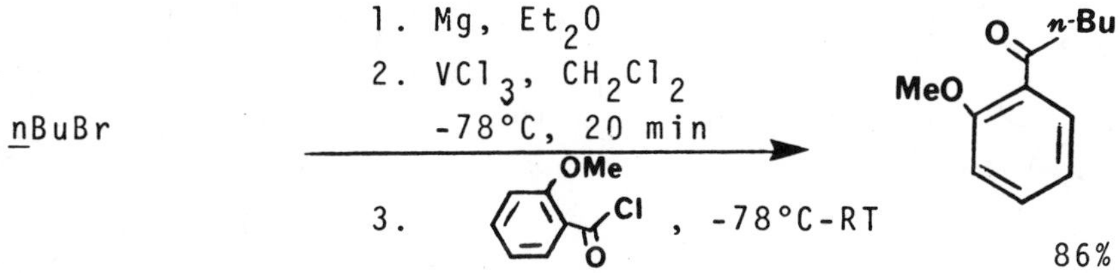
nBuBr
1. Mg, Et2O
2. VCl3, CH2Cl2
-78°C, 20 min
3.
OMe
Cl
O
, -78°C-RT
O
n-Bu
MeO

86%

Hirao, T.*; Misu, D.; Yao, K.; Agawa, T.
Tetrahedron Lett, (1986), 27, 929

54%   20%

Miura, M.[*]; Akase, F.; Nomura, M.   **JCS Chem Comm**, (1986), 241

1. Ac_2O, DMF, e^-, Mg anode
   st. steel cathode

PhCH$_2$Cl ──────────────────────────► PhCH$_2$CCH$_3$

        Bu_4NI, 10°C
2. aq. H_2SO_4 80%

d'Incan, E.; Sibille, S.; Perichon, J.
                Tetrahedron Lett, (1986), **27**, 4175

Related Methods: Ketones from Ketones (Section 177)
                 Aldehydes from halides (Section 55)

SECTION 176: Ketones from Hydrides

This section lists examples of the replacement of hydrogen by
ketonic groups, RH → RCOR'. For the oxidation of methylenes,
R_2CH_2 → R_2CO, see section 170 (Ketones from Alkyls and
Methylenes).

SeO_2, dioxane

48h, reflux

71%

Bestmann, H.J.[*]; Schobert, R.
                Angew Chem Int Ed Engl, (1985), **24**, 791

SECTION 177: Ketones from Ketones

This section contains alkylations of ketones and protected
ketones, ketone transpositions and annelations, ring expansions
and ring openings, and dimerizations. Conjugate reductions and
Michael alkylations of enones are listed in Section 74 (Alkyls
from Olefins.

For the preparation of enamines or imines from ketones see
Section 356 (Amine - Olefin).

Shimizu, I.; Ohashi, Y.; Tsuji, J.*
                    Tetrahedron Lett, (1983), **24**, 3865

McMurry, J.E.*; Miller, D.D.   **J Am Chem Soc**, (1983), **105**, 1660

Snider, B.B.*; Cartaya-Marin, C.P.   **J Org Chem**, (1984), **49**, 153
Snider, B.B.*; Kirk, T.C.   **J Am Chem Soc**, (1983), **105**, 2364

$$ClCH_2\overset{\overset{\displaystyle O}{\|}}{C}CH_2Cl$$

1. nBuMgBr, MgBr$_2$
   -20 - 20°C, 4.5h
———————————————————→
2. Li, -20 - 20°C
3. aq. HCl
4. 180°C, 2h

$$CH_3CH_2\overset{\overset{\displaystyle O}{\|}}{C}\underline{n}Bu$$

25%

Barluenga, J.[*]; Florez, J.; Yus, M.   **Synthesis**, (1983), 647

2 SmI$_2$, THF/MeOH
———————————————————→
-78°C - RT, 10 min

quant.

Molander, G.A.[*]; Hahn, G.   **J Org Chem**, (1986), **51**, 1135

$$Ph\overset{\overset{\displaystyle O}{\|}}{C}CH_2Br$$

1. LiN(TMS)$_2$, THF, -78°C
2. TMS-Cl, -78°C - RT
———————————————————→
3. 2 nBuLi, -78°C
4. aq. NH$_4$Cl

$$Ph\overset{\overset{\displaystyle O}{\|}}{C}CH_2SiMe_3$$

51%

Sampson, P.; Hammond, G.B.; Wiemer, D.F.[*]
              J Org Chem, (1986), **51**, 4342

$$Ph\overset{\overset{\displaystyle O}{\|}}{C}CH_2CH_3$$

1. KH, THF, tBuMe$_2$SiCl/NEt$_3$
2. tBuLi·TMEDA, nC$_6$H$_{14}$,
   -10°C
———————————————————→
3. nBuI, -5°C, 45 min
4. HOH

$$Ph\overset{\overset{\displaystyle O}{\|}}{C}CH_2CH_2\underline{n}Bu$$

62%

Trimitsis, G.[*]; Beers, S.; Ridella, J.; Carlon, M.; Cullin, D.;
High, J.; Brutts, D.
              JCS Chem Comm, (1984), 1088

$$CH_3\overset{O}{\underset{||}{C}}CH_2^-, FeSO_4$$

$$NH_3, DMSO, 10\ min$$

60%

Galli, C.; Bunnett, J.F.[*]   **J Org Chem**, (1984), **49**, 3041

$$CH_3\overset{O}{\underset{||}{C}}CH_3 \xrightarrow[\text{AcOH, reflux}]{PhCH_3,\ Mn(OAc)_7}$$

51%

($\underline{o}$:$\underline{m}$:$\underline{p}$ = 68:20:12)

Gardrat, C.   **Bull Chem Soc Belg**, (1984), **93**, 897

$$CH_3CH_2CH_2\overset{NMe_2}{\underset{||}{C}}CH_2CH_3 \xrightarrow[\text{CH}_2\text{Cl}_2,\ \text{reflux}]{\substack{K-10\ clay \\ Fe(NO)_3}} CH_3CH_2CH_2\overset{O}{\underset{||}{C}}CH_2CH_3$$

78%

Laszlo, P.[*]; Polla, E.   **Tetrahedron Lett**, (1984), **25**, 3309

$$\xrightarrow[\text{CH}_2\text{Cl}_2,\ 2.5h,\ RT]{K-10\ clay,\ Fe(NO)_3}$$

98%

Balogh, M.; Cornelis, A.[*]; Laszlo, P.[*]
                    Tetrahedron Lett, (1984), **25**, 3313

$$NO^+BF_4^-, CH_2Cl_2$$

RT, 2h

88%

Olah, G.A.[*]; Arvanaghi, M.; Ohannesian, L.; Prakash, G.K.S.
Synthesis, (1984), 785

$$\underset{PhCPh}{\overset{S}{\overset{\|}{}}} \quad \xrightarrow[\text{3N NaOH, RT, 1h}]{nBu_4N^+HSO_4^-, CH_2Cl_2} \quad \underset{PhCPh}{\overset{0}{\overset{\|}{}}}$$

90%

Alper, H.; Kwiatkowska, C.; Petrignani, J.-F.; Sibtain, F.
Tetrahedron Lett, (1986), **27**, 5449

$$PhC\overset{S}{\overset{\|}{}}\!\!-\!\!\langle\;\rangle\!\!-\!\!NO_2 \quad \xrightarrow[\text{Fe(NO}_3)_3, \text{ RT, 30 min}]{K\text{-10 clay, } CH_2Cl_2} \quad PhC\overset{0}{\overset{\|}{}}\!\!-\!\!\langle\;\rangle\!\!-\!\!NO_2$$

94%

Chalais, S.; Cornelis, A.; Laszlo, P.[*]; Mathy, A.
Tetrahedron Lett, (1985), **26**, 2327

$$CH_3\overset{0}{\overset{\|}{C}}CH_2COOH \quad \xrightarrow[\substack{\text{2. BnBr, THF, 3d} \\ 0°C - RT}]{\text{1. nBuLi, THF, 0°C}} \quad CH_3\overset{0}{\overset{\|}{C}}CH_2CH_2Ph$$

92%

Kjonaas, R.A.[*]; Patel, D.D.   **Tetrahedron Lett**, (1984), **25**, 5467

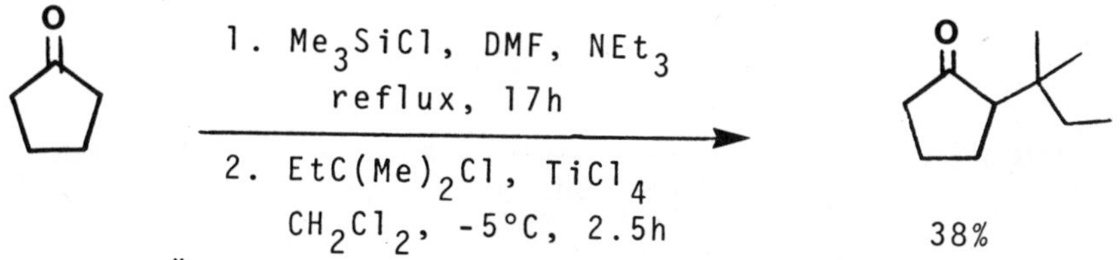
1. Me3SiCl, DMF, NEt3
reflux, 17h
2. EtC(Me)2Cl, TiCl4
CH2Cl2, -5°C, 2.5h
38%

Reetz, M.T.*; Chatzüosifidis, I.; Hubner, F.; Heimbach, H. **Org Syn**, (1984), **62**, 95

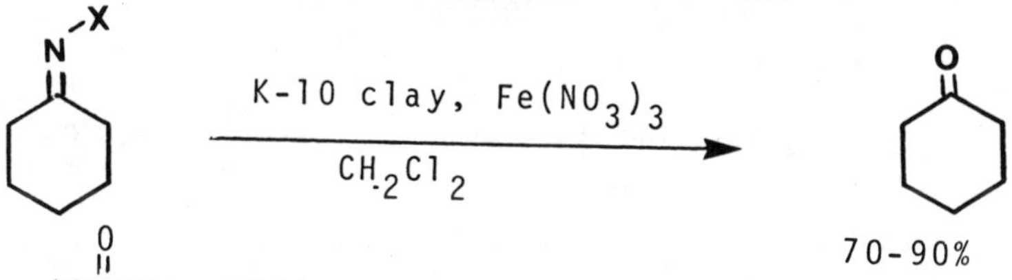
N
X
K-10 clay, Fe(NO3)3
CH2Cl2
70-90%

X = NHC(O)NH$_2$, NHSO$_2$Tol, NhPh

Laszlo, P.*; Polla, E. **Synthesis**, (1985), 439

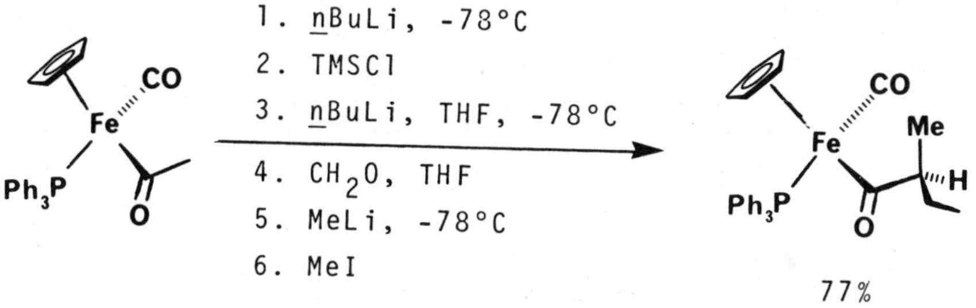
1. nBuLi, -78°C
2. TMSCl
3. nBuLi, THF, -78°C
4. CH2O, THF
5. MeLi, -78°C
6. MeI
CO
Fe
Ph3P
O
Me
H
77%

Davies, S.G.*; Walker, J.C. **JCS Chem Comm**, (1985), 209

$Cp(CO)_2FeC(O)CH_3$ → $Cp(CO)_2FeC(O)CH_2CH_2CH{=}CH_2$

1. LiN(TMS)$_2$, THF, -63°C
2. $CH_2{=}CHCH_2Br$, -63 - -20°C

54%

Brinkman, K.; Helquist, P.* **Tetrahedron Lett**, (1985), **26**, 2845

$$Me_2CHCCHN_2 \quad \xrightarrow[\substack{2.\ CF_3COOH,\ CH_2Cl_2 \\ 2h,\ RT}]{\substack{1.\ PhH,\ Rh(OCCF_3)_2 \\ RT,\ 1h}} \quad Me_2CHCCH_2Ph$$

86%

McKervey, M.A.[*]; Russell, D.N.; Twohig, M.F.
JCS Chem Comm, (1985), 491

PhBr, nBu₃SnOMe, 100°C

PdCl₂(P⟨MeO-C₆H₄⟩₃)₂, PhCH₃
5h

54%

Kosugi, M.; Hagiwara, I.; Sumiya, T.; Migita, T.[*]
Bull Chem Soc Jpn, (1984), **57**, 242

$$\begin{array}{c} O \\ \| \\ C(CH_2)_5CH_3 \\ | \\ CH_2 \\ | \\ COCH_2CH=CH_2 \\ \| \\ O \end{array} \quad \xrightarrow[\substack{NH_4OCHO,\ 35\ min}]{\substack{Pd(OAc)_2,\ PPh_3 \\ dioxane,\ reflux}} \quad CH_3(CH_2)_5CCH_3$$

99%

Tsuji, J.[*]; Nisar, M.; Shimizu, I.
J Org Chem, (1985), **50**, 3416

1. CH₃MgBr, THF
   -70°C, 10 min

2. DDQ, 0°C, 3h

80%

Bartoli, G.[*]; Bosco, M.; Dalpozzo, R.
Tetrahedron Lett, (1985), **26**, 115

$$\underset{PhCCH_2SiMe_3}{\overset{\overset{O}{\|}}{}} \quad \xrightarrow[\text{RT, THF, 12h}]{\text{BnBr, CsF}} \quad \underset{PhCCH_2CH_2Ph}{\overset{\overset{O}{\|}}{}}$$

70%

Fiorenza, M.[*]; Mordini, A.; Papaleo, S.; Pastorelli, S.; Ricci, A.

Tetrahedron Lett, (1985), **26**, 787

$$\underset{CH_3CCH_3}{\overset{\overset{N-NMe_2}{\|}}{}} \quad \xrightarrow{\begin{array}{l}1.\ \underline{n}BuLi,\ -5°C \\ 2.\ \underline{n}C_5H_{11}I,\ RT \\ 3.\ \underline{n}BuLi,\ -5°C \\ \hline 4.\ BnBr,\ RT \\ 5.\ aq.\ HCl\end{array}} \quad \underset{CH_3(CH_2)_5CCH_2Bn}{\overset{\overset{O}{\|}}{}}$$

94%

Yamashita, M.[*]; Matsumiya, K.; Tanabe, M.; Suemitsu, R.
Bull Chem Soc Jpn, (1985), **58**, 407

$$\underset{I(CH_2)_7CCH=CH_2}{\overset{\overset{O}{\|}}{}} \quad \xrightarrow[\text{reflux, 3h}]{\begin{array}{c}\underline{n}Bu_3SnH,\ PhH \\ AIBN\end{array}}$$

$$\left(\begin{array}{c}\overset{O}{\overset{\|}{C}}-CH_2CH_2 \\ (CH_2)_7\end{array}\right) \quad 63\%$$

+

$$\underset{I(CH_2)_6CCH_2CH_2}{\overset{\overset{O}{\|}}{}}(CH_2)_7$$
$$H_2C=CHC\lesssim O \quad 22\%$$

Porter, N.A.[*]; Magnin, D.R.; Wright, B.T.
J Am Chem Soc, (1986), **108**, 2787

$$\underset{\underset{OH}{\overset{|}{PhCCHPh}}}{\overset{\overset{O}{\|}}{}} \quad \xrightarrow[\text{2. AcOH, MeI}]{\text{1. Ph}_2\text{PLi, THF}} \quad \underset{PhCCH_2Ph}{\overset{\overset{O}{\|}}{}}$$

76%

Leone-Bay, A.[*]   **J Org Chem**, (1986), **51**, 2378

$$\underline{n}C_6H_{13}-\overset{\overset{\displaystyle SeMe}{|}}{\underset{\underset{\displaystyle Me}{|}}{C}}-SeMe \quad \xrightarrow[\text{pentane, RT, 90 min}]{\text{K-10 clay, Fe(NO}_3)_3} \quad \underline{n}C_6H_{13}\overset{\overset{\displaystyle O}{||}}{C}CH_3$$

90%

Laszlo, P.*; Pennetreau, P.; Krief, A.
Tetrahedron Lett, (1986), **27**, 3153

$$CH_2=C\overset{\displaystyle OSiMe_3}{\underset{\displaystyle Ph}{<}} \quad \xrightarrow[\text{2. aq. H}^+]{\substack{1. \quad Cl-\bigcirc-N_2^+BF_4^- \\ \text{Py, 0°C, 3.5h}}}$$

73% Cl

Sakakura, T.; Hara, M.; Tanaka, M.*
JCS Chem Comm, (1985), 1545

Reviews:

"Homoenolate Anions"

Werstiuk, N.H.*   **Tetrahedron**, (1983), **39**, 205

"1,2-Carbonyl Transpositions"

Kane, V.V.*; Singh, V.; Martin, A.; Doyle, D.L.
Tetrahedron, (1983), **39**, 345

"Alkylation of Ketones and Aldehydes via Nitrogen Derivatives"

Whitesell, J.K.; Whitesell, M.A.   **Synthesis**, (1983), 517

"Reactive Enolates from Enol Silyl Ethers"

Kuwajima, I.; Nakamura, E.   **Accts Chem Res**, (1985), **18**, 181

Ketones may also be alkyated or homologated via olefinic ketones (Section 374).

Related Methods: Aldehydes from Aldehydes (Section 49)

SECTION 178: Ketones from Nitriles

1. LDA, THF, -78°C
2. O_2, -78°C, 30 min
3. $SnCl_2$, aq. HCl, 0°C
4. 1M NaOH

65%

Freerksen, R.W.; Selikson, S.J.; Wroble, R.R.; Kyler, K.S.; Watt, D.S.*

J Org Chem, (1983), **48**, 4087

DMSO, O_2, HOH

K_2CO_3, overnight

93%

Kulp, S.S.*; McGee, M.J.   **J Org Chem**, (1983), **48**, 4097

nPrCH(OH)
|
nPrCHCH₂CH₂CN

1. PhH, Ph_3SnH, 15h
   AIBN, reflux
2. 80% aq. AcOH
   1h

66%

Clive, D.L.J.*; Beaulieu, P.L.; Set, L.
J Org Chem, (1984), **49**, 1313

SECTION 179: Ketones from Olefins

$CH_3(CH_2)_7CH=CH_2$

O_2, $PdCl_2$, 80°C
$CuCl_2 \cdot 2H_2O$, PhH
Et_3NcetylBr, HOH

$CH_3(CH_2)_7\overset{O}{\overset{\|}{C}}CH_3$

73%

Januskiewicz, K.; Alper, H.*
Tetrahedron Lett, (1983), **24**, 5159

$$Co_2(CO)_8, \text{ 10 atm CO}$$
$$\text{10 atm } H_2, \text{ 20h}$$
$$\overline{120°C, \text{ hexane}}$$

76%

Eilbracht, P.[*]; Acker, M.; Totzauer, W.
          Chem Ber, (1983), **116**, 238
Eilbracht, P.[*]; Balss, E.; Acker, M.
          Tetrahedron Lett, (1984), **25**, 1131

$\underline{n}C_8H_{17}CH=CH_2$

1. $PdCl_2$, CuCl, RT
   O_2, DMF:H_2O, 24h
2. 3N HCl

$\underline{n}C_8H_{17}\overset{\displaystyle O}{\overset{\|}{C}}CH_3$

73%

Tsuji, J.; Nagashima, H.; Nemoto, H.[*]   **Org Syn**, (1984), **62**, 9

1. 3 $CH_3CH_2CH_2CH=CH_2$
   PhH, 30°C, 2h
2. CO, -78°C, 6h
3. AcOH

68%

Yasuda, H.; Nagasuna, K.; Akita, M.; Lee, K.; Nakamura, A.[*]
          Organometallics, (1984), **3**, 1470

1. $NaBH_3OAc$, THF, 50°C
2. NaOMe, $CHCl_3$, 50°C
3. H_2O_2, NaOH

59%

Narayana, C.; Periasamy, M.[*]
          Tetrahedron Lett, (1985), **26**, 6361

Review: "Synthetic Applications of the Palladium Catalyzed
      Oxidation of Olefins to Ketones"

Tsuji, J.[*]   **Synthesis**, (1984), 369

See also: Ethers and Epoxides from Olefins (Section 134) and
Ketones from Ethers and Epoxides (Section 174).

SECTION 180: Ketones from Miscellaneous Compounds

Conjugate reductions and reductive alkylations of enones are
listed in Section 74 (Alkyls from Olefins).

$$PhCH=C \begin{smallmatrix} CH_3 \\ \\ NO_2 \end{smallmatrix} \xrightarrow[\substack{aq. \ NaH_2PO_2 \cdot xH_2O \\ 2h, \ 40-60°C}]{Ni(R), \ EtOH, \ pH \ 5} PhCH_2 \overset{\overset{\displaystyle O}{\|}}{C} CH_3$$

88%

Monti, D.[*]; Gramatica, P.; Speranza, G.; Manitto, P.
                    Tetrahedron Lett, (1983), **24**, 417

$$\underline{n}C_6H_{13}CH=C \begin{smallmatrix} CH_3 \\ \\ NO_2 \end{smallmatrix} \xrightarrow[\substack{2. \ 3\% \ aq. \ HCHO}]{\substack{1. \ e^-, \ aq. \ HClO_4 \\ Pb \ electrodes \\ CH_2Cl_2/dioxane}} \underline{n}C_6H_{13}CH_2 \overset{\overset{\displaystyle O}{\|}}{C} CH_3$$

81%

Torii, S.[*]; Tanaka, H.; Katoh, T.   **Chem Lett**, (1983), 607

1. LiBHsecBu$_3$, THF
   THF, RT

2. 4N H$_2$SO$_4$, -10°C

81%

Mourad, M.S.; Varma, R.S.; Kabalka, G.W.[*]
                    Synthesis, (1985), 654

HN—N=CHEt
 |
 CMe$_3$

1. nBuLi, 0°C, 1h
2. BnBr, -78 - 20°C, 2d
——————————————————→
3. TFA, 20°C, 6h
4. (COOH)$_2$, aq. Et$_2$O, 14h

$$CH_3CH_2\overset{O}{\overset{\|}{C}}Bn$$

67%

Adlington, R.M.; Baldwin, J.E.[*]; Bottaro, J.C.; Perry, M.W.D.
JCS Chem Comm, (1983), 1040

CN
|
PhCHNMe$_2$
(**S**)

1. LDA, THF
2. nC$_6$H$_{13}$CH(Me)Br, HMPT
——————————————————→
3. aq. H$_2$SO$_4$
4. AgNO$_3$

$$PhC\overset{O}{\overset{\|}{C}}CH \begin{smallmatrix} nC_6H_{13} \\ \\ CH_3 \end{smallmatrix}$$

70%
(76% ee)

Hebert, E.[*]; Maigrot, N.; Welvart, Z.
Tetrahedron Lett, (1983), **24**, 4683

$$Me_2N-\overset{N-tBu}{\overset{\|}{C}}-NMe_2$$

IO$_2$ / COOH (C$_6$H$_4$)
CH$_2$Cl$_2$ RT

95%

Barton, D.H.R.[*]; Motherwell, W.B.; Zard, S.Z.
Tetrahedron Lett, (1983), **24**, 5227

CH$_3$(CH$_2$)$_4$CH$_2$NO$_2$

1. nBuLi, THF
   -90 - -30°C
——————————————————→
2. ClHC=NMe$_2$$^+Cl^-$
3. BnCH$_2$MgBr, CuI
   THF, -30°C, 2h

$$CH_3(CH_2)_4\overset{N-OH}{\overset{\|}{C}}CH_2Bn$$

97%

Fujisawa, T.[*]; Kurita, Y.; Sato, T.   **Chem Lett**, (1983), 1537

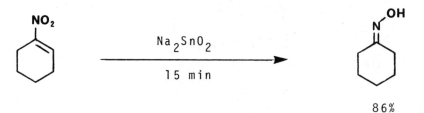

acetone, $\underline{t}$BuOK

$NH_{3(1)}$, $-10°C$, 1.5h

59%

Iwasaki, G.; Saeki, S.; Hamana, M.[*]   **Chem Lett**, (1986), 31

$\underline{n}C_7H_{15}CH$ (NO$_2$) CH$_2$CH$_2$CCH$_3$ (O)

HCO_2Na, MeOH
e^-, 8h

Pt electrodes
$20°C$

$\underline{n}C_7H_{15}CCH_2$ (O) CH$_2$ CH$_3$C (O)   90%

Nokami, J.[*]; Sonoda, T.; Wakabayashi, S.
Synthesis, (1983), 763

$CH_3(CH_2)_5HgBr$

1. $Ni(CO)_4$, KI
   DMF, $80°C$,
2. aq. HCl

$CH_3(CH_2)_5C(CH_2)_5CH_3$ (O)

96%

Ryu, I.; Ryang, M.; Rhee, I.; Omura, H.; Murai, S.[*]; Sonoda, N.
Syn Commun, (1984), **14**, 1175

Na_2SnO_2

15 min

86%

Varma, R.S.; Varma, M.; Kabalka, G.W.[*]
Tetrahedron Lett, (1985), **26**, 6014

Me
$\underset{\underset{Ph}{\overset{|}{\underset{-CHCN}{N}}}}{\overset{\overset{O}{\parallel}}{C}Cl}$

1. $O_2N-\langle\ \rangle-OH$, K_2CO_3
   RT, 24h, MeCN

2. $AgNO_3$, aq. THF
   30 min

78%

$\overset{O\underset{}{\diagup}Ph}{\underset{NO_2}{\bigcirc}}$

Kay, I.T.; Glue, S.E.J.   **Tetrahedron Lett**, (1986), **27**, 113

$\underset{\underline{n}C_5H_{11}}{\overset{Et}{\overset{|}{CHSO_2Ph}}}$

1. $\underline{n}BuLi$, THF, -78°C
2. TMS-OO-TMS, -78°C-RT
   overnight
3. aq. $NaHCO_3$

$\underset{\underline{n}C_5H_{11}}{\overset{O}{\overset{\parallel}{C}Et}}$

76%

Hwu, J.R.[*]   **J Org Chem**, (1983), **48**, 4432

SECTION 180A: Protection of Ketones

See Section 362 (Ester-Olefin) for the formation of enol ethers and Section 367 (Ether-Olefins) for the formation of enol ethers. May of the methods in Section 60A (Protection of Aldehydes) are also applicable to ketones.

1. 2 $HSCH_2CH_2SH$
   ZnOTf, 23°C

2. reflux, 2h

85%

Corey, E.J.[*]; Shimoji, K.   **Tetrahedron Lett**, (1983), **24**, 169

1. LDA, THF/hexane
2. $ZnBr_2$, RT - reflux
3. BnBr, 0°C
4. H_3O^+

79%
(82% ee S)

Saigo, K.[*]; Kasahara, A.; Ogawa, S.; Nohira, H.
   Tetrahedron Lett, (1983), **24**, 511

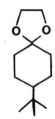

1. $iBu_2AlSCH_2CH_2SAliBu_2$
   PhH, RT, 30 min

2. 3M HCl

3. 10% NaOH

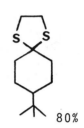

80%

Satoh, T.; Uwaya, S.; Yamakawa, K.*   **Chem Lett**, (1983), 667

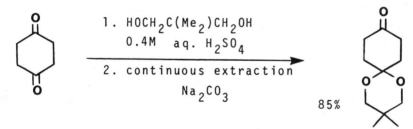

$LiClO_4$, e^-, CH_2Cl_2

Pt electrodes

$TMSOCH_2CH_2OTMS$

91%

Torii, S.*; Inokuchi, T.   **Chem Lett**, (1983), 1349

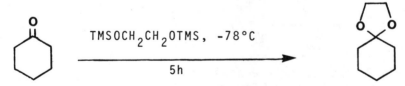

1. $HOCH_2C(Me_2)CH_2OH$
   0.4M   aq. H_2SO_4

2. continuous extraction
   Na_2CO_3

85%

Babler, J.H.*; Spina, K.P.   **Syn Commun**, (1984), **14**, 39

$TMSOCH_2CH_2OTMS$, -78°C

5h

86%

Sakurai, H.*; Sassaki, K.; Hayashi, J.; Hosomi, A.
         J Org Chem, (1984), **49**, 2808

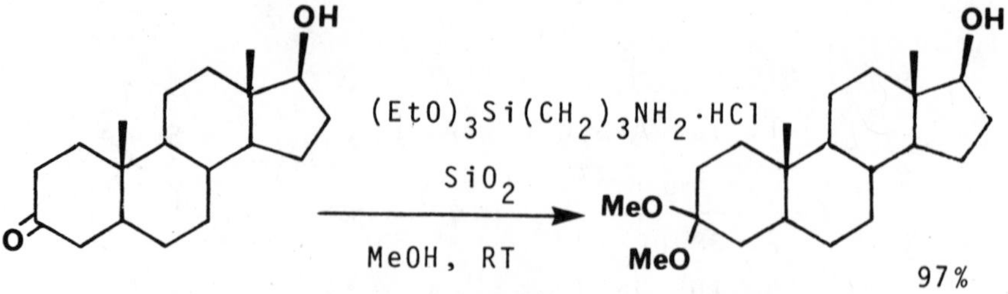
OH
$(EtO)_3Si(CH_2)_3NH_2 \cdot HCl$
SiO_2
MeOH, RT
MeO
MeO
OH
97%

Gasparrini, F.; Giovannoli, M.; Misiti, D.; Palmieri, G.
Tetrahedron, (1984), **40**, 1491

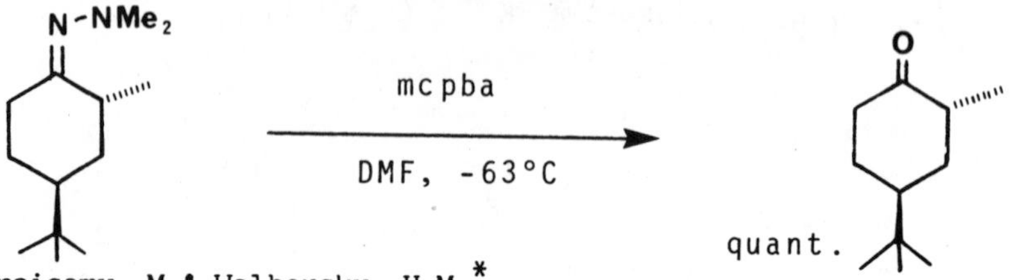
$N-NMe_2$
mcpba
DMF, -63°C
O
quant.

Duraisamy, M.; Walborsky, H.M.*
J Org Chem, (1984), **49**, 3410

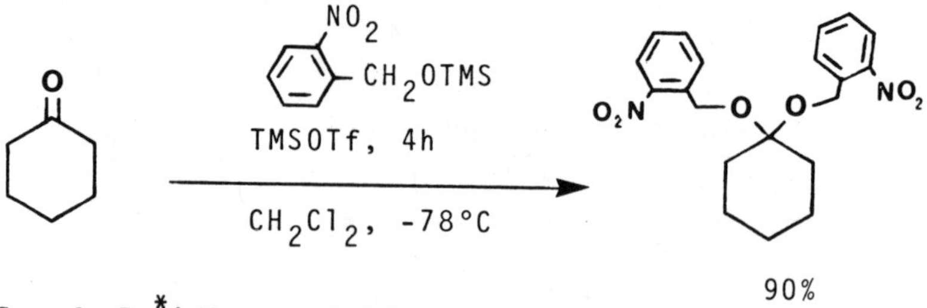
O
NO_2
CH_2OTMS
TMSOTf, 4h
CH_2Cl_2, -78°C
O_2N
O
O
NO_2
90%

Gravel, D.*; Murray, S.; Ladouceur, G.
JCS Chem Comm, (1985), 1828

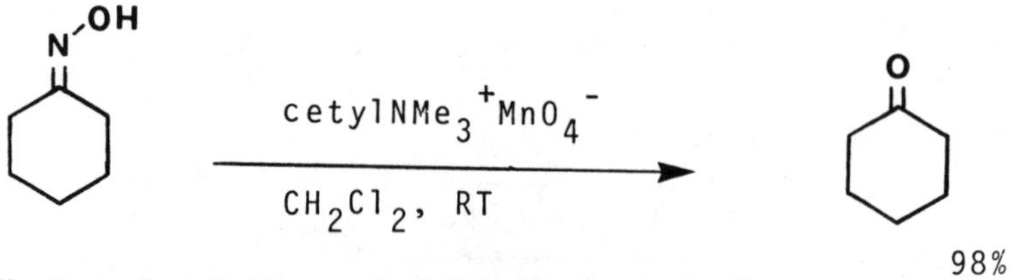
N
OH
$cetylNMe_3^+MnO_4^-$
CH_2Cl_2, RT
O
98%

Vankar, P.; Rathore, R.; Chandrasekaran, S.
J Org Chem, (1986), **51**, 3063

$PhI(OAc)_2$, CH_2Cl_2

Moriarty, R.M.*; Prakash, O.; Vavilikolanu, P.R.
Syn Commun, (1986), **16**, 1247

CHAPTER 13
PREPARATION OF NITRILES

<u>SECTION</u> <u>181</u>: <u>Nitriles</u> <u>from</u> <u>Acetylenes</u>

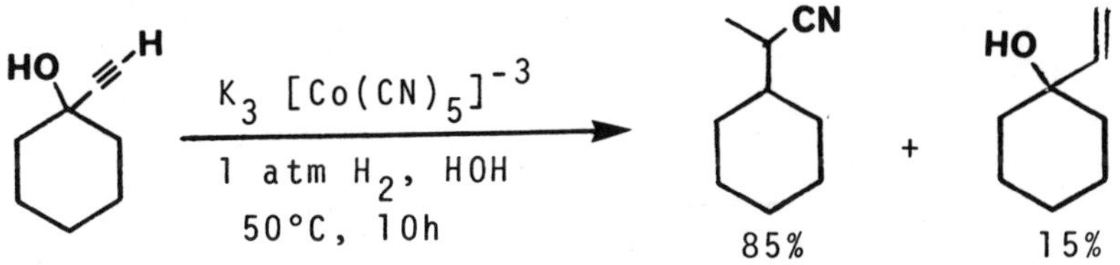

Funabiki, T.[*]; Yamazaki, Y.; Sato, Y.; Yoshida, S.
                 <u>JCS</u> <u>Perkin</u> <u>II</u>, (1983), 1915

<u>SECTION</u> <u>182</u>: <u>Nitriles</u> <u>from</u> <u>Acid</u> <u>Derivatives</u>

No Additional Examples

<u>SECTION</u> <u>183</u>: <u>Nitriles</u> <u>from</u> <u>Alcohols</u> <u>and</u> <u>Thiols</u>

No Additional Examples

<u>SECTION</u> <u>184</u>: <u>Nitriles</u> <u>from</u> <u>Aldehydes</u>

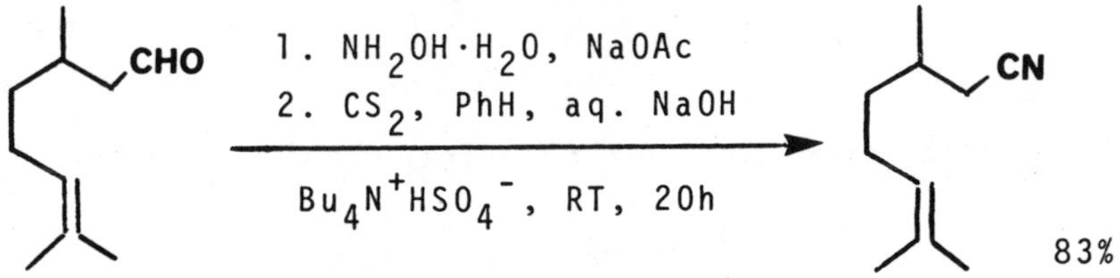

Shinozaki, H.[*]; Imaizumi, M.; Tajima, M.
                 <u>Chem</u> <u>Lett</u>, (1983), 929

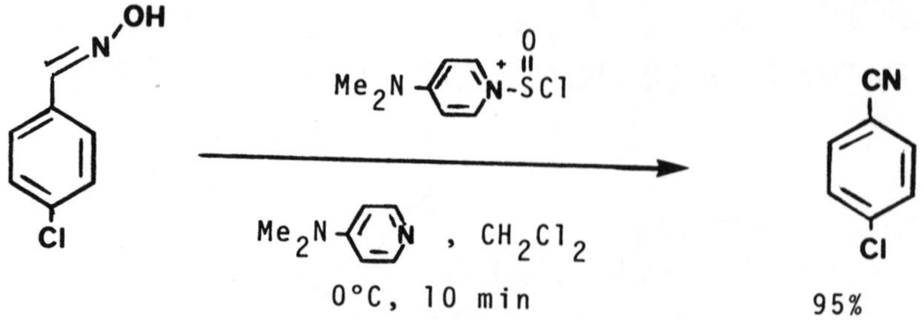

$$Me_2N-\!\!\!\underset{}{\bigcirc}\!\!\!-N^+\!\!\!-\overset{\overset{O}{\|}}{S}Cl$$

$$Me_2N-\!\!\!\underset{}{\bigcirc}\!\!\!N \quad , \quad CH_2Cl_2$$

0°C, 10 min   95%

Arrieta, A.; Palomo, C.[*]   **Synthesis**, (1983), 472

$$\underset{Ph-C-H}{\overset{NOH}{\|}}$$ $$Me_3SiO(\overset{\overset{O}{\|}}{P}O)_n \text{ , DCE}$$

reflux, 95 min   PhCN

92%

Aizpurua, J.M.; Palomo, C.   **Nouv J Chem**, (1983), **7**, 465

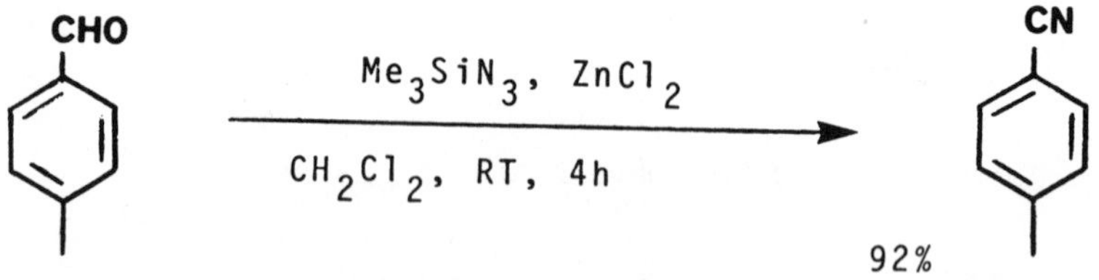

$$Me_3SiN_3, \; ZnCl_2$$

$$CH_2Cl_2, \; RT, \; 4h$$

92%

Nishiyama, K.[*]; Watanabe, A.   **Chem Lett**, (1984), 773

PhCHO $$\xrightarrow[\text{100°C, 4h}]{S_8, \; NaNO_2, \; NH_3}$$ PhCN

74%

Sato, R.[*]; Itoh, Kao.; Itoh, Kaz.; Nishina, H.; Goto, T.; Saito, M.

Chem Lett, (1984), 1913

$$\underset{PhCH}{\overset{\underset{\|}{NOH}}{}} \quad \xrightarrow[\substack{2.\ excess\ PhNH_2,\ DCE \\ RT\ -\ 60°C}]{\substack{1.\ Cl_2C=NMe_2{}^+Cl^- \\ DCE,\ RT\ -\ reflux}} \quad PhCN$$

95%

Kokel, B.[*]; Menichi, G.; Hubert-Habart, M.
                    Synthesis, (1985), 201

$$\xrightarrow[\substack{AcOH,\ reflux \\ 4h}]{EtNO_2,\ NaOAc}$$

80%

Karmarkar, S.N.; Kelkar, S.L.; Wadia, M.S.[*]
                    Synthesis, (1985), 510

SECTION 185: Nitriles from Alkyls, Methylenes, and Aryls

No Additional Examples

SECTION 186: Nitriles from Amides

$$CH_3(CH_2)_{14}\overset{\overset{O}{\|}}{C}NH\underline{t}Bu \quad \xrightarrow[reflux,\ 6h]{POCl_3,\ PhH} \quad CH_3(CH_2)_{14}CN$$

97%

Perni, R.B.; Gribble, G.W.[*]   **Org Prep Proc Int,** (1983), **15,** 297

$$
\begin{array}{c}
CO_2Et \\
| \\
CH_2 \\
| \\
NHCHO
\end{array}
\quad
\xrightarrow[\substack{2.\ Na_2CO_3,\ HOH,\ RT \\ 1\ h}]{\substack{1.\ POCl_3,\ iPr_2NH \\ CH_2Cl_2,\ 0°C}}
\quad
\begin{array}{c}
CO_2Et \\
| \\
CH_2 \\
| \\
N\equiv C \\
+ \ -
\end{array}
$$

84%

Obrecht, R.; Herrmann, R.; Ugi, I.*   **Synthesis**, (1985), 400

$$
CH_3(CH_2)_6\overset{\overset{\displaystyle S}{\|}}{C}NH_2
\quad
\xrightarrow[\substack{30\%\ aq.\ NaOH \\ Bu_4NBr}]{2\ BnCl,\ PhH}
\quad
CH_3(CH_2)_6CN
$$

80%

Funakoshi, Y.; Takido, T.*; Itabashi, K.
          Syn Commun, (1985), **15**, 1299

$$
PhC\overset{\overset{\displaystyle O}{\|}}{N}H_2
\quad
\xrightarrow{\substack{1.\ Cl_3C\overset{\overset{\displaystyle O}{\|}}{C}Cl,\ NEt_3 \\ CH_2Cl_2,\ 5°C \\ 2.\ HOH \\ 3.\ NaOH \\ 4.\ H_2SO_4 \quad 5.\ HOH}}
\quad
PhCN
$$

90%

Saednya, A.*   **Synthesis**, (1985), 184

93%

Mai, K.*; Patil, G.   **Tetrahedron Lett**, (1986), **27**, 2203

SECTION 187: Nitriles from Amines

No Additional Examples

SECTION 188: Nitriles from Esters

No Additional Examples

SECTION 189: Nitriles from Ethers, Epoxides, and Thioethers

No Additional Examples

SECTION 190: Nitriles from Halides and Sulfonates

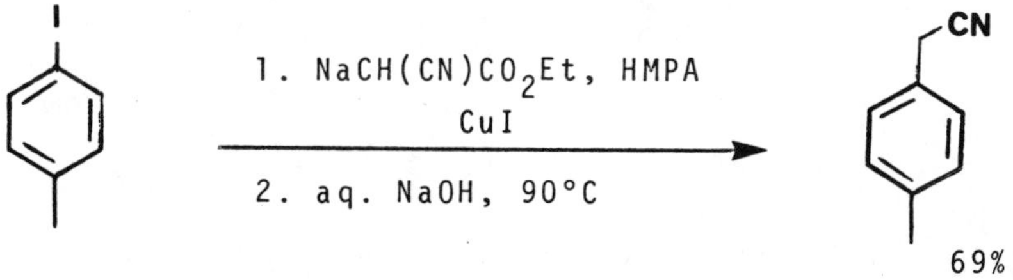

1. NaCH(CN)CO$_2$Et, HMPA
   CuI
2. aq. NaOH, 90°C

69%

Suzuki, H.[*]; Kobayashi, T.; Yoshida, Y.; Osuka, A.[*]
                    Chem Lett, (1983), 193

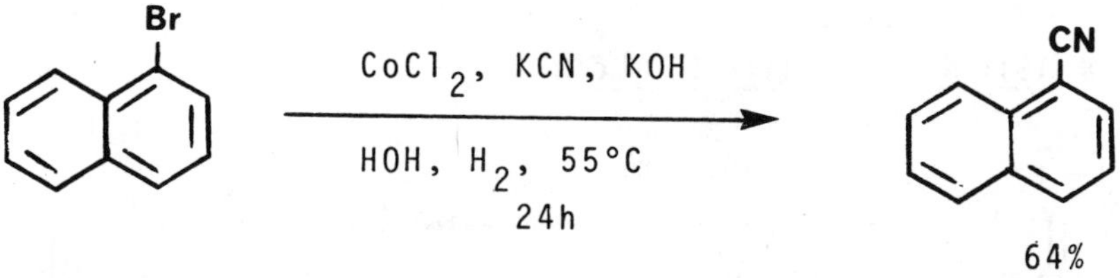

CoCl$_2$, KCN, KOH

HOH, H$_2$, 55°C
24h

64%

Funabiki, T.[*]; Nakamura, H.; Yoshida, S.
                    J Organomet Chem, (1983), 243, 95

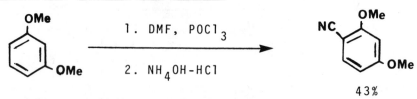

Br
→ CN

nBu$_3$SnCH$_2$CN, xylene
120°C, 3h

PdCl$_2$[P-$\underline{o}$Tol$_3$]$_2$

78%

Kosugi, M.; Ishiguro, M.; Negishi, Y.; Sano, H.; Migita, T.[*]
Chem **Lett**, (1984), 1511

Cl
→ CN

Me$_3$SiCN, AlBr$_3$

CH$_2$Cl$_2$, reflux
23h

78%

Olah, G.A.[*]; Farooq, O.; Surya Prakash, G.K.
Synthesis, (1985), 1140

I
→ CN

1. NaCH(CN)$\overset{\text{O}}{\overset{\|}{\text{P}}}(OMe)_2$
170°C - 200°C

2. aq. HCl

59%

Suzuki, H.[*]; Watanabe, K.; Yi, Q.   **Chem** **Lett**, (1985), 1779

SECTION 191: Nitriles from Hydrides

OMe

1. DMF, POCl$_3$

2. NH$_4$OH-HCl

NC → OMe, OMe

43%

Liebscher, J.; Bechstein, U.   **Zeit** **Chemie**, (1983), **23**, 214

SECTION 192: Nitriles from Ketones

Reetz, M.T.[*]; Müller-Starke, H.

30%

Tetrahedron Lett, (1984), **25**, 3301

SECTION 193: Nitriles from Nitriles

Conjugate reductions and Michael alkylations of olefinic nitriles are found in Section 74D (Alkyls from Olefins).

$$PhCH_2CN \xrightarrow[\text{PhH, } 40°C, \text{ 4h}]{Al_2O_3/KOH, \text{ EtBr}} \underset{PhCHCH_2CH_3}{\overset{CN}{|}}$$

91%

Sukata, K.[*]   **Bull Chem Soc Jpn**, (1983), **56**, 3306

74%

Kornblum, N.[*]; Singh, H.K.; Boyd, S.D.

J Org Chem, (1984), **49**, 358

$$BrCH_2CN \xrightarrow[\text{2. aq. HCl}]{\substack{\text{1. } PhCH_2Cl, \text{ Ni[*], glyme} \\ 85°C, \text{ 30h}}} PhCH_2CH_2CN$$

57%

Ni[*] = activated nickel

Inaba, S.; Rieke, R.D.[*]   **Synthesis**, (1984), 842

$$\begin{array}{c} CN \\ | \\ Ph-C-SEt \\ | \\ CH_3 \end{array} \quad \xrightarrow[\begin{array}{c} 2. \ BnBr, \ THF, \ -78°C \ - \ RT \\ 3. \ 1N \ NaOH \end{array}]{\begin{array}{c} 1. \ \underline{n}Bu_3SnLi, \ -78°C \\ HMPA, \ THF, \ 30 \ min \end{array}} \quad \begin{array}{c} CN \\ | \\ Ph-C-CH_2Ph \\ | \\ CH_3 \end{array}$$

87%

Takeda, T.*; Ando, K.; Mamada, A.; Fujiwara, T.
Chem Lett, (1985), 1149

SECTION 194: Nitriles from Olefins

No Additional Examples

SECTION 195: Nitriles from Miscellaneous Compounds

$$\text{adamantyl-}NO_2 \quad \xrightarrow[25°C, \ 16h]{Me_3SiI, \ CH_2Cl_2} \quad \text{adamantyl-}I$$

98%

Olah, G.A.*; Narang, S.C.; Field, L.D.; Fung, A.P.
J Org Chem, (1983), **48**, 2766

$$\xrightarrow[-20°C, \ 12h]{\begin{array}{c} 1. \ NaCN, \ 25°C, \ 5 \ min \\ 2. \ Pb(OAc)_4 \end{array}} \quad \text{cyclohexyl-CN}$$

88%

Masuda, Y.; Hoshi, M.; Yamada, T.; Arase, A.*
JCS Chem Comm, (1984), 398

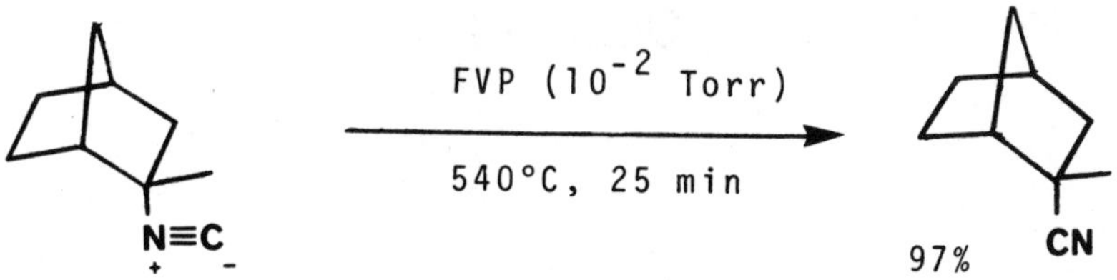

$$\text{FVP} \ (10^{-2} \ \text{Torr})$$
$$540°C, \ 25 \ \text{min}$$

97%

Meier, M.; Rüchardt, C.*   **Tetrahedron Lett**, (1984), **25**, 3441

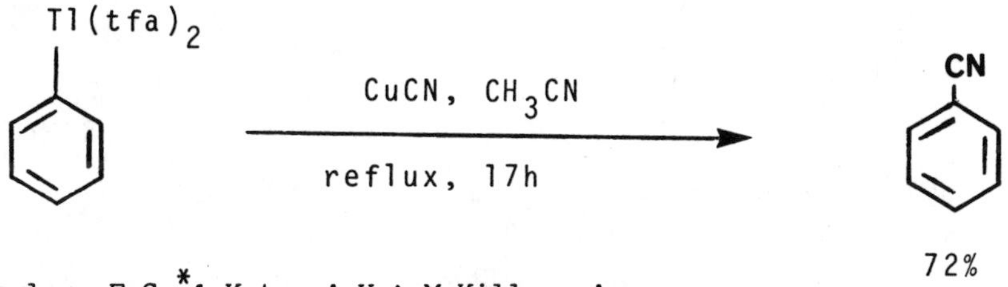

Tl(tfa)$_2$

$$\text{CuCN, } CH_3CN$$
$$\text{reflux, 17h}$$

72%

Taylor, E.C.*; Katz, A.H.; McKillop, A.
                    Tetrahedron Lett, (1984), **25**, 5473

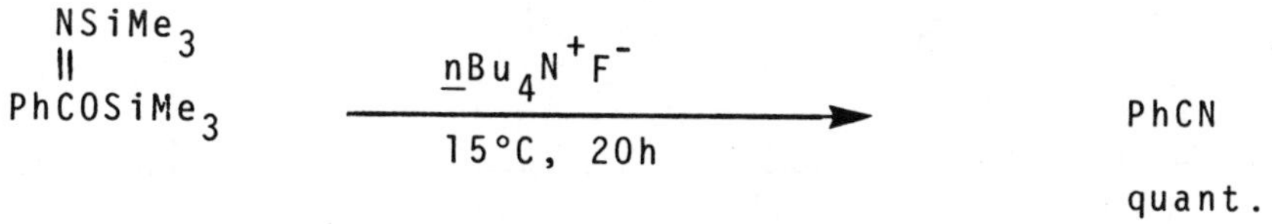

NSiMe$_3$
‖
PhCOSiMe$_3$

$$\underline{n}Bu_4N^+F^-$$
$$15°C, \ 20h$$

PhCN

quant.

Rigo, B.*; Lespagnol, C.; Pauly, M.
                    Tetrahedron Lett, (1986), **27**, 347

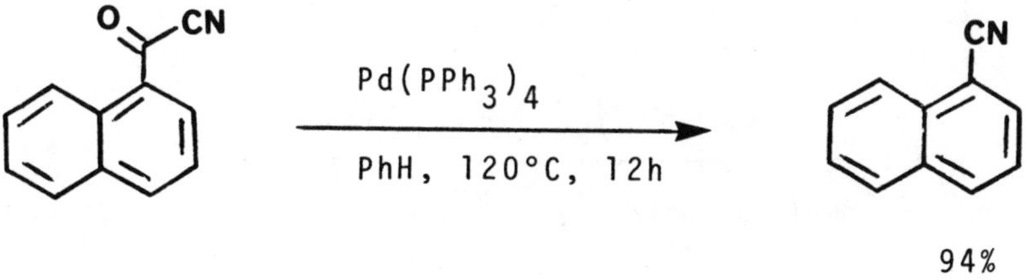

$$\text{Pd(PPh}_3)_4$$
$$\text{PhH, } 120°C, \ 12h$$

94%

Murahashi, S.-I.*; Naota, T.; Nakajima, N.
                    J Org Chem, (1986), **51**, 898

CHAPTER 14
PREPARATION OF OLEFINS

SECTION 196: Olefins from Acetylenes

$CH_3(CH_2)_4C{\equiv}CSiMe_3$ $\xrightarrow[\substack{2. \ CH_2=CHCH_2 \\ CuI, \quad CH_2I}]{\substack{1. \ iBuMgBr \\ Cp_2TiCl_2}}$

$CH_3(CH_2)_4$ $SiMe_3$
C=C
HCH$_2$
H$_2$C=CHCH$_2$

86%

Sato, F.*; Watanabe, H.; Tanaka, Y.; Yamaji, T.; Sato, M.
Tetrahedron Lett, (1983), **24**, 1041

$Ph-C{\equiv}C-Ph$ $\xrightarrow[\substack{PEG, \ CH_2Cl_2, \ 30 \ min}]{NaBH_4-PdCl_2}$ $PhCH=CHPh$

(E)

PEG = polyethylene glycol

91%

Suzuki, N.*; Tsukanaka, T.; Nomoto, T.; Ayaguchi, Y.; Izawa, Y.
JCS Chem Comm, (1983), 515

$Ph-C{\equiv}C-Ph$ $\xrightarrow[\substack{FeCl_2, \ \overset{S-S}{\diagdown}(CH_2)_4\overset{O}{\overset{\|}{C}}NH_2}]{NaBH_4, \ EtOH, \ 35°C, \ 8h}$

PhPh

75%

Kijima, M.; Nambu, Y.*; Endo, T.***Chem Lett**, (1985), 1851

$$HC \equiv C - \underset{\underset{CO_2Me}{|}}{\overset{\overset{Ph}{|}}{C}} H \quad \xrightarrow[\substack{Et_2O \\ 3\% \ Pd(PPh_3)_4}]{PhZnCl, \ THF} \quad \underset{H}{\overset{Ph}{>}} C = C = C \underset{H}{\overset{Ph}{<}}$$

80%

Elsevier, C.J.; Stehouwer, P.M.; Westmijze, H.; Vermeer, P.[*]
J Org Chem, (1983), **48**, 1103

$$CH_3(CH_2)_4 C \equiv CEt \quad \xrightarrow[\substack{EDA, \ -25°C \\ 4h}]{MeNH_2 \cdot Ca} \quad CH_3(CH_2)_4 CH = CHEt$$

79%

Benkeser, R.A.[*]; Belmonte, F.G.   **J Org Chem**, (1984), **49**, 1662

$$PhC \equiv CPh \quad \xrightarrow[\substack{THF/EtOH \\ quinoline, \ H_2}]{NaH/\underline{t}AmOH/Pd(OAc)_2} \quad$$

Ph   Ph

97%

Brunet, J.-J.; Caubere, P.[*]   **J Org Chem**, (1984), **49**, 4058

$$\underline{n}C_6H_{13}C \equiv CSiMe_3 \quad \xrightarrow{\substack{1. \ Dibal-H, \ Et_2O \\ \quad -20°C - reflux \\ 2. \ MeLi, \ 0°C \\ 3. \ CuI, \ P(OEt)_3 \\ \quad THF, \ -78°C \\ 4. \ EtI, \ -78°C - RT \\ 5. \ ice/ \ dil. \ HCl}} \quad \underset{}{\overset{\underline{n}C_6H_{13} \qquad SiMe_3}{>=<}}$$

78%

Ziegler, F.E.[*]; Mikami, K.   **Tetrahedron Lett**, (1984), **25**, 131

$HC{\equiv}C(CH_2)_4OTs$ $\xrightarrow[\text{THF, 0°C - 25°C}]{PhMe_2SiMgMe, CuI}$

91%

Okuda, Y.; Morizawa, Y.; Oshima, K.[*]; Nozaki, H.
Tetrahedron Lett, (1984), **25**, 2483

$CH_3C{\equiv}C\underset{\overset{|}{OH}}{C}HEt$ $\xrightarrow[\begin{array}{c}2.\ PhCH_2CH_2MgBr,\ CuI\\ -30°C,\ 30\ min\end{array}]{\begin{array}{c}1.\quad\text{(pyrrolidine enamine with Cl)}\quad,\ THF-CH_2Cl_2\\ 0°C,\ 5h\end{array}}$ $\underset{CH_2CH_2Ph}{\overset{Me}{C}}=C=CHEt$

84%

Fujisawa, T.[*]; Iida, S.; Sato, T.
Tetrahedron Lett, (1984), **25**, 4007

$\underset{HO}{\quad}{\equiv}-SiMe_3$ (cyclohexane) $\xrightarrow[\begin{array}{c}PhH,\ reflux,\ 15\ min\\ 2.\ PhH,\ 80°C,\ 48h\end{array}]{\begin{array}{c}1.\ MeMgBr,\ Et_2O\\ dppp,\ NiCl_2\end{array}}$ (cyclohexylidene)=C=CH–SiMe_3

95%

Wenkert, E.[*]; Leftin, M.H.; Michelotti, E.L.
J Org Chem, (1985), **50**, 1122

$HC{\equiv}C(CH_2)_4OH$ $\xrightarrow[\begin{array}{c}-60°C,\ THF/MeOH\\ 12h\end{array}]{2\ Me_3SnCu \cdot SMe_2}$ $CH_2{=}\underset{\overset{|}{SnMe_3}}{C}(CH_2)_4OH$

85%

Piers, E.[*]; Chong, J.M.   **JCS Chem Comm**, (1983), 934

$$CH_3C{\equiv}CCH_2OH \xrightarrow[\text{CuCN, 25°C, THF}]{PhMe_2SiZnEt_2Li} CH_3CH=CHCH_2OH$$

$$SiMe_2Ph \quad 85\%$$

Okuda, Y.; Wakamatsu, K.; Tückmantel, W.; Oshima, K.[*]; Nozaki, H.

Tetrahedron Lett, (1985), **26**, 4629

$$\text{2\% Pd}(PPh_3)_4, \text{ RT}$$
$$SmI_2, \text{ THF, 2h}$$

90%

Tabuchi, T.; Inanaga, J.[*]; Yamaguchi, M.
Tetrahedron Lett, (1986), **27**, 5237

SECTION 197: Olefins from Acid Derivatives

1. $CH_3CH_2CH_2MgBr$, THF
   $-78°C - -60°C$

2. $CH_2=CHCH_2MgBr$, Et_2O

3. Li, $-60°C - 20°C$

$$\underset{\displaystyle CH_3CH=CCH=CH_2}{\overset{\displaystyle CH_2CH_2CH_3}{|}}$$

87%

Barluenga, J.[*]; Yus, M.; Concellón, J.M.; Bernad, P.
J Org Chem, (1983), **48**, 609

$$EtO_2C\underset{N}{\overset{+PPh_3}{\underset{|}{\overset{|}{N}}}}NCO_2Et$$

$iPrCH=CHPh$

($\underline{Z}:\underline{E}$ = 98:2)

73%

Mulzer, J.[*]; Lammer, O.
Angew Chem Int Ed Engl, (1983), **22**, 628

$$
\begin{array}{c}
O \\
\parallel \\
iPrCH_2CCl
\end{array}
\quad
\begin{array}{l}
\text{1. } CH_2N_2, \ Et_2O, \ -20°C \\
\text{2. HCl, } Et_2O, \text{ overnight} \\
\qquad 0°C - RT \\
\text{3. } EtMgBr/MgBr_2, \text{ THF} \\
\text{4. Li, RT, overnight}
\end{array}
\quad
\begin{array}{c}
Et \\
\mid \\
iPrCH_2C=CH_2 \\
75\%
\end{array}
$$

Barluenga, J.[*]; Yus, M.; Concellon, J.M.; Bernad, P.
J Org Chem, (1983), **48**, 3116

$$
\xrightarrow[\text{DMF-AcOH, RT, 3h}]{Pb(OAc)_4, \ Cu(OAc)_2 \cdot H_2O}
$$

72%

Nishiyama, H.[*]; Matsumoto, M.; Arai, H.; Sakaguchi, H.; Itoh, K.
Tetrahedron Lett, (1986), **27**, 1599

SECTION 198: Olefins from Alcohols and Thiols

$$
\xrightarrow[\substack{Me_3SiCl, \ RT \\ 10 \ min}]{CH_3CN/NaI}
$$

60%

Sarma, D.N.; Sarma, J.C.; Barua, N.C.; Sharma, R.P.[*]
JCS Chem Comm, (1984), 813

$$
\xrightarrow[\text{reflux, 8h}]{TsOH/SiO_2}
$$

97%

D'Onofrio, F.; Scettri, A.[*]   **Synthesis**, (1985), 1159

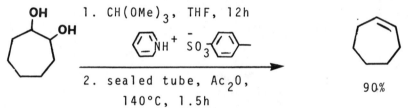

Fadel, A.; Salaun, J.[*]   **Tetrahedron**, (1985), **41**, 1267
                            Tetrahedron, (1985), **41**, 413

Araki, S.; Hatano, M.; Butsugan, Y.[*]
                    J Org Chem, (1986), **51**, 2126

Ando, M.[*]; Ohhara, H.; Takase, K.   **Chem Lett**, (1986), 879

SECTION 199: Olefins from Aldehydes

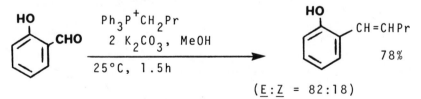

$(E:Z = 82:18)$

Le Bigot, Y.; Delmas, M.; Gaset, A.
                    Tetrahedron Lett, (1983), **24**, 193

$$Ph_3 \overset{+}{P} \overset{-}{C} H \overset{O}{\overset{\|}{C}} CH_2 CH_2 CO_2 Me, \quad DMF, \quad 250°C, \quad 22h \longrightarrow \qquad 70\%$$

Ronald, R.C.; Wheeler, C.J.[*]   **J Org Chem**, (1983), **48**, 138

$$2 \ ClMo \overset{\diagup O}{\underset{CH_2}{\diagdown CH_2}}, \quad THF, \quad -70°C - RT \longrightarrow \qquad 84\%$$

Kauffmann, T.[*]; Fiegenbaum, P.; Wieschollek, R.
                    Angew Chem Int Ed Engl, (1984), **23**, 531

$$CH_3(CH_2)_5 CHO \quad \frac{1. \ Cl_3 Mo = CH_2, \ THF}{Et_2 O, \ -70 - 20°C} \quad CH_3(CH_2)_5 CH = CH_2$$
2. HOH   65%

Kauffmann, T.; Enner, B.; Sandler, J.; Wieschollek, R.
                    Angew Chem Int Ed Engl, (1983), **22**, 244

1. $(iPrO)_3 TiCH_2 CH = CHPPh_2$
   -78°C - 0°C

2. MeI, 0°C - RT   86%

($\underline{E}:\underline{Z}$ = 4:96)

Ukai, J.; Ikeda, Y.; Ikeda, N.; Yamamoto, H.[*]
                    Tetrahedron Lett, (1983), **24**, 4029

$$2 \ PhCHO \quad \frac{WCl_6, \ e^-, \ THF}{4h} \quad PhCH = CHPh$$
($\underline{E}:\underline{Z}$ = 88:9)   97%

Petit, M.; Mortreux, A.; Petit, F.[*]   **JCS Chem Comm**, (1984), 341

$$PhCHO \quad \frac{Cp_2 Zr(PPh_3) = CHCH_2 \underline{t}Bu}{PhCH_3, \ 3h} \quad PhCH = CHCH_2 \underline{t}Bu$$
($\underline{E}:\underline{Z}$ = 1:2)   76%

Clift, S.M.; Schwartz, J.[*]   **J Am Chem Soc**, (1984), **106**, 8300

$$CH_3(CH_2)_5CHO \xrightarrow[\substack{RT, \ 20 \ min \\ \text{))))}}]{CH_2I_2, \ Zn, \ THF} CH_3(CH_2)_5CH=CH_2$$

71%

Yamashita, J.[*]; Inoue, Y.; Kondo, T.; Hasimoto, H.
Bull Chem Soc Jpn, (1984), **57**, 2335

$$CH_2=CHPPh_3 \overset{+}{} Br^- \xrightarrow[\substack{2. \ \underline{n}C_5H_{11}CHO \\ -50 - -20°C \\ 3. \ aq. \ NH_4Cl}]{\substack{1. \ \underline{n}Bu_2CuLi, \ THF \\ HMPA, \ -50°C, \ 9h}} \underline{n}BuCH_2CH=CH\underline{n}C_5H_{11}$$

(Z:E = 1:4)   58%

Just, G.; O'Connor, B.   **Tetrahedron Lett**, (1985), **26**, 1799

PhCHO

$$\text{(arene)(}\underline{t}Bu\text{)}(O)_3\overset{O}{\overset{\|}{W}}=CH\underline{t}Bu$$

$$\xrightarrow[25°C]{} PhCH=CH\underline{t}Bu$$

95%

(Z:E = 1:4)

Freudenberger, J.H.; Shrock, R.R.[*]
                  Organometallics, (1986), **5**, 398

PhCHO

$$\xrightarrow[PhH, \ 25:C, \ 5 \ min]{(cyclopentylidene)WI_2(OCH_2\underline{t}Bu)_2}$$

(cyclopentylidene)=CHPh

88%

Aguero, A.; Kress, J.; Osborn, J.A.   **JCS Chem Comm**, (1986), 531

Related Methods: Olefins from Ketones (Section 207)

SECTION 200: Olefins from Alkyls, Methylenes, and Aryls

This section contains dehydrogenations to form olefins and unsaturated ketones, esters, and amides. It also includes the conversion of aromatic rings to olefins. Reduction of aryls to dienes is found in Section 377 (Olefin-Olefin). Hydrogenation of aryls to alkanes and dehydrogenations to form aryls are included in Section 74 (Alkyls, Methylenes, and Aryls from Olefins).

$$CH_3(CH_2)_6CH_3 \quad \xrightarrow[\text{2. } (MeO)_3P]{\begin{array}{c} \text{1. } CH_2=CH\underline{t}Bu \\ (Ph_3P)_2ReH_2 \end{array}} \quad CH_3(CH_2)_4CH=CH_2$$

$$95\%$$

Baudry, D.[*]; Ephritikhine, M.[*]; Felkin, H.; Zakrzewski, J.
Tetrahedron Lett, (1984), **25**, 1283
Baudry, D.[*]; Ephritikhine, M.[*]; Felkin, H.; Holmes-Smith, R.
JCS Chem Comm, (1983), 788

$$\xrightarrow[\text{2. } Et_2O, \text{ aq. } NH_4Cl]{\begin{array}{c} \text{1. } CH_3NH_2, \text{ DEA, Ca} \\ -25°C \end{array}}$$

76%   +   20%

Benkeser, R.A.[*]; Belmonte, F.G.; Kang, J.
J Org Chem, (1983), **48**, 2796

SECTION 201: Olefins from Amides

No Additional Examples

Related Methods: Alkyls and Aryls from Alkyls and Aryls
(Section 65)
Alkyls and Aryls from Olefins (Section 74)

SECTION 202: Olefins from Amines

No Additional Examples

SECTION 203: Olefins from Esters

$$Ph_2\overset{\overset{\displaystyle O}{\displaystyle \|}}{P}CH_2CH_2iPr \quad \xrightarrow[\begin{array}{l}3. \ NaBH_4, \ EtOH\\4. \ NaH. \ DMF\end{array}]{\begin{array}{l}1. \ \underline{n}BuLi\\2. \ PhCO_2Et\end{array}} \quad \begin{array}{c}PhCH=CHCH_2iPr\\49\%\end{array}$$

Buss, A.D.; Mason, R.; Warren, S.[*]
                    Tetrahedron Lett, (1983), **24**, 5293

$$\underline{n}C_8H_{17}CO_2Et \quad \xrightarrow[\begin{array}{l}3. \ LiAlH_4\\4. \ BF_3 \cdot OEt_2, \ CH_2Cl_2\end{array}]{\begin{array}{l}1. \ LDA\\2. \ Ph_2MeSiCl\end{array}} \quad \begin{array}{c}\underline{n}C_8H_{17}CH=CH_2\\85\%\end{array}$$

De Maldonado, V.C.; Larson, G.L.[*]   **Syn Commun**, (1983), **13**, 1163

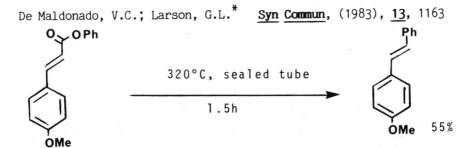

Chamchaang, W.; Chantarasiri, N.; Chaona, S.; Thebtaranonth,
C.[*]; Thebtaranonth, Y.
                    Tetrahedron, (1984), **40**, 1727

$nC_8H_{17}\overset{\displaystyle CO_2Et}{\underset{\displaystyle SiPh_2Me}{CH}}$
$\xrightarrow[\begin{array}{l}3.\ 9\ \underline{t}BuO^-K^+,\ RT\\ \qquad\qquad 1h\end{array}]{\begin{array}{l}1.\ 2\ MeMgI,\ THF\\ \quad reflux,\ 24h\\ 2.\ 3\ MeLi,\ reflux\end{array}}$
$nC_8H_{17}CH=CMe_2$

54%

Hernández, D.; Larson, G.L.*   J Org Chem, (1984), 49, 4285

$CH_3CH=CHCO_2Et$
$\xrightarrow[\begin{array}{l}\qquad\qquad\qquad NO_2\\ 3.\ \text{(arene)} SeCN,\ Bu_3P\\ 4.\ H_2O_2\end{array}]{\begin{array}{l}1.\ PhMe_2Si)_2CuLi\\ 2.\ LiAlH_4\end{array}}$
$PhMe_2Si\overset{\displaystyle CH_3}{\underset{}{CH}}CH=CH_2$

50%

Fleming, I.*; Waterson, D.   JCS Perkin I, (1984), 1809

$PhCO_2Et$
$\xrightarrow[\begin{array}{l}2.\ NaBH_4,\ CeCl_3,\ MeOH\\ \qquad\qquad -78°C\\ 3.\ base\end{array}]{1.\ \text{(dibenzophosphole)} ,\ THF}$
$PhCH=CHCH_3$
(E)

96%

Elliott, J.; Warren, S.*   Tetrahedron Lett, (1986), 27, 645

SECTION 204: Olefins from Ethers, Epoxides, and Thioethers

$nC_{12}H_{25}\overset{\displaystyle S}{\overset{\displaystyle \diagup\!\diagdown}{CHCH_2}}$
$\xrightarrow[80°C,\ 12h]{P_2I_4,\ DMF}$
$nC_{12}H_{25}CH=CH_2$

97%

Schauder, J.R.; Denis, J.N.; Krief, A.*
                    Tetrahedron Lett, (1983), 24, 1657

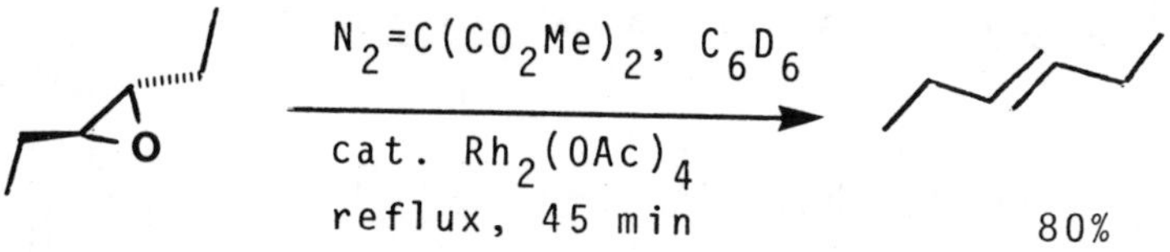

$$N_2=C(CO_2Me)_2, \quad C_6D_6$$

cat. $Rh_2(OAc)_4$

reflux, 45 min

80%

Martin, M.G.; Ganem, B.[*]   **Tetrahedron** **Lett**, (1984), **25**, 251

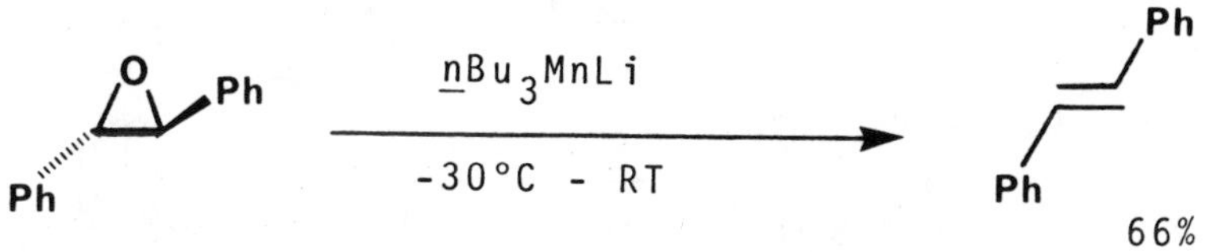

$\underline{n}Bu_3MnLi$

$-30°C$ - RT

66%

Kauffmann, T.[*]; Bisling, M.   **Tetrahedron** **Lett**, (1984), **25**, 293

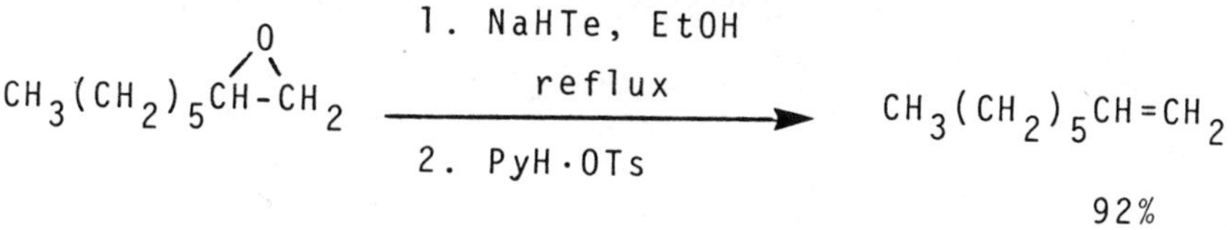

$CH_3(CH_2)_5CH-CH_2$

1. NaHTe, EtOH
   reflux

2. PyH·OTs

$CH_3(CH_2)_5CH=CH_2$

92%

Barton, D.H.R.[*]; Fekih, A.; Lusinchi, X.

Tetrahedron **Lett**, (1985), **26**, 6197

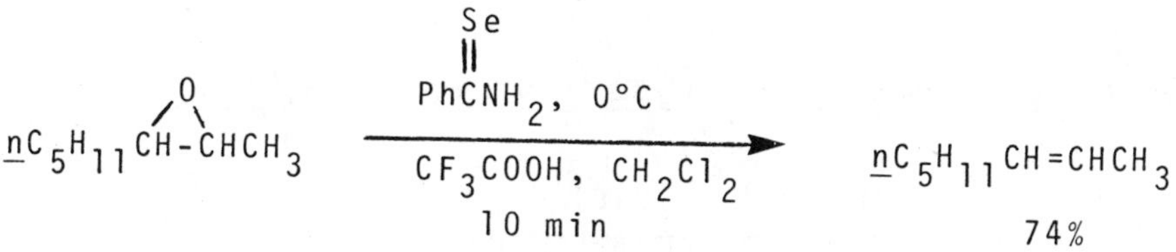

$\underline{n}C_5H_{11}CH-CHCH_3$

$$\overset{Se}{\overset{\|}{PhCNH_2}}, \quad 0°C$$

$CF_3COOH, \quad CH_2Cl_2$

10 min

$\underline{n}C_5H_{11}CH=CHCH_3$

74%

Ogawa, A.; Miyake, J.; Murai, S.; Sonoda, N.[*]

Tetrahedron **Lett**, (1985), **26**, 669

SECTION 205: Olefins from Halides and Sulfonates

$$CH_2=CH_2, \ NiI_2(PPh_3)_2$$
$$\overline{e^-, \ THF/HMPA}$$

80%

Rollin, Y.; Meyer, G.; Troupel, M.; Fauvarque, J.-F.; Perichon, J.
JCS Chem Comm, (1983), 793

$$\begin{array}{c} MeOH, \ 1\% \ dvb, \ 50°C \\ \hline \textcircled{P}CH_2S^-Na^+, \ 3.5h \end{array}$$

95%

$\textcircled{P}$ = poly[styrene (99) co-divinylbenzene]

Janout, V.; Čefelín, P.
Tetrahedron Lett, (1983), **24**, 3913

1. (PhTe)$_2$, NaBH$_4$, EtOH
   reflux
2. Br$_2$, CCl$_4$, 0°C
3. aq. NaOH, 25°C
   3h

62%

Uemura, S.*; Fukuzawa, S.   **J Am Chem Soc**, (1983), **105**, 2748

$$\begin{array}{c} Cl \\ | \\ PhCHCH_3 \end{array}$$
1. CH$_2$=CHBr, NiCl$_2$, Et$_2$O
   Me$_2$N-CH$\begin{array}{c}CH_2PPh_2\\CH_2SMe\end{array}$, RT
2. 10% HCl, 0°C
$$\begin{array}{c} CH_3 \\ | \\ PhCHCH=CH_2 \end{array}$$
90%

(23% ee, S)

Griffin, J.H.; Kellogg, R.M.*   **J Org Chem**, (1985), **50**, 3261

1. Mg, Et_2O
2. $ZnCl_2$, THF/Et_2O
3. $CH_2=CHBr$, THF/Et_2O
   cat., 0°C

95%

(85% ee, S)

cat. =

Hayashi, T.; Hagihara, T.; Katsuro, Y.; Kumada, M.[*]
Bull Chem Soc Jpn, (1983), **56**, 363
Hayashi, T.[*]; Konishi, M.; Okamoto, Y.; Kabeta, K.; Kumada, M.
J Org Chem, (1986), **51**, 3772

$$\underset{\substack{| \\ PhCHCH_2Br}}{Br} \qquad
\begin{array}{l}
\text{1. } C_8K, \text{ PhH, RT, 8h} \\
\text{2. EtOH} \\
\hline
\text{3. HOH}
\end{array}
\qquad PhCH=CH_2$$

80%

C_8K = graphite powder/200°C/K

Rabinovitz, M.[*]; Tamarkin, D.   **Syn Commun**, (1984), **14**, 377

$$\underset{\substack{| \\ Br}}{\overset{\substack{Br \\ |}}{PhCHCHPh}} \qquad
\begin{array}{c}
\text{aq. } Na_2S \cdot 9H_2O, \text{ PhCH}_3 \\
\hline
Bu_3P^+C_{16}H_{33} \text{ } Br^- \\
25°C, \text{ 95 min}
\end{array}
\qquad PhCH=CHPh$$

quant.

Landini, D.[*]; Milesi, L.; Quadri, M.L.; Rolla, F.
J Org Chem, (1984), **49**, 152

1. $\underline{n}Bu_2CuLi$, Et_2O
   -20°C
2. HOH

n-Bu

80%

($\underline{E}$:$\underline{Z}$ = 2:1)

Hrubiec, R.T.; Smith, M.B.[*]   **J Org Chem**, (1984), **49**, 385

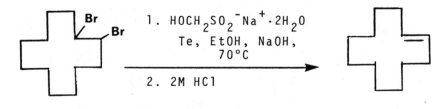

$$\underset{t\text{-Bu}}{\overset{\text{OTs}}{\bigcirc}} \quad \xrightarrow[\substack{2 \text{ LiCl, THF, 70°C} \\ 41h}]{nBu_4Sn, \ Pd(PPh_3)_4} \quad \underset{t\text{-Bu}}{\overset{n\text{-Bu}}{\bigcirc}}$$

80%

Scott, W.J.; Crisp, G.T.; Stille, J.K.*
                              J Am Chem Soc, (1984), **106**, 4630

$$\underset{\underset{Br}{|}}{\underset{|}{nPrCHCHCH_3}} \quad \xrightarrow[15 \text{ min}]{Cp_2TiCl_2, \ THF} \quad \underline{n}PrCH=CHCH_3$$

                                                                95%

Davies, S.G.*; Thomas, S.E.   **Synthesis**, (1984), 1027

$$\underset{CH_3C(CH_2)_5Cl}{\overset{\overset{NNHTRIS}{\|}}{}} \quad \xrightarrow[\substack{2. \ 0°C, \ 10 \ min}]{\substack{1. \ \underline{t}BuLi, \ THF, \ -78°C \\ Li_2CuCl_4}} \quad \text{(methylenecyclohexane)}$$

                                                                66%

Chamberlin, A.R.*; Bloom, S.H.
                              Tetrahedron Lett, (1984), **25**, 4901

$$\text{(cross, Br, Br)} \quad \xrightarrow[\substack{2. \ 2M \ HCl}]{\substack{1. \ HOCH_2SO_2^-Na^+ \cdot 2H_2O \\ Te, \ EtOH, \ NaOH, \\ 70°C}} \quad \text{(cross with double bond)}$$

                                                                77%

Suzuki, H.*; Inouye, M.   **Chem Lett**, (1985), 225

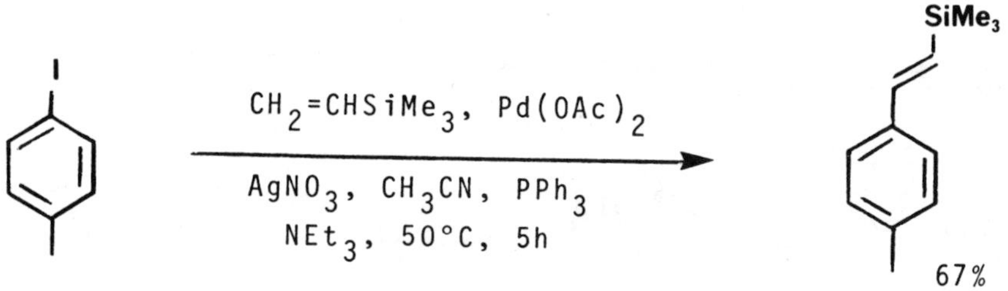
SiMe3
I
CH2=CHSiMe3, Pd(OAc)2
AgNO3, CH3CN, PPh3
NEt3, 50°C, 5h
67%

Karabelas, K.; Hallberg, A.*
Tetrahedron Lett, (1985), **26**, 3131

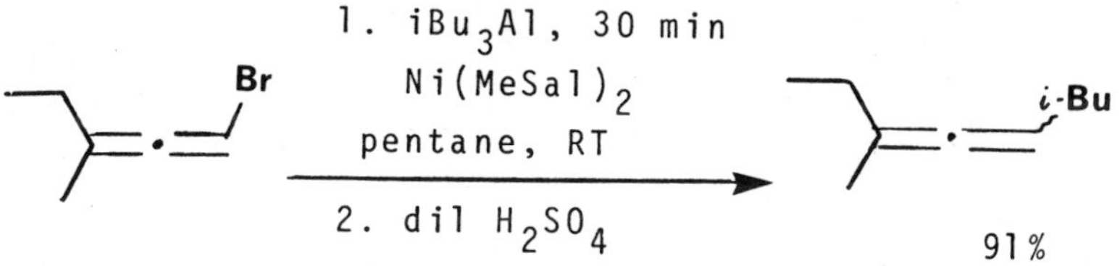
1. iBu3Al, 30 min
Ni(MeSal)2
pentane, RT
2. dil H2SO4
Br
i-Bu
91%

MeSal = N-methylsalycilaldimine

Caporusso, A.M.*; Da Settimo, F.; Lardicci, L.
Tetrahedron Lett, (1985), **26**, 5101

BrCh=CHPh (Z) $\xrightarrow[Et_2O/THF,\ -78°C,\ 3h]{(Me_3Si)_3MnMgMe}$ PhCH=CHSiMe$_3$. (Z) 70%

Fugami, K.; Oshima, K.*; Utimoto, K.; Nozaki, H.
Tetrahedron Lett, (1986), **27**, 2161

SECTION 206: Olefins from Hydrides

No Additional Examples

For conversions of methylenes to olefins (RCH$_2$R' → RR'C=CH$_2$, see Section 200.

SECTION 207: Olefins from Ketones

$$\underset{\text{PhCPh}}{\overset{\displaystyle\text{O}}{\|}} \xrightarrow{\quad (MeS)_2\overset{+\ -}{B}CH_2 \quad} Ph_2C=CH_2$$

75%

Pelter, A.; Singaram, B.; Wilson, J.W.
Tetrahedron Lett, (1983), 24, 635

$$\underset{\text{PhCPh}}{\overset{\displaystyle\text{O}}{\|}} \xrightarrow[\text{PhCH}_3, \ -10°C]{\quad Cp_2TiCH_2 \cdot ZnI_2 \quad} Ph_2C=CH_2$$

78%

Eisch, J.J.*; Piotrowski, A.
Tetrahedron Lett, (1983), 24, 2043

$$\xrightarrow[\text{8h}]{\quad Cp_2Ti \quad \text{(cyclic)} \quad}{Et_2O, \ 0°C - RT}$$

84%

Clawson, L.; Buchwald, S.L.; Grubbs, R.H.*
Tetrahedron Lett, (1984), 25, 5733

$$\xrightarrow[\underset{\displaystyle 0}{CH_3\overset{\|}{S}CH_2Na}, \ -60°C]{\quad \text{(P)}-PPh_2Me, \ DMSO \quad}$$

99%

Bernard, M.; Ford, W.J.*   J Org Chem, (1983), 48, 326

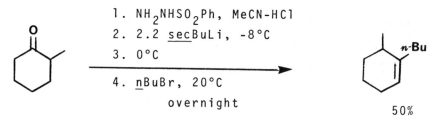

1. NH_2NHSO_2Ph, MeCN-HCl
2. 2.2 <u>sec</u>BuLi, -8°C
3. 0°C

4. <u>n</u>BuBr, 20°C
     overnight

50%

Chamberlin, A.R.[*]; Liotta, E.L.; Bond, F.T.
                    Org Syn, (1983), **61**, 141

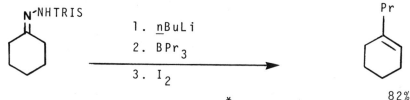

1. <u>n</u>BuLi
2. BPr_3
3. I_2

82%

Baba, T.; Avasthi, K.; Suzuki, A.[*]
                    Bull Chem Soc Jpn, (1983), **56**, 1571

O
‖
PhCiPr

1. H_2N-N△Ph

2. decalin, 160°C
     2h

$PhCH=CMe_2$

92%

Mohamadi, F.; Collum, D.B.[*]   **Tetrahedron Lett**, (1984), **25**, 271

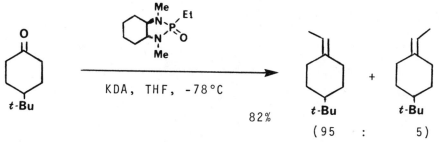

KDA, THF, -78°C

82%

(95   :   5)

Hanessian, S.[*]; Delorme, D.; Beaudoin, S.; Leblanc, Y.
                    J Am Chem Soc, (1984), **106**, 5754

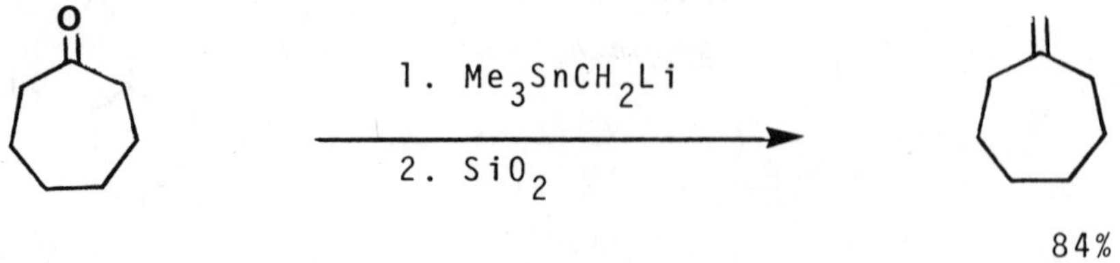

84%

Murayama, E.; Kikuchi, T.; Sasaki, K.; Sootome, N.; Sato, T.*
Chem Lett, (1984), 1897

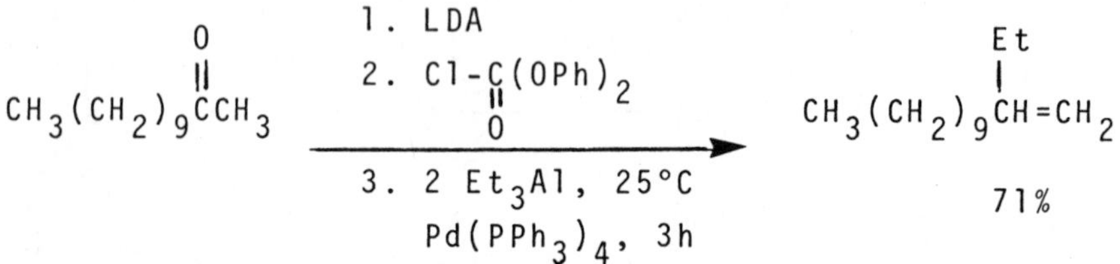

Takai, K.; Sato, M.; Oshima, K.*; Nozaki, H.
Bull Chem Soc Jpn, (1984), **57**, 108

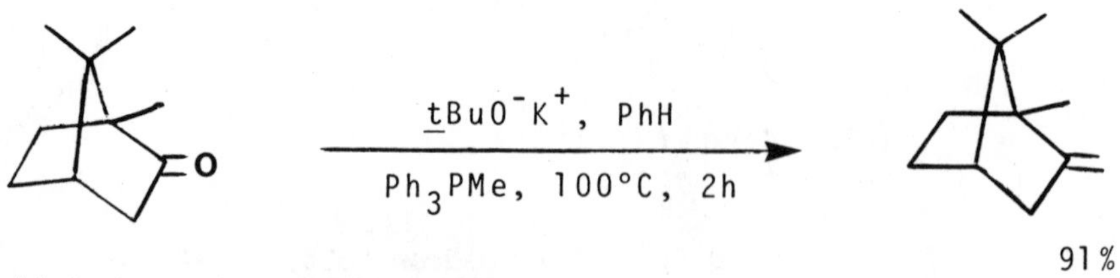

91%

add ketone to mixture

Fitjer, L.*; Quabeck, U.   **Syn Commun**, (1985), **15**, 855

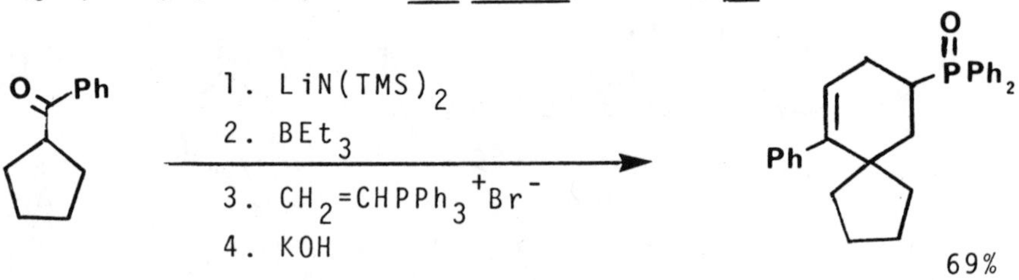

69%

Posner, G.H.*; Lu, S.-B.   **J Am Chem Soc**, (1985), **107**, 1424

$$\text{1. ClCH}_2\text{I, MeLi}$$
$$-78 - -60°\text{C}$$
$$\text{2. Li, -60 - 20°C}$$

80%

Barluenga, J.[*]; Fernandez-Simon, J.L.; Concellon, J.M.; Yus, M.
           JCS Chem Comm, (1986), 1665

Review: "Titanium Induced Dicarbonyl-Coupling Reactions"

McMurry, J.E.[*]   **Accts Chem Res**, (1983), **16**, 405

Related Methods: Olefins from Aldehydes (Section 199)

SECTION 208: Olefins from Nitriles

1. LDA, THF, -78°C

$\underline{n}C_{12}H_{25}CHCH_3$ (with CN)

2. $\underline{n}Bu_3SnCH_2I$, -78 - 0°C

3. MeLi, LiBr/Et$_2$O
   THF, -20°C

4. aq. NaOH

$\underline{n}C_{12}H_{25}C=CH_2$ (with Me)

83%

Pearlman, B.A.[*]; Putt, S.R.; Fleming, J.A.
           J Org Chem, (1985), **50**, 3625

SECTION 209: Olefins from Olefins

$PhSO_2CH=CH\underline{n}C_8H_{17}$

1. $\underline{n}Bu_3SnLi$

2. CHCl$_3$, SiO$_2$
   60°C, 17h

$\underline{n}C_8H_{17}CH=CH_2$

78%

Ochiai, M.; Ukita, T.; Fujita, E.[*]   **JCS Chem Comm**, (1983), 619
                                         Chem Lett, (1983), 1457

$$\begin{array}{c} Fp^{+} \\ | \\ CH \ddagger CH_2 \\ | \\ (CH_2)_4 \\ | \\ CH \ddagger CH_2 \\ | \\ Fp^{+} \end{array}$$

1. CH_2Cl_2, $HBF_4 \cdot OEt_2$, 0°C
2. NaI, acetone

3. HCl, CH_2Cl_2, RT, 4h
4. $\underline{n}BuNI$, CH_2Cl_2, RT

52%

Lennon, P.; Rosenblum, M.[*]   J Am Chem Soc, (1983), 105, 1233

$Mo(CO)_6$, $PhCH_3$, 2h

$(Me_3Si)_3Al \cdot OEt_2$
reflux

72%

$(\underline{E}:\underline{Z} = 2:1)$

Trost, B.M.[*]; Yoshida, J.; Lautens, M.
                    J Am Chem Soc, (1983), 105, 4494

Cp_2TiCl_2, THF

$iPrMgBr$, 20°C
1h

76%   +   16%

Lehmkuhl, H.[*]; Tsien, Y.-L.[*]   Chem Ber, (1983), 116, 2437

1. Cp_2Ti

2. PhCPh, PhH, RT, 12h
      ‖
      O

Ph   Ph

80%

Buchwald, S.L.; Grubbs, R.H.[*]   J Am Chem Soc, (1983), 105, 5490

$$CH_2=CH-\underset{\underset{CH_3}{|}}{\overset{\overset{NO_2}{|}}{C}}-CH_2CH_2CO_2Et \xrightarrow[\substack{Et_2O \\ -30°C - RT \\ 15h}]{nBu_2CuLi} nBuCH_2CH=\underset{\overset{|}{CH_3}}{C}CH_2CH_2CO_2Et$$

76%

$(\underline{E}:\underline{Z} = 88:12)$

Ono, N.[*]; Hamamoto, I.; Kaji, A.   **JCS Chem Comm**, (1984), 274

$$PhCH_2CH=CHCH_3 \xrightarrow[60°C, \ 30 \ min]{10 \ KOH, \ DME} PhCH=CHCH_2CH_3$$

95%

D'Incan, E.[*]; Viout, P.   **Tetrahedron**, (1984), **40**, 4321

$\xrightarrow[0°C, \ 12 \ min]{\substack{PhMe_2SiMgMe \\ CuI}}$

SiMe$_2$Ph   SiMe$_2$Ph

(95 : 5)

74%

Morizawa, Y.; Oda, H.; Oshima, K.[*]; Nozaki, H.
                Tetrahedron Lett, (1984), **25**, 1163

1. $O=S=NSO_2Ph$, Et_2O
   $0 - 20°C$, 10h

2. EtMgCl, CuBr/Me$_2$S

3. aq. H[+]

80%

Deleris, G.[*]; Dunogues, J.; Gadras, A.
                Tetrahedron Lett, (1984), **25**, 2135

$(PhSO_2)_2C(CH_2CH=CH_2)_2$ →
(PhHgCl, CuCl$_2$ / 10% PdCl$_2$, THF RT)

PhSO$_2$ / PhSO$_2$ —— Bn   50%

Trost, B.M.*; Burgess, K.   **JCS Chem Comm**, (1985), 1084

(structure with NO$_2$ on cyclopentene) →
PhSO$_2$Na·2 H$_2$O, DMF / 5% Pd(PPh$_3$)$_4$, 20°C / 10h
→ (structure with SO$_2$Ph on cyclopentene)   70%

Ono, N.*; Hamamoto, I.; Kawai, T.; Kaji, A.; Tamura, R.*;
Kakihana, M.

Bull Chem Soc Jpn, (1986), **59**, 405

Tamura, R.*; Hayashi, K.; Kakihana, M.; Tsuji, M.; Oda, D.;
Scholz, D.*

Tetrahedron Lett, (1985), **26**, 851

(structure SiMe$_3$, SO$_2$Ar) →
1. $\underline{n}$BuLi, THF, -78 - 0°C
2. aq. NH$_4$Cl
→ (structure SiMe$_3$, n-Bu)   90%

Padwa, A.*; Wannamaker, M.W.
Tetrahedron Lett, (1986), **27**, 5817

SECTION 210: Olefins from Miscellaneous Compounds

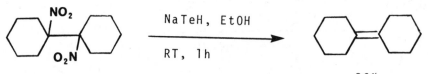

(structure with NO$_2$ and O$_2$N dicyclohexyl) →
NaTeH, EtOH / RT, 1h
→ (dicyclohexylidene)   86%

Osuka, A.*; Shimizu, H.; Suzuki, H.*   **Chem Lett**, (1983), 1373

$$PhCCH_2SO_2Me \quad \begin{array}{l} \text{1. NaH, } \underline{n}BuLi \\ \text{2. } \underline{n}C_8H_{17}I \\ \text{3. NaOH, } Br_2 \\ \text{4. NaOEt, EtOH, reflux} \end{array} \longrightarrow \quad \underline{n}C_8H_{17}CH=CH_2$$

(O double bond above PhCC)

75%

Scholz, D.[*]   **Liebigs Ann Chem**, (1983), 98

$$PhCH_2SPh \quad \begin{array}{l} \text{1. } \underline{n}BuLi, \text{ THF, } 0°C \\ \text{2. 24\% Te, RT, 3h} \\ \text{3. 50\% aq. AcOH} \end{array} \longrightarrow \quad PhCH=CHPh$$

(O double bonds on S)

90%

($\underline{Z}$:$\underline{E}$ = 2:8)

Engman, L.[*]   **J Org Chem**, (1984), **49**, 3559

$$\begin{array}{c} NNHTs \\ \| \\ PhC(CH_2)_4OH \end{array} \quad \begin{array}{l} \text{1. 2.2 KH, diglyme} \\ \\ \text{2. 80°C, 6h} \end{array} \longrightarrow$$

74%

Harada, T.; Akiba, E.; Oku, A.[*]   **Tetrahedron Lett**, (1985), **26**, 651

Reviews:

"Carbene Complexes in Organic Synthesis"

Dötz, K.H.   **Angew Chem Int Ed Engl**, (1984), **23**, 587

"The Peterson Reaction"

Ager, D.J.[*]   **Synthesis**, (1984), 384

"Recent Developments in Alkene Chemistry"

Pasto, D.J.[*]   **Tetrahedron**, (1984), **40**, 2805

CHAPTER 15

PREPARATION OF OXIDES

This Chapter contains reactions which prepare the oxides of nitrogen, sulfur, and selenium. Included are N-oxides, nitroso, nitro compounds, nitrile oxides, sulfoxides, selenoxides, and sulfones. Oximes are found in Sections 60A (Protection of Aldehydes) and 180A (Protection of Ketones). Preparation of sulfonic acid derivatives are found in Chapter Two and the preparation of sulfonates in Chapter Ten.

SECTION 211: Oxides from Acetylenes

No Additional Examples

SECTION 212: Oxides from Acid Derivatives

No Additional Examples

SECTION 213: Oxides from Alcohols and Thiols

$$PhCH_2CH_2S^-Na^+ \xrightarrow[\begin{array}{c}\text{3. } NaBH_4, \text{ MeOH}\\ \text{4. } BnBr\end{array}]{\begin{array}{c}\text{1. } \\ \text{2. } KMnO_4, \text{ aq. AcOH}\end{array}} BnSO_2Bn$$

84%

Ueno, Y.[*]; Kojima, A.; Okawara, M.   **Chem Lett**, (1984), 2125

SECTION 214: Oxides from Aldehydes

No Additional Examples

SECTION 215: Oxides from Alkyls, Methylenes, and Aryls

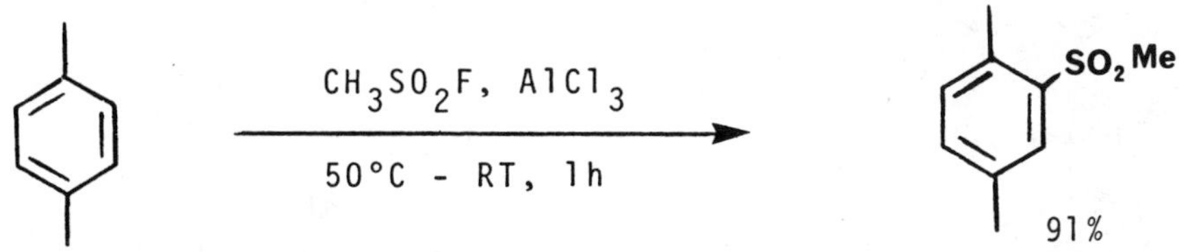

$$CH_3SO_2F, \ AlCl_3$$
$$50°C - RT, \ 1h$$

SO_2Me

91%

Hyatt, J.A.[*]; White, A.W.   **Synthesis**, (1984), 214

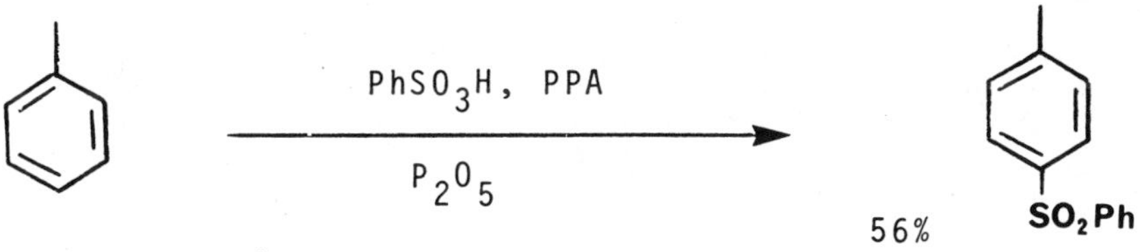

$$PhSO_3H, \ PPA$$
$$P_2O_5$$

56% SO_2Ph

Sipe Jr., H.J.[*]; Clary, D.W.; White, S.B.
              Synthesis, (1984), 283

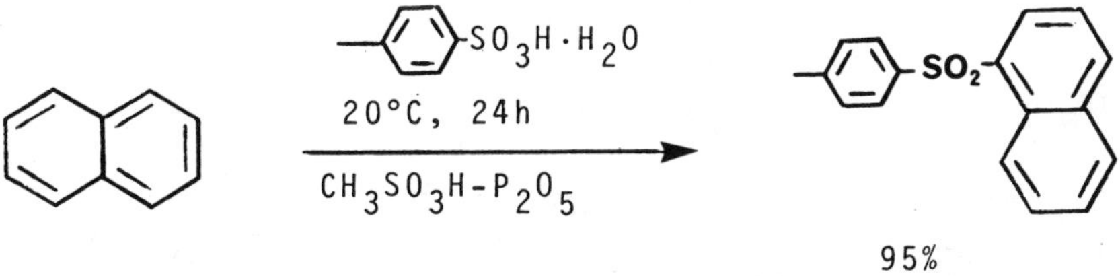

$$SO_3H \cdot H_2O$$
$$20°C, \ 24h$$
$$CH_3SO_3H-P_2O_5$$

SO_2

95%

Ueda, M.[*]; Uchiyama, K.; Kano, T.
              Synthesis, (1984), 323

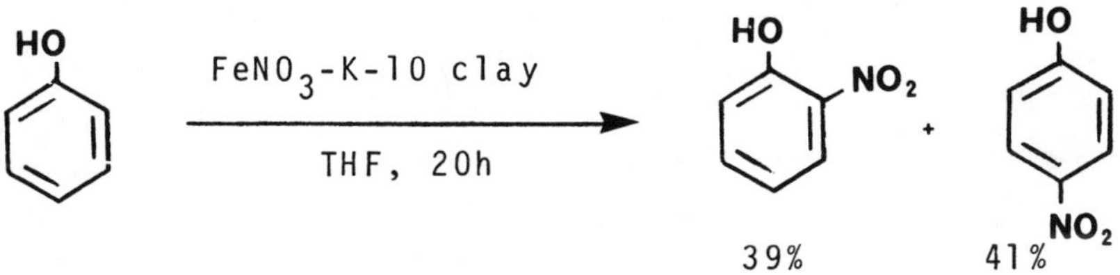

$$FeNO_3-K-10 \ clay$$
$$THF, \ 20h$$

NO_2

+

NO_2

39%   41%

Cornelis, A.; Laszlo, P.[*]; Pennetreau, P.
              Bull Chem Soc Belg, (1984), **93**, 961

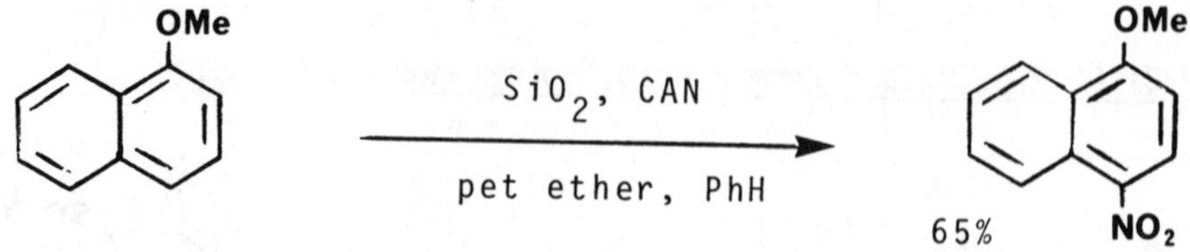

65%

Chawla, H.M.[*]; Mittal, R.S.   **Synthesis**, (1985), 70

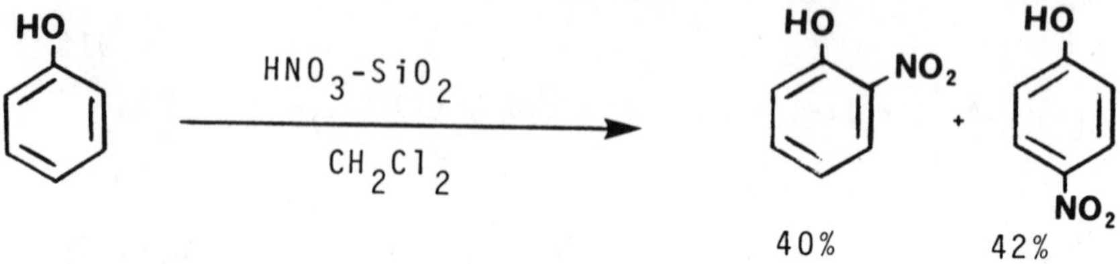

40%   42%

Tapia, R.[;] Torres, G.[;] Valderrama, J.A.
Syn Commun, (1986), **16**, 681

SECTION 216: Oxides from Amides

No Additional Examples

SECTION 217: Oxides from Amines

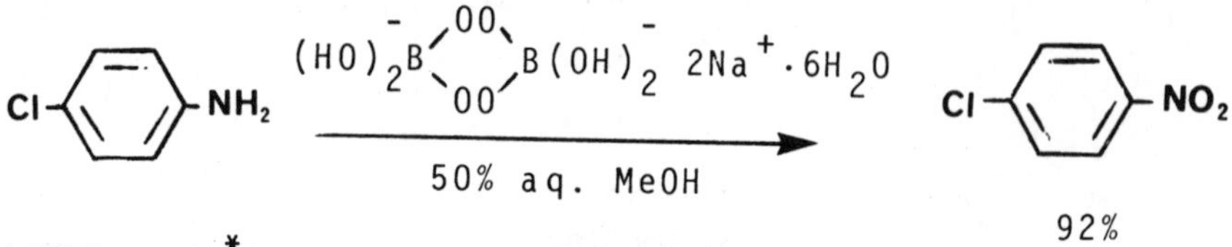

50% aq. MeOH

92%

McKillop, A.[*]; Tarbin, J.A   **Tetrahedron Lett**, (1983), **24**, 1505

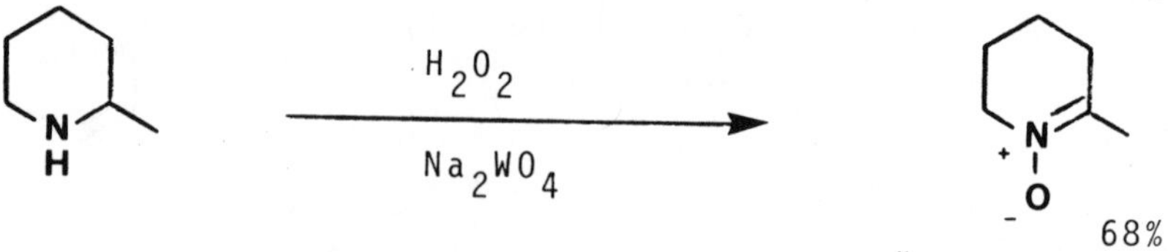

68%

Mitsui, H.[;] Zenki, S.[;] Shiota, T.[;] Murahashi, S.[*]
JCS Chem Comm, (1984), 874

SECTION 218: Oxides from Esters

No Additional Examples

SECTION 219: Oxides from Ethers, Epoxides, and Thioethers

$$PhSCH_3 \xrightarrow[\text{50\% aq. MeOH}]{(HO)_2B\overset{OO}{\underset{OO}{\diagup\diagdown}}B(OH)_2 \cdot 6H_2O} \underset{\underset{74\%(1\ eq)}{O}}{\overset{O}{\underset{\parallel}{PhSCH_3}}} + \underset{\underset{94\%(excess)}{O}}{\overset{O}{\underset{\parallel}{PhSCH_3}}}$$

McKillop, A.[*]; Tarbin, J.A   **Tetrahedron Lett**, (1983), **24**, 1505

$$PhC{\equiv}CSMe \xrightarrow[RuCl_2(PPh_3)_3]{\overset{\displaystyle COOH}{\underset{}{\bigcirc\text{--}IO\ ,\ 4h}}} \overset{O}{\underset{\parallel}{PhC{\equiv}CSMe}}$$

Müller, P.; Godoy, J.   **Helv Chim Acta**, (1983), **66**, 1790

$$BnSCH_2CH_2CH_2Br \xrightarrow[\underset{4h}{Bu_4N^+\ AuClO_4^-}]{CH_3NO_2,\ aq.\ HNO_3} \underset{89\%}{\overset{O}{\underset{\parallel}{BnSCH_2CH_2CH_2Br}}}$$

Gasparrini, F.; Giovannoli, M.; Misiti, D.; Natile, G.;
Palmieri, G.
                    Tetrahedron, (1983), **39**, 3181

OH
|
CH$_3$CHOOOH

Pr-S-Pr $\xrightarrow{}$

-40°C, 30 min

O
‖
Pr-S-Pr

no yield

Shereshovets, V.V.; Komissarov, V.D.; Shafikov, N.Ya.;
Bachanova, L.A.; Kolosnitsin, V.S.; Nikitin, Yu.E.; Tolstikov,
G.A.

J Org Chem USSR, (1983), **19**, 207

1. CH$_3$NO$_2$, aq. HNO$_3$

Ph-S-iPr $\xrightarrow{\text{aq. H}_2\text{SO}_4\text{, RT, 1h}}$

2. HOH

O
‖
Ph-S-iPr

93%

Gasparrini, F.[*]; Giovannoli, M.; Maresca, L.; Natile, G.;
Palmieri, G.

Syn Commun, (1984), **14**, 1111

PhCH$_2$-S-CH$_3$ $\xrightarrow[\text{(+)-diethyl tartrate}]{\text{tBuOOH, Ti(OiPr)}_4}$

CH$_2$Cl$_2$, -77°C, 24h

O
PhCH$_2$-S-CH$_3$

70%

(46% ee)

DiFuria, F.; Modena, G.; Seraglia, R.   **Synthesis**, (1984), 325

Ti(OiPr)$_4$, tBuOOH

Ph-S-CH$_3$ $\xrightarrow[\text{HOH, -20°C}]{\text{(R,R)-diethyl tartrate}}$

O
Ph-S-CH$_3$

80%

(89% ee, R)

Pitchen, P.; Duñach, E.; Deshmukh, M.N.; Kagan, H.B.[*]
[*]**J Am Chem Soc**, (1984), **106**, 8188
Pitchen, P.; Kagan, H.B.[*] **Tetrahedron Lett**, (1984), **25**, 1049

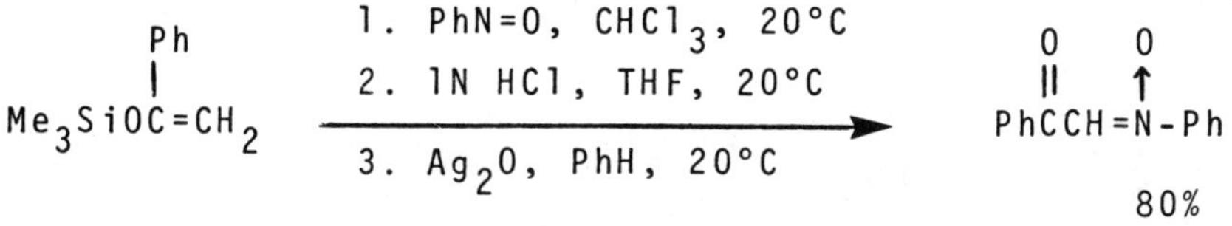
Ph
Me3SiOC=CH2
1. PhN=O, CHCl3, 20°C
2. 1N HCl, THF, 20°C
3. Ag2O, PhH, 20°C
O O
PhCCH=N-Ph
80%

Sasaki, T.; Mori, K.; Ohno, M.* **Synthesis**, (1985), 279

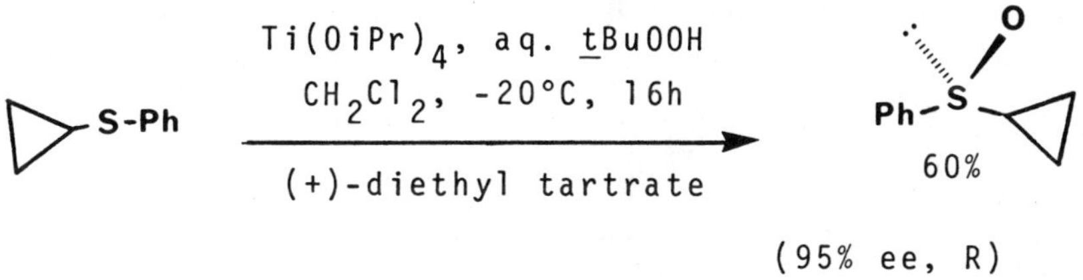
S-Ph
Ti(OiPr)4, aq. tBuOOH
CH2Cl2, -20°C, 16h
(+)-diethyl tartrate
O
Ph-S
60%
(95% ee, R)

Duñach, E.; Kagan, H.B.* **Nouv J Chem**, (1985), **9**, 1

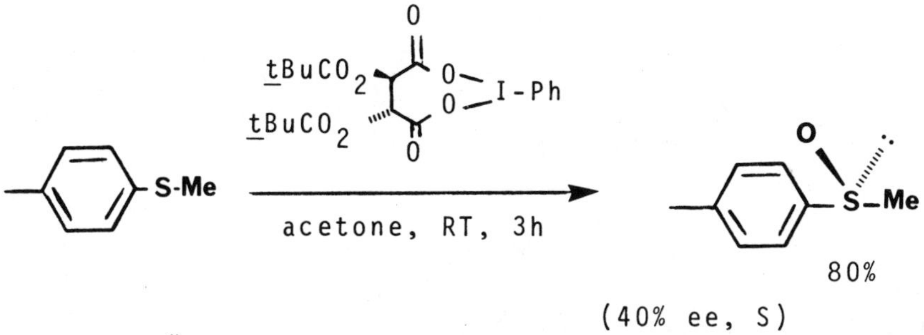
O
tBuCO2
O
I-Ph
O
tBuCO2
O
S-Me
acetone, RT, 3h
O
S-Me
80%
(40% ee, S)

Imamoto, T.*; Koto, H. **Chem Lett**, (1986), 967

SECTION 220: Oxides from Alkyl Halides and Sulfonates

No Additional Examples

SECTION 221: Oxides from Hydrides

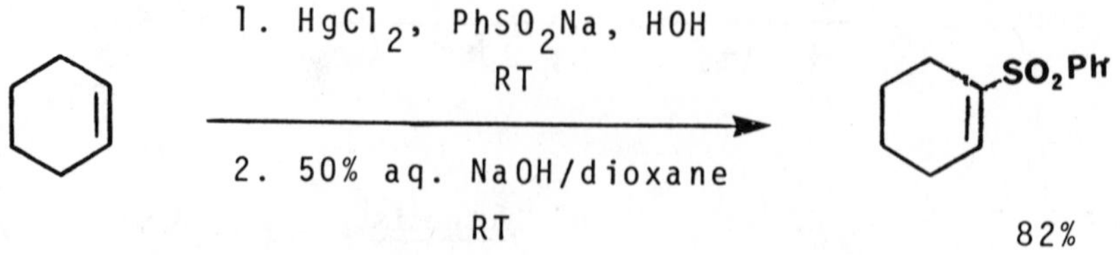

1. HgCl$_2$, PhSO$_2$Na, HOH
   RT

2. 50% aq. NaOH/dioxane
   RT

SO$_2$Ph

82%

Sas, W.   **JCS Chem Comm**, (1984), 862

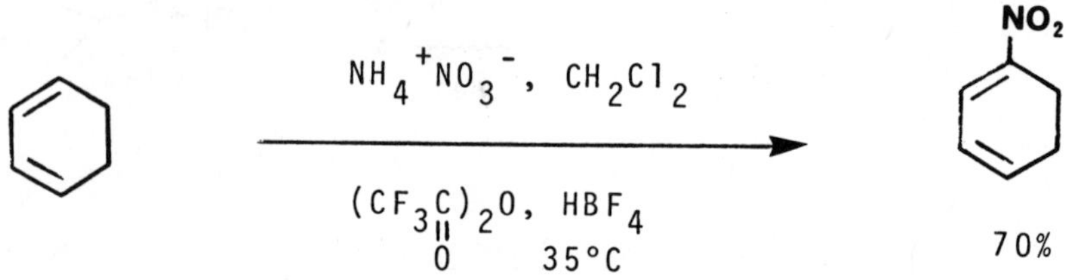

NH$_4$$^+NO_3$$^-$, CH$_2Cl_2$

(CF$_3$C)$_2$O, HBF$_4$
   O   35°C

NO$_2$

70%

Bloom, A.J.; Mellor, J.M.[*]   **Tetrahedron Lett**, (1986), **27**, 873

SECTION 222: Oxides from Ketones

No Additional Examples

SECTION 223: Oxides from Nitriles

No Additional Examples

SECTION 224: Oxides from Olefins

No Additional Examples

SECTION 225: Oxides from Miscellaneous Compounds

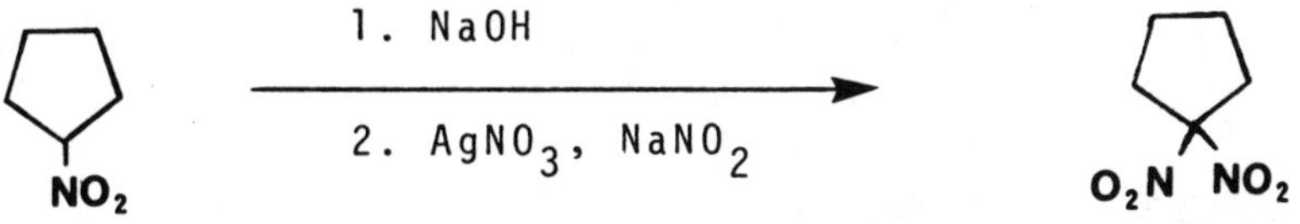

85%

Kornblum, N.[*]; Singh, H.K.; Kelly, W.J.
J Org Chem, (1983), 48, 332

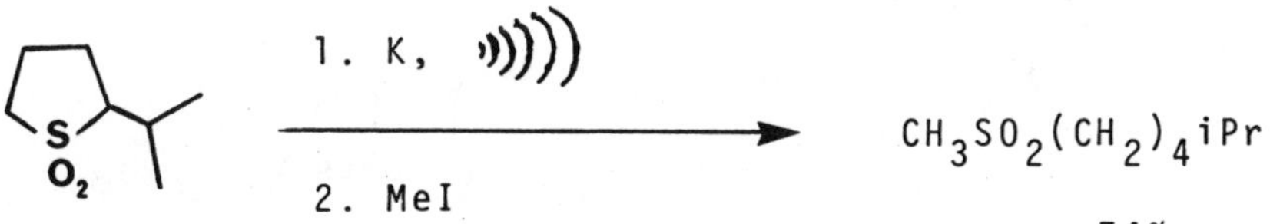

$CH_3SO_2(CH_2)_4 iPr$

74%

Chou, T.[*]; You, M.   Tetrahedron Lett, (1985), 26, 4495

CHAPTER 16
PREPARATION OF DIFUNCTIONAL COMPOUNDS

SECTION 300: Acetylene - Acetylene

$$HC \equiv C\overset{\overset{\displaystyle CH_3}{|}}{\underset{\underset{\displaystyle Co_2(CO)_6}{|}}{C}}HOAc \xrightarrow[\text{2. CAN, acetone}]{\begin{array}{c} \text{1. } \underline{n}BuC\equiv C-)_3Al, \text{ 3h} \\ CH_2Cl_2, -78 - 0°C \end{array}} HC \equiv C\overset{\overset{\displaystyle CH_3}{|}}{C}HC\equiv C\underline{n}Bu$$

55%

Padmanabhan, S.; Nicholas, K.M.[*]
Tetrahedron Lett, (1983), 24, 2239

$$PhCH = C \overset{\displaystyle Br}{\underset{\displaystyle Br}{<}} \xrightarrow[\text{BnNEt}_3{}^+Cl^-]{\begin{array}{c} Ph(Ph_2PCH_2CH_2PPh_2)_2 \\ PhH, CO, 5N NaOH \end{array}} PhC \equiv C - C \equiv CPh$$

54%

Galamb, V.; Gopal, M.; Alper, H.[*]
Organometallics, (1983), 2, 801

$$CH_3(CH_2)_5C\equiv CH \xrightarrow[\begin{array}{c} Pd(PPh_3)_4 \\ \text{4. NaNH}_2/NH_{3(1)} \end{array}]{\begin{array}{c} \text{1. } \underline{n}BuLi \\ \text{2. ZnCl}_2 \\ \text{3. ICH=CHCl, THF} \end{array}} CH_3(CH_2)_5C\equiv C - C\equiv CCH_3$$

67%

Negishi, E.[*]; Okukado, N.; Lovich, S.F.; Luo, F.T.
J Org Chem, (1984), 49, 2629

$$CH_3C\equiv CC=CHCH_3 \text{ (OTf)} \xrightarrow[\text{55°C, 30 min}]{\begin{array}{c} tBu \\ \text{(2,6-di-tBu-phenyl)}-O^-K^+/glyme \\ tBu \end{array}} CH_3C\equiv C-C\equiv CCH_3$$

85%

Stang, P.J.[*]; Dixit, V.   **Synthesis**, (1985), 962

$$\underline{n}BuC\equiv CH \xrightarrow[\substack{CH_3\overset{O}{\underset{\|}{C}}CH_2Cl, \; NEt_3, \; PhH}]{Pd(PPh_3)_4 \; , \; CuI} \underline{n}BuC\equiv C-C\equiv C\underline{n}Bu$$

54%

Rossi, R.; Carpita, A.; Bigelli, C.
                    Tetrahedron Lett, (1985), **26**, 523

SECTION 301: Acetylene - Carboxylic Acid

$$Me_2C=CH\overset{O}{\underset{\|}{C}}Cl \xrightarrow[\substack{TiCl_4, \; CH_2Cl_2 \\ 3. \; HOH}]{\substack{1. \; Cu_2CN_2, \; MeCN \\ 2. \; Me-C\equiv C-SiMe_3}} MeC\equiv C\overset{Me}{\underset{Me}{C}}CH_2\overset{O}{\underset{\|}{C}}CN$$

55%

Jellal, A.; Zahra, J.-P.; Santelli, M.[*]
                    Tetrahedron Lett, (1983), **24**, 1395

$$\begin{array}{c} C\equiv CH \\ | \\ (CH_2)_{10} \\ | \\ C\equiv CH \end{array} \xrightarrow[\substack{2. \; IC\equiv C(CH_2)_7Me, \; MeOH/Et_2O \\ NH_2OH\cdot HCl}]{\substack{1. \; aq. \; KOH, \; CuCl, \; EtNH_2 \\ EtNHOH\cdot HCl}} \begin{array}{c} CH_3(CH_2)_7 \\ C \\ \| \\ C \\ | \\ C \\ \| \\ C \\ / \\ HOOC(CH_2)_{10} \end{array}$$

36%

Singh, A.[*]; Schnur, J.M.   **Syn Commun**, (1986), **16**, 847

SECTION 302: Acetylene - Alcohol, Thiol

$$CH_2=C=C\begin{smallmatrix}OMe\\SiMe_3\end{smallmatrix}$$

1. LDA

2. [cyclopentanone]

$\longrightarrow$

[structure: cyclopentane ring with HO and C(OMe)=C=CH–SiMe$_3$ substituent]

89%

Kuwajima, J.*; Sugahara, S.; Enda, J.
Tetrahedron Lett, (1983), **24**, 1061

$CH_3C\equiv CSiMe_3$

1. [1,3-dioxane ring with C_8H_{17} and H], $TiCl_4$
   CH_2Cl_2, -78°C

2. PCC

3. KOH, aq. MeOH

$\longrightarrow$

[structure: nC_8H_{17}, HO, H substituents on carbon bearing $C\equiv CMe$] 76%

(87% ee)

Johnson, W.S.*; Elliott, R.; Elliott, J.D.
J Am Chem Soc, (1983), **105**, 2904

$$nBuC\equiv C\overset{O}{\overset{\|}{C}}Et$$

[bicyclic]–CH_2AlCl_2, Et_2O

RT, 4h

$\longrightarrow$

$$nBuC\equiv C\overset{OH}{\overset{|}{C}}HEt$$

78%

(64% ee, R)

Giacomelli, G.*; Lardicci, L.; Palla, F.
J Org Chem, (1984), **49**, 310

$nC_6H_{13}CH=C=CH_2$

1. tBuLi, THF
   -90°C

2. ZnCl$_2$, -74°C

3. EtCHO

4. H$_3$O$^+$

$\longrightarrow$

[structure: Et–CH(OH)–CH(nC_6H_{13})–C≡CH]   95%

(erythro:threo = 4:96)

Zweifel, G.*; Hahn, G.   **J Org Chem**, (1984), **49**, 4565
Hahn, G.; Zweifel, G.*   **Synthesis**, (1983), 883

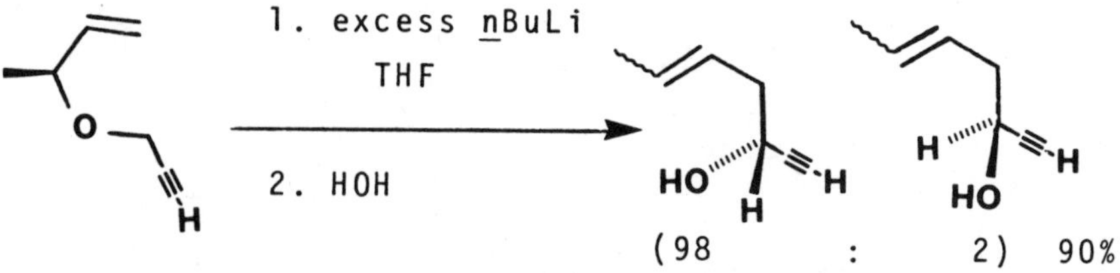
1. excess nBuLi
THF
2. HOH
HO
H
H
(98 : 2) 90%

Sayo, N.; Shirai, F.; Nakai, T.*; Kitahara, E.
Chem Lett, (1984), 255, 259

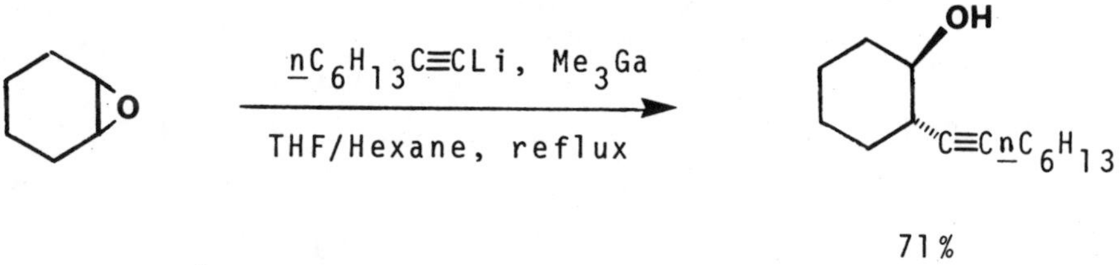
O
$\underline{n}C_6H_{13}C{\equiv}CLi$, Me_3Ga
THF/Hexane, reflux
OH
$C{\equiv}C\underline{n}C_6H_{13}$
71%

Utimoto, K.*; Lambert, C.; Fukuda, Y.
Tetrahedron Lett, (1984), **25**, 5423

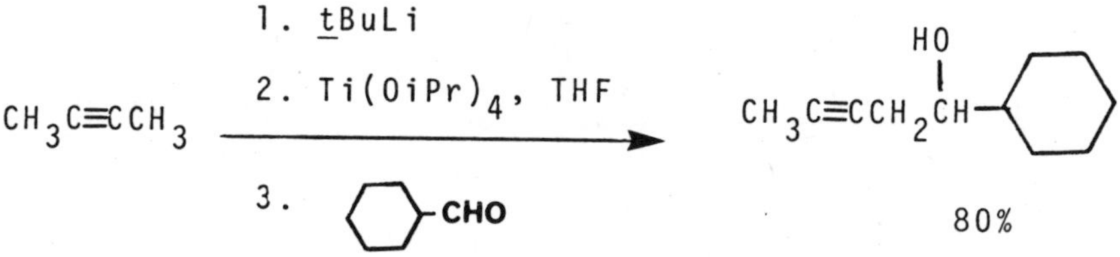
$CH_3C{\equiv}CCH_3$
1. tBuLi
2. $Ti(OiPr)_4$, THF
3. CHO
HO
$CH_3C{\equiv}CCH_2CH$
80%

Furuta, K.; Ishiguro, M.; Haruta, R.; Ikeda, N.; Yamamoto, H.*
Bull Chem Soc Jpn, (1984), **57**, 2768

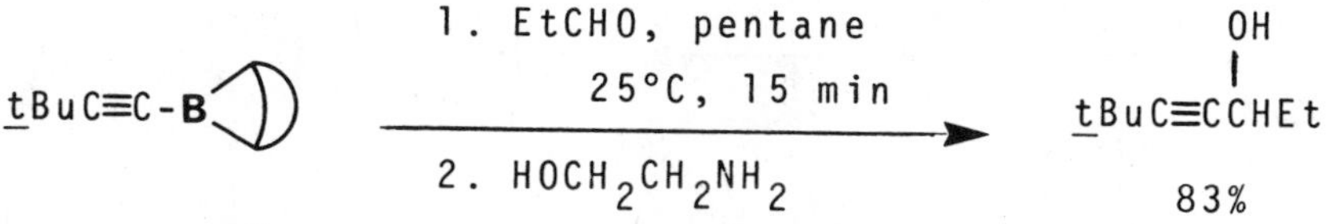
$\underline{t}BuC{\equiv}C$-B
1. EtCHO, pentane
25°C, 15 min
2. $HOCH_2CH_2NH_2$
OH
$\underline{t}BuC{\equiv}CCHEt$
83%

Brown, H.C.*; Molander, G.A.; Singh, S.M.; Rocherla, U.S.
J Org Chem, (1985), **50**, 1577

$$\underline{n}BuC{\equiv}CI \xrightarrow[\text{25°C, 2h}]{\text{PhCHO, DMF, CrCl}_2} \underset{\underline{n}BuC{\equiv}CCHPh}{\overset{OH}{|}}$$

82%

Takai, K.[*]; Kuroda, T.; Nakatsukasa, S.; Oshima, K.[*]; Nozaki, H.
Tetrahedron Lett, (1985), **26**, 5585

$$\underset{\overset{|}{OH}}{PhCHCH_2}\overset{\overset{O}{\|}}{C}CH_3 \xrightarrow[\text{2. H}_2O_2/NaOH]{\substack{\text{1. } CH_2{=}C{=}CHB(OH)_2 \\ \text{5 Å sieve, Et}_2O}}$$

95%

(99:1)

Ikeda, N.; Omori, K.; Yamamoto, H.[*]
Tetrahedron Lett, (1986), **27**, 1175

$$PhCH_2CH_2CHO \xrightarrow[\substack{\text{TiCl}_4, \text{ CH}_2Cl_2 \\ -78°C}]{CH_2{=}C{=}C{\overset{SiMe_3}{\underset{Me}{\big\langle}}}} \underset{BnCH_2CHCH_2C{\equiv}CCH_3}{\overset{OH}{|}}$$

85%

Danheiser, R.L.[*]; Carini, D.J.; Kwasigroch, C.A.
J Org Chem, (1986), **51**, 3870

$$\xrightarrow[\text{3. NaOH/H}_2O_2]{\substack{\text{1. } LiC{\equiv}C\underline{n}C_6H_{13} \\ \text{2. } I_2}} HO(CH_2)_5C{\equiv}C\underline{n}C_6H_{13}$$

71%

Brown, H.C.[*]; Basavaiah, D.; Bhat, N.G.
J Org Chem, (1986), **51**, 4518

$$PhCHO \xrightarrow[\begin{array}{l} \text{TMSCl, DMF} \\ \text{3. aq. HCl, AcOEt} \end{array}]{\begin{array}{l} \text{1. 2 HC} \equiv \text{CCH}_2\text{Br} \\ \text{2. Pb, Bu}_4\text{NBr} \end{array}}$$

$$\overset{OH}{PhCHCH=C=CH_2} \quad (4)$$
$$+$$
$$\underset{OH}{PhCHCH_2C\equiv CH} \quad (1)$$
$$\quad\quad\quad\quad 95\%$$

Tanaka, H.; Hamatani, T.; Yamashita, S.; Torii, S.[*]
Chem Lett, (1986), 1461

SECTION 303: Acetylene - Aldehyde

No Additional Examples

SECTION 304: Acetylene - Amide

No Additional Examples

SECTION 305: Acetylene - Amine

PhC≡CH

1. nBuLi
2. BF₃·OEt₂, THF, -78°C
3. (pyrrolidinone) N-CH₃ , -78°C
4. LiAlH₄, -78 - -10°C

80%

Yamaguchi, M.[*]; Hirao, I.   **Tetrahedron Lett**, (1983), **24**, 1719

$$PhC\equiv C-)_3CuLi_2 \xrightarrow{Me_2N-SO_3Me} PhC\equiv CNMe_2$$
$$\quad\quad\quad\quad\quad\quad\quad 83\%$$

Boche, G.[*]; Bernheim, M.; Niessner, M.
Angew Chem Int Ed Engl, (1983), **22**, 53

$$N-Bn$$
$$\parallel$$
$$H-C-\underline{n}Pr$$

1. $[PhC{\equiv}CLi/BF_3 \cdot OEt_2]$
   THF, $-78°C$ - RT
   ———————————————→
2. aq. NaOH

NHBn
|
$\underline{n}PrCHC{\equiv}CPh$

60%

Wada, M.; Sakurai, Y.; Akiba, K.[*]
                    Tetrahedron Lett, (1984), 25, 1083

SECTION 306: Acetylene - Ester

No Additional Examples

SECTION 307: Acetylene - Ether, Epoxide, Thioether

$$Ph_3P{=}CHSMe$$

$$\overset{O}{\overset{\parallel}{}}$$
1. $PhCCl$, PhH, RT
   ———————————————→
2. $230°C$, 0.005 mmHg

$PhC{\equiv}CSMe$

49%

Braga, A.L.; Comasseto, J.V.[*]; Petragnani, N.
                    Tetrahedron Lett, (1984), 25, 1111

1. $MeS{-}\overset{+}{S}Me_2$ BF_4^-
   RT, DCE, 15 min
   ———————————————→
2. $[LiC{\equiv}C\underline{n}C_5H_{11}/Et_2AlCl]$
   $80°C$, 2h

SMe

$C{\equiv}C\underline{n}C_5H_{11}$

88%

Trost, B.M.[*]; Martin, S.J.   J Am Chem Soc, (1984), 106, 4263

$$\underline{n}BuC\equiv CCCH_3 \quad (C=O)$$

1. ⎡HO⏜OH⎤ , TsOH, Py

2. Dibal-H, CH_2Cl_2

   0°C, 1h

3. aq. HCl, 0°C

→ [structure: n-Bu-C≡ with O-linked ring, OH]

90%

($\underline{R}:\underline{S}$ = 96:4)

Ishihara, K.; Mori, A.; Arai, I.; Yamamoto, H.*
Tetrahedron Lett, (1986), **27**, 983

$HC\equiv CCH_2Br$

1. PBr_3, Py, THF

2. $BnOCH_2Cl$, Al

   THF

→ $BnOCH_2CH_2C\equiv CH$

53%

Guedin-Vuong, D.; Nakatani, Y.   **Bull Chem Soc Fr**, (1986), II245

SECTION 308: Acetylene - Halide

$$CF_3CF_2CF_2CCl \quad (C=O)$$

1. $P(OEt_3)_3$

2. $\underline{n}BuLi$, CuI

3. TBAF, THF

   RT

→ $CF_3CF_2C\equiv CH$

89%

Ishihara, T.*; Maekawa, T.; Ando, T.
Tetrahedron Lett, (1984), **25**, 1377

$CH_3CH=C=CHHgI$

1. I_2, -40°C

2. 3M $Na_2S_2O_3$

3. 1N HCl

→ $CH_3CHC\equiv CH$ (with I substituent)

85%

Larock, R.C.*; Chow, M.-S.   **Organometallics**, (1986), **5**, 603

SECTION 309: Acetylene - Ketone

$$\underset{\underset{CH_3}{|}}{\underset{nPrCHCSEt}{\overset{O}{\|}}} \xrightarrow[\substack{2 \text{ AgBF}_4, \text{ CH}_2\text{Cl}_2 \\ \text{RT, 30 min}}]{\underline{n}C_6H_{13}C\equiv CSiMe_3} \underset{\underset{Me}{|}}{\underset{nPrCHCC\equiv C\underline{n}C_6H_{13}}{\overset{O}{\|}}}$$

91%

Kawanami, Y.; Katsuki, T.; Yamaguchi, M.*
Tetrahedron Lett, (1983), <u>24</u>, 5131

$$\underline{n}C_5H_{11}C\equiv CH \xrightarrow[\substack{3. \text{ PhCNMePh, THF} \\ \overset{\|}{O} \qquad -78°C}]{\substack{1. \ \underline{n}BuLi, \text{ THF, } -78°C \\ 2. \ BF_3\cdot OEt_2, \ -78°C}} \underset{}{\underline{n}C_5H_{11}C\equiv CCPh}$$

(with C=O on CPh) 84%

Yamaguchi, M.*; Waseda, T.; Hirao, I.   **Chem Lett**, (1983), 35

$$CH_3(CH_2)_5C\equiv CH \xrightarrow[\substack{3. \text{ Ac}_2O, \text{ THF, } -78°C}]{\substack{1. \ \underline{n}BuLi, \text{ THF, } -78°C \\ 2. \ BF_3\cdot OEt_2, \ -78°C}} CH_3(CH_2)_5C\equiv CCCH_3$$

(with C=O) 72%

Brown, H.C.*; Rocherla, U.S.; Singh, S.M.
Tetrahedron Lett, (1984), <u>25</u>, 2411

$$\underline{n}BuC\equiv CBr \xrightarrow[\substack{2. \text{ LiC}\equiv C\underline{n}Bu, \ -78°C-RT \\ 3. \text{ NaOH/H}_2O_2}]{\substack{1. \quad \text{BH}\cdot\text{HCl}\cdot\text{SMe}_2 \\ \text{BBr}_3/\text{pentane}}} \underline{n}BuC\equiv CCCH_2\underline{n}Bu$$

(with C=O) 62%

Brown, H.C.*; Bhat, N.G.; Basavaiah, D.   **Synthesis**, (1983), 885
Israel J Chem, (1984), <u>24</u>, 72

nBuC≡CH

1. $\underline{n}$BuLi, CH_2Cl_2, -78°C
2. VCl_3, CH_2Cl_2, -78°C
3. PhCHO, -78°C - reflux

$\underset{\substack{\| \\ O}}{nBuC\equiv CCPh}$

71%

Hirao, T.*; Misu, D.; Agawa, T.
Tetrahedron Lett, (1986), **27**, 933

SECTION 310: Acetylene - Nitrile

No Additional Examples

SECTION 311: Acetylene - Olefin

$\underset{\substack{| \\ C\equiv CSiMe_3}}{C\equiv C(CH_2)_9 OTHP}$

1. $Li(AlHiBu_2\underline{n}Bu)$
   DME/hexane, 25°C
2. 3N HCl
3. $KF \cdot 2H_2O$, DMF
4. $\underline{n}$BuLi, hexane/diglyme
   $\underline{n}C_3H_7Br$

$\underset{\substack{| \\ \underline{n}C_3H_7C\equiv C}}{THPO(CH_2)_9 CH=CH}$

81%

Miller, J.A.; Zweifel, G.*   **J Am Chem Soc**, (1983), **105**, 1383

$\langle \rangle$—)$_2$BCH=CH$\underline{n}$Bu

(**E**)

1. aq. NaOH, THF
   0°C
2. $Cu(acac)_2$
3. $BrC\equiv C\underline{n}Bu$, -15°C

$\underline{n}$BuC≡CCH=CH$\underline{n}$Bu

(**E**)

75%

Hoshi, M.; Masuda, Y.; Arase, A.*
Bull Chem Soc Jpn, (1983), **56**, 2855

$$MeOCH_2C \equiv CCH_2OMe \xrightarrow[\substack{\text{-78 - -45°C} \\ \text{2. } (\underline{n}C_7H_{15})_3B, \text{ THF} \\ \text{-78°C-RT} \\ \text{3. AcOH, RT, 2h} \\ \text{4. 3N NaOH}}]{\text{1. 2 } \underline{n}BuLi, \text{ } Et_2O} \underline{n}C_7H_{15}C \equiv CCH = CH_2$$

88%

Koshino, J.; Sugawara, T.; Suzuki, A.[*]
Syn Commun, (1984), **14**, 245

$$\begin{array}{c} C \equiv CTMS \\ | \\ C \equiv C\underline{n}C_6H_{13} \end{array} \xrightarrow[\text{2. NaOH, MeOH}]{\text{1. DBE, Zn, EtOH}} \begin{array}{c} HC \equiv C \\ | \\ \underline{n}C_6H_{13}CH = CH \end{array}$$

70%

Aerssens, M.H.P.J.; Brandsma, L.[*]   **JCS Chem Comm**, (1984), 735

$$CH_3(CH_2)_5C \equiv CH \xrightarrow[\substack{RuH_2(PBu_3)_4 \\ PhH, 80°C, 2h}]{CH_2 = CHCH = CH_2} CH_3(CH_2)_5C \equiv CCH = CHEt$$

88%

Mitsudo, T.[*]; Nakagawa, Y.; Watanabe, K.; Hori, Y.; Misawa, H.;
Watanabe, H.; Watanabe, Y.[*]
J Org Chem, (1985), **50**, 565

$$Me_2C = CHCH_2SO_2Ph \xrightarrow[\substack{DCE, AlCl_3 \\ Et_2O, \text{ reflux}}]{\substack{\text{1. } \underline{n}BuLi, \underline{n}C_6H_{13}Br \\ \text{2. } Et_2AlC \equiv C\underline{n}C_5H_{11}}} \begin{array}{c} CMe_2 \\ \| \\ HCCHC \equiv C\underline{n}C_5H_{11} \\ | \\ \underline{n}C_6H_{13} \end{array}$$

71%

Trost, B.M.[*]; Ghadiri, M.R.   **J Am Chem Soc**, (1986), **108**, 1098

SECTION 312: Carboxylic Acid - Carboxylic Acid

$$nBuCH_2CO_2H \xrightarrow{\begin{array}{c} \text{1. 2 LDA, THF} \\ -78°C - RT \\ \hline \text{2. } I_2, \text{ THF, } -65°C-RT \\ \text{overnight} \end{array}} \begin{array}{c} nBuCHCOOH \\ | \\ nBuCHCOOH \\ 82\% \end{array}$$

Belletire, J.L.[*]; Spletzer, E.G.; Pinhas, A.R.
Tetrahedron Lett, (1984), 25, 5969

1. $^1O_2$, MeCN, 10°C
   9,ω-dicyanoanthracene

   Ph-Ph
2. H_2O_2

88%

Schaap, A.P.[*]; Siddiqui, S.; Prasad, G.; Palomino, E.; Sandison, M.

Tetrahedron, (1985), 41, 2229

1. Cl_2CHCCl, NEt_3 (with O above the C)
2. nBuLi, THF, -78°C
3. Ac_2O, -78°C - RT

4. $NaIO_4$, RuO_2, MeCN
   CCl_4, HOH
5. aq. NaOH
6. aq. H^+

82%

Depres, J.-P.; Coelho, F.; Greene, A.E.[*]
J Org Chem, (1985), 50, 1972

SECTION 313: Carboxylic Acid - Alcohol, Thiol

1. $\bar{O}-N\equiv\overset{+}{C}-CN$, 0°C

2. H_2, Ni(R)

3. H_5IO_6

45%

Kozikowski, A.P.*; Adamczyk, M.   **J Org Chem**, (1983), **48**, 366

1. $CH_2=CHPh$

2. H_2, Ni(R)

   $AlCl_3$, aq. MeOH

3. $NaIO_4$, $NaHCO_3$

4. CH_2N_2

HOOC — OH   Ph H

90%

($\underline{R}:\underline{S}$ = 1:1)

Kozikowski, A.P.*; Kitagawa, Y.; Springer, J.P.
                    JCS Chem Comm, (1983), 1460

HO

5.76 NaOH, 230.4g HOH

15 mmol Cu, 80°C

0.66 CCl_4, 6h   87%

HO — COOH

Sasson, Y.*; Razintsky, M.   **JCS Chem Comm**, (1985), 1134

1. $TiCl_4$, CH_2Cl_2

2. $CF_3COOH \cdot HOH$

3. Jone's Oxidation

4. KOH/HOH

$\underline{n}C_8H_{17}$

$\underline{n}C_8H_{17}$ — COOH   H OH

76%

Elliott, J.D.*; Steele, J.; Johnson, W.S.
                    Tetrahedron Lett, (1985), **26**, 2535

1. nBuLi, THF, -78°C

2. Et$_2$AlCl, -40°C

3. EtCHO, -10°C

4. 2 nBuLi

5. MeI

6. oxidation

no yield

Davies, S.G.[*]; Dordor-Hedgecock, I.M.; Warner, P.; Ambler, P.W.
          Tetrahedron Lett, (1985), **26**, 2125, 2129
Davies, S.G.[*]; Dordor, I.M.; Walker, J.C.; Warner, P.
          Tetrahedron Lett, (1984), **25**, 2709
Davies, S.G.[*]; Dordor, I.M.; Warner, P.
          JCS Chem Comm, (1984), 956

1. LDA

2. MX

3. EtCHO

4. oxidation

MX = iBu$_2$AlCl   (5.2   :   1)   73%

MX = SnCl$_2$ (1   :   11.6) 66%

Liebeskind, L.S.[*]; Welker, M.E.
                    Tetrahedron Lett, (1984), **25**, 4341

nC$_7$H$_{15}$ 3 Me$_2$CuLi

CO$_2$H   Et$_2$O

0°C, 3h

(8   :   1)

91%

Chong, J.M.; Sharpless, K.B.[*]
               Tetrahedron Lett, (1985), **26**, 4683

Me
O
||
O—CH(Me)(H)
|
O   Ph
(butenyl ester)

1. KN(TMS)$_2$, TMSCl
2. CH$_2$N$_2$

3. KOH, MeOH
4. H$_2$, Pd-C, MeOH

HO$_2$C—CH(OH)—CH(Me)—
82%

(2S,3R = major)

Kallmerten, J.[*]; Gould, T.J.   **J Org Chem**, (1986), **51**, 1152

CH$_3$CHCO$_2$SiMe$_3$
|
SiMe$_3$

1. PhCHO, THF
   CsF, reflux

2. H$_3$O$^+$

Ph—CH(HO)—CH(Me)—CO$_2$H   Ph—CH(HO)—CH(Me)—CO$_2$H

(74   :   26)

84%

Bellasoued, M.; Dubois, J.-E.; Bertounesque, E.
Tetrahedron Lett, (1986), **27**, 2623

OH
|
CH$_3$CH$_2$CCH$_2$CO$_2$Me
|
Me

(S)

PLE

1M phosphate
3d

OH
|
CH$_3$CH$_2$CCH$_2$COOH
|
Me

(S)   51%

PLE = Pig Liver Esterase

Wilson, W.K.; Baca, S.B.; Barber, Y.J.; Scallen, T.J.; Morrow, C.J.[*]

J Org Chem, (1983), **48**, 3960

SECTION 314: Carboxylic Acid - Aldehyde

No Additional Examples

SECTION 315: Carboxylic Acid - Amide

1. $\underline{t}BuO_2C)_2O$, NEt_3
   DMAP, 25°C, 8h

2. 3 LiOH, aq. THF
   RT

COOH
NH-BOC
96%

Flynn, D.L.; Zelle, R.E.; Grieco, P.A.[*]
J Org Chem, (1983), **48**, 2424

1. 3 BnO_2CNH_2, 6 NEt_3
   3 NCS, MeOH, 0°C

2. O_3, CH_2Cl_2/MeOH, -78°C

3. Me_2S, -78 - 25°C

4. CrO_3, aq. H_2SO_4 (79% ee)
   acetone

HO_2C NH-CBZ
40%

Fitzner, J.N.; Shea, R.G.; Fankhauser, J.E.; Hopkins, P.B.[*]
J Org Chem, (1985), **50**, 419

$Ph_2CHCOOH$

1. 2 Na^+, THF

2. $NCPh$, -15°C-RT, 1d

3. H^+

$\overset{O}{\overset{\|}{PhCNHCH_2}}CH_2\overset{Ph}{\underset{Ph}{\overset{|}{\underset{|}{C}}}}COOH$

58%

Stamm, H.[*]; Weiss, R.   **Synthesis**, (1986), 395

1. LDA, THF, -78°C

2. BOC-N=N-BOC

3. gl. AcOH

4. LiOH, aq. THF   (99% ee)

Bn-CH-COOH
  N-BOC
  NH-BOC
75%

Evans, D.A.[*]; Britton, T.C.; Dorow, R.L.; Dellaria, J.F.
J Am Chem Soc, (1986), **108**, 6395

1. nBuLi
2. PhCCl
   ‖
   O
3. LDA
4. BOC-N=N-BOC
5. LiSH
6. AcOOH

Bn-CH-COOH
   |
   N-BOC
   |
   NH-BOC

85%

($\underline{R}$:$\underline{S}$ = 17:83)

Trimble, L.A.; Vederas, J.C.*   J Am Chem Soc, (1986), 108, 6397

Related Methods: Acid-Amine (Section 316)
                 Amide-Ester (Section 344)
                 Amine-Ester (Section 351)

SECTION 316: Carboxylic Acid - Amine

$CH_3CH=CHCHO$

1. $BnNH_2$, 4Å sieve
2. TMS-CN, CH_2Cl_2
3. 6N HCl, reflux
4. Dowex 50W-X4
   HOH/Py/MeOH

$CH_3CH=CHCHNHBn$

COOH
|

43%

Greenlee, W.J.*   J Org Chem, (1984), 49, 2632

1. EtOH, Me_2NH, RT
2. tBuCHO, pentane
3. $PhCO_2)O$, 30°C
4. LDA, THF, -78 - 0°C
   $BrCH_2$—⟨⟩—OMe
                OMe
5. 6N HCl, 185°C
6. Dowex 50W-X8(H^+)

43%

Seebach, D.*; Aebi, J.D.; Naef, R.; Weber, T.
              Helv Chim Acta, (1985), 68, 144

1.

tBu

$\text{—}\langle\text{ring}\rangle\text{—O}^-\text{K}^+$, Ph-**B**$\langle\rangle$

$Ph_2C=N$ $\overset{CO_2Et}{\underset{OAc}{\big|}}$

tBu , THF, 0°C

2. $HCl-Et_2O$

3. 6M HCl, heat

$\overset{NH_2}{\underset{|}{}}$
PhCHCOOH

31%

O'Donnell, M.J.[*]; Falmagne, J.-B.   **JCS Chem Comm**, (1985), 1168

$(MeS)_2C=NCH_2\overset{\overset{O}{\|}}{C}-N\langle\text{ring}\rangle$ OMOM (top) OMOM (bottom)

1. LDA, THF, -78°C

2. MeI, -78 - -20°C

3/ 1N HCl

$\overset{Me}{\underset{}{}}$
$H_2N\overset{|}{C}HCOOH$

82%
(95% ee)

Ikegami, S.; Hayama, T.; Katsuki, T.[*]; Yamaguchi, M.
Tetrahedron Lett, (1986), **27**, 3403

$\overset{O}{\triangle}$COOH

$\underline{n}C_7H_{15}$

excess Et$_2$NH

RT, 4h

1.5 Ti(OiPr)$_4$

$\overset{Et_2N}{\underset{\underline{n}C_7H_{15}}{}}$ COOH HO

+

$\overset{OH}{\underset{\underline{n}C_7H_{15}}{}}$ COOH NEt$_2$

(20 : 1)
92%

Chong, J.M.; Sharpless, K.B.[*]   **J Org Chem**, (1985), **50**, 1560

$\langle\text{camphor}\rangle$ CH$_2$SO$_2$N$\langle$cyclohexyl$\rangle)_2$ O$_2$CCH=CHMe

1. EtCu·BF$_3$

2. LDA, TMSCl, NBS

3. NaN$_3$

4. Ti(OBn)$_4$

5. H$_2$/Pd

$\overset{Me}{H\text{—}C\text{—}Et}$ 97.8% ee

$HO_2C\underset{H}{\overset{}{}}$ NH$_2$

99.3% ee

37%

Oppolzer, W.[*]; Pedrosa, R.; Moretti, R.
Tetrahedron Lett, (1986), **27**, 831

Ph

Ph — [structure]
CBZ

1. NBS, CCl$_4$, reflux

2. CH$_2$=C$\underset{OSiMe_2tBu}{\overset{OEt}{<}}$

   ZnCl$_2$, THF

3. HOH

4. H$_2$/PdCl$_2$, EtOH
   20 psi , 25°C   (97% ee)

$H_3\overset{+}{N}$ — $\overset{CO_2^-}{\underset{H}{C}}$

CH$_2$CO$_2$Et

71%

Sinclair, P.J.; Zhai, D.; Reibenspies, J.; Williams, R.M.[*]
J Am Chem Soc, (1986), **108**, 1103

PhCHO

1. (TMS)$_2$NCH=C(OTMS)$_2$
   Me$_3$SiOTf, -78°C, 3h

2. H$_3$O$^+$

PhCHCHCOOH
with NH$_2$ and OH substituents

95%

(<u>threo</u>:<u>erythro</u> = 2:1)

Hvidt, T.; Martin, O.R.; Szarek, W.A.[*]
Tetrahedron Lett, (1986), **27**, 3807

[structure with Ph, Me, NMe$_2$, OTMS, O]

1. BOC-N=N-BOC
   TiCl$_4$, CH$_2$Cl$_2$, -78°C
2. CF$_3$COOH, RT

3. LiOH
4. Dowex 50W
5. H$_2$/PtO$_2$
6. Dowex 50W   (98% ee R)

$\underset{HO_2C}{}$ — C — $\underset{NH_2}{Me, H}$

50%

Gennari, C.[*]; Colombo, L.; Bertolini, G.
J Am Chem Soc, (1986), **108**, 6394

1. $[Au(Cy-NC)_2]^+BF_4^-$
   cat., MeO_2CCH_2NC

PhCHO   ──────────────────────────────────────►

2. conc. HCl/MeOH, 50°C

3. 6N HCl, 80°C, 6h

cat. = (Fe)⟍Me / N⟍NEt₂ / PPh₂ / PPh₂

90%

Ito, Y.*; Sawamura, M.; Hayashi, T.*
          J Am Chem Soc, (1986), 108, 6405

Related Methods: Acid-Amide (Section 315)
                 Amide-Ester (Section 344)
                 Amine-Ester (Section 351)

SECTION 317: Carboxylic Acid ⇌ Ester

COOH
│
│ $NaNO_2$/aq. H_2SO_4
│   ──────────────────────────►
H₂N ᴴ   RT, 15h
     COOH

70%

Gringore, O.H.; Rouessac, F.P.   Org Syn, (1984), 63, 121

nBu O $NaIO_4-RuO_4 \cdot 2H_2O$
              ──────────────────────────►
              CCl_4-HOH, 10h

nBu   COOH
         OCHO

93%

Torii, S.*; Inokuchi, T.; Kondo, K.
          J Org Chem, (1985), 50, 4980

```
  ┌─COOH              ⬡  ,  Al₂O₃                    ┌─CO₂Me
  │                  ─────────────────────▶          │
  └─COOH                   CH₂N₂                      └─COOH
```
$$\text{quant.}$$

Ogawa, H.; Chihara, T.; Taya, K.*
J Am Chem Soc, (1985), **107**, 1365

```
        Me                                              Me
        │          PLE, pH 8, NaOH                      │
  nBu ──┼── CO₂Me  ────────────────────▶          nBu ──┼── CO₂H
        │          0.1N phosphate                        │
        CO₂Me                                            CO₂Me
```
PLE = Pig Liver Esterase   ($\underline{R}$:$\underline{S}$ = 25:75)
Schneider, M.*; Engel, N.; Boensmann, H.
Angew Chem Int Ed Engl, (1984), **23**, 66

```
      Me                                              Me
      │                  PLE, pH 7                    │
  H ──┼── CH₂CO₂Me  ────────────────────▶       H ──┼── CH₂CO₂Me
      │              aq. KH₂PO₄                       │
      CH₂CO₂Me                                        CH₂COOH
```
PLE = Pig Liver Esterase   (79% ee R)   95%
Lam, L.K.P.; Hui, R.A.H.F.; Jones, J.B.*
J Org Chem, (1986), **51**, 2047

SECTION 318: Carboxylic Acid - Ether, Epoxide, Thioether

```
                   1. Na₂WO₄·2H₂O, 3h
                      HOH, 1N KOH, 65°C              O
  nPrCH=CHCOOH     ────────────────────▶       nPrCH-CHCOOH
                      pH 6
                   2. aq. H₂O₂(pH 5.8-6.8)       82%
```

Kirshenbaum, K.S.; Sharpless, K.B.*
J Org Chem, (1985), **50**, 1979

$$PhCH=CHCOOH \xrightarrow[\begin{array}{c} 3. \text{ aq. HCl} \end{array}]{\begin{array}{c} 1. \text{ NaHCO}_3, \text{ aq. acetone} \\ 2. \text{ KHSO}_5, \text{ Na}_2\text{EDTA} \\ 27°C, \text{ 2h} \end{array}} \begin{array}{c} O \\ \diagup \diagdown \\ PhCH-CHCOOH \\ 92\% \end{array}$$

Corey, P.F.[*]; Ward, F.E.   **J Org Chem**, (1986), **51**, 1925

$$PhCH_2COOH \xrightarrow[\begin{array}{c} 2. \text{ BrCH}_2\text{SeBn}, \text{ 1h} \\ -78°C - RT \end{array}]{\begin{array}{c} 1. \text{ 2 LDA, THF/HMPA} \\ -78°C \end{array}} \begin{array}{c} CH_2SeBn \\ | \\ PhCHCOOH \\ 81\% \end{array}$$

Reich, H.J.[*]; Jasperse, C.P.; Renga, J.M.
                    J Org Chem, (1986), **51**, 2981

SECTION 319: Carboxylic Acid - Halide, Sulfonate

$$CH_3(CH_2)_{16}COOH \xrightarrow[\text{TCNQ, 150°C}]{Cl_2, \ ClSO_3H} \begin{array}{c} Cl \\ | \\ CH_3(CH_2)_{15}CHCOOH \\ 88\% \end{array}$$

TCNQ = 7,7,8,8-tetracyanoquinodimethane

Crawford, R.J.[*]   **J Org Chem**, (1983), **48**, 1364

$$HC\equiv CCOOH \xrightarrow[\text{2. AcOH}]{\begin{array}{c} 1. \text{ MeMgI, THF} \\ reflux \end{array}} \begin{array}{c} ICH=CHCOOH \\ (\underline{Z}) \\ 80\% \end{array}$$

Jung, M.E.[*]; Hagenak, J.A.; Long-Mei, Z.
                    Tetrahedron Lett, (1983), **24**, 3973

$$\underset{\underset{iPrCHCOOH}{|}}{\overset{NH_2}{|}} \quad \xrightarrow[\text{HF/Py}]{\text{NaNO}_2, \text{ KBr}} \quad \underset{\underset{iPrCHCOOH}{|}}{\overset{Br}{|}}$$

84%

Olah, G.A.; Shih, J.; Surya Prakash, G.K.
    Helv Chim Acta, (1983), **66**, 1028

$$CH_3COOH \quad \xrightarrow[\text{reflux, 40h}]{I_2, \text{ Cu(OAc)}_2} \quad ICH_2COOH$$

80%

Horiuchi, C.A.[*]; Satoh, J.Y.   **Chem Lett**, (1984), 1509

$$\underset{\underset{nBu}{|}}{\overset{O}{\overset{\|}{Ph_2AsCHCOOH}}} \quad \xrightarrow[\text{2. Br}_2]{\text{1. LiAlH}_4} \quad \underset{\underset{nBuCHCOOH}{}}{\overset{Br}{|}}$$

56%

Kauffmann, T.[*]; Joussen, R.; Woltermann, A.
    Chem Ber, (1986), **119**, 2135

SECTION 320: Carboxylic Acid - Ketone

$$BrMgO_2C\underset{}{\overset{S}{\underset{N}{\diagdown}}}N\,CO_2MgBr \quad \xrightarrow[\text{2. H}_3O^+]{\overset{\overset{O}{\|}}{\text{1. PhCCH}_3, \text{ DMSO}} \atop 150°C} \quad \overset{O}{\overset{\|}{PhCCH_2COOH}}$$

4 eq.   74%

Matsumura, N.[*]; Asai, N.; Yoneda, S.[*]
    JCS Chem Comm, (1983), 1487

Ph $\triangle$ MeO   COOH

$\xrightarrow[\text{70°C, 30 min}]{\text{HOH/MeOH}}$

$\overset{\text{O}}{\overset{\|}{\text{PhCCH}_2\text{CH}_2\text{COOH}}}$

quant.

Kunz, H.[*]; Lindig, M.   **Chem Ber**, (1983), **116**, 220

$CH_3(CH_2)_4COOH$

1. 2 LDA, THF, -50°C
2. HMPT, RT, 30 min

$\xrightarrow{}$

3. CH_2=CMeNO$_2$, 4h
    -100 - 0°C
4. aq. HCl, RT, overnight

$\overset{\text{O}}{\overset{\|}{\text{CH}_3}}\overset{}{\text{CCH}_2}\overset{n\text{Bu}}{\overset{|}{\text{CHCOOH}}}$

65%

Miyashita, M.; Yamaguchi, R.; Yoshikoshi, A.[*]
          J Org Chem, (1984), **49**, 2857

OSiMe$_3$

$\xrightarrow[\text{2. HOH}]{\begin{array}{c}\text{1. } \overset{\text{O}}{\overset{\|}{\text{MeCH-CHCCN}}}\text{, TiCl}_4 \\ \text{CH}_2\text{Cl}_2\text{, -70 - -30°C}\end{array}}$

$\overset{\text{O}}{\Large\triangleleft}\overset{\text{CH}_3}{\overset{|}{\text{CHCH}_2\text{COOH}}}$

65%

El-Abed, D.; Jellal, A.; Santelli, M.[*]
          Tetrahedron Lett, (1984), **25**, 4503

$\overset{\text{O}}{\overset{\|}{\text{PhCH=CHCCH}_3}}$

$\xrightarrow[\begin{array}{c}\text{Hg electrodes} \\ \text{Et}_4\text{N}^+\text{ClO}_4^-\end{array}]{\text{e}^-, \text{CO}_2, \text{MeCN}}$

$\overset{\text{COOH}}{\overset{|}{\text{PhCHCH}_2}}\overset{\text{O}}{\overset{\|}{\text{CCH}_3}}$

82%

Harada, J.; Sakakibara, Y.; Kunai, A.[*]; Sasaki, K.
          Bull Chem Soc Jpn, (1984), **57**, 611

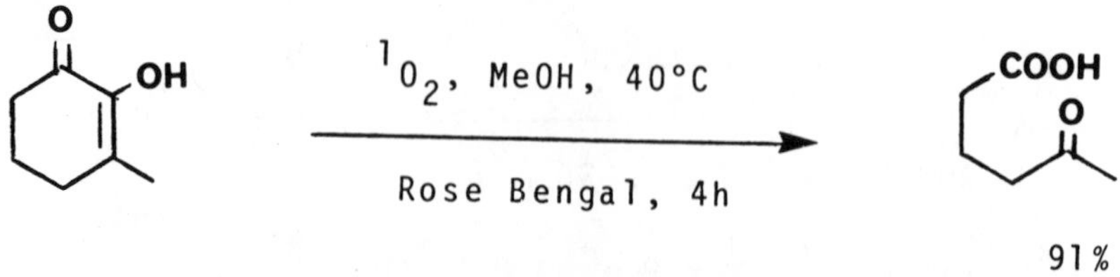

$^1O_2$, MeOH, 40°C

Rose Bengal, 4h

91%

Utaka, M.; Nakatani, M.; Takeda, A.*
Tetrahedron, (1985), **41**, 2163

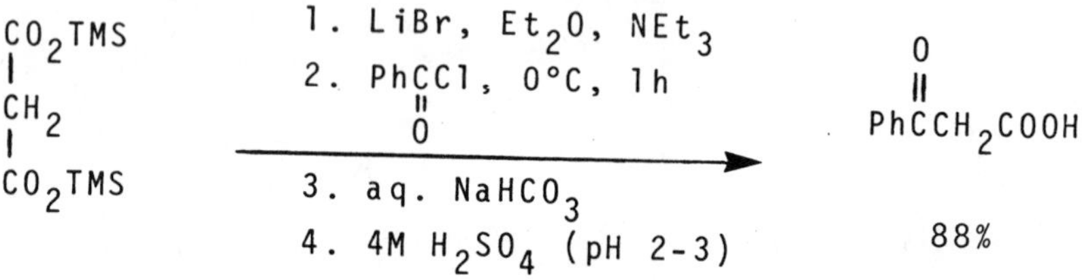

CO_2TMS
|
CH_2
|
CO_2TMS

1. LiBr, Et_2O, NEt_3
2. PhCCl, 0°C, 1h
      ‖
      O
3. aq. $NaHCO_3$
4. 4M H_2SO_4 (pH 2-3)

$PhCCH_2COOH$
with O double bond above

88%

Rathke, M.W.*; Nowak, M.A.   **Syn Commun**, (1985), **15**, 1039

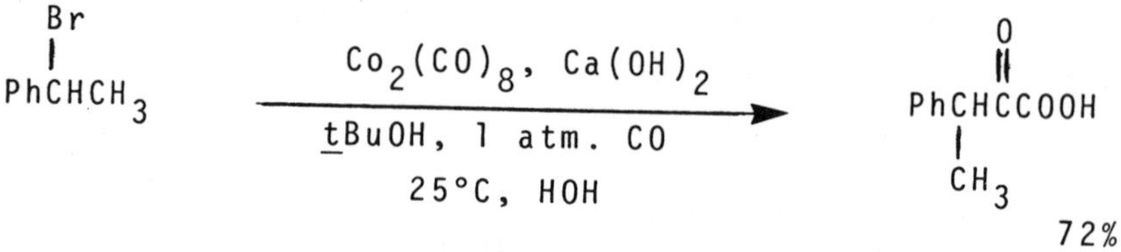

Br
|
$PhCHCH_3$

$Co_2(CO)_8$, $Ca(OH)_2$
_____
<u>t</u>BuOH, 1 atm. CO
25°C, HOH

O
‖
PhCHCCOOH
|
CH_3

72%

Francalanci, F.; Bencini, E.; Gardano, A.; Vincenti, M.; Foà, M.
J Organomet Chem, (1986), **301**, c027

Review: "Synthesis and Properties of α-Keto-Acids"

Cooper, A.J.L.; Ginos, J.Z.; Meister, A.
Chem Rev, (1983), **83**, 321

Also via: Ketoesters (Section 360)

SECTION 321: Carboxylic Acid - Nitrile

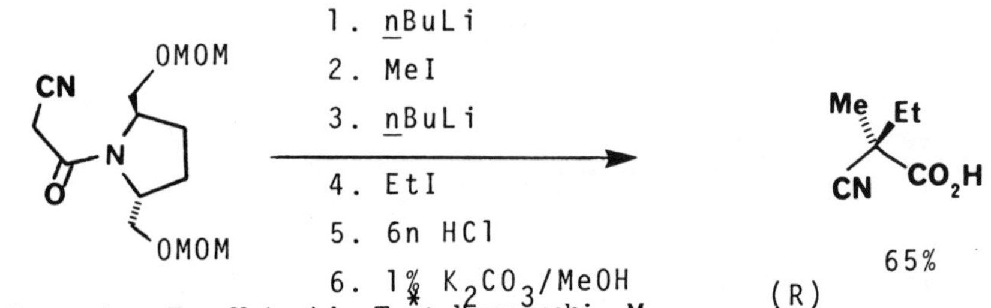

OMOM
CN
N
O
OMOM
1. nBuLi
2. MeI
3. nBuLi
4. EtI
5. 6n HCl
6. 1% K_2CO_3/MeOH
Me
Et
CN
CO_2H
(R)
65%

Hanamoto, T.; Katsuki, T.*; Yamaguchi, M.
Tetrahedron Lett, (1986), 27, 2463

Also via: Cyanoesters (Section 361)

SECTION 322: Carboxylic Acid - Olefin

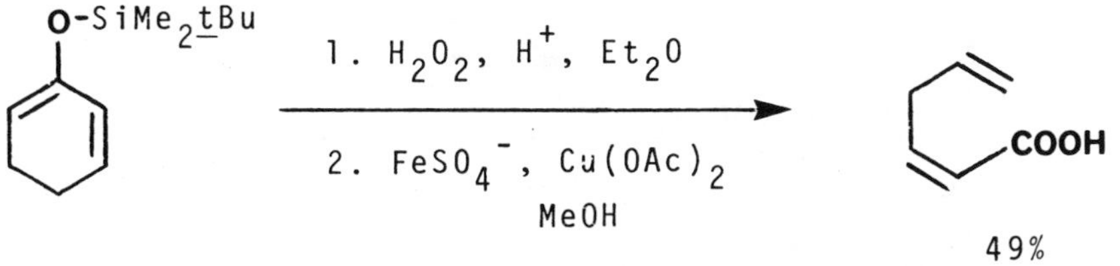

O-SiMe$_2$tBu
1. H_2O_2, H^+, Et_2O
2. $FeSO_4^-$, $Cu(OAc)_2$
MeOH
COOH
49%

Saito, I.*; Nagata, R.; Yuba, K.; Matsuura, T.
Tetrahedron Lett, (1983), 24, 1737, 4439

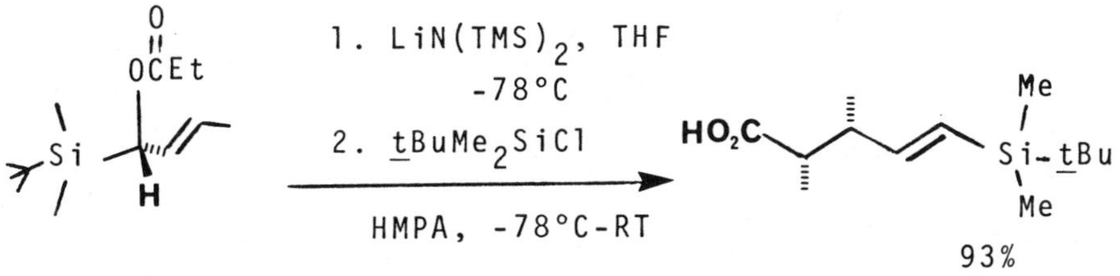

O
OCEt
Si
H
1. LiN(TMS)$_2$, THF
-78°C
2. tBuMe$_2$SiCl
HMPA, -78°C-RT
HO_2C
Me
Si-tBu
Me
93%

Ireland, R.E.*; Varney, M.D. **J Am Chem Soc**, (1984), 106, 3668

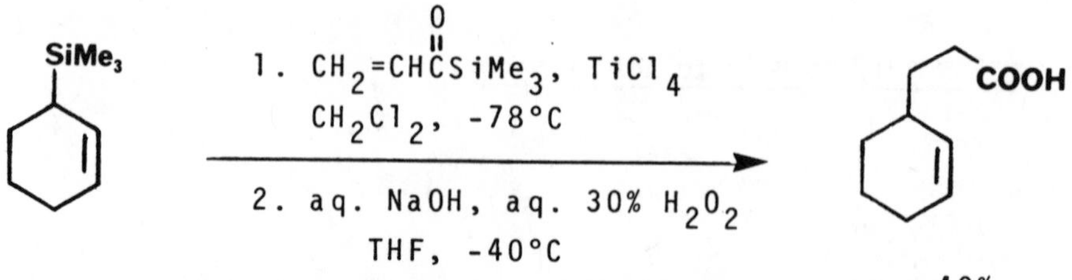

1. $CH_2=CH\overset{\overset{O}{||}}{C}SiMe_3$, $TiCl_4$
   CH_2Cl_2, -78°C

2. aq. NaOH, aq. 30% H_2O_2
   THF, -40°C

49%

Danheiser, R.L.*; Fink, D.M.
Tetrahedron Lett, (1985), **26**, 2509, 2513

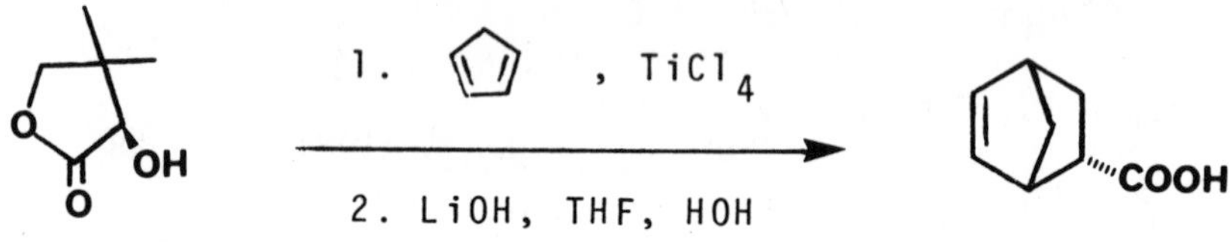

1. [cyclopentadiene] , $TiCl_4$

2. LiOH, THF, HOH

(99.9% ee)   79%

Poll, T.; Sobczak, A.; Hartmann, H.; Helmchen, G.*
Tetrahedron Lett, (1985), **26**, 3095

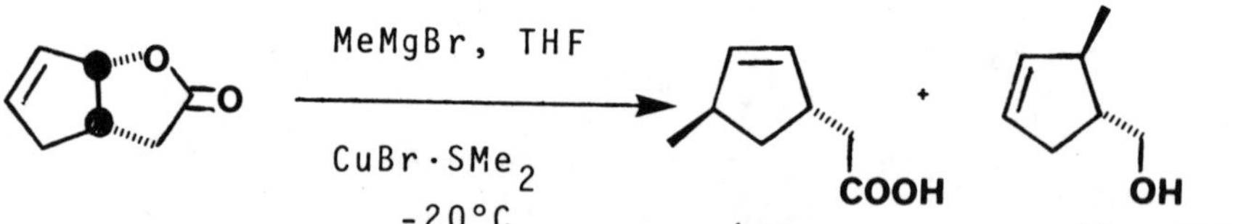

MeMgBr, THF

CuBr·SMe$_2$
-20°C

(98   :   2)   97%

Curran, D.P.*; Chen, M.-H.; Leszczweski, D.; Elliott, R.L.;
Rakiewicz, D.M.
J Org Chem, (1986), **51**, 1612

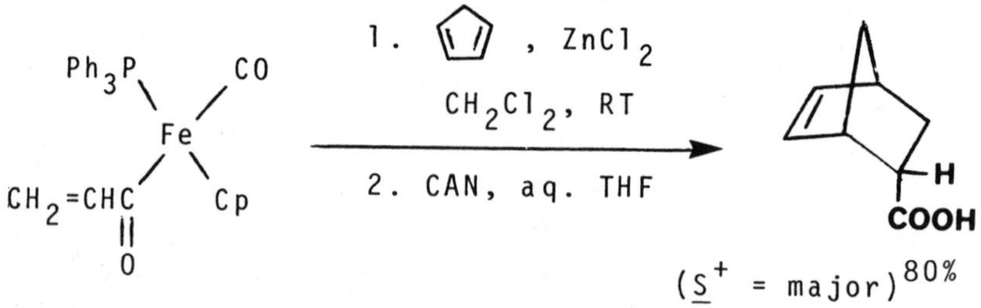

1. [cyclopentadiene] , $ZnCl_2$
   CH_2Cl_2, RT

2. CAN, aq. THF

($\underline{S}^+$ = major) 80%

Davies, S.G.*; Walker, J.C.   **JCS Chem Comm**, (1986), 609

Also via: Hydroxy acids (Section 313); Olefinic amides (Section 349); Olefinic esters (Section 362); Olefinic nitriles (Section 376).

SECTION 323: Alcohol, Thiol - Alcohol, Thiol

PhCHO $\xrightarrow{\text{1. 2 SmI}_2, \text{ 0.5 min}}_{\text{2. H}_3\text{O}^+}$

$$\begin{array}{c} \text{OH} \\ | \\ \text{PhCHCHPh} \\ | \\ \text{OH} \quad 95\% \end{array}$$

Namy, J.L.; Souppe, J.; Kagan, H.B.[*]
Tetrahedron Lett, (1983), **24**, 765

$$\xrightarrow[\text{Et}_2\text{O, 0°C}]{\text{Zn(BH}_4)_2}$$

(87 : 13) no yield

Nakata, T.[*]; Tanaka, T.; Oishi, T.[*]
Tetrahedron Lett, (1983), **24**, 2653

R - iPr
$$\text{Me}_4\text{NH}^+\text{B(OAc)}_3{}^-, \text{ AcOH}$$
MeCN, -20°C, 5h   86%(anti:syn = 96:4)

Evans, D.A.[*]; Chapman, K.T.   **Tetrahedron Lett**, (1986), **27**, 5939

R = Ph
1. Dibal-H, THF
2. 10% HCl   87% (syn:anti = 93:7)

Kiyooka, S.[*]; Kuroda, H.; Shimasaki, Y.
Tetrahedron Lett, (1986), **27**, 3009

R = Me
1. iPr$_2$SiHCl, Py, PhH
2. SnCl$_4$, CH$_2$Cl$_2$, -80°C   61%
3. aq. HF, MeCN

Anwar, S.; Davis, A.P.[*]   **JCS Chem Comm**, (1986), 831

1. $\vdash\!\!\!-BH_2$

THF

2. $NaOH/H_2O_2$

(15 : 1)

91%

Still, W.C.[*]; Barrish, J.C.   **J Am Chem Soc**, (1983), **105**, 2487

1. $\underline{n}Bu_3SnH$, AIBN
   PhH, reflux

2. 30% H_2O_2, KF
   DMF, 60°C, 8h

(84 : 16)

85%

Nishiyama, H.[*]; Katajima, T.; Matsumoto, M.; Itoh, K.
              J Org Chem, (1984), **49**, 2298

1. $\underline{n}BuLi$

2. $\underset{Li}{\text{naphthyl}}$, THF, -78°C

3. $Et\overset{O}{\underset{||}{C}}Bn$

Et 55%

($\underline{SR}:\underline{SS}$ = 1:1)

Nájera, C.; Yus, M.; Seebach, D.[*]
              Helv Chim Acta, (1984), **67**, 289

1. $MeS_2B^+\text{-}\bar{C}HMe$

2. $NaOH/H_2O_2$

Ph 72%

(erythro:threo = 4:1)

Pelter, A.[*]; Bugden, G.; Rosser, R.
              Tetrahedron Lett, (1985), **26**, 5097

PhCH=CHPh

(E)

OsO$_4$, CH$_2$Cl$_2$, -100°C

αnaph

H

→

HO
Ph   Ph
      OH

71%

(90% ee RR)

Yamada, T.; Narasaka, K.   **Chem Lett**, (1986), 131

CH$_3$(CH$_2$)$_4$CH=CH$_2$

NMe$_2$

NMe$_2$

OsO$_4$, RT

→

CH$_3$(CH$_2$)$_4$-C-CH$_2$OH

H   OH

(86% ee R)   75%

Tokles, M.; Snyder, J.K.*   **Tetrahedron Lett**, (1986), **27**, 3957

OH

t-Bu

1. Hg(OCCF$_3$)$_2$, CH$_2$Cl$_2$
   RT, 12h

2. LiAlH$_4$, THF, 0°C

3. HOH

→

HO   Me   OH

t-Bu   58%

(inversion:retention = 8:1)

Collum, D.B.; Still, W.C.; Mohamadi, F.*
         J Am Chem Soc, (1986), **108**, 2094

OH

1. TMS$_2$NH, RT, 1h

2. H$_2$PtCl$_6$·6H$_2$O

60°C, 10 min

3. 30% H$_2$O$_2$, NaHCO$_3$

→

OH   OH

69%

Tamao, K.*; Tanaka, T.; Nakajima, T.; Sumiya, R.; Arai, H.; Ito, Y.*

         Tetrahedron Lett, (1986), **27**, 3377

Tamao, K.*; Nakajima, T.; Sumiya, R.; Arai, H.; Higuchi, N.; Ito, Y.*

         J Am Chem Soc, (1986), **108**, 6090

Review: "Stereoselective Reduction of β-Hydroxy Ketones
       to 1,3-Diols"

Narasaka, K.[*]; Pai, F.-C.   **Tetrahedron**, (1984), **40**, 2233

Also via: Hydroxyesters (Section 327); Diesters (Section 357)

SECTION 324: Alcohol, Thiol - Aldehyde

1. $ClCH_2SO_2Ph$
   tBuOH/tBuOK
   THF, 10°C

2. tBuOH/tBuOK
   THF
   tr. HOH, 25°C

64%

Adamczyk, M.; Dolence, E.K.; Watt, D.S.[*]; Christy, M.R.;
Reibenspies, J.H.; Anderson, O.P.
                J Org Chem, (1984), **49**, 1378

1. tBuLi, HMPA/THF
   -78°C, 30 min

2. PhCHO, 20°C
3. Na/Hg
4. Cu^{+2}

40%

(34% ee)

Annunziata, R.; Cozzi, F.; Cinquini, M.[*]; Colombo, L.; Gennari,
C.; Poli, G.; Scolastico, C.
                JCS Perkin I, (1985), 251

1. NaH, DMSO
   $ClCH_2SO_2CH_2Cl$

2. $TiClO_4$, HOH

32%

Nagashima, E.; Suzuki, K.; Ishikawa, M.; Sekiya, M.[*]
                Heterocycles, (1985), **23**, 1873

$$\text{CH}_3\text{CHCH}_2\text{CH}_2\text{OH} \xrightarrow[\text{Cp}_2\text{ZrH}_2,\ 150°\text{C},\ 8\text{h}]{} \overset{\text{OH}}{\underset{}{\text{CH}_3\text{CHCH}_2\text{CHO}}}$$

OH (on left reactant)

98%

Nakano, T.; Terada, T.; Ishi, Y.[*]; Ogawa, M.
Synthesis, (1986), 774

Related Methods: Alcohol⇌Ketone (Section 330)

SECTION 325: Alcohol, Thiol - Amide

1. $\underline{t}\text{BuMgBr}$, THF
   $-78°\text{C}$

2. iPrCH_2CHO

3. Na/Hg

$$\text{Tol-}\overset{\text{O}}{\underset{}{\text{S}}}\text{-CH}_2\overset{\text{O}}{\underset{}{\text{C}}}\text{NMe}_2 \longrightarrow \text{iPrCHCH}_2\overset{\text{O}}{\underset{}{\text{C}}}\text{NMe}_2$$

OH (on product)

73%

(89% ee S)

Annunziata, R.; Cinquini, M.; Cozzi, F.; Montanari, F.; Restelli A.

JCS Chem Comm, (1983), 1138

$$\text{Ph}\overset{\text{O}}{\underset{}{\text{C}}}\overset{\text{Me}}{\underset{}{\text{CH}}}\overset{\text{O}}{\underset{}{\text{C}}}\text{NEt}_2$$

PhMe$_2$SiH, 0°C

CF$_3$COOH

(99% erythro)

$$\text{Ph}\overset{\text{HO}}{\underset{}{\text{CH}}}\overset{\text{Me}}{\underset{}{\text{CH}}}\overset{\text{O}}{\underset{}{\text{C}}}\text{NEt}_2$$ 98%

PhMe$_2$SiH, TASF

0°C, DMPU   (99% threo)

$$\text{Ph}\overset{\text{HO}}{\underset{}{\text{CH}}}\overset{\text{Me}}{\underset{}{\text{CH}}}\overset{\text{O}}{\underset{}{\text{C}}}\text{NEt}_2$$ 98%

TASF = tris(diethylamino)sulfonium difluorotrimethyl
       silicate
DMPU = 1,3-dimethyl-3,4,5,6-tetrahydro-2(1H)
       pyrimidinone
Fujita, M.; Hiyama, T.[*]   **J Am Chem Soc**, (1985), **107**, 8294

$$\underset{\underset{Me}{|}}{\overset{O\ \ \ O}{\overset{||\ \ \ ||}{RCCHCNR^1R^2}}} \longrightarrow \underset{\underset{Me}{|}}{\overset{OH\ \ \ O}{\overset{|\ \ \ \ ||}{RCHCHCNR^1R^2}}}$$

R = Me, R^1 = H
R^2 = Bn

1. $Zn(BH_4)_2$, Et_2O, -78°C   97%
2. 3% aq. H_2PO_4 (syn:anti = 97:3)

Ito, Y.; Yamaguchi, M.*   **Tetrahedron Lett**, (1983), **24**, 5385

R = iPr, R^1 = Me
R^2 = $CH_2CH_2OSiMe_2tBu$

$KBEt_3H$, Et_2O 95%
0°C   (syn:anti = 1:80)

Ito, Y.; Katsuki, T.; Yamaguchi, M.*
Tetrahedron Lett, (1985), **26**, 4643

$$CH_3(CH_2)_4CHO \xrightarrow[\text{2. 2N HCl, RT}]{\overset{\overset{NMe}{||}}{1.Cl_3Ti\text{-}C\text{-}Cl,\ 0°C}} CH_3(CH_2)_4\underset{\underset{OH}{|}}{\overset{\overset{O}{||}}{CHCNHMe}}$$

96%

Schiess, M.; Seebach, D.   **Helv Chim Acta**, (1983), **66**, 1618

PhCH=CHPh
(E)

$$\xrightarrow[\begin{array}{c}1\%\ OsO_4,\ MeCN,\ RT\\ HOH,\ 5\ min\end{array}]{\overset{\overset{O}{||}}{EtOCNClNa,\ AgNO_3}}$$

69%

Ph   OH

NH   Ph
CO_2Et

Herranz, E.; Sharpless, K.B.*   **Org Syn**, (1983), **61**, 93

1. nBuLi
2. Cp_2ZrCl_2
3. DMAP
4. PhCHO

94%
(R:S = 4.4:1)

Katsuki, T.; Yamaguchi, M.*   **Tetrahedron Lett**, (1985), **26**, 5807

1. $\underline{n}Bu_2BOSO_2Me$
   CH_2Cl_2, NEt_3, $-78°C$
2. EtCHO, $-78°C$
3. aq. $NaHSO_4$
4. H_2O_2, pH 7   94%

Evans, D.A.[*]; Sjogren, E.B.; Bartroli, J.; Dow, R.L.
Tetrahedron Lett, (1986), **27**, 4957

1. $CH_3N=C=O$, BCl_3
   CH_2Cl_2, $PhCH_3$
   reflux, 12h
2. 5% HCl, RT, 2h   75%

Piccolo, O.[*]; Filippini, L.; Tinucci, L.; Valoti, E.; Citterio, A.[*]
Tetrahedron, (1986), **42**, 885

1. $\underline{n}BuLi$, THF
   $-100°C$
2. Cp_2ZrCl_2
   $-100 - -70°C$
3. aq. KF   4. SiO_2 (92% de)   75%

Uchikawa, M.; Hanamoto, T.; Katsuki, T.[*]; Yamaguchi, M.
Tetrahedron Lett, (1986), **27**, 4577 4581

SECTION 326: Alcohol, Thiol - Amine

1. Me_3SnNEt_2, CH_2Cl_2
   reflux 1d
2. $HO_2CCH_2CO_2H$, 150°C
   1.5 mmHg   55%

Taddei, M.; Papini, A.; Fiorenza, M.; Ricci, A.
Tetrahedron Lett, (1983), **24**, 2311

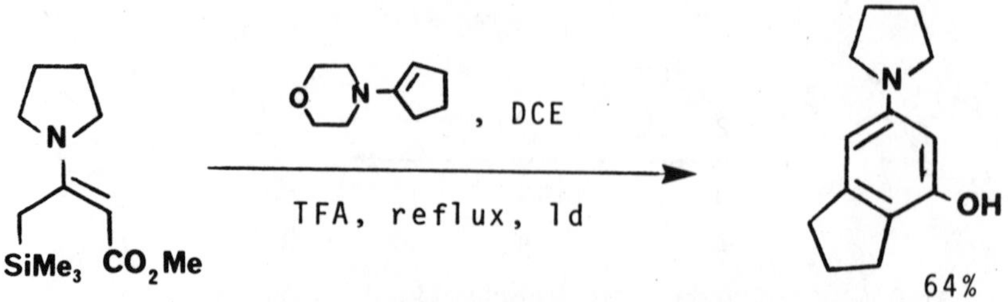

, DCE
TFA, reflux, 1d
SiMe₃ CO₂Me
OH
64%

Chan, T.H.; Kang, G.J. Tetrahedron Lett, (1983), 24, 3051

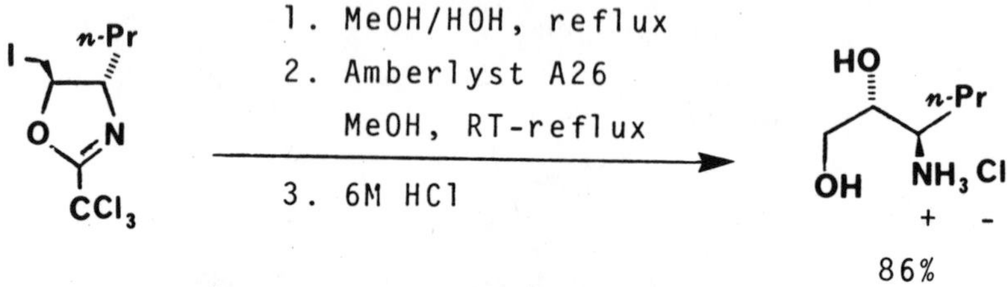

1. MeOH/HOH, reflux
2. Amberlyst A26
MeOH, RT-reflux
3. 6M HCl
n-Pr
CCl3
HO
OH
NH3 Cl
86%

Cardillo, G.*; Orena, M.; Sandri, S.
JCS Chem Comm, (1983), 1489

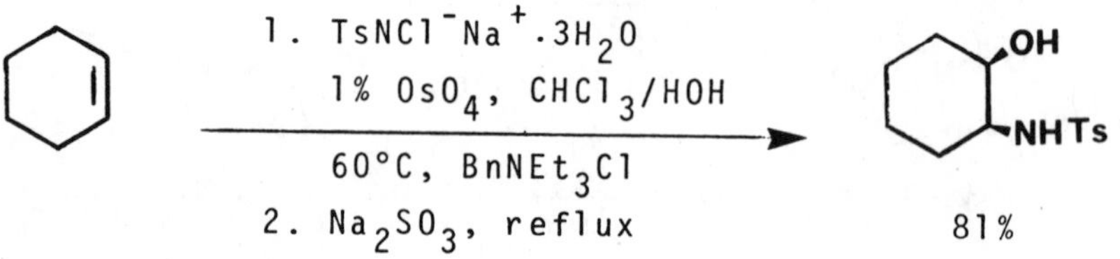

1. TsNCl⁻Na⁺·3H2O
1% OsO4, CHCl3/HOH
60°C, BnNEt3Cl
2. Na2SO3, reflux
OH
NHTs
81%

Herranz, E.; Sharpless, K.B.* Org Syn, (1983), 61, 85

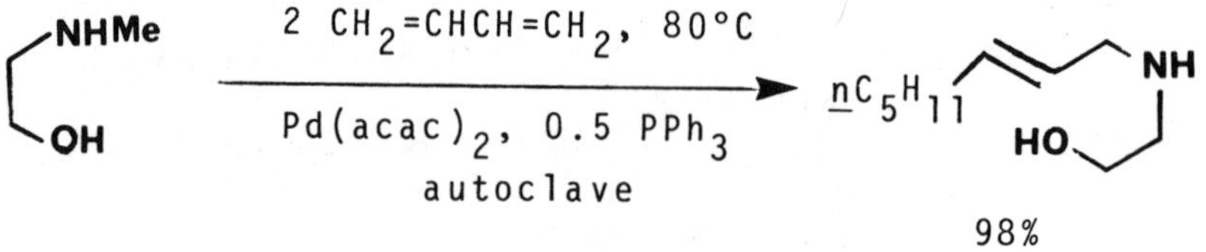

NHMe
OH
2 CH2=CHCH=CH2, 80°C
Pd(acac)2, 0.5 PPh3
autoclave
nC5H11
NH
HO
98%

Groult, A.; Guy, A. Tetrahedron, (1983), 39, 1543

O
PhCH-CH₂

1. Me₃Sn-N⟨ ⟩ , CH₂Cl₂

reflux, 16h

2, HO₂CCH₂CO₂H, Et₂O
reflux, 2h

$$PhCH-CH_2$$

1. Me_3Sn-N⟨ ⟩ , CH_2Cl_2, reflux, 16h

2. $HO_2CCH_2CO_2H$, Et_2O, reflux, 2h

83%

Fiorenza, M.; Ricci, A.*; Taddei, M.; Tassi, D.
Synthesis, (1983), 640

HO   O

n-Bu n-Bu

1. $\underline{n}BuONH_2 \cdot HCl$
   Py, MeOH
   reflux
2. $LiAlH(OMe)_3$, 0°C
   NaOMe, Et_2O

OH   NH₂

n-Bu n-Bu

+

OH   NH₂

n-Bu n-Bu

(96   :   4)

92%

Narasaka, K.*; Ukaji, Y.; Yamazaki, S.
                    Bull Chem Soc Jpn, (1986), **59**, 525
Narasaka, K.*; Ukaji, Y. **Chem Lett**, (1984), 147
Narasaka, K.; Yamazaki, S.; Ukaji, Y. **Chem Lett**, (1984), 2065

NH₂

1. SbF_5-HF, 70% H_2O_2
   -20°C, 15 min

2. $NaHCO_3$, ice/HOH

NH₂

OH

($\underline{o}:\underline{p}:\underline{m}$ = 5:36:21%)

Jacquesy, J.-C.; Jouannetaud, M.-P.; Morellet, G.; Vidal, Y.
                    Tetrahedron Lett, (1984), **25**, 1479

O

Ph   NMe₂
      Me

$PhMe_2SiH$, RT

TBAF, HMPA

12h

HO

Ph   NMe₂
      Me   83%

($\underline{threo}:\underline{erythro}$ = 99:1)

Fujita, M.; Hiyama, T.* **J Am Chem Soc**, (1984), **106**, 4629

$$CH_2=C \overset{OTMS}{\underset{Ph}{\big<}} \quad \xrightarrow[\text{2. LiAlH}_4, \text{ Et}_2O]{\begin{array}{c}\text{1. Ph-N=O, PhH}\\20°C\end{array}} \quad \overset{OH}{\underset{}{PhCHCH_2NHPh}}$$

78%

Sasaki, T.; Mori, K.; Ohno, M.[*]   **Synthesis**, (1985), 280

$$EtCH-CHCH_2OH \quad \xrightarrow[\text{1.5 Ti(OiPr)}_4]{\begin{array}{c}\text{excess Et}_2\text{NH}\\\text{RT, 5h}\end{array}}$$

Et—CH—CH—CH₂OH with NEt₂ and OH   +   Et with OH and OH, NEt₂

(20   :   1)

90%

Caron, M.; Sharpless, K.B.[*]   **J Org Chem**, (1985), **50**, 1557

$$\xrightarrow[\begin{array}{c}CH_2Cl_2,\\25°C, 1.5h\end{array}]{2 \text{ AlCl}_3}$$

n-Bu   structure with Me OH   +   structure with *n*-Bu, Me OH

(3   :   1)

91%

Overman, L.E.[*]; Lesuisse, D.
                    Tetrahedron Lett, (1985), **26**, 4167

$$\square\text{O} \quad \xrightarrow[\begin{array}{c}\text{2. HCl/MeOH}\\\text{3. NaOH, HOH}\end{array}]{\begin{array}{c}\text{1. TMS-CN, CH}_2\text{Cl}_2\\\text{ZnI}_2, \text{12h}\end{array}} \quad \begin{array}{c}\text{OH}\\\text{NH}_2\end{array}$$

61%

Gassman, P.G.[*]; Haberman, L.M.
                    Tetrahedron Lett, (1985), **26**, 4971

$$\underset{\substack{| \\ CH_3 \\ |}}{CH_3\overset{\displaystyle CN}{\underset{\displaystyle C}{C}}OSiMe_3} \quad \xrightarrow[\substack{2.\ NaBH_4,\ MeOH \\ RT}]{\substack{1.\ PhMgBr,\ Et_2O \\ RT}} \quad \underset{\substack{| \\ CH_3}}{CH_3\overset{\displaystyle HO\ Ph}{\underset{\displaystyle C}{C}}-CHNH_2}$$

84%

Krepski, L.R.[*]; Jensen, K.M.; Heilmann, S.M.[*]; Rasmussen, J.K.[*]
Synthesis, (1986), 301

$$\underset{(\underline{R})}{PhCH\overset{\displaystyle O}{-}CH_2} \quad \xrightarrow[\substack{44h}]{\substack{NHTMS \\ | \\ PhCHCH_3\ (\underline{S})\ ,\ 65°C}} \quad$$

80%

Atkins, R.K.; Frazier, J.; Moore, L.L.; Weigel, L.O.[*]
Tetrahedron Lett, (1986), **27**, 2451

$$\xrightarrow[\substack{37\%\ aq.\ HCHO}]{\substack{\underline{n}BuNH_2\cdot TFA,\ 35°C \\ HOH,\ 48h}}$$

94%

Larsen, S.D.; Grieco, P.A.[*]; Fobare, W.F.
J Am Chem Soc, (1986), **108**, 3512

$$\xrightarrow[\substack{2.\ NaOH/HOH}]{\substack{1.\ (-)\ IpC_2BH,\ -15°C}}$$

89%

IpC_2BH = isopinocamphenylborane (84% ee)

Brown, H.C.[*]; Prasad, J.V.N.V.; Gupta, A.K.
J Org Chem, (1986), **51**, 4296

SECTION 327: Alcohol, Thiol - Ester

R = CH$_2$CH=CMe$_2$ Baker's Yeast, MgSO$_4$ 73%
R' = Me   D-glucose, KH$_2$PO$_4$, HOH   (92% ee)

Hirama, M.[*]; Shimizu, M.; Iwashita, M.
            JCS Chem Comm, (1983), 599

R = Et   Thermoanaerobium brockii
R' = Et   Trypton, Na$_2$S, yeast extract   80%
          D-glucose   (84% ee S)
Seebach, D.[*]; Züger, M.F.; Giovannini, F.; Sonnleitner, B.;
Fieehter, A.
       Angew Chem Int Ed Engl, (1984), **23**, 151

R = Me
E' = Et   Baker's Yeast, HOH
        sucrose, 30°C   (S:R = 93:7)
Seebach, D.[*]; Sutler, M.A.; Weber, R.H.; Züger, M.F.
         Org Syn, (1984), **63**, 1
Seebach, D.[*]; Giovannini, F.; Lamatsch, B.
        Helv Chim Acta, (1985), **68**, 958

R = Ph   LiBH$_4$, tBuOH, THF   94%
R' = Et   (RR')NN'dibenzoyl
        cystine, -78 - -30°C   (87% ee R)
Soai, K.[*]; Yamanoi, T.; Hikima, H.; Oyamada, H.
        JCS Chem Comm, (1985), 138

Yamamoto, Y.[*]; Maeda, N.; Maruyama, K.
        JCS Chem Comm, (1983), 774

BrCH₂CO₂Et, 0°C
$$BrCH_2CO_2Et, \ 0°C$$

Zn-graphite

88%

Boldrini, G.P.; Savoia, D.; Tagliavini, E.; Trombini, C.*;
Umani-Ronchi, A.*

J Org Chem, (1983), **48**, 4108

PhCH₂CH₂CHO $\xrightarrow{\text{Cl}_3\text{Ti}^{\nearrow O}, \ 0-20°C}$ BnCH₂CHCH₂CH₂CO₂iPr

$$Cl_3Ti^{\nearrow O\overset{OiPr}{\diagdown}}, \ 0-20°C$$

CH₂Cl₂, 2h

OH
|
BnCH₂CHCH₂CH₂CO₂iPr

·80%

Nakamura, E.; Kuwajima, I.*   **J Am Chem Soc**, (1983), **105**, 651

2.8 CH₂=C $\overset{OSiMe_2tBu}{\underset{OMe}{}}$

CH₂Cl₂, BF₃·OEt₂
-78°C, 4h

$$Ph\overset{Me}{\underset{HO}{\diagdown}}\diagdown CO_2Me \ + \ Ph\overset{Me}{\underset{OH}{\diagdown}}\diagdown CO_2Me$$

(15 : 1)

Heathcock, C.H.*; Flippin, L.A.   81%
J Am Chem Soc, (1983), **105**, 1667

K9-0-DIPGF9-BBNH

THF, -78°C

$$\overset{H}{\diagup}\overset{OH}{\underset{}{\diagup}}CO_2Et$$

80%

(92% ee S)

K⁺   = K9-DIPGF9-BBNH

Brown, H.C.*; Cho, B.T.; Park, W.S.
J Org Chem, (1986), **51**, 3396

PhCCO$_2$Et $\xrightarrow[\substack{\text{SnCl}_2/\text{Dibal-H, }-100°C \\ 20 \text{ min}}]{\text{[chiral amine]}, \text{ CH}_2\text{Cl}_2}$ PhCHCO$_2$Et
90%
(85% ee S)

Mukaiyama, T.; Tomimori, K.; Oriyama, T.
Chem Lett, (1985), 813, 1359

CH$_3$CCO$_2$Et $\xrightarrow[0°C, 24h]{\text{[chiral borane]}}$ CH$_3$CHCO$_2$Et
81%
(89% ee S)

Brown, H.C.[*]; Pai, G.G.; Jadhav, P.K.
J Am Chem Soc, (1984), **106**, 1531

[cyclopropane with SiMe$_3$ and CO$_2$Me] $\xrightarrow[\text{CH}_3\text{CHO}]{\text{TBAF}}$ [cyclopropane with CHCH$_3$/OH and CO$_2$Me]
90%

Paquette, L.A.[*]; Blankenship, C.; Wells, G.J.
J Am Chem Soc, (1984), **106**, 6442

R*OCCH$_3$ $\xrightarrow[\substack{2. \text{ClSiMe}_2\text{tBu} \\ \text{HMPT/THF} \\ 3. \text{EtCHO, TiCl}_4 \\ \text{CH}_2\text{Cl}_2}]{\substack{1. \text{LICA, THF} \\ -80°C}}$ R*O...Et (HO) + R*O...Et (HO)
(93 : 7)
67%

R* = [bornane sultam structure] N SO$_2$ Ar

Ar= [dimethylphenyl]

Helmchen, G.[*]; Leikauf, U.; Taufer-Knöpfel, I.
Angew Chem Int Ed Engl, (1985), **24**, 874

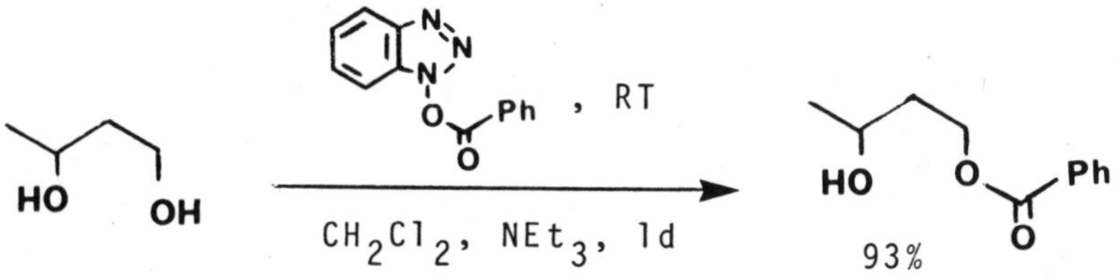

RT
CH_2Cl_2, NEt_3, 1d
HO
OH
Ph
93%

Kim, S.*; Chang, H.; Kim, W.J. J Org Chem, (1985), 50, 1751

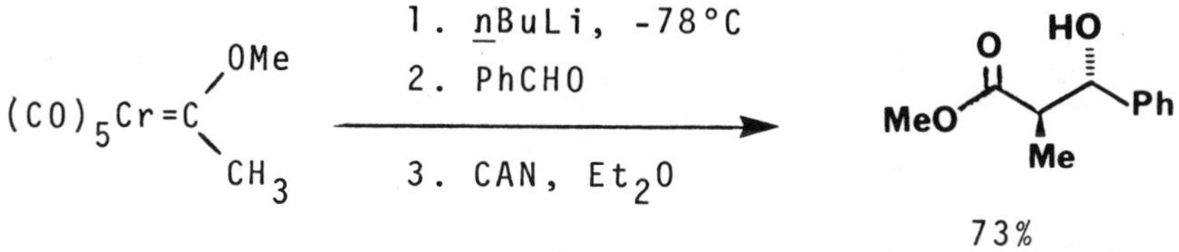

$(CO)_5Cr=C(OMe)CH_3$
1. nBuLi, -78°C
2. PhCHO
3. CAN, Et_2O
MeO
HO
Ph
Me
73%

Wulff, W.D.*; Gilbertson, S.R. J Am Chem Soc, (1985), 107, 503

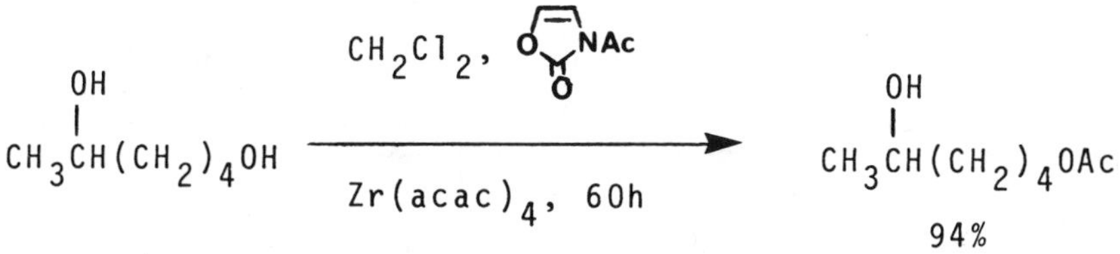

CH_2Cl_2,
NAc
$CH_3CH(OH)(CH_2)_4OH$
Zr(acac)$_4$, 60h
$CH_3CH(OH)(CH_2)_4OAc$
94%

Kunieda, T.*; Mori, T.; Higuchi, T.; Hirobe, M.* Tetrahedron Lett, (1985), 26, 1977

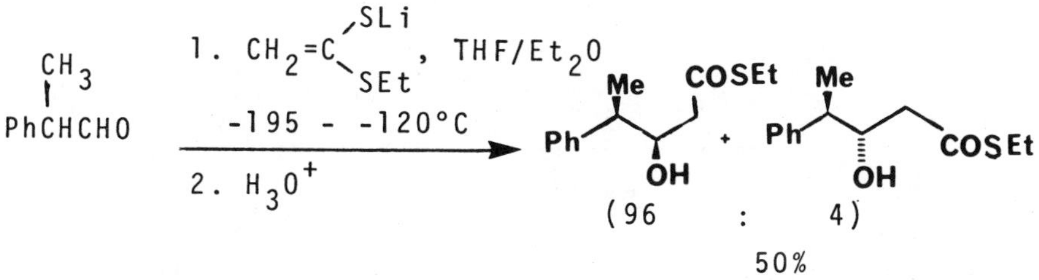

$PhCH(CH_3)CHO$
1. $CH_2=C(SLi)(SEt)$, THF/Et_2O
-195 - -120°C
2. H_3O^+
Me
COSEt
Ph
OH
(96 : 4)
50%

Meyers, A.I.*; Walkup, R.D. Tetrahedron, (1985), 41, 5089

OH
|
$CH_3CH(CH_2)_3CO_2\underline{t}Bu$

1. 2 $LiNEt_2$
THF/HMPA

-100 - -78°C

2. Me_2SO_4
-100°C

HO ⟶ $CO_2\underline{t}Bu$ Me (92)

HO + $CO_2\underline{t}Bu$ Me (8)

71%

Narasaka, K.; Ukaji, Y.   **Chem Lett**, (1986), 81

HO   O   **CO₂Me**

$FeCl_2$, MeOH

50 psi H_2
18h

HO OH
n-Pr ⟶ **CO₂Me**
90%

(syn:anti = 9:1)

Kathawala, F.G.; Prager, B.; Prasad, K.*; Repič, O.; Shapiro,
M.J.; Stabler, R.S.; Widler, L.
Helv Chim Acta, (1986), **69**, 803

O   CO₂Et

TMS

$TiCl_4$, CH_2Cl_2

-78°C, 2h

Me   OH CO₂Et
Me

88%

Molander, G.A.*; Andrews, S.W.
Tetrahedron Lett, (1986), **27**, 3115

$CH_2=C$
O-B
SCEt₃

nPrCHO

pentane
-78°C, 6h

OH   O
| ||
nPrCHCH₂CSCEt₃
82%

(91.2% ee R)

Masamune, S.*; Sato, T.; Kim, B.M.; Wollmann, T.A.
J Am Chem Soc, (1986), **108**, 8279

$$\underset{R*OCCH_2CH_2Ph}{\overset{O}{\parallel}} \xrightarrow[\substack{2 \ KN(TMS)_2 \\ 8 \ CH_3CH_2\overset{|}{\underset{\underset{K^+}{O^-}}{C}}HCH_3}]{MoOPH, \ 4h} \underset{\underset{OH}{\overset{|}{\ }}}{\overset{O}{\parallel}}R*OCCHCH_2Ph$$

73%

(R:S = 1:99)

R* =

Gamboni, R.; Tamm, C.*   **Tetrahedron Lett**, (1986), **27**, 3999

$$\underset{CH_3-\overset{\overset{HO\quad H}{|\quad|}}{C}-CO_2ET}{} \xrightarrow[\substack{\text{overnight} \\ 2. \ 3\% \ HCl/EtOH}]{1. \ CsSAc, \ DMF} \underset{CH_3-\overset{\overset{H\quad SH}{|\quad|}}{C}-CO_2Et}{}$$

86%

(92% ee R)

Strijtveen, B.; Kellogg, R.M.*   **J Org Chem**, (1986), **51**, 3664

68%

Molander, G.A.*; LaBelle, B.E.; Hahn, G.
          J Org Chem, (1986), **51**, 5259

Also via: Hydroxyacids (Section 313)

SECTION 328: Alcohol, Thiol - Ether, Epoxide, Thioether

Heathcock, C.H.[*]; Kiyooka, S.; Blumenkopf, T.A.
J Org Chem, (1984), 49, 4214
Kiyooka, S.; Heathcock, C.H.
Tetrahedron Lett, (1983), 24, 4765

Tius, M.A.[*]; Thurkauf, A.   J Org Chem, (1983), 48, 3839

(E:Z = 13:87)

Takai, K.; Oshima, K.[*]; Nozaki, H.
Bull Chem Soc Jpn, (1983), 56, 3791

Ph
Ph
OH

(A) HO—$\overset{O}{\underset{}{C}}$—NH-Bn
    HO—C—NH-Bn
           $\overset{O}{}$, tBuOOH

(B) Ti(OiPr)$_4$, -20°C
(A:B = 2.4:2   1:2.0)
               96% ee   82% ee

Ph $\triangle$ O
Ph
OH

Ph $\triangle$ O
Ph
OH

Lu, L.D.L.; Johnson, R.A.; Finn, M.G.; Sharpless, K.B.[*]
J Org Chem, (1984), **49**, 728

CH$_3$(CH$_2$)$_7$CH=CHCH$_2$OH
(E)

$\xrightarrow{\text{(+)diethyl tartrate}}$
4Å sieve, -20°C
Ti(OiPr)$_4$, tBuOOH
CH$_2$Cl$_2$

nC$_8$H$_{17}$$\overset{O}{\overset{\triangle}{CH}}$-CHCH$_2$OH
79%
(95% ee 2S,3S)

Hanson, R.M.; Sharpless, K.B.[*]   **J Org Chem**, (1986), **51**, 1922
Hill, J.G.; Sharpless, K.B.[*]; Exon, C.M.; Regenye, R.
Org Syn, (1984), **63**, 66

1. BH$_2$Cl·OEt$_2$, Et$_2$O
   0°C, 1h

2. NaOH, H$_2$O$_2$
   EtOH

90%

Borders, R.J.; Bryson, T.A.[*]   **Chem Lett**, (1984), 9

(CO)$_5$Cr=C$\overset{OMe}{\underset{C≡CSiMe_3}{}}$

1. $\overset{OTMS}{\diagup\!\!\diagup}$, THF
   nPrC≡CH, 50°C

2. SiO$_2$, air

58%

Wulff, W.D.[*]; Yang, D.C.   **J Am Chem Soc**, (1984), **106**, 7565

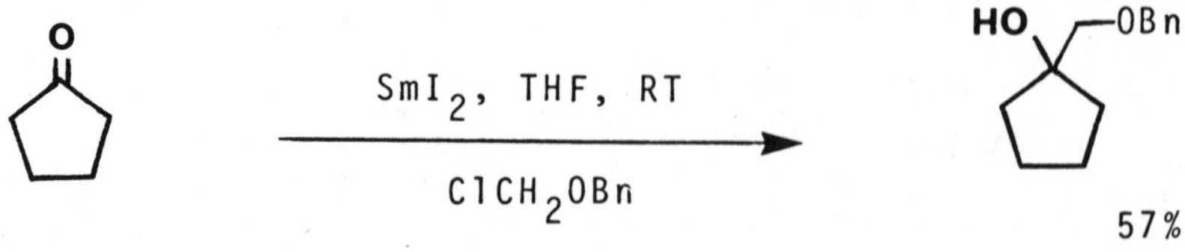

57%

Imamoto, T.[*]; Takeyama, T.; Yokoyama, M.
                      Tetrahedron **Lett**, (1984), **25**, 3225

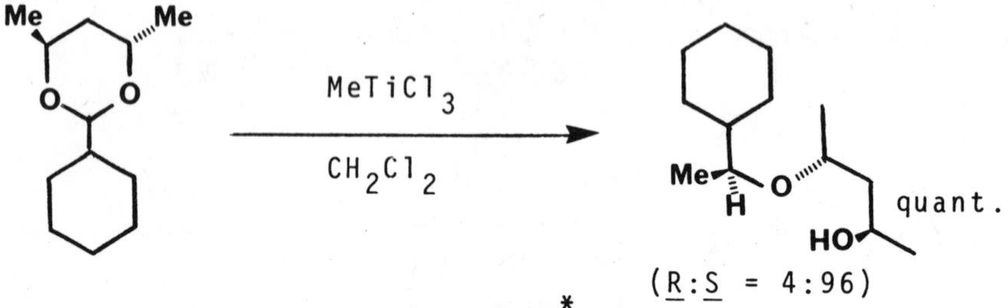

quant.

($\underline{R}$:$\underline{S}$ = 4:96)

Mori, A.; Maruoka, K.; Yamamoto, H.[*]
                      Tetrahedron **Lett**, (1984), **25**, 4421

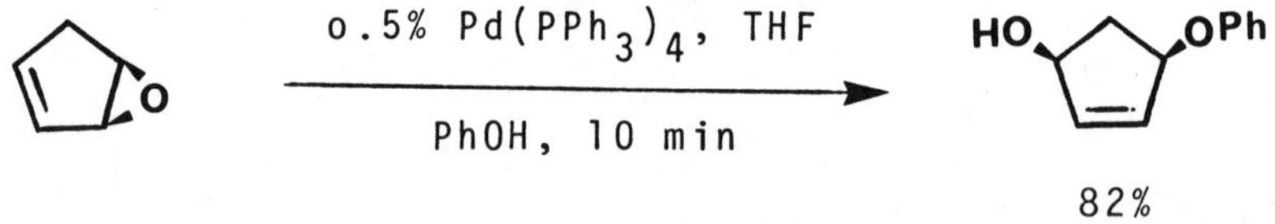

82%

Deardorff, D.R.[*]; Myles, D.C.; MacFerrin, K.D.
                      Tetrahedron **Lett**, (1985), **26**, 5615

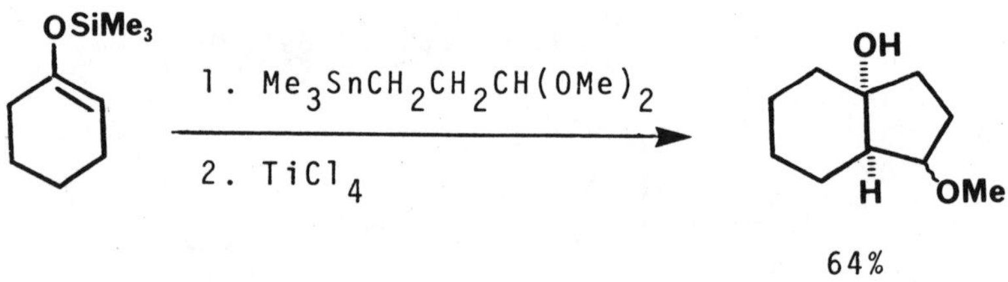

64%

Lee, T.V.[*]; Richardson, K.A.
                      Tetrahedron **Lett**, (1985), **26**, 3629

$$CF_3COOH, NaBH_4$$
$$THF, 20°C$$

$$PhCH_2OCH_2CH_2OH$$

83%

Nutaitis, C.F.; Gribble, G.W.[*]
Org Prep Proc Int, (1985), **17**, 11

PhCH-CH_2

$$\underline{n}Bu_2SnO, RT$$
$$5h$$
$$P(O\underline{n}Bu)_3$$

OH
|
PhCHCH_2OMe

OMe
|
+ PhCHCH_2OH

(98   :   2)

77%

Otera, J.[*]; Yoshinaga, Y.; Hirakawa, K.
Tetrahedron Lett, (1985), **26**, 3219

1. NCS, CCl_4
2. MeLi, THF
   -80°C - RT

49%

Harada, T.; Akiba, E.; Tsujimoto, K.; Oku, A.[*]
Tetrahedron Lett, (1985), **26**, 4483

$$CH_3CH=C \begin{matrix} SPh \\ \\ CH_3 \end{matrix}$$

1. PhCHO, Et_3SiH, -78°C
   CH_2Cl_2, 1.5h
2. ZnBr_2, CH_2Cl_2
   -78 - -40°C, 4h
3. HOH

SPh OH
| |
CH_3CHCHCHPh
|
CH_3

77%

Takeda, T.[*]; Tsuchida, T.; Nakagawa, I.; Ogawa, S.; Fujiwara, T.
Tetrahedron Lett, (1985), **26**, 5313

iPrCH=CHiPr

(E)

$$\xrightarrow[\text{Ti(OiPr)}_4, \text{ CH}_2\text{Cl}_2]{^3\text{O}_2, \text{ TPP, h}\nu, 0°\text{C}}$$

Me
Me $>$ C-CHCHiPr (with epoxide O)

75%

TPP = tetraphenyl porphine

Adam, W.[*]; Griesbeck, A.; Staab, E.

(erythro:threo = 9:1)

Tetrahedron Lett, (1986), **27**, 2839

$$\overset{\text{OH}}{\underset{|}{CH_3CH=CHCH\underline{n}C_6H_{13}}}$$

(E)

$$\xrightarrow[\text{60°C, 6h}]{\text{nBu}_2\text{SnO, } \underline{t}\text{BuOOH}}$$

$$CH_3\overset{O}{CH-}\overset{OH}{\underset{|}{CHCH\underline{n}C_6H_{13}}}$$

75%

(threo:erythro = 4:6)

Kanemoto, S.; Nonaka, T.; Oshima, K.[*]; Utimoto, K.

Tetrahedron Lett, (1986), **27**, 3387

PhCHO

$$\xrightarrow[\substack{\text{CrCl}_2, \text{ TMEDA, 25°C} \\ \text{THF, 6h}}]{\overset{\text{Cl}}{\underset{|}{\text{PhSCHCH}_3}} , \text{ LiI}}$$

Ph $-$ Me $-$ OH   +   Ph $-$ Me $-$ OH

(88   :   12)

96%

Nakatsukasa, S.; Takai, K.[*]; Utimoto, K.

J Org Chem, (1986), **51**, 5045

$$CH_2=C\overset{CH_3}{\underset{CH_2OH}{<}}$$

1. Ti(OiPr)$_4$, -20°C
   (+)diisopropyl tartrate
   cumene hydroperoxide
   3Å sieve
2. P(OMe)$_3$
3. PhSH, Ti(OiPr)$_4$

$\xrightarrow{\hspace{3cm}}$

HO Me

PhS $-$ **OH**

quant.

(92% ee)

Ko, S.Y.; Sharpless, K.B.[*]   **J Org Chem**, (1986), **51**, 5413

Reviews:

"Chelation or Non-chelation Control in Addition Reactions of
Chiral α and β-Alkoxy Carbonyl Compounds"

Reetz, M.T.[*]   **Angew Chem Int Ed Engl**, (1984), **23**, 556

"Asymmetric Epoxidation of Allylic Alcohols: The Sharpless
Reaction"

Pfenniger, A.[*]   **Synthesis**, (1986), 89

SECTION 329: Alcohol, Thiol - Halide, Sulfonate

$$
\begin{array}{ccc}
O & & OH \\
\parallel & \xrightarrow[\text{2. NaOH, } H_2O_2]{\text{1. } \quad \text{, THF, 4d}} & \mid \\
PhCCH_2Br & & PhCHCH_2Br \\
& & 95\%
\end{array}
$$

(86% ee R)

Brown, H.C.[*]; Pai, G.G.   **J Org Chem**, (1983), **48**, 1784

$$
\xrightarrow[25°C, 18h]{Li_2NiBr_4, \ THF}
$$

96%

Dawe, R.D.; Molinski, T.F.; Turner, J.V.[*]
                          Tetrahedron Lett, (1984), **25**, 2061

$$
nC_7H_{15}CH=CHCH_2OMe \xrightarrow[\substack{2. \ Na_2SO_4, \ Et_2O \\ RT, \ 1h}]{\substack{1. \ TiCl_4, \ tBuOOH \\ CH_2Cl_2, \ -78°C}} nC_7H_{15} \quad HO \quad OMe
$$

(E)

92%

Klunder, J.M.; Caron, M.; Uchiyama, M.; Sharpless, K.B.[*]
                          J Org Chem, (1985), **50**, 912

HO—⟨C₆H₄⟩ $\xrightarrow[\underline{t}BuOCl]{NaI,\ CH_3CN,\ 0°C}$ HO—⟨C₆H₄⟩—I   97%

Kometani, T.*; Watt, D.S.*; Ji, T.; Fitz, T.
J Org Chem, (1985), 50, 5384

cyclohexanone $\xrightarrow[\substack{MeLi/LiBr/Et_2O \\ 30\ min}]{CF_3CF_2I,\ -78°C}$ HO, CF₂CF₃ cyclohexane   quant.

Gassman, P.G.*; O'Reilly, N.J.
Tetrahedron Lett, (1985), 26, 5243

$\underset{PhCCH_2Cl}{\overset{O}{\|}}$ $\xrightarrow{LiBH_4,\ \underline{t}BuOH,\ THF}$ $\underset{PhCHCH_2Cl}{\overset{OH}{|}}$

76%

$$\underset{\substack{| \\ COOH}}{\overset{\substack{CH_2SSCH_2 \\ | \\ PhCNHCH \qquad CHNHCPh}}{}}$$
$$\underset{O}{\overset{\|}{}} \quad (\underline{S},\underline{S}) \quad \underset{HOOC}{} \quad \overset{\|}{O}$$ (72% ee S)

Soai, K.*; Yamanoi, T.; Hikima, H.
J Organomet Chem, (1985), 290, c023

$\underset{EtC(CH_2)_3Cl}{\overset{O}{\|}}$ $\xrightarrow[\substack{TRIS·HCl,\ HOCH_2CH_2SH \\ 37°C}]{TBADH,\ NADP,\ aq.\ iPrOH}$ $\underset{EtCH(CH_2)_3Cl}{\overset{OH}{|}}$

82%

TBADH = Thermoanaerobium brockii   (99% ee S)
        alcohol dehydrogenase

Keinan, E.*; Seth, K.K.; Lamed, R.
J Am Chem Soc, (1986), 108, 3474

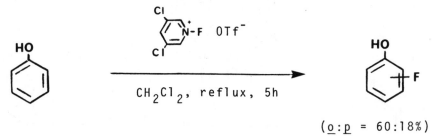

n = 1   81%

Imamoto, T.[*]; Takeyama, T.; Koto, H.
                    Tetrahedron Lett, (1986), <u>27</u>, 3243

n = 0   82%

Tabuchi, T.; Inanaga, J.[*]; Yamagkuchi, M.
                    Tetrahedron Lett, (1986), <u>27</u>, 3891

1. Li_2CuCl_4, THF
        4.75 h

2. HOH

                                                    98%

Ciaccio, J.A.; Addess, K.J.; Bell, T.W.[*]
                    Tetrahedron Lett, (1986), <u>27</u>, 3697

CH_2Cl_2, reflux, 5h

($\underline{o}$:$\underline{p}$ = 60:18%)

Umemoto, T.[*]; Kawada, K.; Tomita, K.
                    Tetrahedron Lett, (1986), <u>27</u>, 4465

SECTION 330: Alcohol, Thiol - Ketone

Reetz, M.T.[*]; Kesseler, K.; Schmidtberger, S.; Wenderoth, B.;
Steinbach, R.

Angew Chem Int Ed Engl, (1983), 22, 989

Ando, W.[*]; Tsumaki, H.   Chem Lett, (1983), 1409

(P) = Amberlyst A26(F⁻)

Cardillo, G.[*]; Orena, M.; Porzi, G.; Sandri, S.; Tomasini, C.
J Org Chem, (1984), 49, 701

Braun, M.[*]; Hild, W.   Angew Chem Int Ed Engl, (1984), 23, 723

Davis, F.A.[*]; Vishwakarma, L.C.; Billmers, J.M.; Finn, J.[*]
J Org Chem, (1984), **49**, 3241

Curran, D.P.[*]; Scanga, S.A.; Fenk, C.J.
J Org Chem, (1984), **49**, 3474

Osuka, A.[*]; Taka-Oka, K.; Suzuki, H.[*]   **Chem Lett**, (1984), 271

Moriarty, R.M.; Hou, K.-C.; Prahash, I.
Org Syn, (1985), **64**, 138
Moriarty, R.M.; Hou, K.-C.   **Tetrahedron Lett**, (1984), **25**, 691

$$\underline{n}C_6H_{13}\overset{O}{\overset{\|}{C}}CH_2Br \xrightarrow[\substack{\underline{n}Bu_3SnAlEt_2 \\ 30\ min}]{PhCHO,\ 0°C} \underline{n}C_6H_{13}\overset{O}{\overset{\|}{C}}CH_2\overset{OH}{\overset{|}{C}}HPh$$

81%

Tsuboniwa, N.; Matsubara, S.; Morizawa, Y.; Oshima, K.[*]; Nozaki, H.

Tetrahedron Lett, (1984), **25**, 2569

$$\underset{\underset{Br}{|}}{tBuC\overset{O}{\overset{\|}{C}}CHCH_3} \xrightarrow[THF,\ RT,\ 24h]{EtCHO,\ CrCl_2} \quad t\text{-}Bu \overset{O}{\diagup}\diagdown\underset{Me}{}\diagdown\overset{OH}{\diagup}\diagdown$$

81%

Dubois, J.-E.[*]; Axiotis, G.; Bertounesque, E.

Tetrahedron Lett, (1985), **26**, 4371

iPrCHO $\xrightarrow[\substack{2.\ N(CH_2CH_2OH)_3}]{\substack{1.\ \overset{CH_3}{\underset{H}{}}C=C\overset{OB\diagdown}{\underset{CH_3}{}}}}$

HO ... Me (91 : 9) HO ... Me

85%

Hoffmann, R.W.[*]; Ditrich, K.

Tetrahedron Lett, (1984), **25**, 1781

Ph Me / O NH $\xrightarrow[\substack{3.\ \underline{n}PrCHO,\ 0°C \\ 4.\ H_3O^+}]{\substack{1.\ 2\ LDA,\ THF \\ 2.\ SnCl_2}}$

O HO ... n-Pr

60%

(58% ee)

Narasaka, K.; Miwa, T.; Hayashi, H.; Ohta, M.

Chem Lett, (1984), 1399

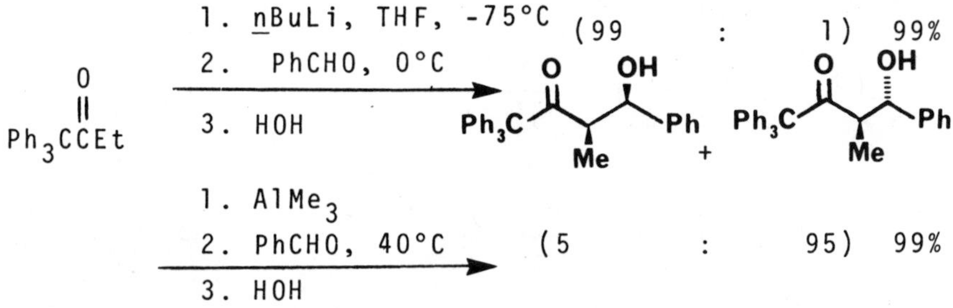

1. nBuLi, THF, -75°C   (99   :   1)   99%
2. PhCHO, 0°C
3. HOH

Ph₃CCEt

1. AlMe₃
2. PhCHO, 40°C   (5   :   95)   99%
3. HOH

Ertas, M.; Seebach, D.*   **Helv Chim Acta**, (1985), **68**, 961

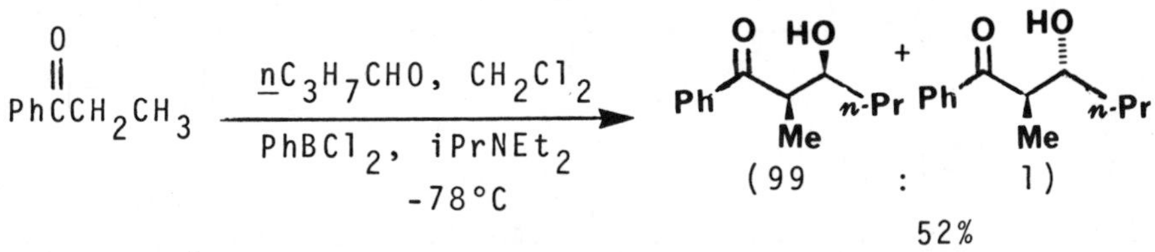

PhCCH₂CH₃

nC₃H₇CHO, CH₂Cl₂
PhBCl₂, iPrNEt₂
-78°C

(99   :   1)

52%

Hamana, H.*; Sasakura, K.; Sugasawa, T.
Chem Lett, (1984), 1729

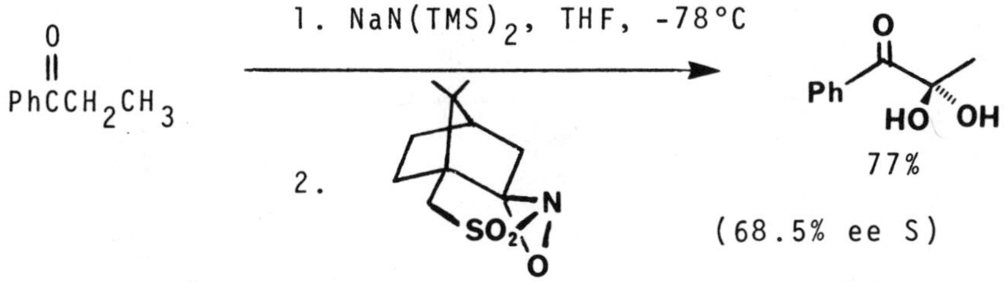

PhCCH₂CH₃

1. NaN(TMS)₂, THF, -78°C

2.

77%

(68.5% ee S)

Davis, F.A.*; Haque, M.S.   **J Org Chem**, (1986), **51**, 4083

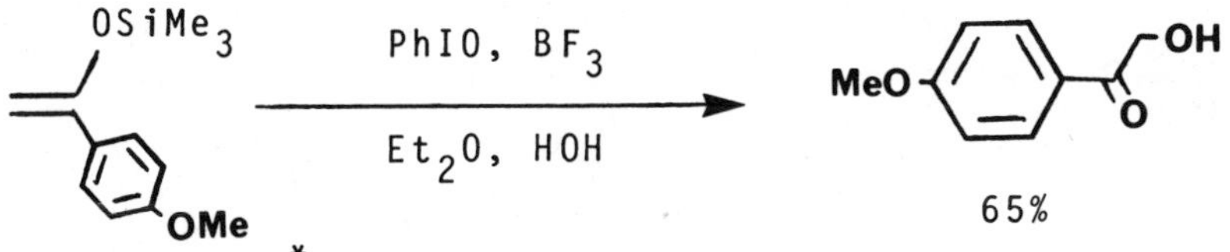

PhIO, BF₃

Et₂O, HOH

65%

Moriarty, R.M.*; Prakash, O.; Duncan, M.P.
Synthesis, (1985), 943

Tamura, Y.[*]; Yakura, T.; Haruta, J.; Kita, Y.
Tetrahedron Lett, (1985), **26**, 3837

DMI = 1,3-dimethylimidazolidin-2-one

Ricci, A.; Degl'Innocenti, A.[*]; Chimichi, S.; Fiorenza, M.;
Rossini, G.; Bestmann, H.J.
J Org Chem, (1985), **50**, 130

Matsumoto, T.; Ohishi, M.; Inoue, S.[*]
J Org Chem, (1985), **50**, 603

Koreeda, M.[*]; Luengo, J.I.   **J Am Chem Soc**, (1985), **107**, 5572

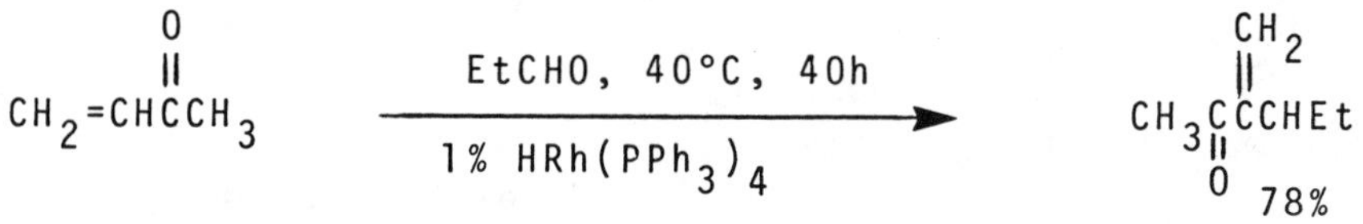

$$CH_2=CHCCH_3 \quad \xrightarrow[\text{1\% HRh(PPh}_3)_4]{\text{EtCHO, 40°C, 40h}} \quad CH_3CCCHEt$$

O (left structure)

CH₂ / O (right structure) 78%

Sato, S.; Matsuda, I.*; Izumi, Y. **Chem Lett**, (1985), 1875

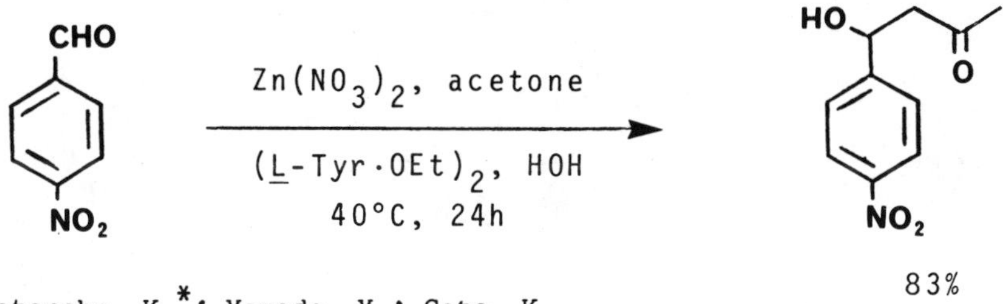

$$\xrightarrow[\substack{(\underline{L}\text{-Tyr}\cdot\text{OEt})_2, \text{ HOH} \\ 40°C, 24h}]{Zn(NO_3)_2, \text{ acetone}}$$

83%

Watanabe, K.*; Yamada, Y.; Goto, K.
                    Bull Chem Soc Jpn, (1985), **58**, 1401

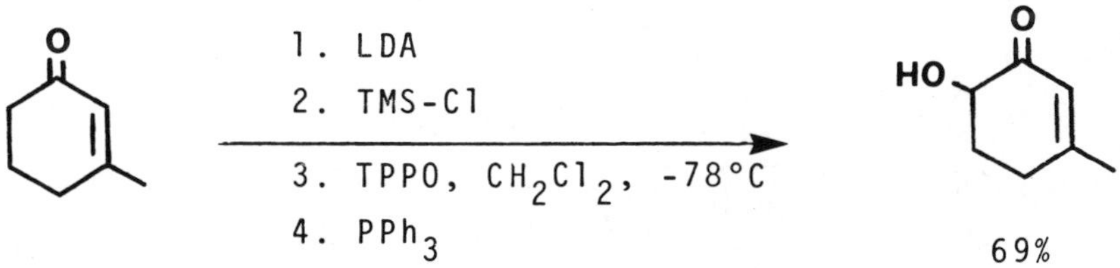

1. LDA
2. TMS-Cl

3. TPPO, CH_2Cl_2, -78°C
4. PPh_3

69%

TPPO = triphenylphosphite ozonide

Iwata, C.*; Takemoto, Y.; Nakamura, A.; Imanishi, T.
                    Tetrahedron Lett, (1985), **26**, 3227

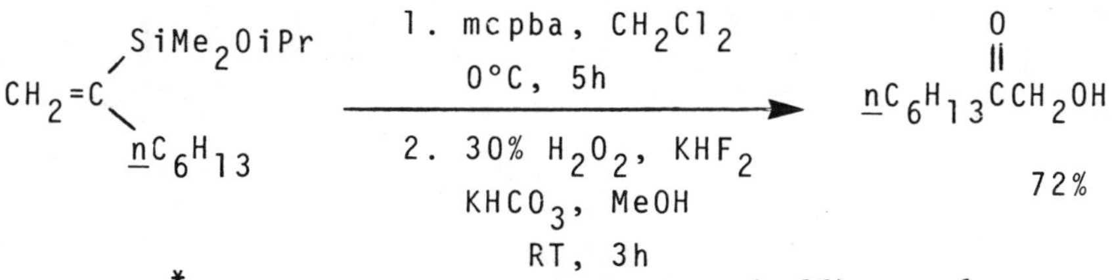

$$CH_2=C\begin{smallmatrix}SiMe_2OiPr\\ \\ \underline{n}C_6H_{13}\end{smallmatrix} \quad \xrightarrow[\substack{2. \ 30\% \ H_2O_2, \ KHF_2 \\ KHCO_3, \ MeOH \\ RT, \ 3h}]{\substack{1. \ mcpba, \ CH_2Cl_2 \\ 0°C, \ 5h}} \quad \underline{n}C_6H_{13}CCH_2OH$$

O

72%

Tamao, K.*; Maeda, K. **Tetrahedron Lett**, (1986), **27**, 65

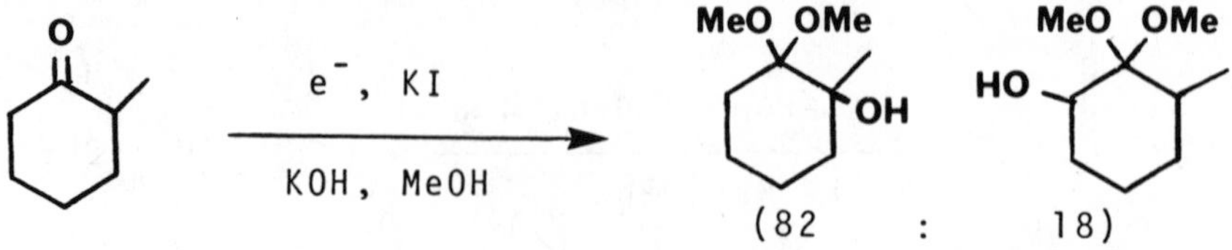

(82   :   18)

77%

Shono, T.[*]; Matsumura, Y.; Inoue, K.; Iwasaki, F.
JCS Perkin I, (1986), 73

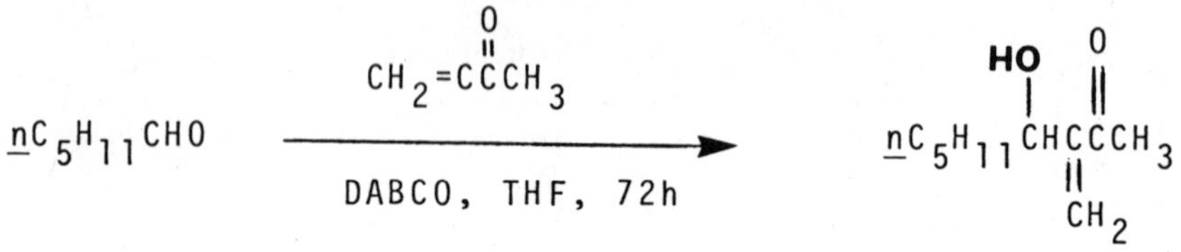

65%

Basavaiah, D.[*]; Gowriswari, V.V.L.
Tetrahedron Lett, (1986), **27**, 2031

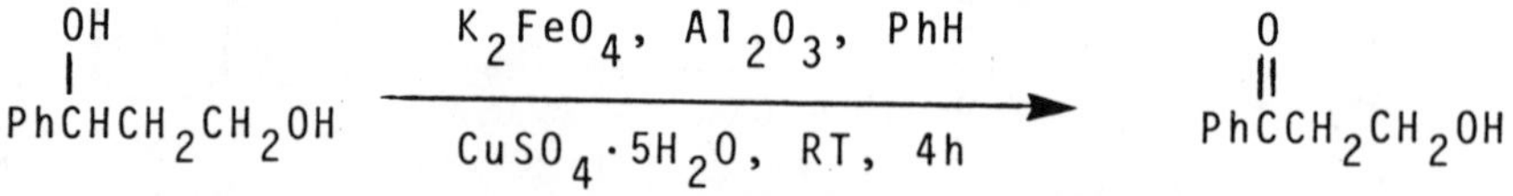

95%

Kim, K.S.[*]; Song, Y.H.; Lee, N.H.; Hahn, C.S.
Tetrahedron Lett, (1986), **27**, 2875

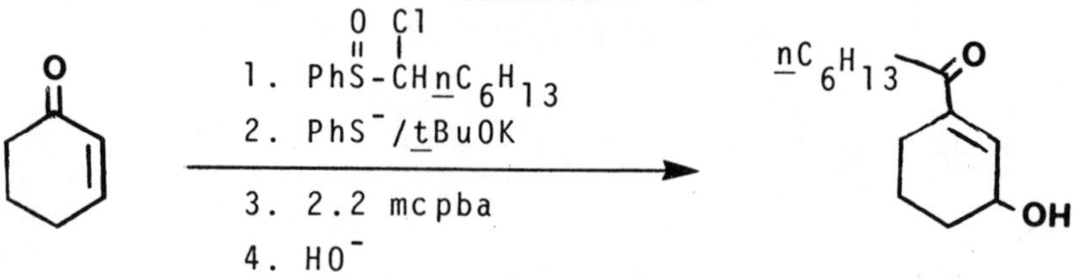

1. PhS−CH$\underline{n}$C$_6$H$_{13}$
2. PhS$^-$/$\underline{t}$BuOK
3. 2.2 mcpba
4. HO$^-$

83%

Satoh, T.; Motohashi, S.; Yamakawa, K.[*]
Tetrahedron Lett, (1986), **27**, 2889

$$CH_3\overset{O}{\overset{||}{C}}CH-CH\underline{n}C_5H_{11} \quad \xrightarrow[\text{THF/MeOH, } -90°C]{Sm/CH_2I_2} \quad CH_3\overset{O}{\overset{||}{C}}CH_2\overset{OH}{\overset{|}{C}}H\underline{n}C_5H_{11}$$

81%

Molander, G.A.[*]; Hahn, G.   **J Org Chem**, (1986), **51**, 2596

$$\xrightarrow[-78°C, \text{ 30 min}]{TiCl_4, \ CH_2Cl_2}$$

81%

Maruoka, K.; Hasegawa, M.; Yamamoto, H.[*]
               J Am Chem Soc, (1986), **108**, 3827

1. EtCHO, <u>t</u>BuMgBr
   THF/Et$_2$O, -78°C

2. Al/Hg, THF, 0°C

67%

(64% ee)

Schneider, F.[*]; Simon, R.   **Synthesis**, (1986), 582

$$PhCH-\overset{O}{\overset{\diagup\diagdown}{C}}HCH_2OH \quad \xrightarrow[\substack{\text{sealed tube, } 140°C \\ 60h}]{4\% \ Pd(PPh_3)_4, \ PhCH_3} \quad PhCH_2\overset{O}{\overset{||}{C}}CH_2OH$$

62%

Vankar, Y.D.[*]; Chaudhuri, N.C.; Singh, S.P.
               Syn Commun, (1986), **16**, 1621

SECTION 331: Alcohol, Thiol - Nitrile

Reetz, M.T.[*]; Kesseler, K.; Jung, A.
Angew Chem Int Ed Engl, (1985), **24**, 989

Ni[*] = NiI/Li/naphthalene/glyme/RT/12h

Inaba, S.; Rieke, R.D.[*]   **Tetrahedron Lett**, (1985), **26**, 155

Review: "Acyl Anion Equivalents from Cyanohydrins,
Protected Cyanohydrins and α-Dialkylaminonitriles"

Albright, J.D.[*]   **Tetrahedron**, (1983), **39**, 3207

SECTION 332: Alcohol, Thiol - Olefin

Allylic and benzylic hydroxylation (C=C-CH → C=C-C-OH, etc. is
listed in Section 41 (Alcohols and Phenols from Hydrides).

no yield
(threo:erythro = 91:9)

Mikami, K.; Kimura, Y.; Kishi, N.; Nakai, T.[*]
J Org Chem, (1983), **48**, 279

$$CH_3CH=CHCH_2SiMe_3 \xrightarrow[\substack{CH_2Cl_2 \\ -78°C, \ 1h}]{\substack{iPrCHO \\ TiCl_4}}$$

(E)

+ (97 : 3)

92%

Hayashi, T.[*]; Kabeta, K.; Hamachi, I.; Kumada, M.[*]
Tetrahedron Lett, (1983), **24**, 2865

$$\xrightarrow[\substack{CH_2CH_2OH \\ H-C-NHMe \\ CH_2NHPh}]{LiAlH_4}$$

95%

(100% ee S)

Sato, T.; Gotoh, Y.; Wakabayashi, Y.; Fujisawa, T.[*]
Tetrahedron Lett, (1983), **24**, 4123

$$\underset{HC(CH_2)_4\underline{C}nBu}{\overset{O \quad\quad\quad O}{\| \quad\quad\quad \|}} \xrightarrow[\substack{DMF, \ 25°C, \ 15 \ min}]{\substack{2 \ CH_2=CIMe \\ 4 \ CrCl_2}} \underset{\underset{OH}{|}}{\overset{\overset{CH_3}{|} \quad\quad\quad \overset{O}{\|}}{CH_2=CCH(CH_2)_4\underline{C}nBu}}$$

94%

Takai, K.[*]; Kimura, K.; Kuroda, T.; Hiyama, T.[*]; Nozaki, H.
Tetrahedron Lett, (1983), **24**, 5281

$$Me_2C=C=CHBr \xrightarrow[\substack{THF/hexane, \ reflux \\ 2h}]{PhCH_2OLi, \ \underline{t}BuOLi} \underset{\underset{OH}{|}}{PhCHCH=C=CMe_2}$$

65%

Harada, T.; Nozaki, Y.; Oku, A.[*]
Tetrahedron Lett, (1983), **24**, 5665

(threo:erythro = 100:1)

Wuts, P.G.M.*; Thompson, P.A.; Callen, G.R.
            J Org Chem, (1983), 48, 5398

R = tBu   X = Br   (ax:eq = 57:43)   89%
Hiyama, T.*; Sawahata, M.; Obayashi, M.
                  Chem Lett, (1983), 1237

R = H   Ca(Hg), THF, 0°C
X = I   4h   59%

Imamoto, T.*; Kusumoto, T.; Tawarayama, Y.; Sugiura, Y.; Mita,
T.; Hatanaka, Y.; Yokoyama, M.
                  J Org Chem, (1984), 49, 3904

R = H   Pb, Bu₄NBr, DMF, RT   86%
X = Br

Tanaka, H.; Yamashita, S.; Hamatani, T.; Ikemoto, Y.; Torii, S.*
                  Chem Lett, (1986), 1611

PhI
        1. Yb, THF, -30°C   OH
              1.5h |
        ──────────────────────→   PhCHCH=CH₂
        2. CH₂=CHCHO, -30°C-RT
        3. aq. NH₄Cl, HCl
                                       66%

Yokoo, K.; Yamanaka, Y.; Fukagawa, T.; Taniguchi, H.; Fujiwara,
Y.*
                  Chem Lett, (1983), 1301

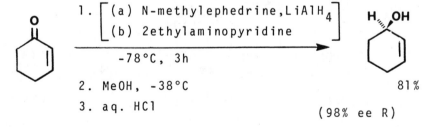

DeRenzi, A.; Panunzi, A.[*]; Saporito, A.; Vitagliano, A.
JCS Perkin II, (1983), 993

Yamamoto, Y.[*]; Yatagai, H.; Saito, Y.; Maruyama, K.
J Org Chem, (1984), **49**, 1096

Mori, I.; Takai, K.; Oshima, K.; Nozaki, H.
Tetrahedron, (1984), **40**, 4013

Kawasaki, M.; Suzuki, Y.; Terashima, S.[*]
Chem Lett, (1984), 239

$$HOCH_2C{\equiv}CCH_2OH \xrightarrow[\text{Et}_2\text{O/THF, RT - reflux}]{4 \quad \text{(p-tolyl)}-MgBr}$$

60%

Ishino, Y.[*]; Wakamoto, K.; Hirashima, T.
Chem Lett, (1984), 765

$$\xrightarrow[\substack{\text{aq. dioxane} \\ \text{cat. AcOH, 24h}}]{\text{Sn, Al, RT}}$$

61%

Nokami, J.[*]; Wakabayashi, S.; Okawara, R.
Chem Lett, (1984), 869

$$RCHO \xrightarrow{R'X} \underset{\underset{RCHR'}{|}}{\overset{OH}{}}$$

R = Ph, X = Br $\xrightarrow[\substack{\text{AcOH, Pt electrodes} \\ \text{55°C, cyclohexane}}]{\text{Sn, MeOH, e}^-}$ 91%

R' = $CH_2CH{=}CH_2$

Uneyama, K.; Matsuda, H.; Torii, S.[*]
Tetrahedron Lett, (1984), **25**, 6017

R = iPr   X = B(   )₂   1. -78°C   86%

R' = $CH_2CH{=}CH_2$ 2. 25°C   (90% ee S)

3. NaOH, H_2O_2

Jadhav, P.K.; Bhat, K.S.; Perumal, P.T.; Brown, H.C.[*]
J Org Chem, (1986), **51**, 432

R = $\underline{n}C_6H_{13}$

R = ($\underline{E}$-$CH_2CH{=}CHPh$) $\xrightarrow[\text{THF, 0°C, 2.5h}]{\text{Pd(PPh}_3)_4, \text{ SmI}_2}$ 63%

X = OAc

Tabuchi, T.; Inanaga, J.[*]; Yamaguchi, M.
Tetrahedron Lett, (1986), **27**, 1195

R = CH_2-Tol
R' = $CH_2CH=CH_2$
X = $AliBu_2$

$-18°C$

$Sn(OTf)_2$, CH_2Cl_2 (87% ee S)

Mukaiyama, T.; Minowa, N.; Oriyama, T.; Narasaka, K.
Chem Lett, (1986), 97

R = Ph
R' = $CH_2CH=CH_2$
X = $OP(O)[OPh]_2$

SnF_2-Et_2AlCl 95%

Matsubara, S.; Wakamatsu, K.; Morizawa, Y.; Tsuboniwa, N.;
Oshima, K.*; Nozaki, H.
Bull Chem Soc Jpn, (1985), **58**, 1196

R = Ph
R' = $CH_2CH=CH_2$
X = Br

1. Bi, DMF , R'X
─────────────────────   94%
2. RCHO

Wada, M.; Akiba, K.*   **Tetrahedron Lett**, (1985), **26**, 4211

R = Ph
R' = $CH_2CH=CH_2$
X = Br

$BiCl_3$, Zn
─────────────   99%
THF, RT

Wada, M.; Ohki, H.; Akiba, K.*
Tetrahedron Lett, (1986), **27**, 4771

R = $PhCH_2CH_2$
R' = $CH_2CH=CH_2$
X = $GePh_3$

$BF_3·OEt_2$, CH_2Cl_2
─────────────────────   55%
$-78°C$ - RT

Sano, H.; Miyazaki, Y.; Okawara, M.; Ueno, Y.
Synthesis, (1986), 776

PhCHO

$(\underline{E})$-PhCH=CHCH$_2$Cl
───────────────────────
$SnCl_2$-Al, aq. THF

OH

Ph

Ph

87%

($\underline{threo}$:$\underline{erythro}$ = 98:2)

Uneyama, K.; Nanbu, H.; Torii, S.*
Tetrahedron Lett, (1986), **27**, 2395

PhCHO

1. CH$_2$=CHCH$\begin{smallmatrix}TMS\end{smallmatrix}$ (pinanediol boronate)

$\xrightarrow{}$

Et$_2$O, Py

2. 120°C

PhCHCH$_2$CH=CHTMS

91%

($\underline{Z}$:$\underline{E}$ = 78:22)

Tsai, D.J.S.; Matteson, D.S.   **Organometallics**, (1983), **2**, 236

2 Ph $\overset{HO}{\diagup}$

1. 2 $\underline{n}$BuLi
2. TiCl$_4$
3. PhCH$_2$CH$_2$CHO
4. Me$_3$SiCH$_2$CH=CH$_2$
5. TMSI

$\xrightarrow{}$

PhCH$_2$CH$_2$$\overset{OH}{\underset{|}{C}}HCH_2$CH=CH$_2$

75%

Imwinkelried, R.; Seebach, D.*   (80% ee R)

Angew Chem Int Ed Engl, (1985), **24**, 765

$\overset{OH}{\underset{|}{PhCHCH_3}}$

1. TMEDA, $\underline{n}$BuLi, Et$_2$O
   0°C - reflux
2. Cu$_2$(CN)$_2$, CH$_2$=CHCH$_2$Cl
   Et$_2$O
3. NH$_4$Cl/NH$_4$OH/HOH

$\xrightarrow{}$

87%

Taber, D.F.*; Dunn, B.S.; Mack, J.F.; Saleh, S.A.
J Org Chem, (1985), **50**, 1987

$\overset{OH}{\underset{|}{CH_3CHCH_2C\equiv CH}}$

$\xrightarrow[\text{2h}]{\text{TiCl}_4, \text{Me}_3\text{Al} \atop \text{CH}_2\text{Cl}_2, -78°C}$

$\overset{OH}{\underset{|}{CH_3CHCH_2CH=CHCH_3}}$

80%

Ewing, J.C.; Ferguson, G.S.; Moore, D.W.; Schultz, F.W.;
Thompson, D.W.*

J Org Chem, (1985), **50**, 2124

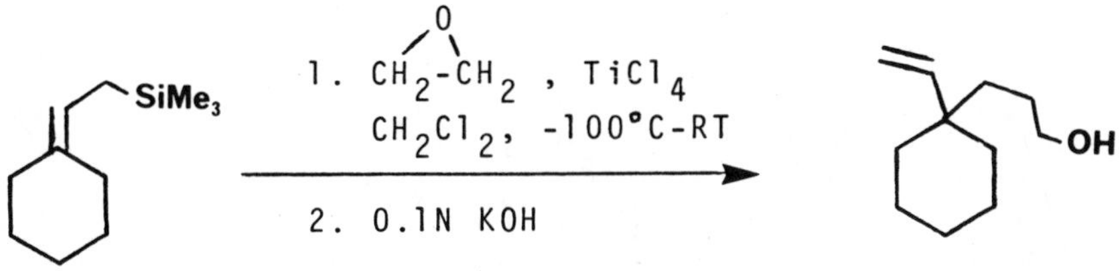
SiMe3
1. CH2-CH2 (O), TiCl4
CH2Cl2, -100°C-RT
2. 0.1N KOH
OH

86%

Carr, S.A.; Weber, W.P.* **J Org Chem**, (1985), **50**, 2782

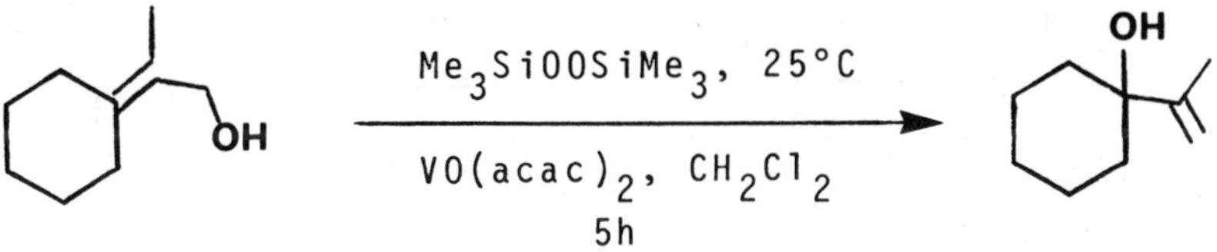
OH
Me3SiOOSiMe3, 25°C
VO(acac)2, CH2Cl2
5h
OH

85%

Matsubara, S.; Okazoe, T.; Oshima, K.; Takai, K.*; Nozaki, H. **Bull Chem Soc Jpn**, (1985), **58**, 844

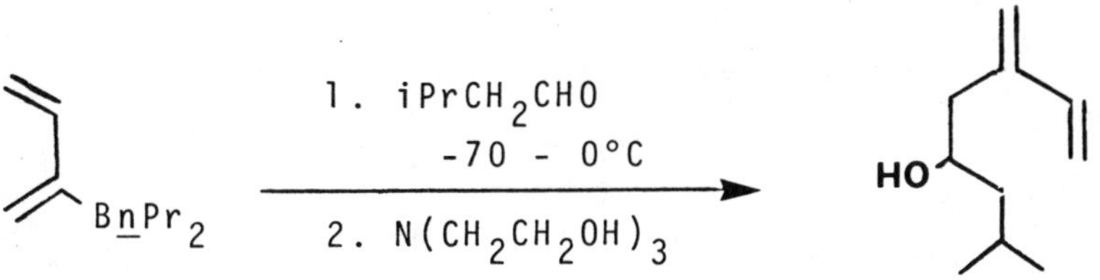
BnPr2
1. iPrCH2CHO
-70 - 0°C
2. N(CH2CH2OH)3
HO

94%

Bubnov, Y.N.*; Etinger, M.Yu. **Tetrahedron Lett**, (1985), **26**, 2797

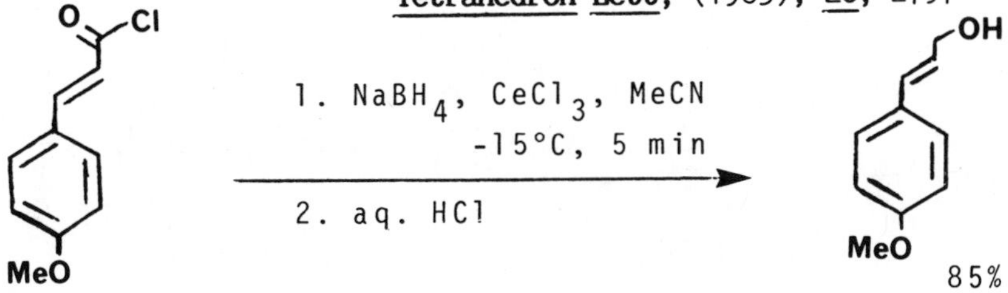
O
Cl
MeO
1. NaBH4, CeCl3, MeCN
-15°C, 5 min
2. aq. HCl
OH
MeO

85%

Lakshmy, K.V.; Mehta, P.G.; Sheth, J.P.; Trivedi, G.K.* **Org Prep Proc Int**, (1985), **17**, 251

$EtCH=CHCH_2CH_2OH$
($\underline{E}$)

1. $AlMe_3$, $TiCl_4$
   Me_2O, pentane
   20°C, 3h
2. HOH

$\begin{array}{c} Et \\ Me \end{array} C=C \begin{array}{c} CH_2CH_2OH \\ H \end{array}$

55%

Moret, E.; Schlosser, M.[*]   **Tetrahedron Lett**, (1985), **26**, 4423

1. PhI, 2% $Pd(PPh_3)_2(OAc)_2$
   Pip, 10h, 60°C, 4% CuI

2. HCOOH

70%

Arcadi, A.; Cacchi, S.[*]; Marinelli, F.
                    Tetrahedron, (1985), **41**, 5121

$MeCeCl_2$

THF/Et_2O, -78°C
30 min

98%

Imamoto, T.[*]; Sugiura, Y.   **J Organomet Chem**, (1985), **285**, c021

EtCHO

1. $\underline{E}$ MeCH=CHCH
   Cl
   Pet ether
2. $N(CH_2CH_2OH)_3$

5%   65%
      (96% ee)

Hoffmann, R.W.[*]; Dresely, S.
                 Angew Chem Int Ed Engl, (1986), **25**, 189

$$\text{4 [CyN}\underline{\text{secBu}}\text{/MeMgBr]}$$

with Li on N

0°C, 20h

85%

Masset, P.; Manna, S.; Viala, J.; Falck, J.R.[*]
Tetrahedron Lett, (1986), **27**, 299

5 PhCHO

1. 4 $CH_2=CHCH_2Br$, -5°C
   (+)diethyl tartrate

   THF, NaH,   5h

2. NaH

3. $SnCl_2$

$$PhCHCH_2CH=CH_2$$
with OH

82%

(62% ee S)

Boldrini, G.P.; Tagliavini, E.; Trombini, C.; Umani-Ronchi, A.
JCS Chem Comm, (1986), 685

$\underline{n}C_{10}H_{21}C=CH_2$ with OTf

PhCHO, $CrCl_2$, DMF

cat. $NiCl_2$, 25°C

$\underline{n}C_{10}H_{21}CCHPh$ with CH_2 and OH

87%

Takai, K.[*]; Tagashira, M.; Kuroda, T.; Oshima, K.; Utimoto, K.;
Nozaki, H.
J Am Chem Soc, (1986), **108**, 6048

$\underline{n}PrCH=CHCCH_2OH$ with CH_3 and NO_2

PPh_3, THF, iPrOH

5% $Pd(PPh_3)_4$, 0°C

$NaBH_4$, 5h

(97) $\underline{n}PrCH=CHCHCH_2OH$ with Me
       (E)

+

86%

$\underline{n}PrCH_2CH=C$ with CH_2OH and CH_3

(3)

Ono, N.[*]; Hamamoto, I.; Kamimura, A.; Kaji, A.
J Org Chem, (1986), **51**, 3734

PhCHO
$$\xrightarrow[\text{K}_2\text{CO}_3, \text{ iPrOH, 12h}]{\overset{+}{\text{Ph}_3}\text{PCH}_2\text{CH}_2\text{CH}_2\text{OH Cl}^-}$$
PhCH=CHCH$_2$CH$_2$OH

84%

Cheik-Rouhou, F.; Le Bigot, Y.; El Gharbi, R.; Delmas, M.[*];
Gaset, A.
Syn Commun, (1986), **16**, 1617

Review: "Double Asymmetric Synthesis and a New Strategy for
Stereochemical Control In Organic Synthesis"

Masamune, S.[*]; Choy, W.; Petersen, J.S.; Suta, L.R.
Angew Chem Int Ed Engl, (1985), **24**, 1

Also via: Acetylenes-Alcohols (Section 302)

SECTION 333: Aldehyde - Aldehyde

(structure)
1. PhPh, ^{1}O$_2$, MeCN
   DCA, 10°C, 30 min
2. Me$_2$S
(structure)

88%

DCA = 9,10-dicyanoanthracene

Schaap, A.P.[*]; Siddiqui, S.; Prasad, G.; Palomino, E.; Sandison,
M.
Tetrahedron, (1985), **41**, 2229

SECTION 334: Aldehyde - Amide

No Additional Examples

SECTION 335: Aldehyde - Amine

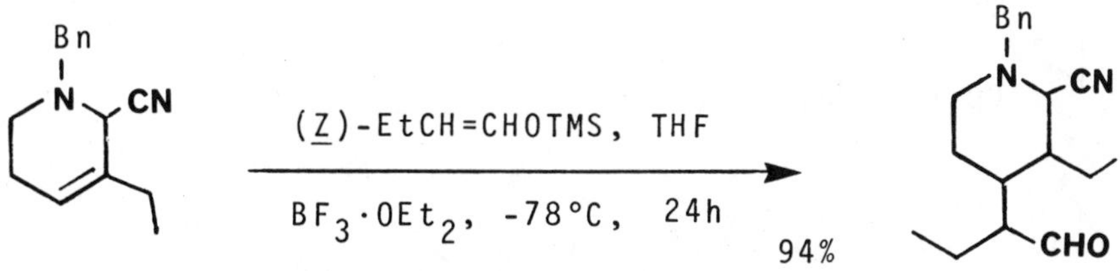

$$(\underline{Z})\text{-EtCH=CHOTMS, THF}$$
$$BF_3 \cdot OEt_2, \ -78°C, \ 24h$$

94%

Koskinen, A.; Lounasmaa, M.[*]   **JCS Chem Comm**, (1983), 821

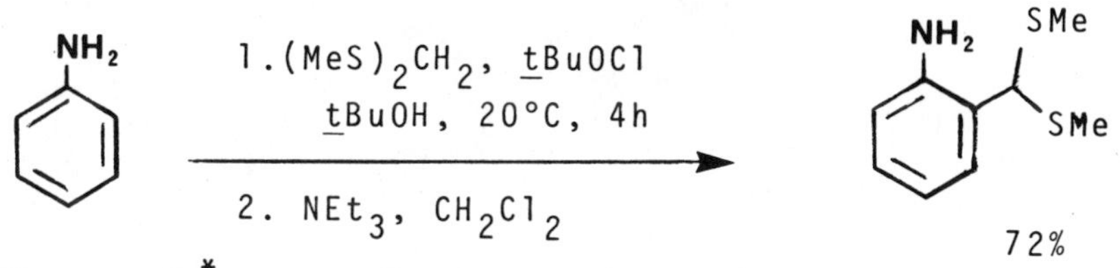

1. $(MeS)_2CH_2$, $\underline{t}BuOCl$
   $\underline{t}BuOH$, 20°C, 4h

2. NEt_3, CH_2Cl_2

72%

Claus, P.K.[*]; Jäger, E.; Setzer, A.
                    Monatsh Chem, (1985), **116**, 1017

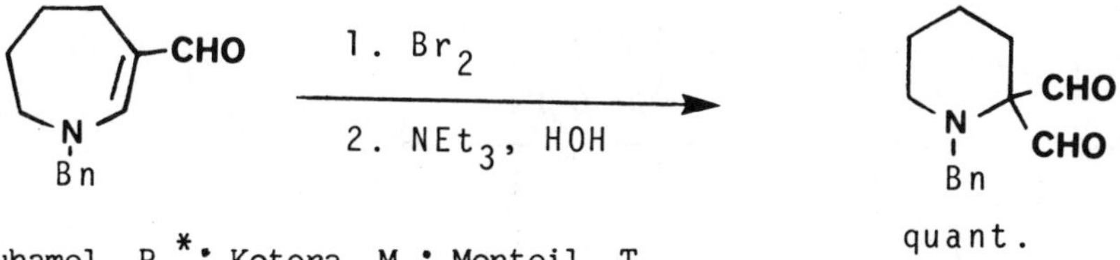

1. Br_2

2. NEt_3, HOH

quant.

Duhamel, P.[*]; Kotera, M.; Monteil, T.
                    Bull Chem Soc Jpn, (1986), **59**, 2353

SECTION 336: Aldehyde - Ester

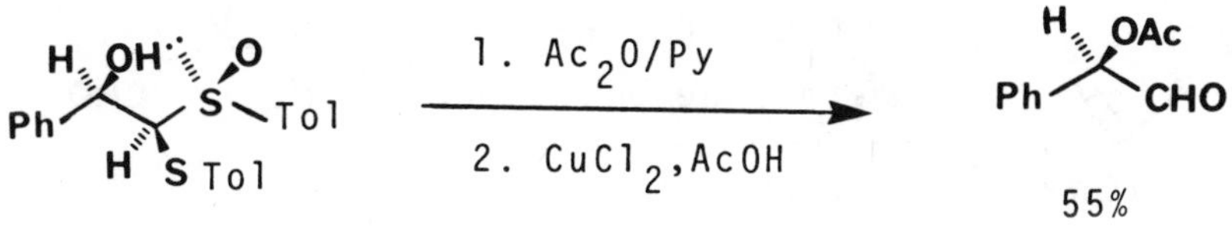

1. Ac_2O/Py

2. $CuCl_2$, AcOH

55%

Ogura, K.[*]; Fujita, M.; Inaba, T.; Takahashi, K.; Iida, H.
                    Tetrahedron Lett, (1983), **24**, 503

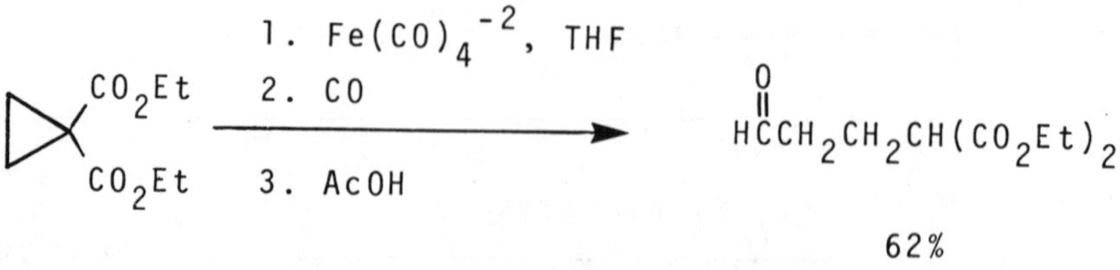

1. $Fe(CO)_4^{-2}$, THF
2. CO
3. AcOH

CO_2Et
CO_2Et

$HCCH_2CH_2CH(CO_2Et)_2$

62%

Tamblyn, W.H.[*]; Waltermire, R.E.
Tetrahedron Lett, (1983), **24**, 2803

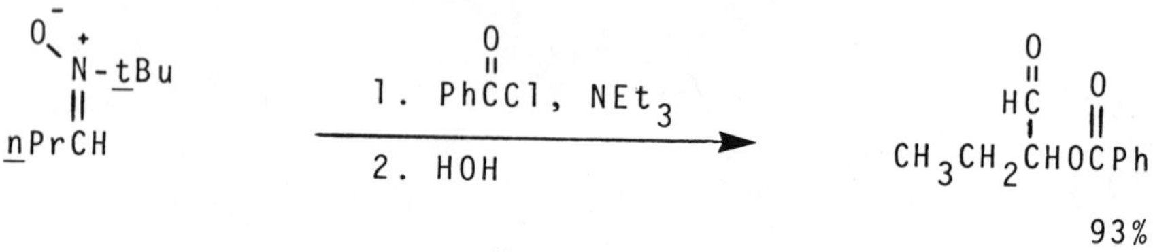

$O^- + N-tBu$
$nPrCH$

1. $PhCCl$, NEt_3
2. HOH

$CH_3CH_2CHOCPh$

93%

Cummins, C.H.; Coates, R.M.[*]   **J Org Chem**, (1983), **48**, 2070

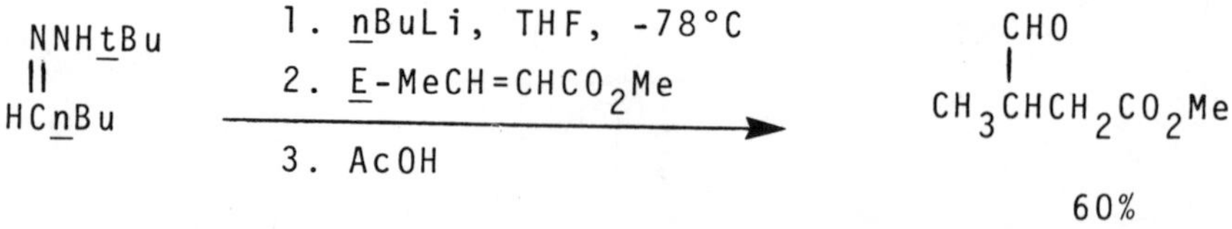

NNHtBu
HCnBu

1. nBuLi, THF, -78°C
2. E-MeCH=CHCO_2Me
3. AcOH

CHO
$CH_3CHCH_2CO_2Me$

60%

Baldwin, J.E.[*]; Adlington, R.M.; Bottaro, J.C.; Jain, A.U.;
Kolhe, J.N.; Perry, M.W.D.; Newington, I.M.
JCS Chem Comm, (1984), 1095

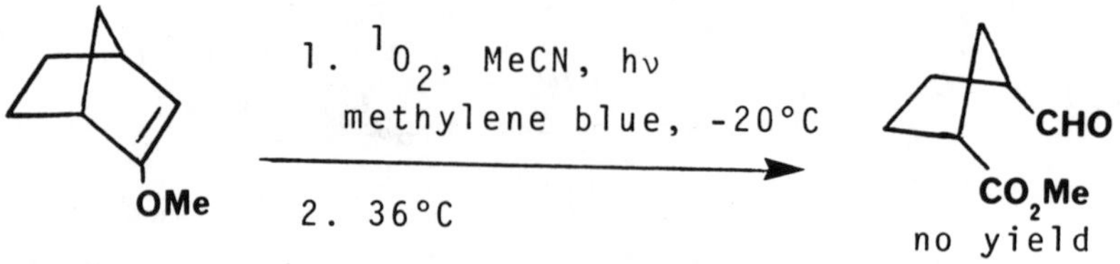

1. $^1O_2$, MeCN, hν
   methylene blue, -20°C
2. 36°C

CHO
CO_2Me
no yield

Jefford, C.W.[*]; Boukouvalas, J.; Kohmoto, S.
Tetrahedron, (1985), **41**, 2081

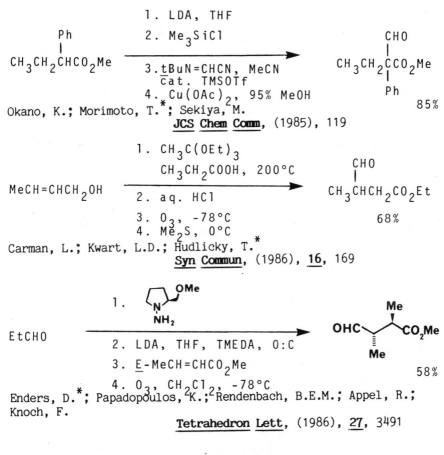

1. LDA, THF

2. Me₃SiCl

Ph
|
CH₃CH₂CHCO₂Me

3. tBuN=CHCN, MeCN
   cat. TMSOTf
4. Cu(OAc)₂, 95% MeOH

CHO
|
CH₃CH₂CCO₂Me
|
Ph
                                                    85%

Okano, K.; Morimoto, T.*; Sekiya, M.
          JCS Chem Comm, (1985), 119

1. CH₃C(OEt)₃
   CH₃CH₂COOH, 200°C

MeCH=CHCH₂OH

2. aq. HCl

3. O₃, -78°C
4. Me₂S, 0°C

CHO
|
CH₃CHCH₂CO₂Et
                68%

Carman, L.; Kwart, L.D.; Hudlicky, T.*
          Syn Commun, (1986), **16**, 169

1.

EtCHO

2. LDA, THF, TMEDA, 0:C

3. E-MeCH=CHCO₂Me

4. O₃, CH₂Cl₂, -78°C

OHC   Me
   \ |
    \ CO₂Me
     |
    Me
              58%

Enders, D.*; Papadopoulos, K.; Rendenbach, B.E.M.; Appel, R.;
Knoch, F.
          Tetrahedron Lett, (1986), **27**, 3491

SECTION 337: Aldehyde - Ether, Epoxide, Thioether

1. E-nBuOCH=CHCO₂Me
   iBuOH

HC≡CO₂Me

2. Dibal-H
3. MnO₂

nBuOCH=CHCHO
      (E)
              61%

Vo Quang, Y.; Marais, D.; Vo Quang, L.; Le Goffic, F.
          Tetrahedron Lett, (1983), **24**, 5209

$$PhCH_2CH_2CHO \xrightarrow[\substack{(CO)_5MnH, \ CD_3CN, \ 50°C \\ 7d}]{(CO)_5MnSiMe_3, \ \underline{t}BuOH}} PhCH_2\overset{\overset{\displaystyle OTMS}{|}}{C}HCHO$$

78%

Brinkman, K.C.; Gladysz, J.A.[*]   **Organometallics**, (1984), **3**, 147

SECTION 338: Aldehyde - Halide, Sulfonate

$$MeCH=CHCHO \atop (\underline{E}) \xrightarrow[\substack{HOCH_2CH_2OH \\ -15°C \ - \ reflux}]{TMS-Cl, \ CH_2Cl_2}}$$

88%

Gil, G.[*]   **Tetrahedron Lett**, (1984), **25**, 3805

$$CH_3(CH_2)_4C{\equiv}CH \xrightarrow[\substack{3. \ 30\% \ H_2O_2}]{\substack{1. \ BBr_3, \ pH \ 5 \\ 2. \ KOAc}} CH_3(CH_2)_4\overset{\overset{\displaystyle Br}{|}}{C}HCHO$$

65%

Satoh, Y.; Tayano, T.; Koshino, H.; Hara, S.; Suzuki, A.
              Synthesis, (1985), 406

$$\underline{n}BuCHO \xrightarrow[\substack{Cl_2}]{\substack{10\% \ [pyridine]N{=}CHOH^+ \ Cl^- \\ CCl_4, \ 70°C}} \underline{n}Pr\overset{\overset{\displaystyle Cl}{|}}{C}HCHO$$

99%

Pews, R.G.[*]; Lysenko, Z.[*]   **Syn Commun**, (1985), **15**, 977

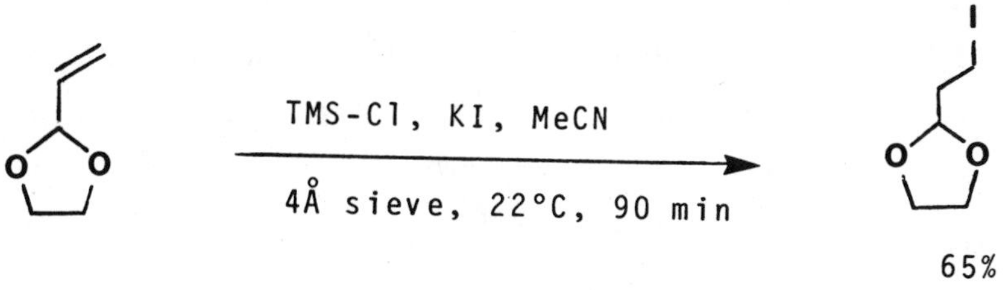

TMS-Cl, KI, MeCN

4Å sieve, 22°C, 90 min

65%

Feringa, B.L.[*] **Syn Commun**, (1985), **15**, 87

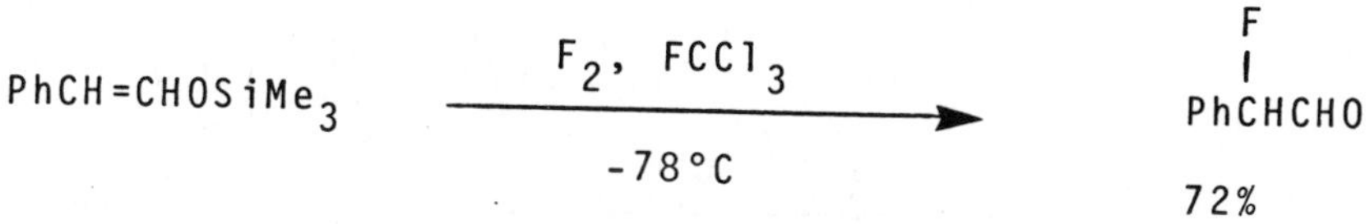

PhCH=CHOSiMe$_3$

F$_2$, FCCl$_3$

-78°C

$$\underset{\text{F}}{\underset{|}{\text{PhCHCHO}}}$$

72%

Purrington, S.T.[*]; Lazaridis, N.V.; Bumgardner, C.L.[*]
                    Tetrahedron Lett, (1986), **27**, 2715

SECTION 339: Aldehyde - Ketone

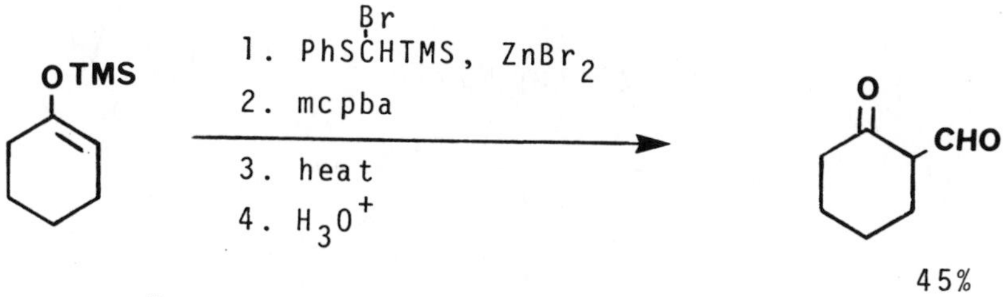

Br
1. PhSCHTMS, ZnBr$_2$
2. mcpba
3. heat
4. H$_3$O$^+$

45%

Ager, D.J.[*]   **Tetrahedron Lett**, (1983), **24**, 419

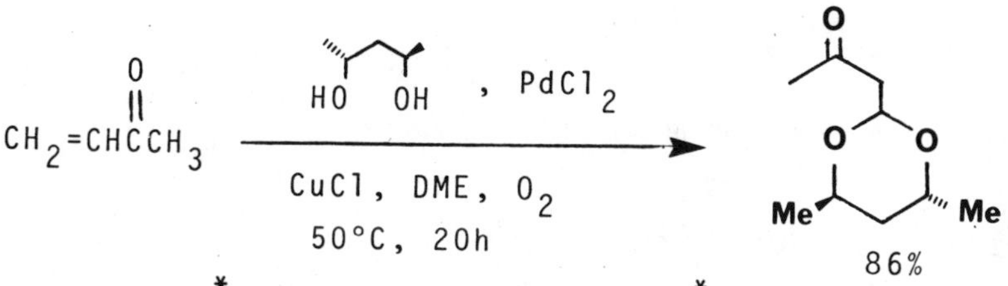

CH$_2$=CHCCH$_3$

HO OH , PdCl$_2$

CuCl, DME, O$_2$
50°C, 20h

86%

Hosokawa, T.[*]; Ohta, T.; Murahashi, S.I.[*]
                    JCS Chem Comm, (1983), 848

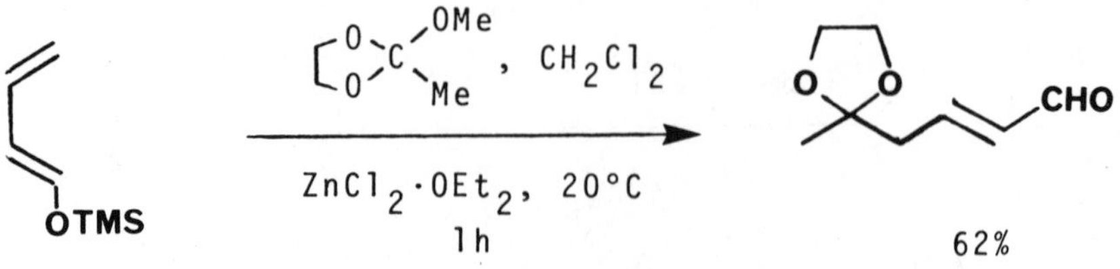
OMe
Me
, CH_2Cl_2
OTMS
$ZnCl_2 \cdot OEt_2$, 20°C
1h
CHO
62%

Akgün, E.; Pindur, U.* **Synthesis**, (1984), 227

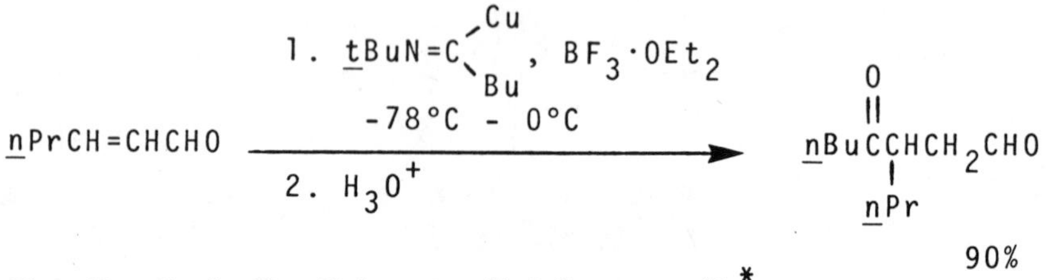
1. tBuN=C(Cu)(Bu), $BF_3 \cdot OEt_2$
-78°C - 0°C
nPrCH=CHCHO
2. H_3O^+
nBuC(=O)CH(nPr)CH$_2$CHO
90%

Ito, Y.; Imai, H.; Matsuura, T.; Saegusa, T.* **Tetrahedron Lett**, (1984), 25, 3091

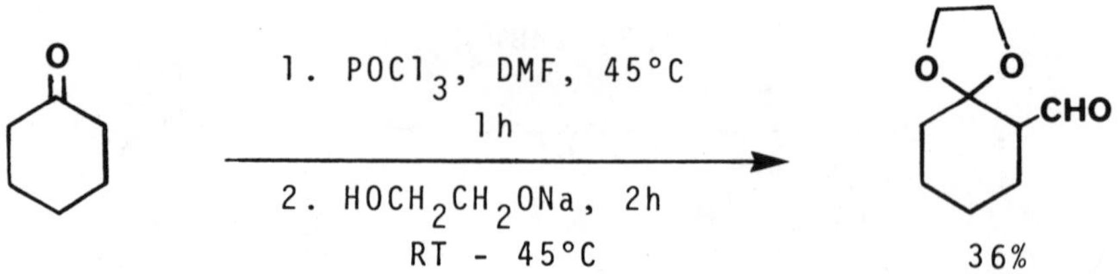
1. $POCl_3$, DMF, 45°C
1h
2. $HOCH_2CH_2ONa$, 2h
RT - 45°C
CHO
36%

Huet, F.* **Synthesis**, (1985), 496

SECTION 340: Aldehyde - Nitrile

No Additional Examples

SECTION 341: Aldehyde - Olefin

For the oxidation of allylic alcohols to olefinic aldehydes see also Section 48 (Aldehydes from Alcohols).

$$
\underset{\substack{| \\ AcOCH_2CH_2C=CH_2}}{CH_2TMS}
\xrightarrow[\substack{BF_3 \cdot OEt_2 \\ dioxane}]{PhIO}
\underset{\substack{| \\ AcOCH_2CH_2C=CH_2}}{CHO}
$$

63%

Ochiai, M.; Fujita, E.[*]   **Tetrahedron Lett**, (1983), **24**, 777

$$
\underset{\substack{| \\ CH_3CH(CH_2)_8CH=CHCH_2OH}}{OH}
\xrightarrow[\substack{RuCl_2(PPh_3)_3}]{TMSOOTMS}
\underset{\substack{| \\ CH_3CH(CH_2)_8CH=CHCHO}}{OH}
$$

70%

Kanemoto, S.; Oshima, K.[*]; Matsubara, S.; Takai, K.
             Tetrahedron Lett, (1983), **24**, 2185

$$
\xrightarrow[\substack{K-10 \ clay/Fe^{+3} \\ CH_2Cl_2}]{CH_2=CHCHO, \ -24°C}
$$

60%

Laszlo, P.[*]; Lucchetti, J.   **Tetrahedron Lett**, (1984), **25**, 2147

CH₂=CHCH₂NHMe

1. tBuOK, THF, -78°C
2. DMAP, THF, -78°C
3. cyclopentanone, THF
   ─────────────────
   -78°C - RT
4. Me₂SO₄, PhH, reflux
5. aq. HCl, RT

49%

Gilbert, J.C.[*]; Senaratne, K.P.A.
             Tetrahedron Lett, (1984), **25**, 2303

1. $\underline{n}C_8H_{17})_3NMeCl$
   KOH, DMF, -15°C

$\underline{n}C_{12}H_{25}Br$

MeSCH=CHCH$_2$SO$_2$Tol $\xrightarrow{}$ $\underline{n}C_{12}H_{25}CH=CHCHO$

2. TiCl$_4$, CuCl$_2$
   RT - 50°C   59%

3. K$_2$CO$_3$, DME, RT

Ogura, K.[*]; Iihama, T.; Takahashi, K.; Iida, H.
Tetrahedron Lett, (1984), **25**, 2671

1. Me$_2$NCH$_2$CN, DMF
   K$_2$CO$_3$, 48h

2. AgNO$_3$, THF
   Et$_2$O, HOH

92%

Stela, L.[*]   **Tetrahedron Lett**, (1984), **25**, 3457

PhCH=CHCH$_2$OH $\xrightarrow[\text{BnNEt}_3\text{Cl, PhH, 25°C}]{\text{K}_2\text{FeO}_4\text{, aq. NaOH}}$ PhCH=CHCHO

1.5h   95%

Kim, K.S.[*]; Chang, Y.K.; Bae, S.K.; Hahn, C.S.
Synthesis, (1984), 866

Et$_2$AlSPh, DCE

25°C, 15 min

84%

(E:Z = 39:61)

Takai, K.; Mori, I.; Oshima, K.[*]; Nozaki, H.
Bull Chem Soc Jpn, (1984), **57**, 446

$$Ph_3\overset{+}{As}-CH_2CHO \ \overset{-}{Br}$$
$$K_2CO_3(trace\ HOH)$$
$$Et_2O/THF, \ RT, \ 3h$$

91%

Huang, Y.[*]; Shi, L.; Yang, J.
Tetrahedron Lett, (1985), **26**, 6447

$$(PhIO)_n, \ DCC$$
$$BF_3 \cdot OEt_2, \ RT$$
$$12h$$

$$\underline{n}C_{10}H_{21}CH=CH(CH_2)_3CHO$$

77%

Ochiai, M.; Ukita, T.; Nagao, Y.; Fujita, E.
JCS Chem Comm, (1985), 637

$$EtOCH=C\underset{Me}{\overset{CHO}{<}}$$

1. EtMgBr, Et$_2$O
   0°C - RT

2. aq. NH$_4$Cl, 0°C

$$EtCH=C\underset{Me}{\overset{CHO}{<}}$$

88%

Spangler, C.W.[*]; Tan, R.P.K.; Gibson, R.S.; McCoy, R.K.
Syn Commun, (1985), **15**, 371

1. CH(OEt)$_3$, CH$_2$Cl$_2$
   BF$_3\cdot$OEt$_2$, iPr$_2$NEt

2. NaBH$_4$, MeOH

3. 6N HCl

76%

DasGupta, R.; Ghatak, U.R.[*]   **Tetrahedron Lett**, (1985), **26**, 1581

Me
|
TESCH=NtBu

OTBS

OHC.
Me

secBuLi

$\longrightarrow$

O-TBS

OHC
Me

77%

Schlessinger, R.H.[*]; Poss, M.A.; Richardson, S.; Lin, P.
Tetrahedron Lett, (1985), **26**, 2391

OMe

CH_2=CHCHO, RT

cat. Yb(fod)$_3$

48h

MeO   CHO

60%

Danishefsky, S.[*]; Bednarski, M.
Tetrahedron Lett, (1985), **26**, 2507

Ph N

1. 5% Pd(PPh$_3$)$_3$, PhH
   2.5% CF$_3$COOH, 50°C
2. 0.5N HCl, RT

$\longrightarrow$

Ph
|
CH_2=CHCH$_2$CCHO
|
Me

78%

Murahashi, S.[*]; Makabe, Y.   **Tetrahedron Lett**, (1985), **26**, 5563

Cl

3 Et-N   O, DMF
    -O

50°C, 4h

$\longrightarrow$

CHO

88%

(E:Z = 7:3)

Suzuki, S.; Onishi, T.; Fujita, Y.; Misawa, H.; Otera, J.[*]
Bull Chem Soc Jpn, (1986), **59**, 3287

Also via: β-hydroxyaldehydes (Section 324)

SECTION 342: Amide - Amide

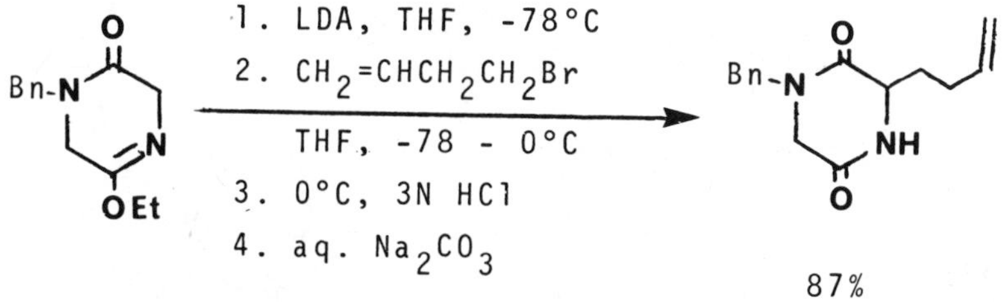

1. LDA, THF, -78°C
2. CH_2=$CHCH_2CH_2Br$

   THF, -78 - 0°C
3. 0°C, 3N HCl
4. aq. Na_2CO_3

87%

Fukuyama, T.[*]; Frank, R.K.; Laird, A.A.
                    Tetrahedron Lett, (1985), **26**, 2955

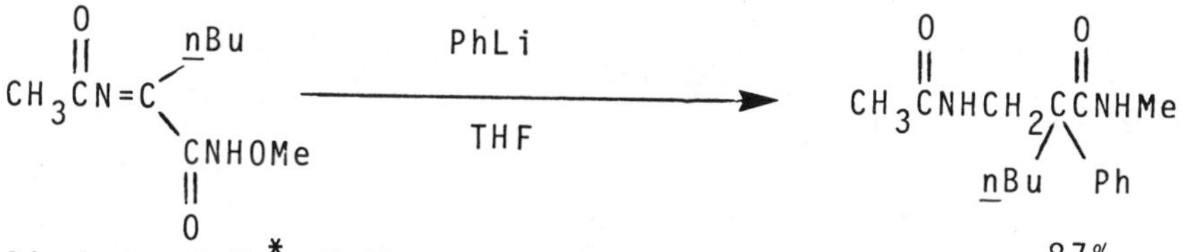

PhLi

THF

87%

Lipshutz, B.H.[*]; Huff, B.; Vaccaro, W.
                    Tetrahedron Lett, (1986), **27**, 4241

Also via: Dicarboxylic acids (Section 312); Diamines (Section 350)

SECTION 343: Amide - Amine

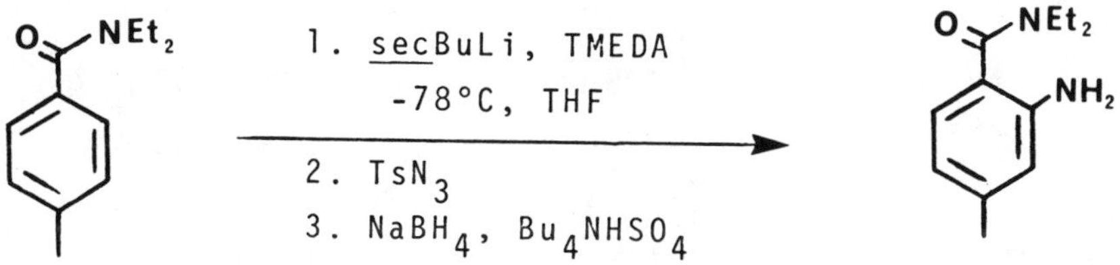

1. secBuLi, TMEDA
   -78°C, THF

2. TsN_3
3. $NaBH_4$, Bu_4NHSO_4

82%

Reed, J.N.; Snieckus, V.[*]   **Tetrahedron Lett**, (1983), **24**, 3795

Kraus, G.A.[*]; Yue, S.   **J Org Chem**, (1983), **48**, 2936

Grigg, R.[*]; Aly, M.F.; Sridharan, V.; Thianpatanagul, S.
JCS Chem Comm, (1984), 182

$$EtCNHCH_2CH_2NHiPr$$

Fazio, M.J.[*]   **J Org Chem**, (1984), **49**, 4889

SECTION 344: Amide - Ester

$$BnNHCH_2CCO_2Me$$

Ikeda, K.; Achiwa, K.; Sekiya, M.
Tetrahedron Lett, (1983), **24**, 913

$$\text{(CH}_2)_3\text{CH(CO}_2\text{Me)}_2 \xrightarrow[\substack{\text{2. TiCl}_4, \text{ CH}_2\text{Cl}_2 \\ \text{RT, 3d}}]{\substack{\text{1. e}^-, \text{ Et}_4\text{NClO}_4 \\ \text{Pt electrodes} \\ \text{MeOH}}}$$

55%

Okita, M.; Wakamatsu, T.; Ban, Y.[*]

Heterocycles, (1983), **20**, 401

$$\underset{\substack{| \\ \text{CO}_2\text{Me}}}{\overset{\substack{\text{NHBOC} \\ |}}{\text{BrCH}_2\text{CH}_2\text{CH}}} \xrightarrow[\text{THF, } -10°\text{C}]{4 \text{ } \underline{n}\text{Bu}_2\text{CuLi}} \underset{\substack{| \\ \text{CO}_2\text{Me}}}{\overset{\substack{\text{NHBOC} \\ |}}{\underline{n}\text{BuCH}_2\text{CH}_2\text{CH}}} \quad 74\%$$

(95% ee)

Bajgrowicz, J.A[*]; El Hallaouri, A.; Jacquier, R.[*]; Pigiere, Ch.;
Viallefont, Ph.[*]

Tetrahedron Lett, (1984), **25**, 2231

$$\underset{\substack{| \\ \underset{\substack{|| \quad || \\ \text{O} \quad \text{O}}}{\text{CCH}_2\text{CCO}_2\text{Me}}}}{\text{N}} \xrightarrow[\text{48h}]{\underline{t}\text{BuOH, h}\nu}$$

70%

Gramain, J.-C.[*]; Remuson, R.; Vallée, D.

J Org Chem, (1985), **50**, 710

$$\text{PhSeCH}_2\text{CH=CHCO}_2\text{Me} \xrightarrow[\substack{3 \text{ NCS, 0}°\text{C}}]{\substack{3 \text{ BnOCNH}_2, \text{ MeOH} \\ 6 \text{ iPr}_2\text{NEt}}} \underset{\substack{| \\ \text{CO}_2\text{Me}}}{\text{CH}_2\text{=CHCHNHCOBn}}$$

63%

Fitzner, J.N.; Pratt, D.V.; Hopkins, P.B.[*]

Tetrahedron Lett, (1985), **26**, 1959

$$CH_2=CHCHCH_2CH \overset{OH \quad NHSO_2Tol}{\underset{Ph}{|}} \xrightarrow[\text{CO, RT, 2d}]{\substack{PdCl_2, CuCl_2 \\ AcONa, AcOH}}$$

80%

Tamaru, Y.; Kobayashi, T.; Kawamura, S.; Ochiai, H.; Yoshida, Z.
Tetrahedron Lett, (1985), **26**, 4479

$$Cl-\overset{CO_2Me}{\underset{H}{\overset{|}{C}}}-NHCBZ \xrightarrow[\text{THF, } -78°C]{CH_2=CHMgBr} CH_2=CH-\overset{CO_2Me}{\underset{H}{\overset{|}{C}}}-NHCBZ$$

65%

Castelhano, A.L.[*]; Horne, S.; Billedeau, R.; Krantz, A.[*]
Tetrahedron Lett, (1986), **27**, 2435

1. LiN(TMS)$_2$, THF
   -78°C
2. (BnOCO)$_2$, -78°C
3. AcOH, 0°C

87%

($\underline{R}:\underline{S}$ = 2:98)

Gore, M.D.; Vederas, J.C.[*] **J Org Chem**, (1986), **51**, 3700

Related Methods: Acid-Amide (Section 315); Acid-Amine (Section 316); Amine-Ester (Section 351)

SECTION 345: Amide - Ether, Epoxide, Thioether

$$(CO)_5Cr=C \overset{OMe}{\underset{Me}{\big\langle}} \quad \xrightarrow[\text{h}\nu,\ Et_2O]{\text{PhCH=NMe}} \quad$$

75%

Hegedus, L.S.[*]; McGuire, M.A.; Schultze, L.M.; Yijun, C.;
Anderson, O.P.
J Am Chem Soc, (1984), **106**, 2680

$$CH_2=CHCH_2\overset{Ph}{\underset{\underset{N\underline{n}Bu}{\|}}{CH-C-OMe}} \quad \xrightarrow[\text{MeCN, 12h}]{\text{PhSeBr, 0°C}} \quad$$

80%

Toshimitsu, A.[*]; Terao, K.; Uemura, S.
JCS Chem Comm, (1986), 530

$$AcNH(CH_2)_3CH=CH_2 \quad \xrightarrow[\text{MeCN, 12h}]{\text{PhSeBr, 20°C}} \quad$$

66%

Toshimitsu, A.[*]; Terao, K.; Uemura, S.
J Org Chem, (1986), **51**, 1724

1. PhOCH$_2$CCl
   NEt$_3$

2. CAN

70%

Antonini, I.; Cardellini, M.; Claudi, F.[*]; Moracci, F.M.
Synthesis, (1986), 379

PhCHO
$$\xrightarrow[\substack{5\% \ CF_3SO_3TMS, \ RT \\ 1h}]{(TMS)_2NCHO, \ CCl_4}$$

OTMS
|
PhCHNHCHO

63%

Johnson, A.P.*; Luke, R.W.A.; Steele, R.W.
JCS Chem Comm, (1986), 1658

SECTION 346: Amide - Halide, Sulfonate

$$Cl_2CHC\overset{\overset{\displaystyle O}{\|}}{N}(CH_2CH=CH_2)_2 \xrightarrow[\substack{diglyme \\ reflux, \ 1h}]{FeCl_2}$$

60%

Tseng, C.K.*; Teach, E.G.; Simons, R.W.
Syn Commun, (1984), **14**, 1027

$$CH_3(CH_2)_6\overset{\overset{\displaystyle O}{\|}}{C}NH_2 \xrightarrow[\substack{RT, \ 3h}]{AcOH, \ NaO_2Br} CH_3(CH_2)_6\overset{\overset{\displaystyle O}{\|}}{C}NHBr$$

81%

Kajigaeshi, S.*; Nakagawa, T.; Fujisaki, S.
Chem Lett, (1984), 2045

$$\xrightarrow[\substack{Et_2O, \ RT, \ 1.5h}]{2 \ I_2, \ K_2CO_3}$$

78%

($\underline{Z}$:$\underline{E}$ = 2.7:1)

Hirama, M.*; Iwashita, M.; Yamazuki, Y.; Ito, S.*
Tetrahedron Lett, (1984), **25**, 4963

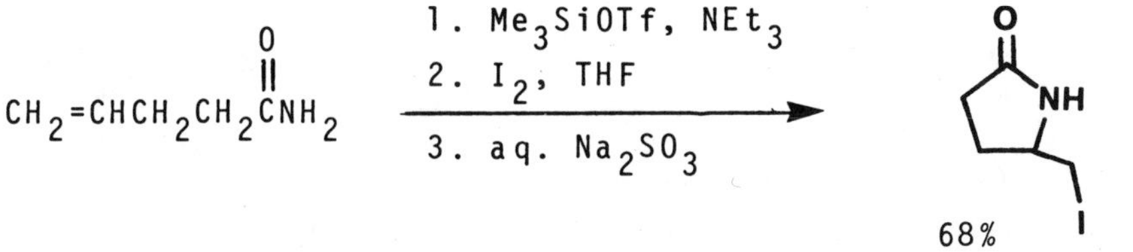

$$CH_2=CHCH_2CH_2\overset{\overset{O}{\|}}{C}NH_2 \quad \xrightarrow{\begin{array}{l}1.\ Me_3SiOTf,\ NEt_3\\ 2.\ I_2,\ THF\\ 3.\ aq.\ Na_2SO_3\end{array}}$$

68%

Knapp, S.[*]; Rodriques, K.E.   **Tetrahedron Lett**, (1985), **26**, 1803

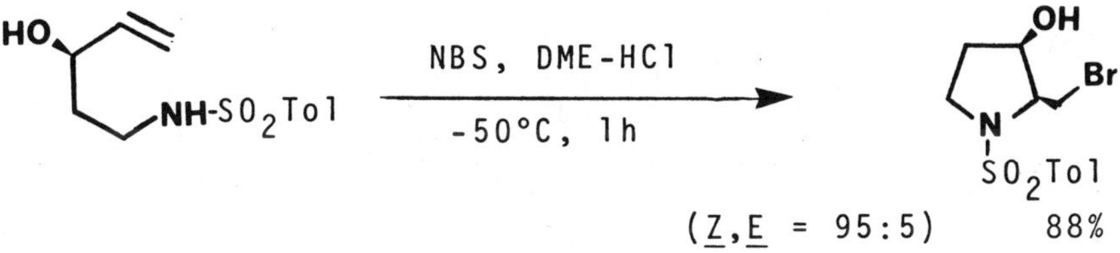

NBS, DME-HCl

-50°C, 1h

($\underline{Z},\underline{E}$ = 95:5)   88%

Tamaru, Y.; Kawamura, S.; Tanaka, K.; Yoshida, Z.[*]
                    Tetrahedron Lett, (1984), **25**, 1063

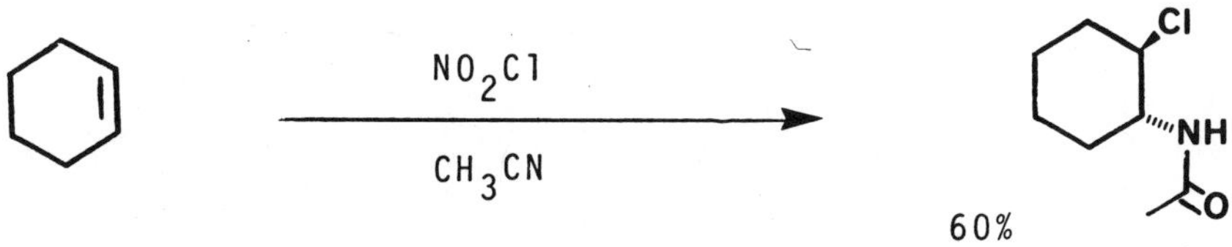

NO_2Cl

CH_3CN

60%

Zyk, N.V.; Nikulin, A.V.; Kolbasenko, S.I.; Kutateladze, A.G.;
Zefirov, N.S.
                    J Org Chem USSR, (1984), **20**, 1209

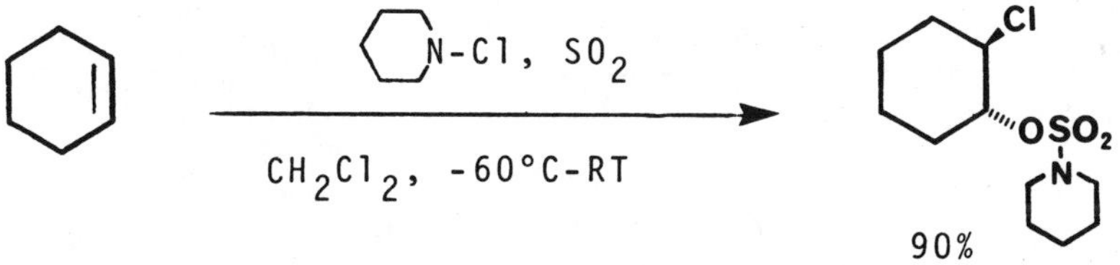

N-Cl, SO_2

CH_2Cl_2, -60°C-RT

90%

Zefirov, N.S.[*]; Zyk, N.V.; Kolbasenko, S.I.;Kutaleladze, A.G.
          J Org Chem, (1985), **50**, 4539

Lambert, C.; Caillaux, B.; Viehe, H.G.*
                    Tetrahedron, (1985), **41**, 3331

Marcos, M.; Castro, J.L.; Castedo, L.; Riguera, R.*
                    Tetrahedron, (1986), **42**, 649

SECTION 347: Amide - Ketone

Lociuro, S.; Pellacani, L.*; Tardella, P.A.
                    Tetrahedron Lett, (1983), **24**, 593

Blechert, S.*   **Helv Chim Acta**, (1983), **68**, 1835

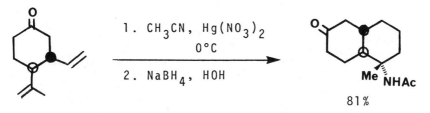

Nishino, H.[*]; Kurosawa, K. **Bull Chem Soc Jpn**, (1983), **56**, 1682

Ahlbrecht, H.[*]; Dietz, M. **Synthesis**, (1985), 417

Akiba, K.[*]; Nakatani, M.; Wada, M.; Yamamoto, Y.
J Org Chem, (1985), **50**, 63

Stevens, R.V.[*]; Albizati, K.F. **J Org Chem**, (1985), **50**, 632

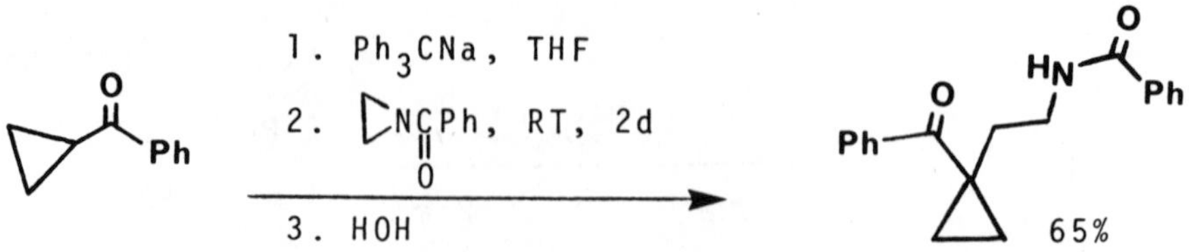

1. Ph_3CNa, THF

2. △NCPh, RT, 2d (N-C(=O)-Ph)

3. HOH

65%

Stamm, H.[*]; Weiss, R.   **Synthesis**, (1986), 392

SECTION 348: Amide - Nitrile

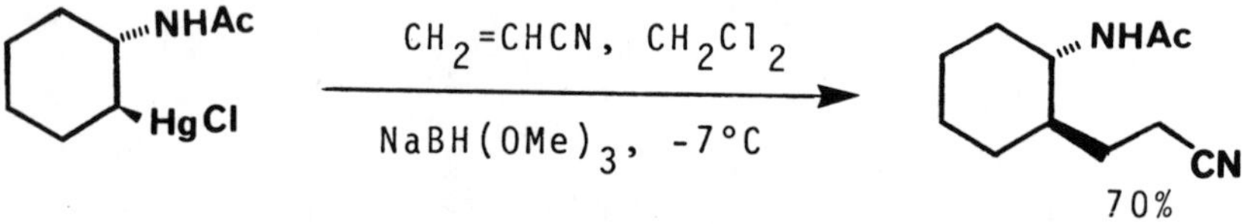

$CH_2=CHCN$, CH_2Cl_2

$NaBH(OMe)_3$, -7°C

70%

($\underline{Z}:\underline{E}$ = 4:6)

Kozikowski, A.P.[*]; Scripko, J.

Tetrahedron Lett, (1983), **24**, 2051

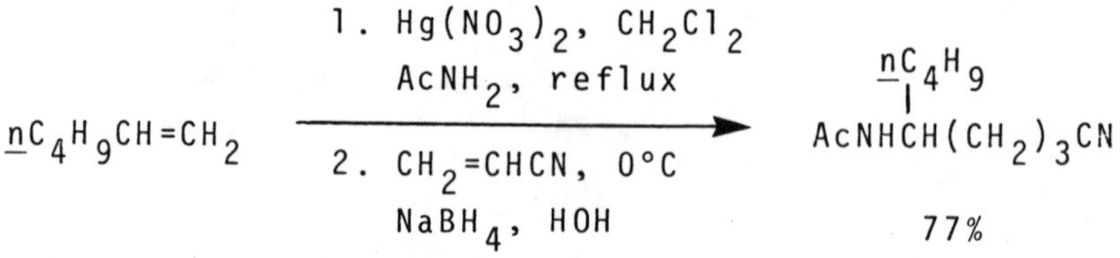

$\underline{n}C_4H_9CH=CH_2$

1. $Hg(NO_3)_2$, CH_2Cl_2
   $AcNH_2$, reflux

2. $CH_2=CHCN$, 0°C
   $NaBH_4$, HOH

$\underset{|}{\overset{\underline{n}C_4H_9}{AcNHCH(CH_2)_3CN}}$

77%

Barluenga, J.[*]; Ferrera, L.; Nájera, C.; Yus, M.
Synthesis, (1984), 831

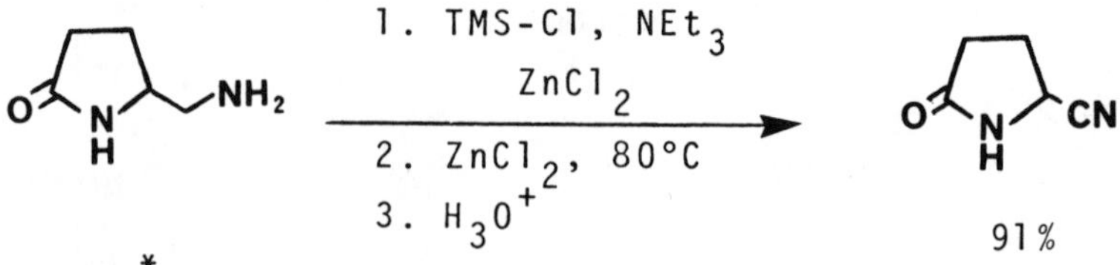

1. TMS-Cl, NEt_3
   $ZnCl_2$

2. $ZnCl_2$, 80°C

3. H_3O^+

91%

Rigo, B.[*]; Lespagnol, C.; Pauly, M.
J Heterocyclic Chem, (1986), **23**, 183

SECTION 349: Amide - Olefin

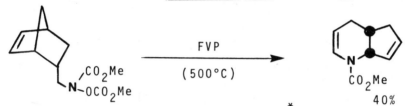

4 CF$_3$COOH, 0°C

CH$_2$Cl$_2$, 2h

90%

Hiemstra, H.[*]; Speckamp, W.N.
Tetrahedron Lett, (1983), 24, 1407

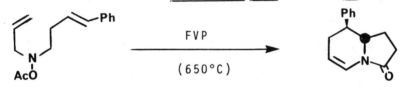

FVP

(500°C)

40%

Chu, M.; Wu, P.-L.; Givre, S.; Fowler, F.W.[*]
Tetrahedron Lett, (1986), 27, 461
Koch, K.; Lin, J.-M.; Fowler, F.W.
Tetrahedron Lett, (1983), 24, 1581

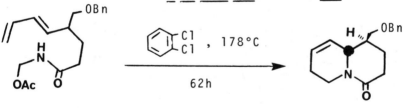

FVP

(650°C)

74%

Cheng, Y.-S.; Lupo Jr., A.T.; Fowler, F.W.[*]
J Am Chem Soc, (1983), 105, 7696

178°C

62h

93%

Bremmer, M.L.; Weinreb, S.M.[*]
Tetrahedron Lett, (1983), 24, 261

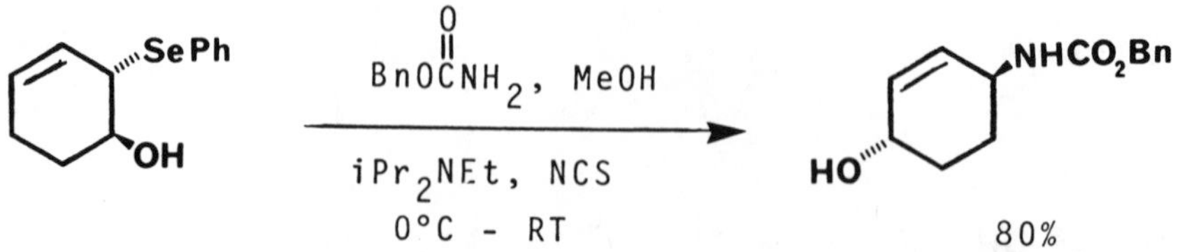

Shea, R.G.; Fitzner, J.N.; Fankhauser, J.E.; Hopkins, P.B.[*]
J Org Chem, (1984), 49, 3647

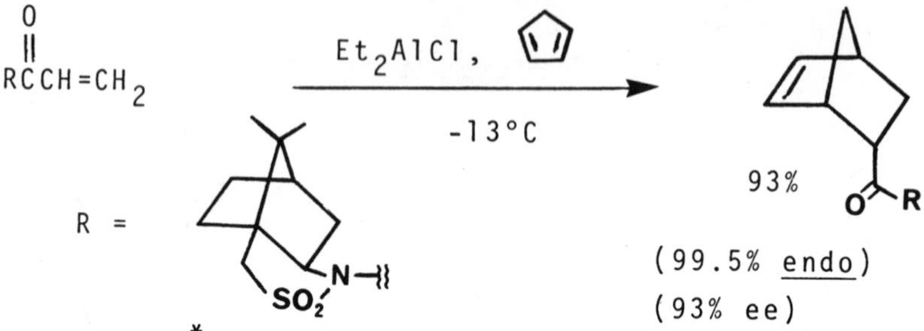

93%

(99.5% endo)
(93% ee)

Oppolzer, W.[*]; Chapuis, C.; Bernardinelli, G.
Helv Chim Acta, (1984), 67, 1397

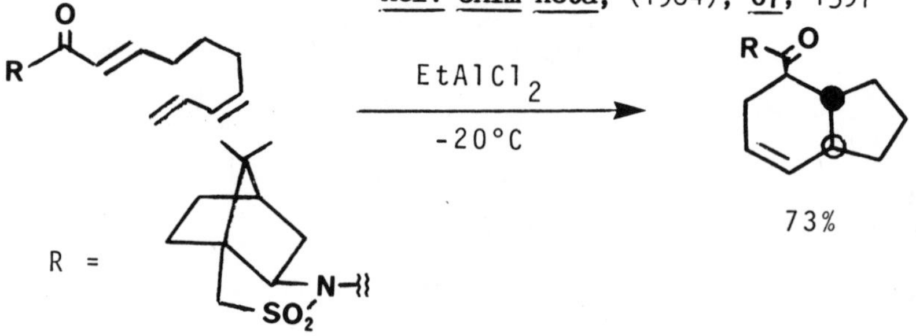

73%

Oppolzer, W.[*]; Dupuis, D.   Tetrahedron Lett, (1985), 26, 5437

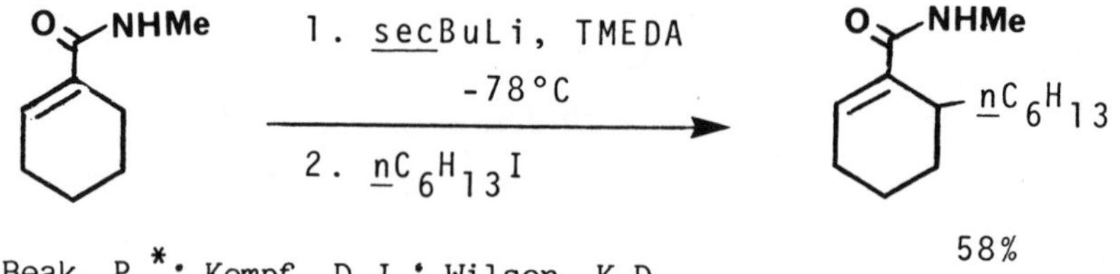

1. secBuLi, TMEDA
   -78°C

2. $\underline{n}C_6H_{13}I$

58%

Beak, P.[*]; Kempf, D.J.; Wilson, K.D.
J Am Chem Soc, (1985), 107, 4745

1. $TiCl_4$, CH_2Cl_2
      0 - 25°C

2. $CH_2=C=CMeSiMe_2\underline{t}Bu$
      0 - 25°C, 45 min
3. NEt_3 4. HOH

67%

Danheiser, R.L.[*]; Kwasigroch, C.A.; Tsai, Y.-M.
J Am Chem Soc, (1985), **107**, 7233

$$\underline{n}PrCH=CHCH_2O\overset{O}{\overset{\|}{P}}(OEt)_2$$
($\underline{E}$)

$HNEt_2$, PhH
20 atm CO
$Rh_6(CO)_{16}$
Bu_4NCl, 50°C

$$\underline{n}PrCH=CHCH_2\overset{O}{\overset{\|}{C}}NEt_2$$
($\underline{E}$)
81%

Murahashi, S.[*]; Imada, Y.   **Chem Lett**, (1985), 1477

$$C\equiv CCH_2SiMe_3$$
$$|$$
$$(CH_2)_2$$
$$|$$
$$N-CH\underline{n}Pr$$
$$| \quad OEt$$
$$Ac$$

HCOOH

68%

Hiemstra, H.[*]; Fortgens, H.P.; Stegenga, S.; Speckamp, W.N.[*]
Tetrahedron Lett, (1985), **26**, 3151, 3155

$$MeCH=C=CH\overset{Me}{\underset{Me}{\overset{|}{C}}}\overset{O}{\overset{\|}{C}}NH_2$$

$AgBF_4$, $CHCl_3$
reflux, 40h

60%

Grimaldi, J.[*]; Cormons, A.   **Tetrahedron Lett**, (1986), **27**, 5089

$$CH_2=C\begin{smallmatrix}CH_2SePh\\\\CH_3\end{smallmatrix} \xrightarrow[\substack{\underset{\displaystyle Me}{|}\\CH_2=COCNH_2\\\quad\ \ \overset{\displaystyle O}{||}}]{iPr_2NEt,\ NCS} CH_2=CCH_2NHCO\underline{t}Bu$$

with Me substituent and O (=CO) group, 92%

Shea, R.G.; Fitzner, J.N.; Fankhauser, J.E.; Spaltenstein, A.;
Carpino, P.A.; Peevey, R.M.; Tenge, B.J.; Hopkins, P.B.*
J Org Chem, (1986), **51**, 5243

$$\underset{\substack{|\ |\\ \underline{n}Bu}}{\underset{\substack{|\ O\\}}{\underline{n}PrCHNCCl}}\overset{\displaystyle CH_2CH=CH_2}{\underset{}{}} \xrightarrow[18\%\ PPh_3]{3\%\ Pd_2(dba)_3CHCl_3}$$

pyrrolidinone products (83 : 17)

69%

Henin, F.*; Muzart, J.; Pete, J.-P.
Tetrahedron Lett, (1986), **27**, 6339

Review: "Alkaloid Total Synthesis by Intramolecular Imino
          Diels Alder Cycloaddiltions"

Weinreb, S.M.*   **Accts Chem Res**, (1985), **18**, 16

Also via: Olefinic acids (Section 322)

SECTION 350: Amine - Amine

$$\underline{n}BuCH=CH_2 \xrightarrow[\substack{3.\ Na_2CO_3\\4.\ aq.\ HCl,\ 120°C}]{\substack{1.\ NH_2CN,\ NBS\\2.\ HCl,\ EtOH}} \underset{\substack{|\\ \displaystyle NH_2}}{\underline{n}BuCHCH_2NH_2}$$

71%

Jung, S.-H.; Kohn, H.*   2d   **J Am Chem Soc**, (1985), **107**, 2931
                              J Am Chem Soc, (1983), **105**, 4106

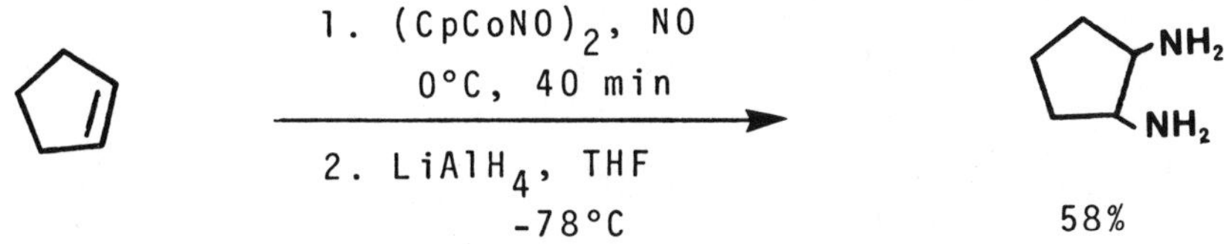

1. $(CpCoNO)_2$, NO
   0°C, 40 min

2. $LiAlH_4$, THF
   -78°C

58%

Becker, P.N.; Bergman, R.G.[*] **Organometallics**, (1983), **2**, 787

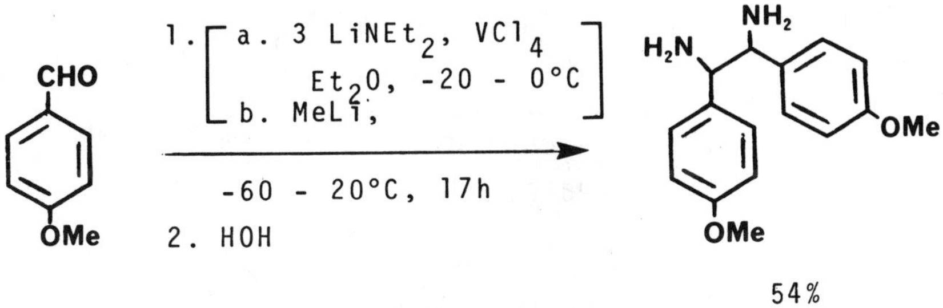

1. ⎡ a. 3 $LiNEt_2$, VCl_4
   ⎢ Et_2O, -20 - 0°C
   ⎣ b. MeLi,   ⎤

   -60 - 20°C, 17h
2. HOH

CHO
OMe

H_2N NH_2
OMe
OMe

54%

Imwinkelried, R.; Seebach, D.[*]
Helv Chim Acta, (1984), **67**, 1496

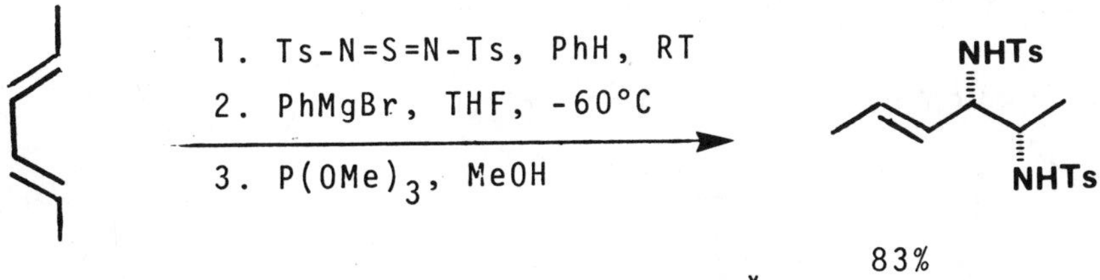

1. Ts-N=S=N-Ts, PhH, RT
2. PhMgBr, THF, -60°C
3. $P(OMe)_3$, MeOH

NHTs
NHTs

83%

Natsugari, H.; Whittle, R.R.; Weinreb, S.M.[*]
J Am Chem Soc, (1984), **106**, 7867

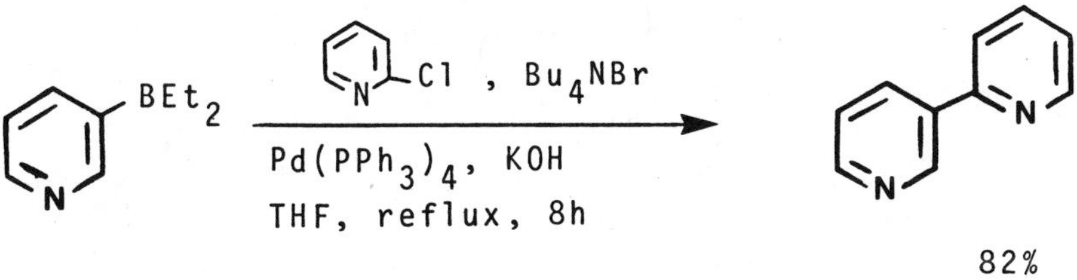

BEt_2 [pyridine]-Cl , Bu_4NBr

$Pd(PPh_3)_4$, KOH
THF, reflux, 8h

82%

Ishikura, M.; Kamada, M.; Terashima, M.[*]
Synthesis, (1984), 937

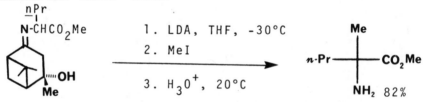

$$\text{pyrrolidine} \xrightarrow[\text{CH}_2\text{Cl}_2/\text{MeOH}]{\text{9 KBar, RT, 2d}} \text{product}$$

50%

Matsumoto, K.[*]; Hashimoto, S.; Ikemi, Y.; Otani, S.
Heterocycles, (1984), **22**, 1417

$$\begin{array}{c} + \\ \text{PPh}_3\text{Br}^- \end{array}$$

1. EDA, CHCl$_3$, 20°C
2. 2 $\underline{t}$BuOK, Et$_2$O

3. $\underline{n}$C$_3$H$_7$CHO, RT
PPh$_3$Br$^-$ 4. H$_2$/Pd-C, MeOH
+

$$\underline{n}\text{C}_5\text{H}_{11} \quad \underline{n}\text{C}_5\text{H}_{11}$$
HN   NH

46%

Cristau, H.-J.[*]; Chiche, L.[*]; Plenát, F.   **Synthesis**, (1986), 56

SECTION 351: Amine - Ester

$\underline{n}$Pr
N-CHCO$_2$Me

OH
Me

1. LDA, THF, -30°C
2. MeI

3. H$_3$O$^+$, 20°C

Me
n-Pr———CO$_2$Me
NH$_2$ 82%

(82% ee)

Bajgrowicz, J.A.[*]; Cossec, B.; Pigiere, C.L.; Jacquier, R.;
Viallefont, Ph.[*]
Tetrahedron Lett, (1983), **24**, 3721

$$\text{PhCH=C} \begin{array}{c} \text{OTMS} \\ \text{OMe} \end{array}$$

CH$_2$=NOBn

10% Me$_3$SiOTf, RT
CH$_2$Cl$_2$, 7h

CO$_2$Me
PhCHCH$_2$NHOBn

93%

Ikeda, K.; Achiwa, K.; Sekiya, M.
Tetrahedron Lett, (1983), **24**, 4707

$$\xrightarrow[\text{2. HOH}(t_{1/2}\ 8\ \text{min})]{\begin{array}{c}1.\ \text{Et}_3\text{O}^+\text{BF}_4^-,\ 0°\text{C}\\ \text{CH}_2\text{Cl}_2\end{array}} \text{EtO}_2\text{C}(\text{CH}_2)_3\overset{+}{\text{N}}\text{H}_3\text{BF}_4^-$$

61%

Smith, M.B.[*]; Shroff, H.N.　**J Org Chem**, (1984), **49**, 2900
　　　　　　　　　　Tetrahedron Lett, (1983), **24**, 2091

$$\xrightarrow[\begin{array}{c}\text{CH}_2\text{Cl}_2,\ 30°\text{C}\\ 5\text{h}\end{array}]{\text{BnN(CH}_2\text{SMe)}_2}$$

68%

Miyazawa, S.; Ikeda, K.; Achiwa, K.; Sekiya, M.[*]
　　　　　　Chem Lett, (1984), 785·

$$\xrightarrow[\text{CsF, THF, }60°\text{C}]{\begin{array}{c}\text{BnN}\overset{\text{SiMe}_3}{\underset{\text{CH}_2\text{OMe}}{}}\\ \text{Me}_3\text{SiOTf}\end{array}}$$

Hosomi, A.[*]; Sakata, Y.; Sakurai, H.[*]　**Chem Lett**, (1984), 1117

$$\xrightarrow[\begin{array}{c}\text{2. 8N HCl, reflux}\\ \text{3. EtOH, reflux}\end{array}]{\begin{array}{c}1.\ 0.1\text{N HCl, THF}\\ \text{RT, 60h}\end{array}}$$

43%

Gull, R.; Schöllkopf, U.[*]　**Synthesis**, (1985), 1052

$PhCH_2CO_2Me$

$\xrightarrow{\begin{array}{l}1.\ LDA\\ 2.\ TMS-Cl\\ \hline 3.\ Py,\ PhN_2{}^+BF_4{}^-,\ 0°C\\ 4.\ 5\ atm\ H_2,\ Pd-C\\ \quad MeOH\end{array}}$

$\overset{NH_2}{\underset{}{PhCHCO_2Me}}$

73%

Sakakura, T.; Tanaka, M.[*]   **JCS Chem Comm**, (1985), 1309

$\xrightarrow{\begin{array}{c}FVP\\ \\ (400°C)\end{array}}$

67%

DeShong, P.[*]; Kell, D.A.; Sidler, D.R.
J Org Chem, (1985), **50**, 2309

$\overset{CO_2Me}{\underset{NHBn}{ICH_2CH_2CH}}$

$\xrightarrow[-60°C,\ 3h]{4\ Et_2CuLi,\ Et_2O}$

$\overset{CO_2Me}{\underset{NHBn}{EtCH_2CH_2CH}}$

70%

Bajgrowicz, J.A.; El Hallaouri, A.; Jacquier, R.; Pigiere, Ch.;
Viallefont, Ph.[*]

Tetrahedron, (1985), **41**, 1833

$\overset{NO_2}{\underset{}{EtCHCO_2Me}}$

$\xrightarrow[MeOH,\ RT]{HCO_2{}^-NH_4{}^+,\ 10\%\ Pd-C}$

$\overset{NH_2}{\underset{}{EtCHCO_2Me}}$

67%

Ram, S.; Ehrenkaufer, R.E.[*]   **Synthesis**, (1986), 133

1. $CH_2=C\diagdown_{CH_2OAc}^{Me}$, THF

$\underset{HC}{\overset{NCH_2CO_2Et}{\underset{||}{\|}}}$ ⬡—Cl

Pd(dppe), 20°C

2. 10% HCl, RT

$\longrightarrow$

$CH_2=CHCH_2\overset{\overset{NH_2}{|}}{C}HCO_2Me$

50%

Ferroud, D.; Genet, J.P.[*]; Kiolle, R.
Tetrahedron Lett, (1986), **27**, 23

$\underset{PhCPh}{\overset{NCH_2CO_2Me}{\underset{||}{}}}$

1. LDA, THF, -78°C
2. $CH_2=CHCH_2OAc$

Pd(dba)$_2$, DIOP

3. 10% HCl
4. K_2CO_3

$\longrightarrow$

$CH_2=CHCH_2CHCO_2Me$

75%

(39% ee R)

DIOP = (4$\underline{S}$, 5$\underline{S}$)-(+)-4,5-$\underline{bis}$-(diphenylphosphinomethyl)
2,$\overline{2}$-dimethyl-1,3-dioxolane

Genet, J.P.[*]; Ferroud, D.; Juge, S.; Montes, J.R.
Tetrahedron Lett, (1986), **27**, 4573

0.1 PdCl$_2$, 3 CuCl$_2$

CO, MeOH, RT, 8h

$\longrightarrow$

67%

Lathbury, D.; Vernon, P.; Gallagher, T.[*]
Tetrahedron Lett, (1986), **27**, 6009

Related Methods: Acid-Amide (Section 315); Acid-Amine (Section 316); Amide-Ester (Section 344)

SECTION 352: Amine - Ether, Epoxide, Thioether

Shi, L.; Narula, C.K.; Mak, K.T.; Kao, L.; Xu, Y.; Heck, R.F.[*]
J Org Chem, (1983), 48, 3894

Ito, Y.[*]; Sawamura, M.; Kominami, K.; Saegusa, T.
Tetrahedron Lett, (1985), 26, 5303

Benati, L.; Montevecchi, P.C.; Spagnolo, P.
Tetrahedron, (1986), 42, 1145

(threo:erythro = 98:2) 95%

Claremon, D.A.[*]; Lumma, P.K.; Phillips, B.T.
J Am Chem Soc, (1986), 108, 8265

SECTION 353: Amine - Halide, Sulfonate

$Cl(CH_2)_4MgBr$

$iPrCH=\overset{+}{N}$ pyrrolidine Cl^-

$\xrightarrow[\text{2h}]{Et_2O, \ 15°C-RT}$

$iPrCH(CH_2)_4Cl$ with N-pyrrolidine

57%

Courtois, G.; Miginiac, P. **Bull Chem Soc Fr**, (1983), II148

$$\underset{\underset{Cl}{|}}{CH_3\overset{\overset{iPrN}{||}}{C}-\overset{\overset{Cl}{|}}{C}-CO_2Me}$$

$\xrightarrow[\text{reflux, 1.7h}]{NaOMe/MeOH}$

$$CH_3\overset{\overset{NiPr}{||}}{C}CHCl_2$$

83%

De Kimpe, N.[*]; Verhe, R.; De Buyck, L.; Schamp, N.
          Tetrahedron Lett, (1985), **26**, 2709

Review: "Reactions of β-Haloenamines"

De Kimpe, N.[*]; Schamp, N. **Org Prep Proc Int**, (1983), **15**, 71

SECTION 354: Amine - Ketone

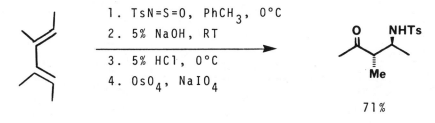

1. TsN=S=O, $PhCH_3$, 0°C
2. 5% NaOH, RT
3. 5% HCl, 0°C
4. OsO_4, $NaIO_4$

71%

Garigipati, R.S.; Morton, J.A.; Weinreb, S.M.[*]
          Tetrahedron Lett, (1983), **24**, 987

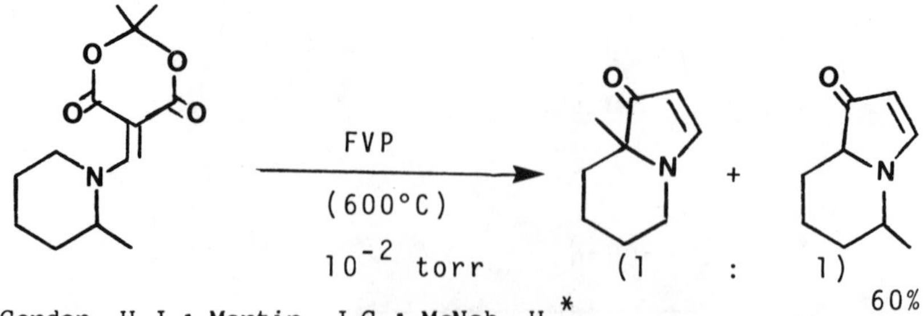

FVP

(600°C)

10^{-2} torr

(1 : 1)

60%

Gordon, H.J.; Martin, J.C.; McNab, H.[*]
JCS Chem Comm, (1983), 957

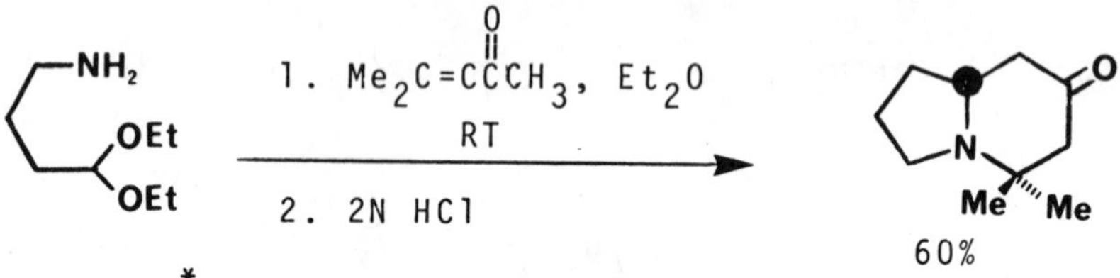

1. $Me_2C=CCCH_3$, Et_2O
   RT

2. 2N HCl

60%

King, F.D.[*] **Tetrahedron Lett**, (1983), **24**, 3281

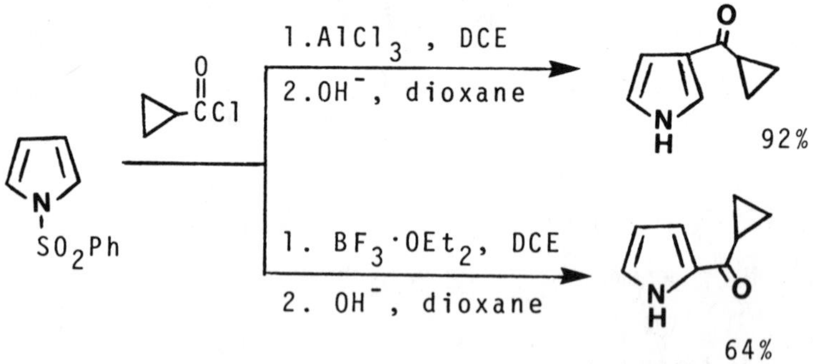

1. $AlCl_3$, DCE

2. OH^-, dioxane

92%

1. $BF_3 \cdot OEt_2$, DCE

2. OH^-, dioxane

64%

Kakushima, M.[*]; Hamel, P.; Frenette, R.; Rokach, J.
J Org Chem, (1983), **48**, 3214

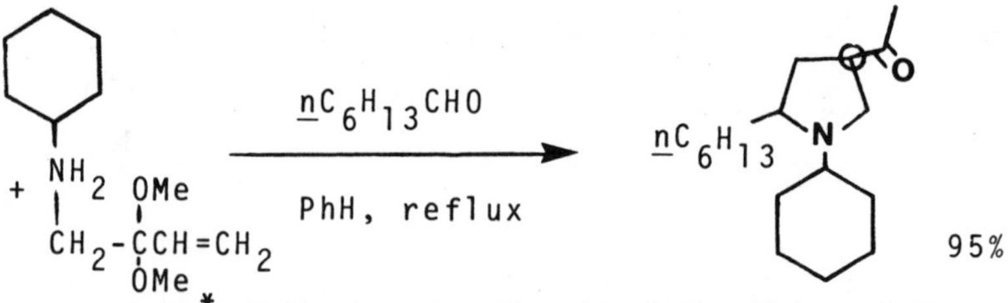

$\underline{n}C_6H_{13}CHO$

PhH, reflux

95%

Overman, L.E.[*]; Kakimoto, M.; Okazaki, M.E.; Meier, G.P.
J Am Chem Soc, (1983), **105**, 6622

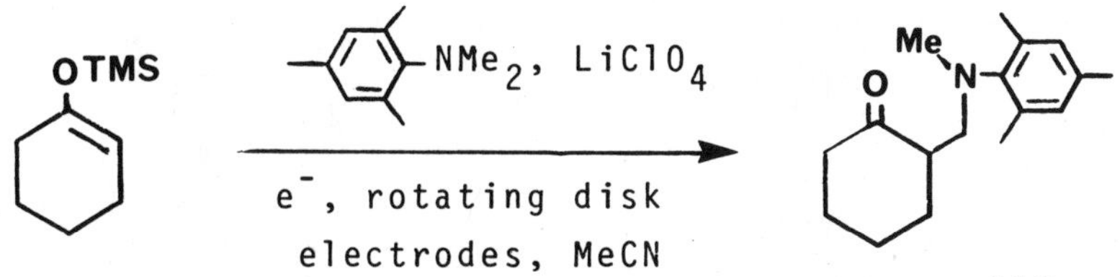

58%

Renaud, R.N.[*]; Bérubé, D.; Stephens, C.J.
Can J Chem, (1983), **61**, 1379

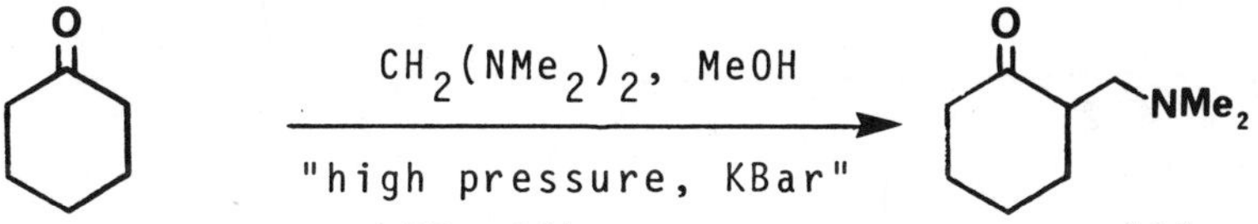

53%

Matsumoto, K.[*]; Hashimoto, S.; Otani, S.; Amita, F.; Osugi, J.
Syn Commun, (1984), **14**, 585

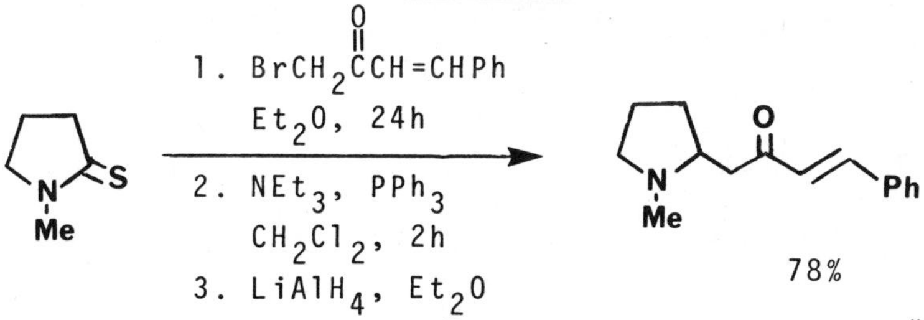

78%

Ghirlando, R.; Howard, A.S.[*]; Katz, R.B.; Michael, J.P.[*]
Tetrahedron, (1984), **40**, 2879

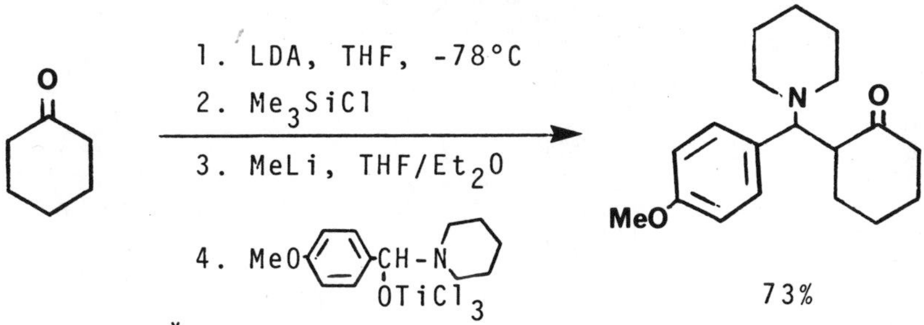

73%

Seebach, D.[*]; Betschart, C.; Schiess, M.
Helv Chim Acta, (1984), **67**, 1593

PhCHO $\xrightarrow[\begin{array}{c}2 \ \pentagon \ , \ 25°C\end{array}]{CH_3NH_2 \cdot HCl, \ HOH}$

exo:endo = 4.2:1) 82%

Grieco, P.A.[*]; Larsen, S.D.; Fobare, W.F.
Tetrahedron **Lett**, (1986), **27**, 1975

$$Me_2CHCCl \ (O) \xrightarrow[\begin{array}{l}2. \ 3N \ HCl \\ 3. \ 3N \ NaOH\end{array}]{\begin{array}{l}1. \ \underline{n}Bu_3SnCH_2NEt_2 \\ \ \ \ 0 - 20°C, \ 15 \ min\end{array}} Me_2CHCCH_2NEt_2 \ (O)$$

64%

Verlhac, J.-B.; Quintard, J.-P.
Tetrahedron **Lett**, (1986), **27**, 2361

$$\begin{array}{c}Ph \\ \ \ \ \diagdown C \diagup NHPh \\ \ \ \ \| \\ \ \ \ CH \\ \ \ \ | \\ \ \ \ C=NH \\ Ph \diagup\end{array} \xrightarrow[\begin{array}{l}2. \ LiAlH_4, \ THF \\ 3. \ H_3O^+\end{array}]{\begin{array}{l}1. \ PhMeC=NPh, \ AlCl_3 \\ \ \ \ dioxane, \ 100°C\end{array}} \begin{array}{c}PhCHNHPh \\ | \\ CH_2 \\ | \\ C=O \\ | \\ Ph\end{array}$$ 56%

Barluenga, J.[*]; Cuervo, H.; Olano, B.; Fustero, S.; Gotor, V.
Synthesis, (1986), 469

$$\begin{array}{c}O \\ \| \\ PhS \quad O \\ \diagdown C - C \diagup H \\ \underline{n}C_6H_{13} \diagup \quad \diagdown Ph\end{array} \xrightarrow[RT, \ 3h]{PiP} \begin{array}{c}O \\ \| \\ \underline{n}C_6H_{13}CCH-N \bigcirc \\ | \\ Ph\end{array}$$

97%

Satoh, T.; Kaneko, Y.; Sakata, K.; Yamakawa, K.[*]
Bull Chem Soc Jpn, (1986), **59**, 457

SECTION 355: Amine - Nitrile

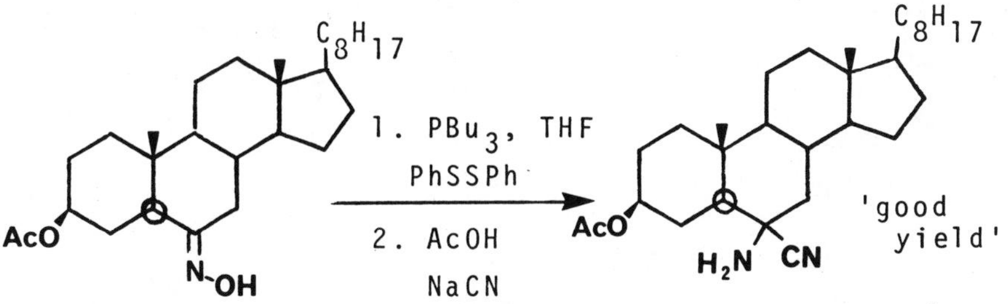

1. PBu$_3$, THF
   PhSSPh

2. AcOH
   NaCN

'good yield'

Barton, D.H.R.; Motherwell, W.B.; Simon, E.S.; Zard, S.Z.[*]
JCS Chem Comm, (1984), 337

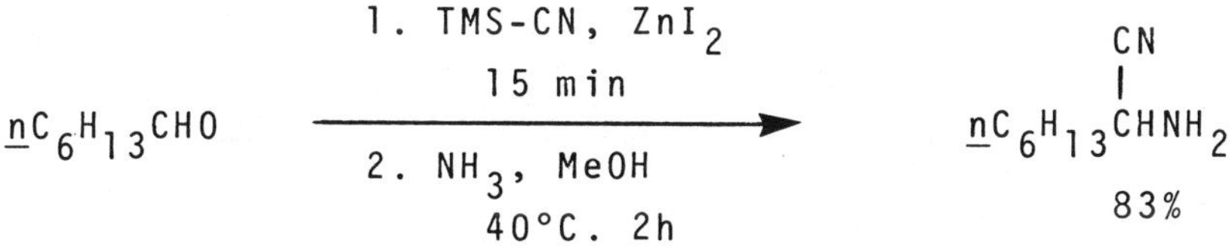

$\underline{n}C_6H_{13}CHO$

1. TMS-CN, ZnI$_2$
   15 min

2. NH$_3$, MeOH
   40°C. 2h

$\underline{n}C_6H_{13}\overset{\displaystyle CN}{\underset{\displaystyle}{C}}HNH_2$

83%

Mai, K.[*]; Patil, G.   **Tetrahedron Lett**, (1984), **25**, 4583

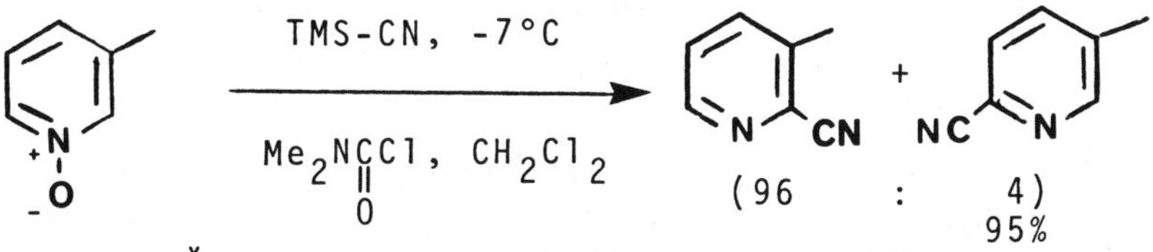

TMS-CN, -7°C

Me$_2$NCCl, CH$_2$Cl$_2$
      ‖
      O

(96   :   4)
95%

Fife, W.K.[*]   **Heterocycles**, (1984), **22**, 93, 1121

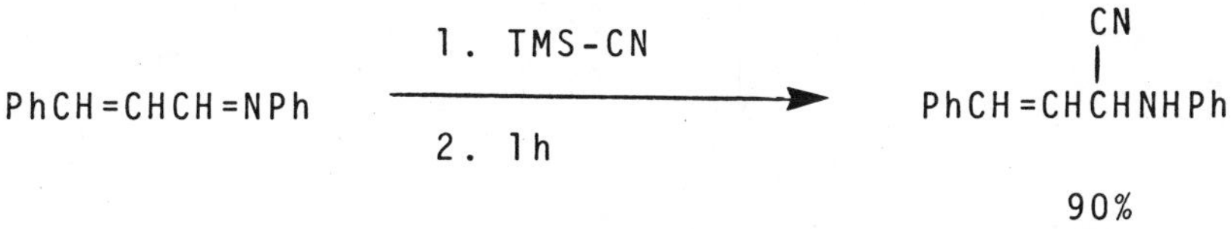

PhCH=CHCH=NPh

1. TMS-CN

2. 1h

PhCH=CHCHNHPh with CN

90%

Prajapati, D.; Borush, R.C.; Sandhu, J.S.[*]; Baruah, J.N.
Ind J Chem B, (1984), **23**, 853

$$CN$$
$$PhCHOTMS , MeOH$$

Me
H''''—NH$_2$
Ph

reflux, 2h

Me H   CN
   \N/
H''''   Ph Ph '''H

97%

$$(\underline{RR}:\underline{RS} = 3:1)$$

Mai, K.*; Patil, G.   **Syn Commun**, (1984), **14**, 1299

O
||
CH$_3$CCH$_3$

1. BnNH$_2$, neat
   100°C, 1 min

2. RT, TMS-CN
3. 100°C, 1 min
4. RT   97%

CN
|
CH$_3$CNHBn
|
CH$_3$

Mai, K.*; Patil, G.   **Syn Commun**, (1985), **15**, 157
                      Org Prep Proc Int, (1985), **17**, 183

1. tBuMe$_2$SiOTf, CH$_2$Cl$_2$, 0°C
2. MeLi, THF, 0°C
3. CH$_2$Cl$_2$, TiCl$_4$
   -78°C, 30 min
4. TMS-CN, -78°C-RT
5. aq. NaOH   63%

Tokitoh, N.; Okazaki, R.*   **Chem Lett**, (1985), 241

1. PsSeSePh, NaCN
   AcOH, 5d, 20°C

2. aq. NaOH   95%

Barton, D.H.R.*; Billion, A.; Boivin, J.
                      Tetrahedron Lett, (1985), **26**, 1229

SECTION 356: Amine - Olefin

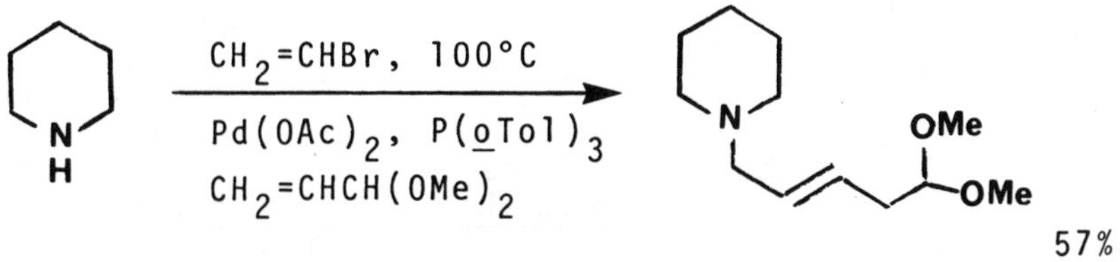
CH2=CHBr, 100°C
Pd(OAc)2, P(oTol)3
CH2=CHCH(OMe)2
N
H
N
OMe
OMe
57%

Fischetti, W.; Mak, K.T.; Stakem, F.G.; Kim, J.-I.; Rheingold, A.L.; Heck, R.F.*
J Org Chem, (1983), 48, 948

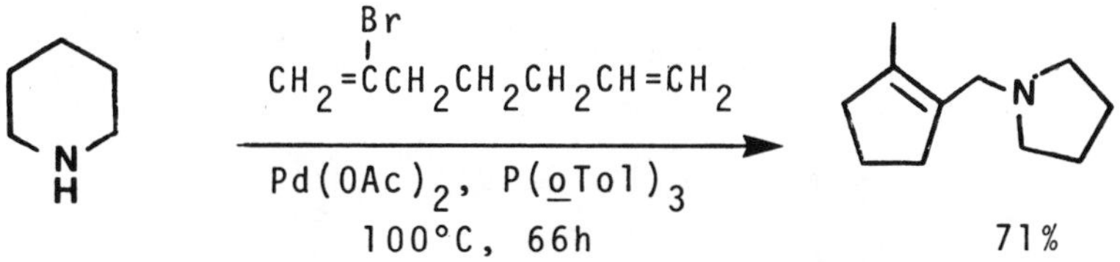
Br
CH2=CCH2CH2CH2CH=CH2
Pd(OAc)2, P(oTol)3
100°C, 66h
N
H
N
71%

Narula, C.K.; Mak, K.T.; Heck, R.F.*
J Org Chem, (1983), 48, 2792

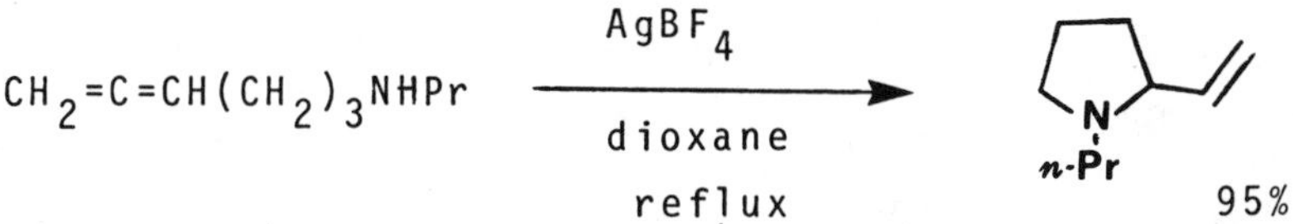
AgBF4
CH2=C=CH(CH2)3NHPr
dioxane
reflux
N
n-Pr
95%

Arseniyadis, S.; Gore, J. Tetrahedron Lett, (1983), 24, 3997

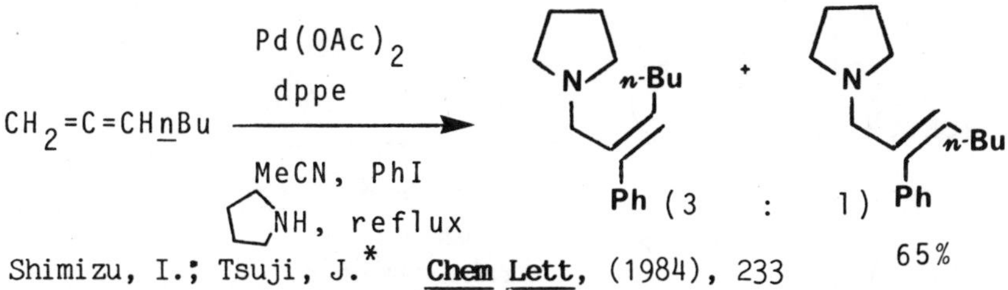
Pd(OAc)2
dppe
CH2=C=CHnBu
MeCN, PhI
NH, reflux
N
n-Bu
Ph (3 : 1) Ph
N
n-Bu
65%

Shimizu, I.; Tsuji, J.* Chem Lett, (1984), 233

$$CH_3CH_2CH=C\begin{smallmatrix}NO_2\\ \\CH_3\end{smallmatrix} \xrightarrow[\text{◯NH, DMF, 75°C}]{Pd(PPh_3)_4, \; dppe}$$

75%

Tamura, R.[*]; Hayashi, K.; Kai, Y.; Oda, D.
Tetrahedron Lett, (1984), **25**, 4437

$$\xrightarrow[\text{2. } CH_2=CHCH_2Br]{\text{1. CuCN}}$$

69%

Ishikura, M.; Kamada, M.; Oda, I.; Terashima, M.[*]
Heterocycles, (1985), **23**, 117
Ishikura, M.; Kamada, M.; Ohta, T.; Terashima, M.[*]
Heterocycles, (1984), **22**, 2475

$$\xrightarrow[\text{CH}_3\text{CN, 82°C}]{\text{CF}_3\text{COOH, 2h}}$$

90%

Overman, L.E.[*]; Burk, R.M.   **Tetrahedron Lett**, (1984), **25**, 5739

$$Me_3SiCH_2CH=CHCH_2CH_2NHBn \xrightarrow[\substack{\text{aq. THF, RT}\\56h}]{\substack{\text{CF}_3\text{COOH}\\\text{HCHO}}}$$

81%

Grieco, P.A.[*]; Fobare, W.F.   **Tetrahedron Lett**, (1986), **27**, 5067

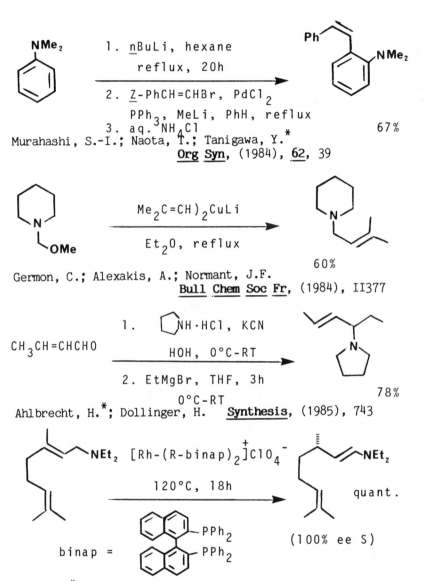

1. nBuLi, hexane
   reflux, 20h

2. Z-PhCH=CHBr, PdCl$_2$
   PPh$_3$, MeLi, PhH, reflux
3. aq. NH$_4$Cl   67%

Murahashi, S.-I.; Naota, T.; Tanigawa, Y.*
Org Syn, (1984), **62**, 39

Me$_2$C=CH)$_2$CuLi

Et$_2$O, reflux

60%

Germon, C.; Alexakis, A.; Normant, J.F.
Bull Chem Soc Fr, (1984), II377

CH$_3$CH=CHCHO

1. ◯NH·HCl, KCN
    HOH, 0°C-RT

2. EtMgBr, THF, 3h
   0°C-RT   78%

Ahlbrecht, H.*; Dollinger, H. **Synthesis**, (1985), 743

NEt$_2$ [Rh-(R-binap)$_2$]$^+$ClO$_4^-$ NEt$_2$

120°C, 18h   quant.

binap =   PPh$_2$ (100% ee S)
                       PPh$_2$

Tani, K.*; Yamagata, T.; Tatsuno, Y.; Yamagata, Y.; Tomita, K.;
Akutagawa, S.; Kumobayashi, H.; Otsuka, S.*
Angew Chem Int Ed Engl, (1985), **24**, 217

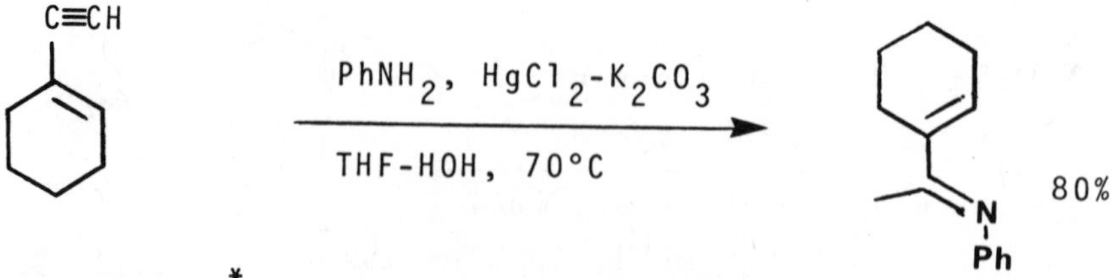
C≡CH
PhNH2, HgCl2-K2CO3
THF-HOH, 70°C
N
Ph
80%

Barluenga, J.*; Aznar, F.; Liz, R.; Cabal, M.-P.
<u>JCS Chem Comm</u>, (1985), 1375

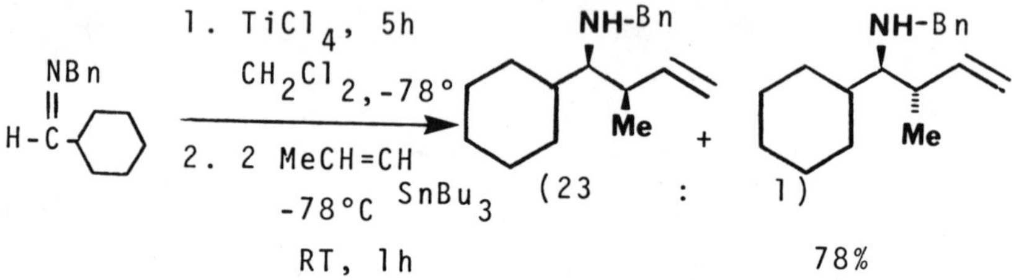
NBn
H-C
1. TiCl4, 5h
CH2Cl2, -78°
2. 2 MeCH=CH
SnBu3
-78°C
RT, 1h
NH-Bn
Me
+
NH-Bn
Me
(23 : 1)
78%

Keck, G.E.*; Enholm, E.J. <u>J Org Chem</u>, (1985), <u>50</u>, 146

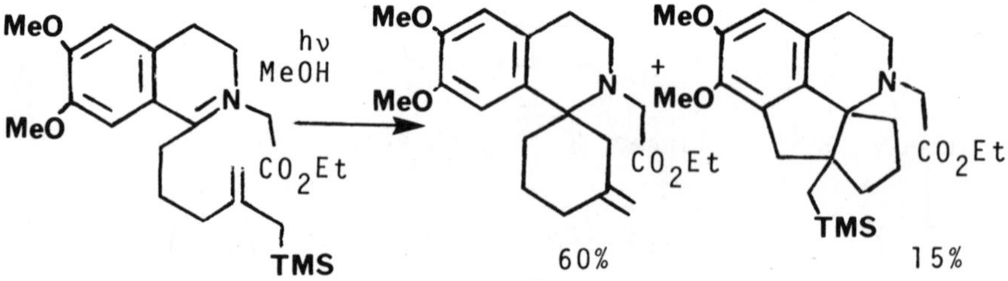
MeO
MeO
N
CO2Et
TMS
hν
MeOH
MeO
MeO
N
CO2Et
60%
+
MeO
MeO
N
CO2Et
TMS
15%

Ahmed-Schofield, R.; Mariano, P.S.*
<u>J Org Chem</u>, (1985), <u>50</u>, 5667

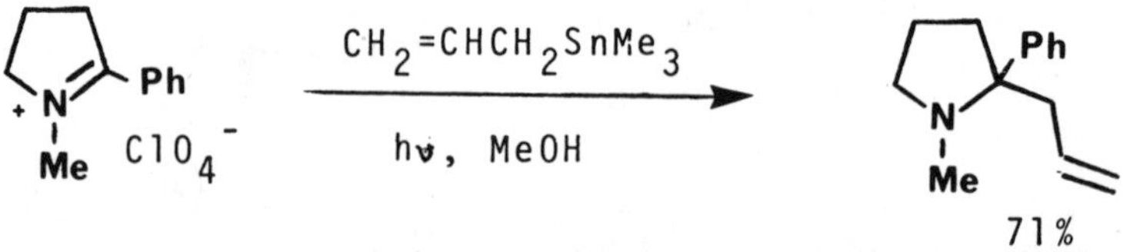
Ph
N
Me
ClO4
CH2=CHCH2SnMe3
hν, MeOH
Ph
N
Me
71%

Borg, R.M.; Mariano, P.S.* <u>Tetrahedron Lett</u>, (1986), <u>27</u>, 2821

PhCH=NPh

1. MeCH=CHCH$_2$-B⟨⟩

   Et$_2$O, -78°C

2. HCl, 0°C - RT

3. aq. NaOH, 0°C

Ph—CH(NHPh)—CH(Me)—CH=CH$_2$

93%

Yamamoto, Y.[*]; Komatsu, T.; Muruyama, K.
J Org Chem, (1985), **50**, 3115

PhNH$_2$

CH$_2$=CHCH$_2$OH, 120-150°C

PtCl$_2$(PPh$_3$)$_2$-SnCl$_2$-2 H$_2$O

CH$_2$=CHCH$_2$NHPh

67%

Tsuji, Y.; Takeuchi, R.; Ogawa, H.; Watanabe, Y.[*]
Chem Lett, (1986), 293

$$\underset{\text{PhCCl}}{\overset{\displaystyle\|}{\overset{\displaystyle NPh}{}}}$$

CH$_2$=CHCH$_2$SnBu$_3$, PhH

PdCl$_2$(PPh$_3$)$_2$, 120°C

10h

$$\underset{\text{PhCCH=CHCH}_3}{\overset{\displaystyle\|}{\overset{\displaystyle NPh}{}}}$$

86%

Kosugi, M.[*]; Koshiba, M.; Atoh, A.; Sano, H.; Migita, T.[*]
Bull Chem Soc Jpn, (1986), **59**, 677

PhCH=NPh

CH$_2$OSO$_2$Me
|
TMSCH$_2$C=CH$_2$

Ni[P(OEt)$_3$]$_4$, PhCH$_3$
reflux, overnight

77%

Jones, M.D.; Kemmitt, R.D.W.   **JCS Chem Comm**, (1986), 1201

$[\underline{o}Tol_3P]_2Pd(OAc)_2$

PhH, 60°C

70%

Trost, B.M.[*]; Chen, S.-F.

J Am Chem Soc, (1986), **108**, 6053

$CH_2CH_2N_3$

$CH=CCH=CH_2$

SPh

CHCl$_3$, 70°C

3d

SPh

62%

Pearson, W.H.[*]; Celebuski, J.E.; Poon, Y.-F.; Dixon, B.R.; Glens, J.H.

Tetrahedron Lett, (1986), **27**, 6301

Reviews:

"Diels Alder Reactions of Azadienes"

Boger, D.L.[*]   **Tetrahedron**, (1983), **39**, 2869

"Synthesis of Primary Allylic Amines"

Cheikh, R.B.; Choabouni, R.; Laurent, A.; Mison, P.; Nafti, A.
Synthesis, (1983), 685

"Recent Advances in the Chemistry of Conjugated Enamines"

Hickmott, P.W.[*]   **Tetrahedron**, (1984), **40**, 2989

SECTION 357: Ester - Ester

OSiMe$_3$

1. Pb(OAc)$_4$, CH$_2$Cl$_2$

RT, 30 min

2. Et$_3$NHF, RT

OAc

66%

Rubottom, G.M.[*]; Gruber, J.M.; Marrero, R.; Juve Jr., H.D.; Kim, C.W.

J Org Chem, (1983), **48**, 4940

$$\underset{PhCCH_2CH_2COOH}{\overset{O}{\parallel}} \quad \xrightarrow[\substack{HC(OMe)_3, \ HOH \\ 2.8h}]{Tl(NO_3)_3 \cdot 3 \ H_2O} \quad \underset{PhCHCH_2CO_2Me}{\overset{CO_2Me}{\mid}}$$

82%

Taylor, E.C.[*]; Conley, R.A.; Katz, A.H.; McKillop, A.H.
J Org Chem, (1984), **49**, 3840

$$\xrightarrow[\substack{FeSO_4 \cdot 7H_2O, \ 80°C \\ 4h}]{(NH_4)_2S_2O_8, \ AcOH}$$

79%

Fristad, W.E.[*]; Peterson, J.R.   **Tetrahedron**, (1984), **40**, 1469

$$\xrightarrow[\substack{AcOH, \ 40°C \\ 24h}]{Mn_3(OAc)_7}$$

64%

Corey, E.J.[*]; Kang, M.   **J Am Chem Soc**, (1984), **106**, 5384

$$\underline{n}PrCO_2Et \quad \xrightarrow[\substack{2. \ MeCH=CHCO_2Et \\ -78°C \\ 3. \ aq. \ NH_4Cl}]{\substack{1. \ LDA, \ -78°C \\ THF/HMPA}}$$

(20   :   1)
87%

Yamaguchi, M.[*]; Tsukamoto, M.; Tanaka, S.; Hirao, I.
Tetrahedron Lett, (1984), **25**, 5661

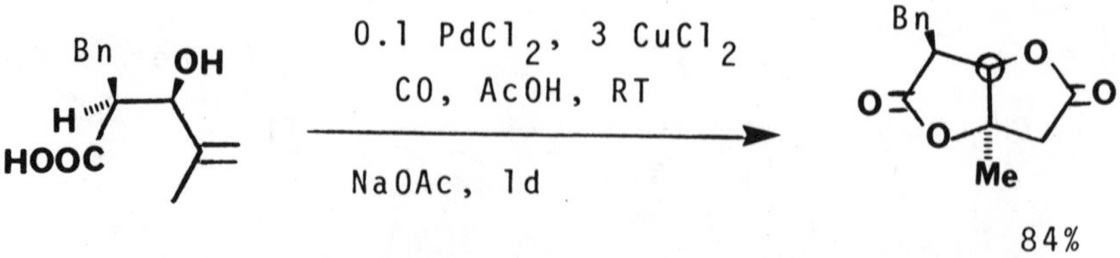

84%

Tamaru, Y.; Higashimura, H.; Naka, K.; Hojo, M.; Yoshida, Z.*
Angew Chem Int Ed Engl, (1985), **24**, 1045

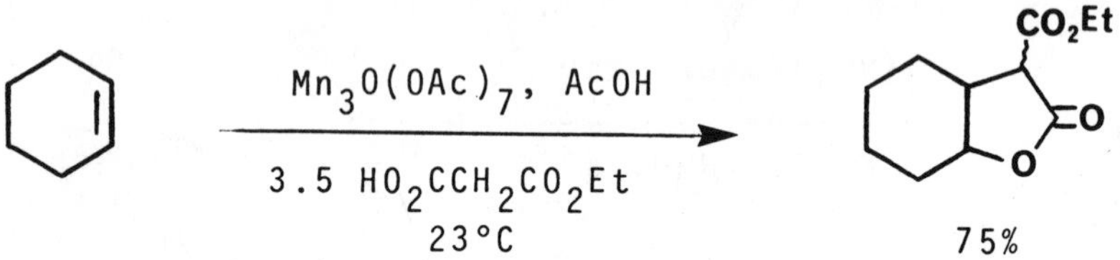

75%

Corey, E.J.*; Gross, A.W.   **Tetrahedron Lett**, (1985), **26**, 4291

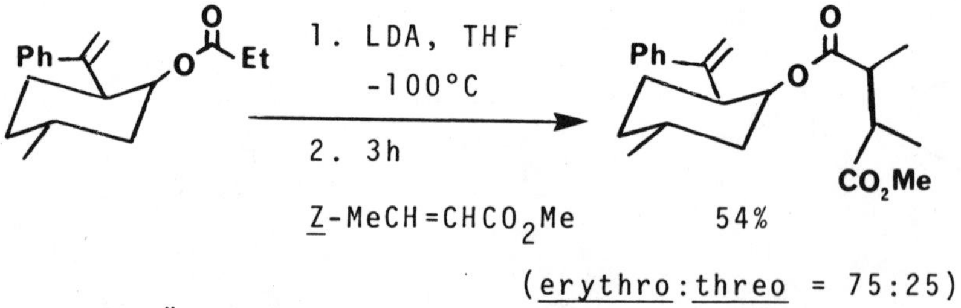

54%

(erythro:threo = 75:25)

Corey, E.J.*; Peterson, R.T.
Tetrahedron Lett, (1985), **26**, 5025

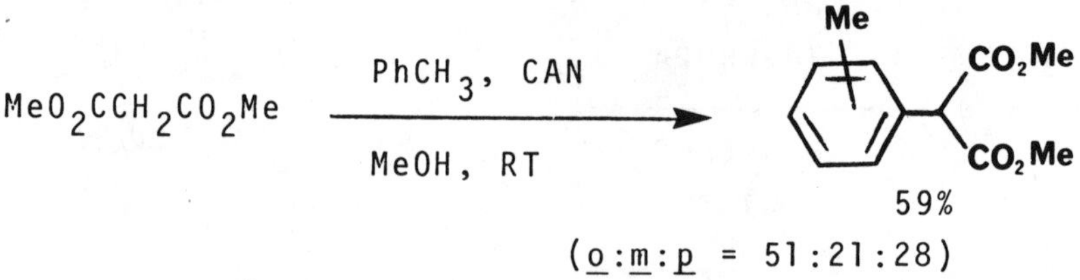

$MeO_2CCH_2CO_2Me$

59%

(o:m:p = 51:21:28)

Baciocchi, E.*; Dell'aira, D.; Ruzziconi, R.
Tetrahedron Lett, (1986), **27**, 2763

O   O
‖   ‖
HC(CH$_2$)$_4$OCCHCH$_3$
                |
                Br

$\xrightarrow{\begin{array}{c} 1. \ SmI_2, \ THF, \ 0°C \\ \hline 2. \ Ac_2O, \ DMAP \end{array}}$

(structure of product)

82%

Tabuchi, T.; Kawamura, K.; Inanaga, J.[*]; Yamaguchi, M.
            Tetrahedron Lett, (1986), **27**, 3889

Also via: Dicarboxylic acids (Section 312); Hydroxyesters
(Section 327; Diols (Section 323)

SECTION 358: Ester - Ether, Epoxide, Thioether

CH$_2$CH$_2$COOH
|
C
‖
C
\
(CH$_2$)$_3$OH

$\xrightarrow{\begin{array}{c} cat. \ HgO \\ \hline DMF, \ reflux \\ 2h \end{array}}$

(structure of product)

73%

Yamamoto, M.[*]; Yoshitake, M.; Yamada, K.
            JCS Chem Comm, (1983), 991

(structure) CO$_2$Me

$\xrightarrow{\begin{array}{c} 5 \ CH_2=C(OMe)_2 \\ \hline PhCH_3, \ 110°C \\ 15h \end{array}}$

(structure) OMe / CO$_2$Me

86%

Boger, D.L.[*]; Mullican, M.D.
            Tetrahedron Lett, (1983), **24**, 4939

Watanabe, Y.[*]; Tsuji, Y.; Takeuchi, R.
Bull Chem Soc Jpn, (1983), **56**, 1428

Bailey, W.F.[*]; Rivera, A.D.   **J Org Chem**, (1984), **49**, 4958

Williams, D.R.[*]; Harigaya, Y.; Moore, J.L.; D'sa, A.
J Am Chem Soc, (1984), **106**, 2641

(erythro:threo = 55:45)

Yamamoto, Y.[*]; Maruyama, K.; Matsumoto, K.
Tetrahedron Lett, (1984), **25**, 1075

PhCH=CHCO$_2$Me $\xrightarrow{\substack{\text{Me} \\ \triangleright\text{-C-OSiMe}_2\text{tBu} \\ \text{OOCPh} \\ \parallel \\ \text{O} \\ \text{Pd(OAc)}_2, \text{ CH}_2\text{Cl}_2, \text{ 8h}}}$ PhCH-CHCO$_2$Me

$\overset{\text{O}}{\overset{\diagup\backslash}{}}$

59%

Nagata, R.; Matsuura, T.; Saito, I.*
Tetrahedron Lett, (1984), **25**, 2691

$\left[\begin{array}{c} \text{S} \diagdown \text{S} \end{array}\right]$ $\xrightarrow{\substack{\text{1. CO, } \underline{n}\text{BuLi, } -110°C \\ \text{THF/pentane/Et}_2\text{O} \\ \text{2. CO, 2h, } -110°C\text{-RT} \\ \text{3. MeI}}}$ $\underline{n}$BuC$\overset{\text{O}}{\overset{\parallel}{\text{S}}}$(CH$_2$)$_4$SMe

55%

Seyferth, D.*; Hui, R.C.   **Organometallics**, (1984), **3**, 327

COOH $\xrightarrow{\substack{\text{1. Me}_2\text{S-SMe BF}_4^- \\ \text{MeCN, 30 min} \\ \text{2. iPr}_2\text{NEt, MeCN} \\ \text{RT, 36h}}}$

SMe

86%

O'Malley, G.J.; Cava, M.P.*   **Tetrahedron Lett**, (1985), **26**, 6159

CHO $\xrightarrow{\substack{\text{CH}_2=\text{C} \diagdown \substack{\text{OSiMe}_2\text{tBu} \\ \text{OMe}} \\ \text{5\% ZnI}_2, \text{ MeCN, 48h} \\ 0°C \text{ - RT } (\underline{\text{anti}}:\underline{\text{syn}} = 96:4)}}$

CO$_2$Me

67%

OSiMe$_2$tBu

Kita, Y.*; Yasuda, H.; Tamura, O.; Itoh, F.; Ke, Y.Y.; Tamura, Y.

Tetrahedron Lett, (1985), **26**, 5777

OBn
|
$\underline{n}C_5H_{11}CHCHO$

1. [diagram: OTMS / OEt], ZnCl$_2$

CH_2Cl_2, 20°C
$\longrightarrow$
2. Py

[product structure] BnO ... $\underline{n}C_5H_{11}$ ... $OSiMe_3$ ... CO_2Et

93%

Oshino, H.; Nakamura, E.[*]; Kuwajima, I.[*]
J Org Chem, (1985), **50**, 2802

[naphthalene structure]

EtO_2CCH_2SMe, CH_2Cl_2
$\longrightarrow$
$SnCl_4$, 0°C - RT

[naphthalene with $MeSCHCO_2Et$ substituent]

95%

Stamos, I.K.[*]   **Tetrahedron Lett**, (1985), **26**, 477

$MeOCH_2OMe$

$Co_2(CO)_8$, γ-picoline
$\longrightarrow$
CO/HOH, MeOH, 160°C
5h   (65% conversion)

$MeOCH_2CO_2Me$

50%

Murota, K.[*]; Matsuda, A.; Masuda, T.
Bull Chem Soc Jpn, (1985), **58**, 2141

Ph [structure] OH   OH

$PdCl_2$, $CuCl_2$
$\longrightarrow$
AcONa, AcOH
CO, RT, 1d

[bicyclic lactone product] Ph

80%

Tamaru, Y.; Kobayashi, T.; Kawamura, S.; Ochiai, H.; Hojo, M.; Yoshida, Z.[*]
Tetrahedron Lett, (1985), **26**, 3207

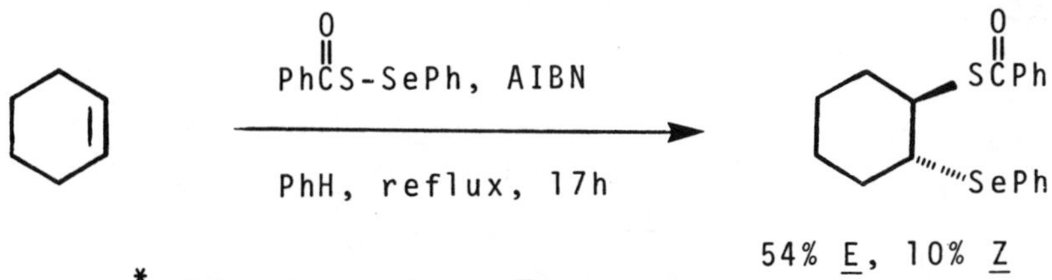
PhCS-SePh, AIBN
PhH, reflux, 17h
SCPh
SePh
54% E, 10% Z

Toru, T.*; Seko, T.; Maekawa, E.
Tetrahedron Lett, (1985), **26**, 3263

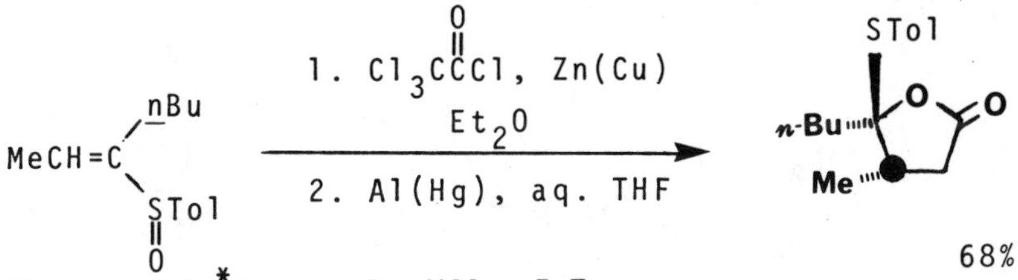
1. Cl3CCCl, Zn(Cu)
Et2O
2. Al(Hg), aq. THF
MeCH=C
nBu
STol
STol
n-Bu
Me
68%

Marino, J.P.*; De la Pradilla, R.F.
Tetrahedron Lett, (1985), **26**, 5381

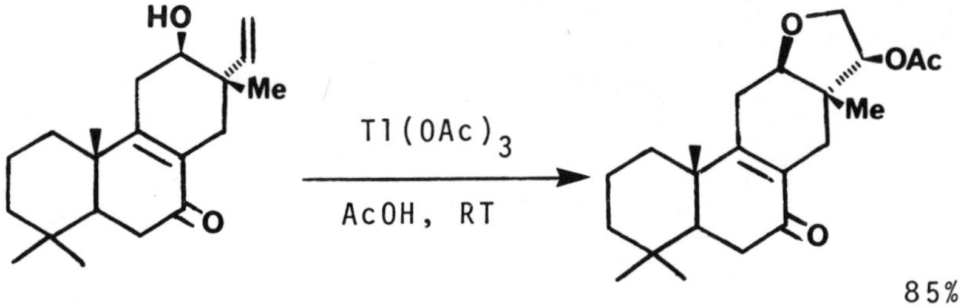
HO
Me
Tl(OAc)3
AcOH, RT
OAc
Me
85%

Ferroz, H.M.C.*; Brocksom, T.J.; Pinto, A.C.; Abla, M.A.; Zocher, D.H.T.
Tetrahedron Lett, (1986), **27**, 811

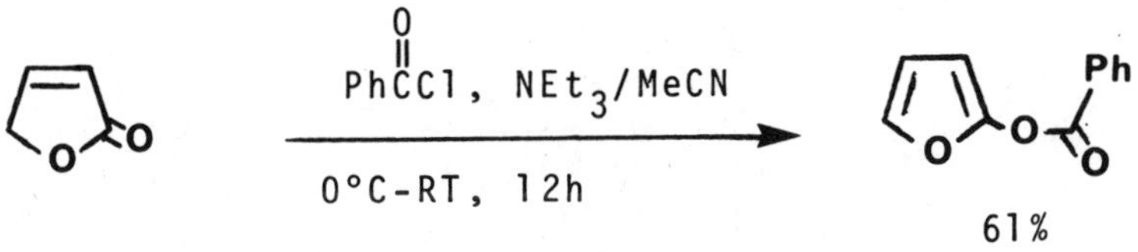
PhCCl, NEt3/MeCN
0°C-RT, 12h
Ph
61%

Hormi, O.E.O.; Näsman, J.H.* **Syn Commun**, (1986), **16**, 69

SECTION 359: Ester - Halide, Sulfonate

$$\underset{\text{PhCHCO}_2\text{Et}}{\overset{\text{CO}_2\text{Et}}{|}} \quad \xrightarrow[\substack{\underline{p}\text{TolSO}_2\text{-N} \\ |\; \text{F}}]{\text{NaH, THF, -20°C}} \quad \underset{\overset{|}{\text{F}}}{\overset{\text{CO}_2\text{Et}}{\underset{|}{\text{PhC-CO}_2\text{Et}}}}$$

81%

Barnette, W.E.[*]   J Am Chem Soc, (1984), 106, 452

I$_2$, DME/HOH

RT, 4d

(9 : 1)

85%

Tamaru, Y.; Mizutani, M.; Furukawa, Y.; Kawamura, S.; Yoshida, Z.[*]; Yanagi, K.; Minobe, M.

J Am Chem Soc, (1984), 106, 1079

$$\underset{\text{PhCCH}_2\text{CH}_2\text{COOH}}{\overset{\overset{\text{O}}{\|}}{}} \quad \xrightarrow[\text{CHCl}_3, \; 30 \text{ min}]{\text{Et}_2\text{NSF}_3, \; 0°C} \quad$$

85%

Patrick, T.B.[*]; Poon, Y.F.   Tetrahedron Lett, (1984), 25, 1019

$$\text{CH}_2\text{=CHCO}_2\text{Me} \quad \xrightarrow[\text{EtOH, 30°C, 30 min}]{\text{PhN}_2^{+}\;\text{Cl}^{-}, \; \text{KI}} \quad \underset{\text{PhCH}_2\text{CHCO}_2\text{Me}}{\overset{\overset{\text{I}}{|}}{}}$$

45%

Ganushchak, N.I.; Obushak, N.D.; Polishchuk, O.P.
J Org Chem USSR, (1984), 20, 595

1. Cl_2, CCl_4, MeCN
2. Bu_4NCl, RT, 8h
3. aq. H_2O_2

82%

Morella, A.M.; Ward, A.D.[*]
Tetrahedron Lett, (1985), **26**, 2899

$(BrCH_2)_2CHCO_2\underline{n}C_6H_{13}$ KF, reflux / MeCN/DMF / 8h   → $CH_2=CCO_2\underline{n}C_6H_{13}$ with CH_2Br

97%

Anzeveno, P.B.[*]; Campbell, J.A.; White, W.L.
Syn Commun, (1986), **16**, 387

1. MeCN, NaOCl
   pH 5-6
2. iPrOH, 48% HBr
   reflux, 3h

83%

Robert, A.[*]; Jaguelin, S.; Guinamant, J.L.
Tetrahedron, (1986), **42**, 2275

Also via: haloacids (Section 319); Halohydrins (Section 329)

SECTION 360: Ester - Ketone

1. $EtNO_2$, e^-, MeCN
   cat. PhN=NPh, Bu_4NBr
2. O_2, e^-, MeCN
   Bu_4NBr

$CH_3CCH_2CHCO_2Me$ with O and CH_3

60%

Monte, W.T.; Baizer, M.M.; Little, R.D.
J Org Chem, (1983), **48**, 803

$$
\begin{array}{c}
\text{OTMS} \\
| \\
\text{PhC=CH}_2
\end{array}
\quad
\xrightarrow[\text{2. Ni/H}_2]{\overset{\text{Cl}}{\underset{}{\text{1. PhSCHCO}_2\text{Me, ZnBr}_2}}}
\quad
\begin{array}{c}
\text{O} \\
\| \\
\text{PhCCH}_2\text{CH}_2\text{CO}_2\text{Me}
\end{array}
$$

92%

Lee, T.V.[*]; Okonkwo, J.O.   **Tetrahedron** **Lett**, (1983), **24**, 323

1. TiCl$_4$/CH$_2$Cl$_2$ 2. Ni(R)   73%
Fleming, I.[*]; Iqbal, J.   **Tetrahedron** **Lett**, (1983), **24**, 327

MeO

1. BrZnCH$_2$CO$_2$Me

2. CAN

90%

Pearson, A.J.[*]; Richards, I.C.
          Tetrahedron **Lett**, (1983), **24**, 2465

$$
\underline{n}\text{C}_5\text{H}_{11}\overset{\overset{\text{O}}{\|}}{\text{C}}\text{CH}_3
$$

1.

2. LDA, THF, -78°C
3. PhCH=CHCO$_2$Me
4. O$_3$, CH$_2$Cl$_2$, -78°C

59%

Enders, D.[*]; Papadopoulos, K.
          Tetrahedron **Lett**, (1983), **24**, 4967

1. LDA, THF, -78-0°C
2. MeOCCN, HMPA, -78°C
3. HOH

86%

Mander, L.N.; Sethi, S.P.   **Tetrahedron** **Lett**, (1983), **24**, 5425

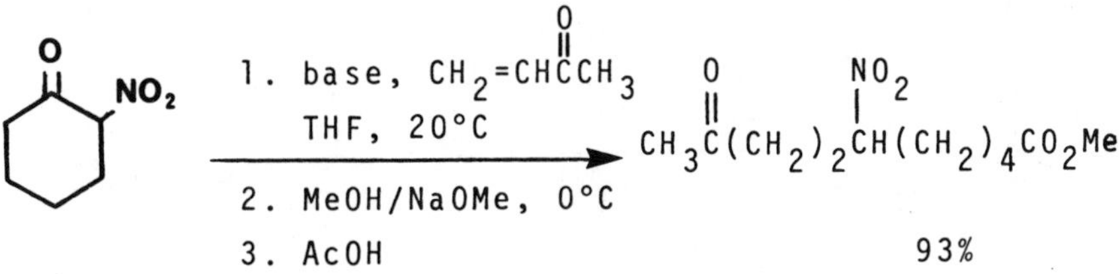

1. base, CH$_2$=CHCCH$_3$

   THF, 20°C

2. MeOH/NaOMe, 0°C

3. AcOH

$$CH_3C(CH_2)_2CH(CH_2)_4CO_2Me$$

93%

Huggenberg, W.; Hesse, M.[*] **Helv Chim Acta**, (1983), **66**, 1519

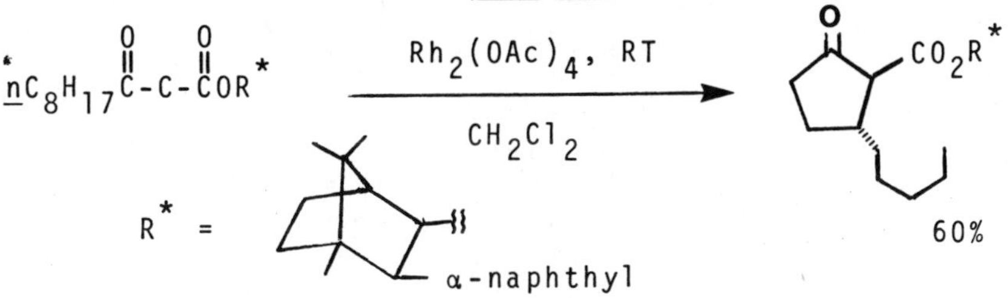

$$\underline{n}C_8H_{17}C-C-COR^*$$

Rh$_2$(OAc)$_4$, RT

CH$_2$Cl$_2$

R* =

α-naphthyl

60%

Taber, D.F.[*]; Raman, K.   **J Am Chem Soc**, (1983), **105**, 5935

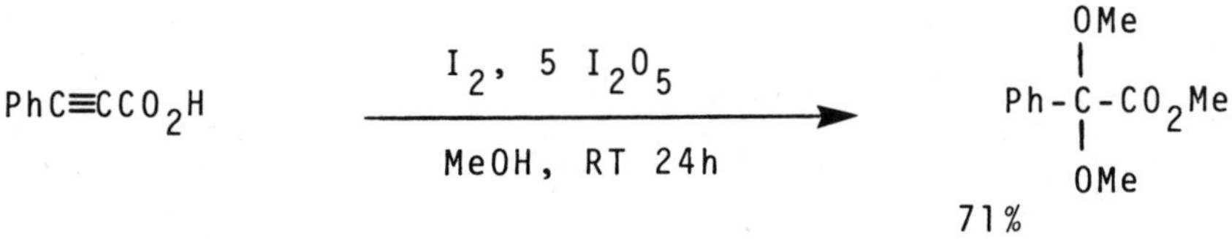

PhC≡CCO$_2$H

I$_2$, 5 I$_2$O$_5$

MeOH, RT 24h

$$Ph-C-CO_2Me$$
(with OMe above and OMe below)

71%

Cohen, M.J.; McNelis, E.[*]   **J Org Chem**, (1984), **49**, 515

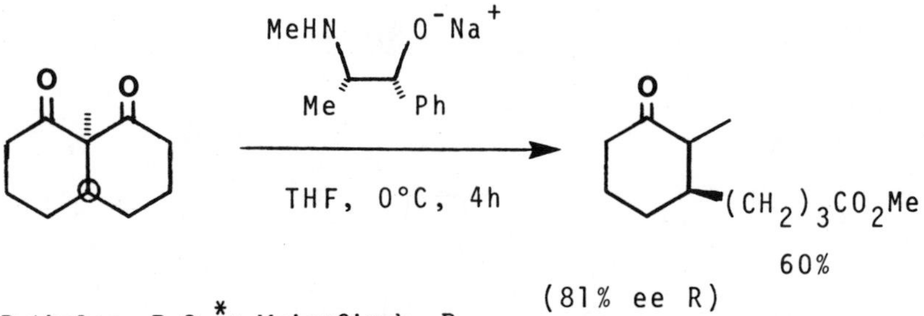

MeHN   O$^-$Na$^+$

Me   Ph

THF, 0°C, 4h

(CH$_2$)$_3$CO$_2$Me

60%

(81% ee R)

Duthaler, R.O.[*]; Maienfisch, P.

Helv Chim Acta, (1984), **67**, 845

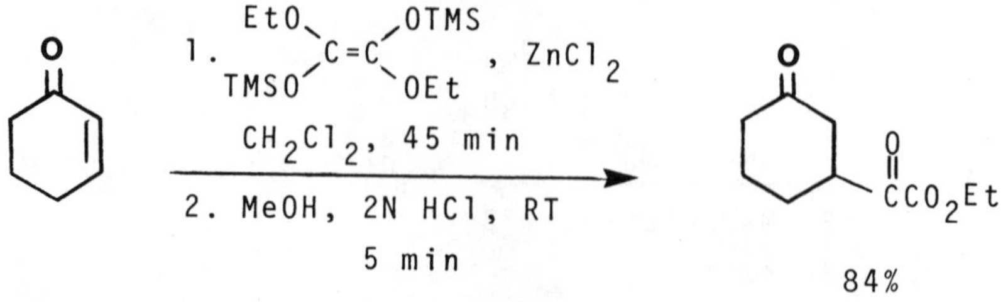

Reetz, M.T.[*]; Heimbach, H.; Schwellnus, K.
Tetrahedron Lett, (1984), 25, 511

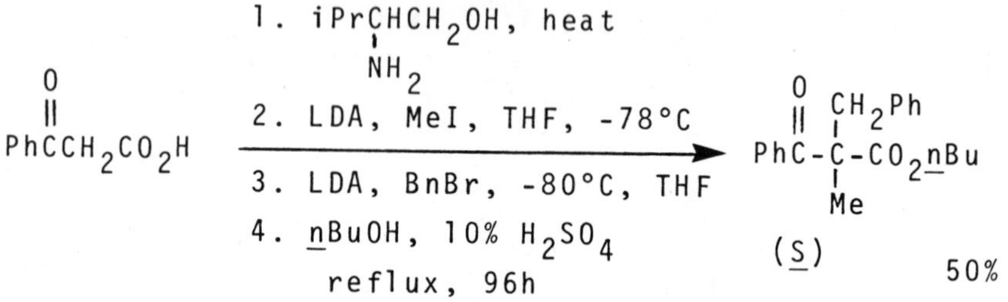

Meyers, A.I.[*]; Harre, M.; Garland, R.
J Am Chem Soc, (1984), 106, 1146

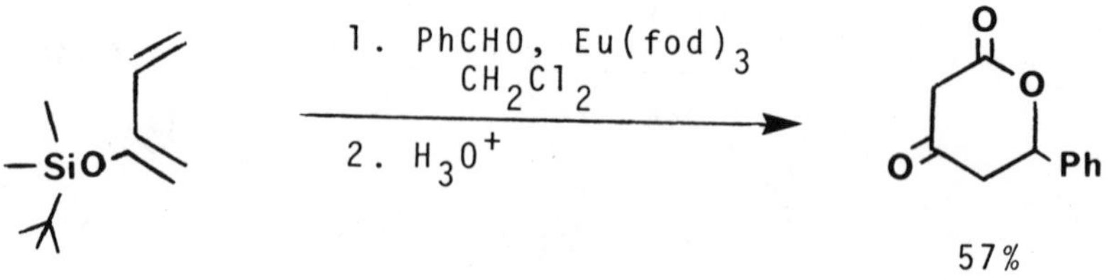

Castellino, S.; Sims, J.J.   Tetrahedron Lett, (1984), 25, 2307

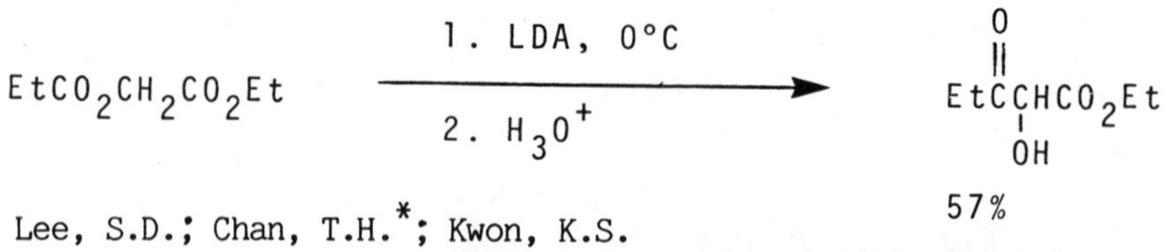

$EtCO_2CH_2CO_2Et$

1. LDA, 0°C

2. H_3O^+

$EtCCHCO_2Et$ with $\overset{O}{\overset{\|}{}}$ and $\underset{OH}{}$

57%

Lee, S.D.; Chan, T.H.[*]; Kwon, K.S.
Tetrahedron Lett, (1984), 25, 3399

$$\underline{n}BuCHO \xrightarrow[\begin{array}{c} \text{2. Zn(dust), 25°C} \\ \text{3. HOH} \end{array}]{\begin{array}{c} \text{1. } (MeO)_2\overset{\overset{\displaystyle O}{\|}}{P}CH\overset{CO_2CMe_2CCl_3}{\underset{CO_2Me}{\diagdown}} \\ \text{LiN(TMS)}_2 \qquad 0°C \end{array}} \underline{n}Bu\overset{\overset{\displaystyle O}{\|}}{C}CO_2Me$$

66%

Horne, D.; Gaudino, J.; Thompson, W.J.*
Tetrahedron **Lett**, (1984), **25**, 3529

$$CH_3\overset{\overset{\displaystyle O}{\|}}{C}\underset{\underset{Me}{|}}{C}HCO_2Me \xrightarrow[\begin{array}{c} \underline{n}Bu_3P, \text{ THF} \\ 5\% \text{ RhH(PPh}_3)_4 \end{array}]{2 \text{ } CH_2=CHCH_2OAc, \text{ } 20°C} CH_2=CHCH_2-\underset{\underset{CO_2Me}{|}}{\overset{\overset{Me}{|}}{C}}-\overset{\overset{\displaystyle O}{\|}}{C}Me$$

93%

Tsuji, J.*; Minami, I.; Shimizu, I.
Tetrahedron **Lett**, (1984), **25**, 5157

$$\xrightarrow[\text{reflux, 21h}]{Mn(OAc)_3, \text{ PhH}}$$

84%

Dunlap, N.K.; Sabol, M.R.; Watt, D.S.*
Tetrahedron **Lett**, (1984), **25**, 5839

$$\xrightarrow[\text{1h}]{\begin{array}{c} (TMS)_2SO_4 \\ DCE, \text{ } 80°C \end{array}}$$

(65   :   35)

71%

Morizawa, Y.; Hiyama, T.*; Oshima, K.; Nozaki, H.
Bull Chem Soc Jpn, (1984), **57**, 1123

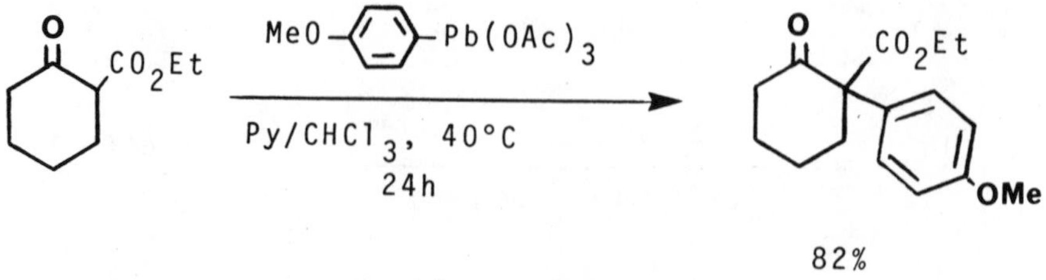

82%

Kozyrod, R.P.; Pinkey, J.T.*   Org Syn, (1984), 62, 24

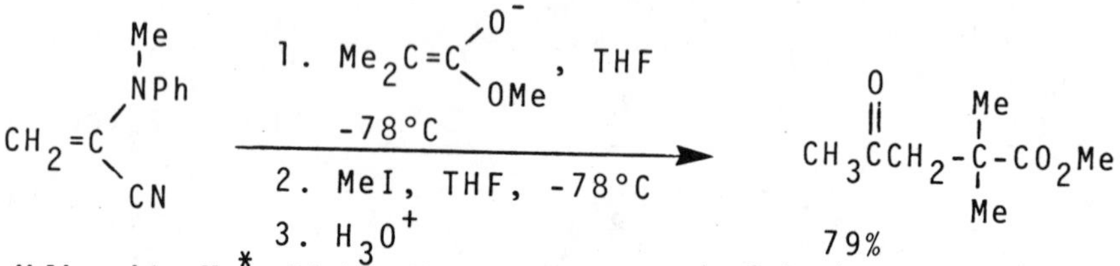

79%

Ahlbrecht, H.*; Dietz, M.   Synthesis, (1985), 417

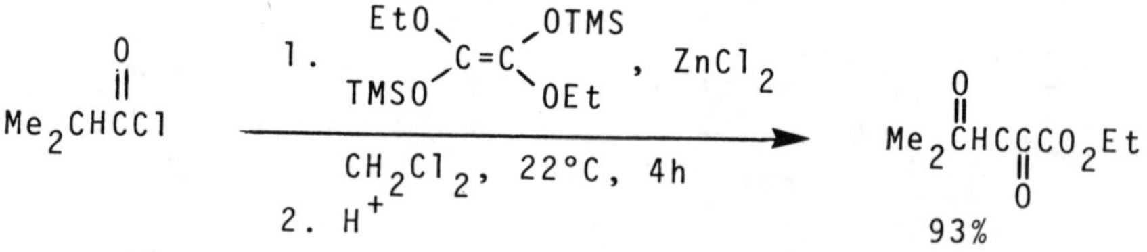

93%

Reetz, M.T.*; Kyung, S.-H.   Tetrahedron Lett, (1985), 26, 6333

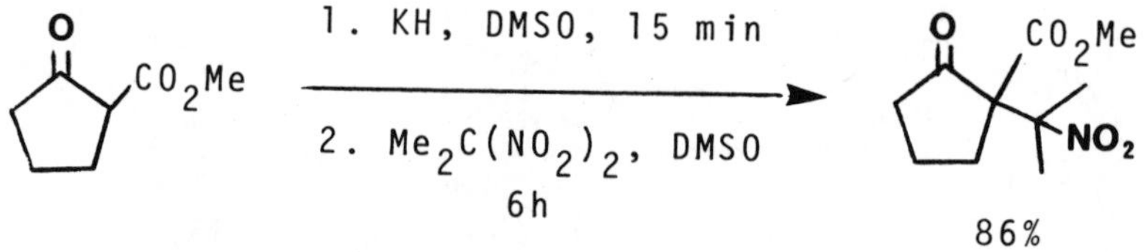

86%

Kornblum, N.*; Kelly, W.J.; Kestner, M.M.
        J Org Chem, (1985), 50, 4720

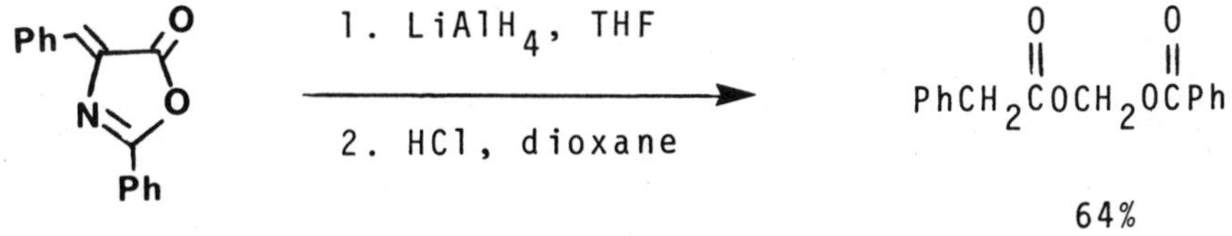
Ph
O
N
O
Ph
1. LiAlH4, THF
2. HCl, dioxane
O O
PhCH2COCH2OCPh
64%

Naim, S.S.; Hussain, M.; Khan, N.H.* **Synthesis**, (1985), 48

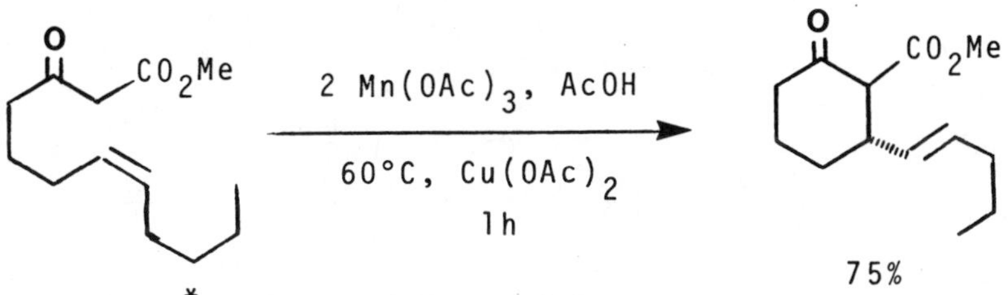
O
CO2Me
2 Mn(OAc)3, AcOH
60°C, Cu(OAc)2
1h
O
CO2Me
75%

Snider, B.B.*; Mohan, R.; Kates, S.A.
J Org Chem, (1985), **50**, 3659

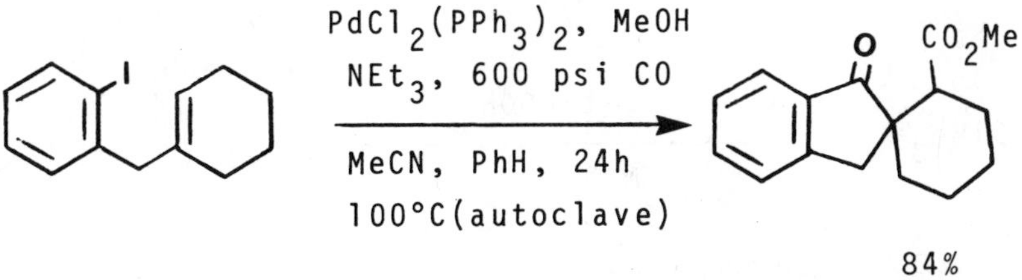
I
PdCl2(PPh3)2, MeOH
NEt3, 600 psi CO
MeCN, PhH, 24h
100°C(autoclave)
O
CO2Me
84%

Tour, J.M.; Negishi, E.* **J Am Chem Soc**, (1985), **107**, 8289

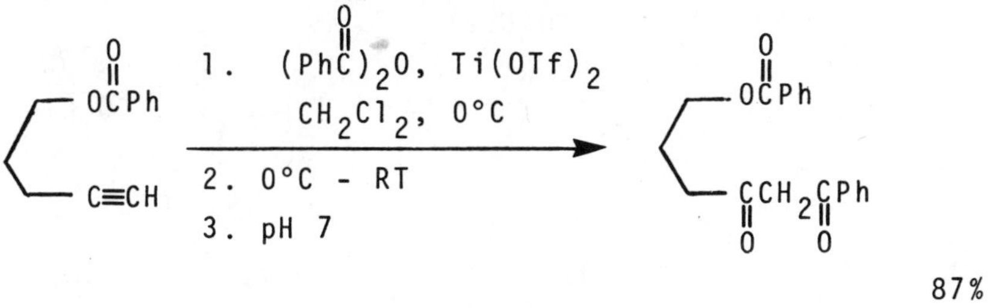
O
OCPh
C≡CH
1. (PhC)2O, Ti(OTf)2
CH2Cl2, 0°C
2. 0°C - RT
3. pH 7
O
OCPh
CCH2CPh
O O
87%

Tanabe, Y.; Mukaiyama, T. **Chem Lett**, (1985), 673

$ICH_2CH_2CO_2Et$

1. Zn/Cu, Me_2NAc, PhH
   RT - 60°C
2. $Pd(PPh_3)_4$, 5h
3. $CH_3CH_2\overset{\text{O}}{\underset{||}{C}}Cl$, 30 min

$Et\overset{O}{\overset{||}{C}}CH_2CH_2CO_2Et$

84%

Tamaru, Y.; Ochiai, H.; Nakamura, T.; Tsubaki, K.; Yoshida, Z.
Tetrahedron Lett, (1985), **26**, 5559

2 $PhCH_2COOH$

1. $(\underline{n}Bu_3Sn)_2O$
2. $BrCH_2\overset{C}{\underset{||}{O}}$⟨⟩-Br

$PhCO\overset{O}{\overset{||}{C}}CH_2\overset{O}{\overset{||}{C}}$⟨⟩-Br

60%

Vijayaraghavan, S.T.; Balasubramanian, T.R.
J Organomet Chem, (1985), **282**, 17

$PhC\equiv CSiMe_3$

$\dfrac{OsO_4, \underline{t}BuOOH}{MeOH}$

$Ph\overset{O}{\overset{||}{C}}CO_2Me$

61%

Page, P.C.B.[*]; Rosenthal, S.
Tetrahedron Lett, (1986), **27**, 1947

$EtOCH_2CH_2\overset{O}{\overset{||}{C}}Cl$

1. 2 $Ph_3P=CHCO_2Et$
   PhH, 5°C-RT, 2h
2. Al(Hg), wet THF

$EtCH_2CH_2\overset{O}{\overset{||}{C}}CH_2CO_2Et$

87%

Sánchez, I.H.[*]; Larraza, M.I.; Breña, F.K.; Crúz, A.; Sotelo, O.; Flores, H.J.
Syn Commun, (1986), **16**, 299

$$PhC(=O)C(Et)(O)-CH(CH_2)_4CH_3 \xrightarrow[\text{2. aq. HCl}]{\begin{array}{c}\text{1. CsOAc/Pb(OAc)}_4 \\ \text{100°C, DMSO}\end{array}} EtCC(=O)CH\underline{n}C_5H_{11}$$

$$\text{OAc}$$

84%

Satoh, T.; Motohashi, S.; Yamakawa, K.[*]
Bull Chem Soc Jpn, (1986), **59**, 946

$$\text{(P)} \quad \text{CH}_2\text{NMe}_3^+ \quad ^-\text{O}_2\text{CPh} \xrightarrow[\text{EtOH, 2h}]{\text{PhCCH}_2\text{Br}} PhC(=O)OCH_2C(=O)Ph$$

99%

Thorat, M.; Mane, R.; Jagdale, M.; Salunkhe, M.
Org Prep Proc Int, (1986), **18**, 203

$$CH_3CH(CO_2Et)_2 \xrightarrow[\text{PdCl}_2\text{(dppf), 33h}]{\text{PhI, CO, 120°C, NEt}_3} PhC(=O)-C(Me)(CO_2Et)-CO_2Et$$

75%

Kobayashi, T.; Tanaka, M.[*] **Tetrahedron Lett**, (1986), **27**, 4745

Also via: Ketoacids (Section 320); Hydroxyketones (Section 330)

SECTION 361: Ester - Nitrile

1. LDA, THF
   0 - 20°C, 12h

2. AcOH, THF
   -70°C

51%

Milenkov, B.; Süsse, M.; Hesse, M.[*]
Helv Chim Acta, (1983), **68**, 2115

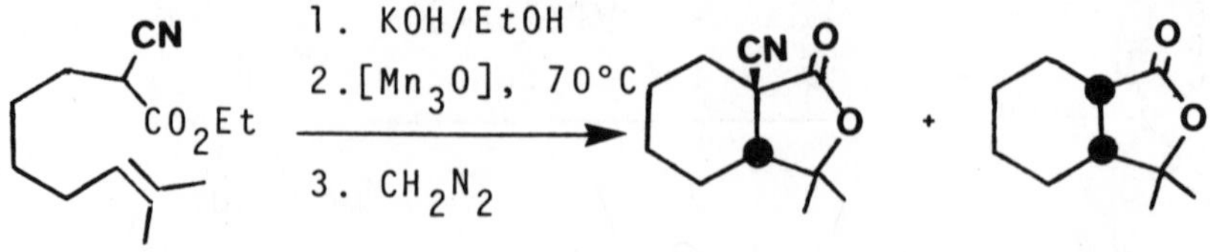

50%   8%

Ernst, A.B.; Fristad, W.E.[*]   **Tetrahedron** **Lett**, (1985), **26**, 3761

$$N{\equiv}CCH_2CO_2Et \xrightarrow[\substack{PdCl_2(PPh_3)_2, \ 70°C \\ monoglyme}]{PhBr, \ \underline{t}BuOK, \ 5h}$$

CN
|
PhCHCO$_2$Et

73%

Uno, M.; Seto, K.; Ueda, W.; Masuda, M.; Takahashi, S.
Synthesis, (1985), 506

SECTION 362: Ester - Olefin

This section contains syntheses of enol esters and esters of
unsaturated acids.

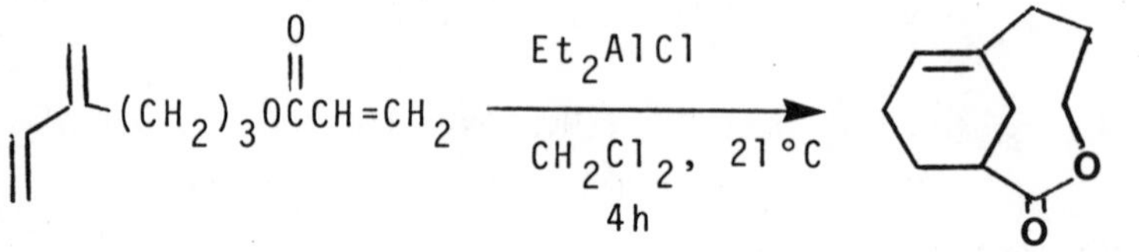

85%

Shea, K.J.; Gilman, J.W.   **Tetrahedron** **Lett**, (1983), **24**, 657

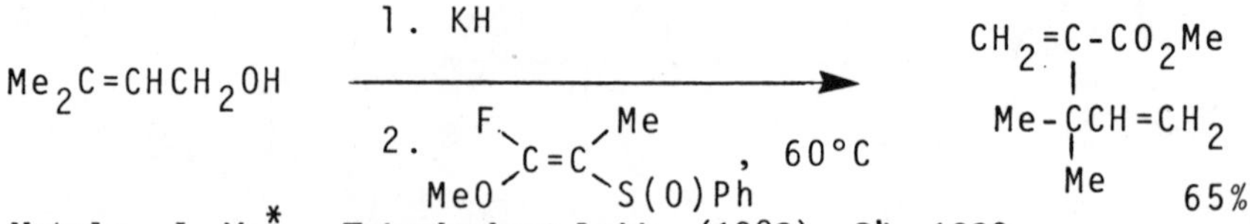

Me$_2$C=CHCH$_2$OH

1. KH

2. (structure: F, Me, MeO, S(O)Ph vinyl), 60°C

CH$_2$=C-CO$_2$Me
|
Me-CCH=CH$_2$
|
Me   65%

Vatele, J.-M.[*]   **Tetrahedron** **Lett**, (1983), **24**, 1239

$$\underset{-20°C, \ 1h}{\overset{\overset{+ \ -}{nBu_2Te\text{-}CHCO_2Et} \ , \ THF}{\xrightarrow{\hspace{3cm}}}}$$

PhCHO

PhCH=CHCO$_2$Et
75%

($\underline{E}$:$\underline{Z}$ = 50:1)

Osuka, A.[*]; Mori, Y.; Shimizu, H.; Suzuki, H.[*]
Tetrahedron **Lett**, (1983), **24**, 2599

$$\xrightarrow[\substack{cat. \ (C_8H_{17})_3NMeCl \\ 1h}]{Na_2S, \ PhH/HOH}$$

88%

Nakayama, J.[*]; Machida, H.; Hoshino, M.
Tetrahedron **Lett**, (1983), **24**, 3001

$$\xrightarrow[25°C, \ 30 \ min]{h\nu, \ PhH}$$

70%

Aryal-Kaloustian, S.; Agosta, W.C.
J **Org** **Chem**, (1983), **48**, 1718

$$\xrightarrow[\substack{18\text{-}crown\text{-}6 \\ (CF_3CH_2O)_2\underset{\underset{CO_2Me}{CH_2}}{P}=O \\ -78°C}]{KN(TMS)_2, \ THF}$$

+

(22 : 1)   81%

Still, W.C.[*]; Gennari, C. **Tetrahedron** **Lett**, (1983), **24**, 4405

56%

Ono, N.[*]; Tamura, R.[*] Eto, H.; Hamamoto, I.; Nakatsuka, T.;
Hayami, J.; Kaji, A.[*]

J Org Chem, (1983), **48**, 3678

84%
(88% ee R)

Oppolzer, W.[*]; Chapuis, C.; Kelly, M.J.
                    Helv Chim Acta, (1983), **66**, 2358
Oppolzer, W.[*]; Chapuis, C.   **Tetrahedron Lett**, (1984), **25**, 5383
Oppolzer, W.[*]; Chapuis, C.; Bernardinelli, G.
                    Tetrahedron Lett, (1984), **25**, 5885

$$CH_2$$
$$TMS-CH_2CCH_2OAc$$

$\underline{n}C_6H_{13}CH=CHCO_2Me$ dppe, THF

($\underline{E}$)

Pd(PPh$_3$)$_4$,
reflux

$\underline{n}C_6H_{13}$ CO$_2$Me

51%

Trost, B.M.[*]; Chan, D.M.T.
                    J Am Chem Soc, (1983), **105**, 2315, 2326
Trost, B.M.[*]; Balkovec, J.M.; Angle, S.R.
                    Tetrahedron Lett, (1986), **27**, 1445

1. $Me_2\overset{\text{Li}}{\underset{}{C}}CO_2Et$, 8 psi CO
   THF-HMPA, -78 - 25°C

2. $EtOSO_2F$, -78 - 25°C

96%

Semmelhack, M.F.[*]; Herndon, J.W.; Springer, J.P.
J Am Chem Soc, (1983), **105**, 2497

$R^*O\overset{\text{O}}{\underset{}{C}}CH=CH_2$

$TiCl_2(OiPr)_2$

-20°C, 4h

$-OCH_2\underline{t}Bu$

CO_2R^*

91%

Oppolzer, W.[*]; Chapuis, C.; Dupuis, D.; Guo, M.
Helv Chim Acta, (1983), **68**, 2100

1. $\underset{CH_2=CH}{\overset{MeO}{\diagdown}}C=Cr(CO)_5$
   PhH, 25°C, 1.5h

2. DMSO, 3.5h, 25°C

CO_2Me

71%

Wulff, W.D.[*]; Yang, D.C.   **J Am Chem Soc**, (1983), **105**, 6726

$PhCH=CHCH_2O\overset{\text{O}}{\underset{}{C}}OMe$

$NaCH(CO_2Me)_2$, THF
THF, 15% bpy

15% $(MeCN)_3W(CO)_3$
4h

$\underset{CO_2Me}{\overset{CH=CH_2}{PhCHCHCO_2Me}}$

91%

Trost, B.M.[*]; Hung, M.-H.   **J Am Chem Soc**, (1983), **105**, 7757

$$CH_3CH=CCNHMe$$

(structure: CH₃CH=C(CH₃)C(=O)NHMe)

1. secBuLi, THF
   TMEDA, -78°C

2. [cyclopentanone] =O, THF
   -78°C-RT

3. pH 3, reflux

[product: spirocyclic lactone with exocyclic methylene]

40%

Ladlow,.M.; Pattenden, G.*   **Syn Commun**, (1984), **14**, 11

[vinylcyclopentane structure]

$$EtO_2CCCO_2Et, 180°C$$

sealed tube, 72h

[product structure with CO₂Et, OH, CO₂Et groups]

90%

Saloman, M.F.; Pardo, S.N.; Saloman, R.G.*
J Org Chem, (1984), **49**, 2446

[cyclopentanone with SMe, SMe substituents]

1. MeLi, THF, 1.5h
   -78 - -10°C

2. 10% aq. HBF₄, RT
   THF/HOH, 24h

[product structure with CSMe group]

54%

Dieter, R.K.*; Lin, Y.J.; Dieter, J.W.
J Org Chem, (1984), **49**, 3183

[alkene boron dioxaborole structure]
$$nC_8H_{17}\text{-}C=C\text{-}H$$

1. Hg(OAc)₂, THF
   NaOAc, -78 - 0°C

2. PdCl₂, LiCl
   MgO, MeOH, CO
   -78°C - RT

$$nC_8H_{17}CH=CHCO_2Me$$

80%

(E:Z = 98:2)

Larock, R.C.*; Narayanan, K.   **J Org Chem**, (1984), **49**, 3411

Larock, R.C.[*]; Harrison, L.W.; Hsu, M.H.
J Org Chem, (1984), **49**, 3662

($\underline{dl}$:$\underline{meso}$ = 6:4)

Bäckvall, J.-E.[*]; Byström, S.E.; Nordberg, R.E.
J Org Chem, (1984), **49**, 4619

Pearson, A.J.[*]; Khan, M.N.I.   **J Am Chem Soc**, (1984), **106**, 1872

Knochel, P.; Normant, J.F.[*]   **Tetrahedron Lett**, (1984), **25**, 1475

$$\text{(cyclohexenyl-CH}_2\text{-CO-CH}_2\text{N}_2) \quad \xrightarrow[\text{BnOH, 4h}]{\text{Cu(OTf)}_2, \text{ reflux}} \quad \text{(methylenecyclohexane-CH}_2\text{CO}_2\text{Bn)}$$

63%

Smith III, A.B.[*]; Toder, B.H.; Branca, S.J.
J Am Chem Soc, (1984), 106, 3995

$$\text{(cyclohexadiene)} \quad \xrightarrow[\substack{5\% \text{ Pd(OAc)}_2, \text{ RT} \\ \text{AcOH}}]{\overset{O}{\overset{\|}{\text{LiOCCF}_3}}, \text{ CF}_3\text{COOH}} \quad \text{(AcO-cyclohexenyl-O-CO-CF}_3)$$

75%

Bäckvall, J.-E.[*]; Vågberg, J.; Nordberg, R.E.
Tetrahedron Lett, (1984), 25, 2717

$$\text{(propenyl lactone)} \quad \xrightarrow[\substack{\text{Me}_3\text{O}^+\text{BF}_4^- \\ 94\text{h}}]{\substack{\text{CH}_2\text{=CHCH}_2\text{SiMe}_3 \\ \text{CH}_2\text{Cl}_2, \text{ RT}}} \quad \text{CH}_2\text{=CHCH}_2\overset{\overset{\displaystyle \text{CH}_3}{|}}{\text{CH}}\text{CH=CHCH}_2\text{CH}_2\text{CO}_2\text{Me}$$

85%

Fujisawa, T.[*]; Kawashima, M.; Ando, S.
Tetrahedron Lett, (1984), 25, 3213

PhCH=CHCH$_2$OH $\xrightarrow[\substack{2. \text{ PdCl}_2(\text{PPh}_3)_2 \\ \text{xylene, reflux} \\ 4\text{h}}]{\substack{1. \text{ CH}_2\text{=C(OEt)}_2, \text{ PhH} \\ \text{cat. PdCl}_2(\text{COD}) \\ 0°\text{C - RT}}}$ PhCH=CHCH$_2$CH$_2$CO$_2$Et

(E)   (E)

86%

Ohshima, M.; Murakami, M.; Mukaiyama, T.
Chem Lett, (1984), 1535

PhCH=CHCH$_2$O$\overset{\overset{\displaystyle O}{\|}}{C}$OMe

dppe

Pd$_2$(dba)$_3$·CHCl$_3$

dioxane, reflux   91%

Tsuji, J.*; Takahashi, K.; Minami, I.; Shimizu, I.
                    Tetrahedron Lett, (1984), **25**, 4783

1. 37% HCHO, PhSH, RT

   cat. TMG, 24h

2. Ac$_2$O, Py

3. $\underline{n}$Bu$_3$SnH, AIBN

   80°C, PhH

TMG = tetramethylguanidine   80%

Ono, N.*; Kamimura, A.; Kaji, A.
                    Tetrahedron Lett, (1984), **25**, 5319

$CH_3CH_2\overset{\overset{\displaystyle O}{\|}}{C}Cl$

Ph$_3$P=CHCO$_2$Et, NEt$_3$

CH$_2$Cl$_2$, RT, 1h

MeCH=C=C$\overset{\text{H}}{\underset{\text{CO}_2\text{Et}}{}}$

                                        64%

Lang, R.W.; Hansen, H.-J.   **Org Syn**, (1984), **62**, 202

PhCH=C$\overset{\text{CH}_2\text{Br}}{\underset{\text{CO}_2\text{Me}}{}}$

(**E**)

1. LiAlH$_4$, CrCl$_3$

   THF, CH$_3$CHO

   25°C, overnight

2. HOH

                                        83%

Drewes, S.E.*; Hoole, R.F.A.   **Syn Commun**, (1985), **15**, 1067
Okuda, Y.; Nakatsukasa, S.; Oshima, K.; Nozaki, H.
                    Chem Lett, (1985), 481

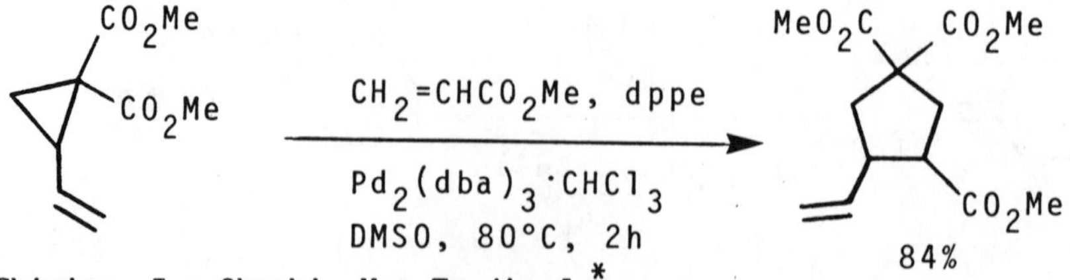
CO2Me
CO2Me
CH2=CHCO2Me, dppe
Pd2(dba)3·CHCl3
DMSO, 80°C, 2h
MeO2C
CO2Me
CO2Me
84%

Shimizu, I.; Ohashi, Y.; Tsuji, J.*
<u>Tetrahedron Lett</u>, (1985), <u>26</u>, 3825

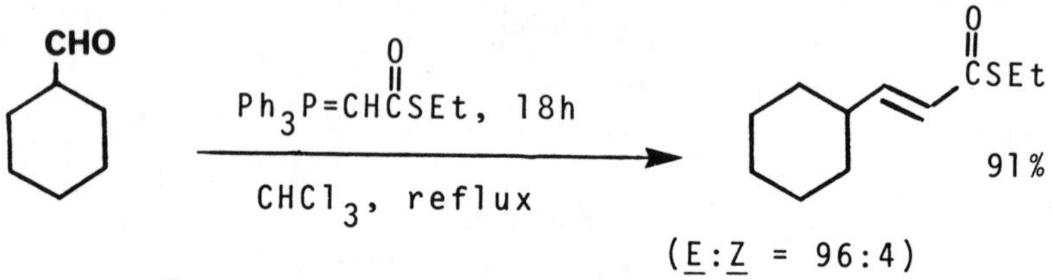
CHO
Ph3P=CHCSEt, 18h
CHCl3, reflux
CSEt
91%
(E:Z = 96:4)

Keck, G.E.*; Boden, E.P.; Mabury, S.A.
<u>J Org Chem</u>, (1985), <u>50</u>, 709

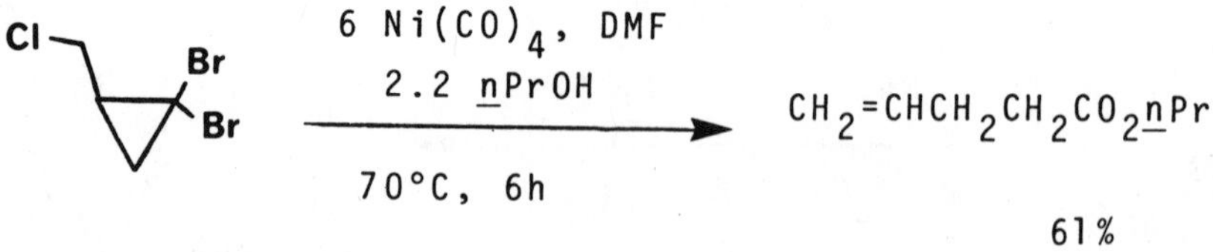
Cl
Br
Br
6 Ni(CO)4, DMF
2.2 nPrOH
70°C, 6h
CH2=CHCH2CH2CO2nPr
61%

Hirao, T.*; Nagata, S.; Agawa, T.
<u>Tetrahedron Lett</u>, (1985), <u>26</u>, 5795

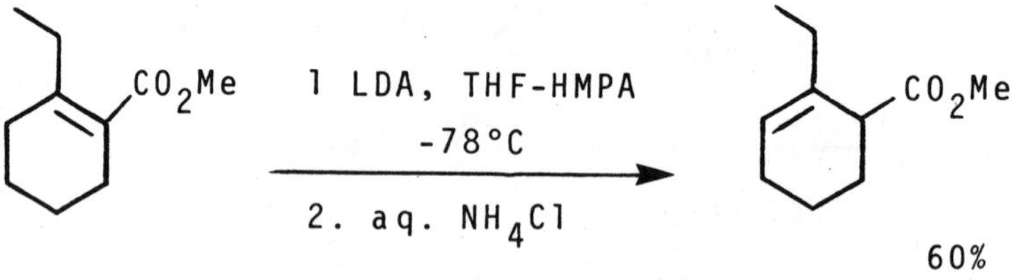
CO2Me
1 LDA, THF-HMPA
-78°C
2. aq. NH4Cl
CO2Me
60%

Harris, F.L.; Weiler, L.* <u>JCS Chem Comm</u>, (1985), 1124
<u>Tetrahedron Lett</u>, (1984), <u>25</u>, 1333

$$Ph_3\overset{+}{A}s-\overset{-}{C}HCH=CHCO_2Et$$

$$Et_2O, \ 200°C, \ 4h$$

85%

Huang, Y.[*]; Shen, Y.; Zheng, J.; Zhang, S.
Synthesis, (1985), 57

$$CH_2=CHCO_2Me, \ DMF$$

$$2\% \ Pd(PPh_3)_2Cl_2$$

$$3 \ NEt_3, \ 75°C$$

91%

Scott, W.J.; Peña, M.R.; Swärd, K.; Stoessel, S.J.; Stille, J.K.[*]

J Org Chem, (1985), **50**, 2302

$$\underline{n}C_8H_{17}CH=CH_2$$

1. AcOH, 120°C
   $ClCH_2COOH$
   $Mn(OAc)_3 \cdot HOH$

2. NaI, acetone
3. NEt_3, THF

$\underline{n}C_8H_{17}$ $\underline{n}C_8H_{17}$

(1   :   2)

72%

Fristad, W.E.[*]; Peterson, J.R.; Ernst, A.B.
J Org Chem, (1985), **50**, 3143

$$CH_2=CHPPh_3Br^-$$

$$KF, \ MeCN$$

$$18\text{-crown-}6$$
$$reflux$$

80%

Marino, J.P.[*]; Laborde, E.   **J Am Chem Soc**, (1985), **107**, 734

MeCH=C(OTMS)(OEt)
$(\underline{E}:\underline{Z} = 85:15)$

1. $CHCl_2Ph$, MeLi

2. $PhCH_3$, heat

MeCH=C(Ph)(CO_2Et)
$(\underline{E}:\underline{Z} = 79:21)$ 79%

Slougui, N.; Rousseau, G.   **Tetrahedron**, (1985), **41**, 2643, 2653

$$\left[\begin{array}{l} Pd(OAc)_2, \; P(OiPr)_3 \\ \underline{n}BuLi, \; THF, \; RT \end{array} \right]$$

40 psi CO_2, THF
4 h

95%

Trost, B.M.[*]; Angle, S.R.   **J Am Chem Soc**, (1985), **107**, 6123

OH
$(CH_2)_7$
CHO

1. $(EtO)_2\overset{O}{\overset{\|}{P}}CH_2CO_2Et$
   K_2CO_3, HOH 20°C

2. KOH, aq. EtOH, 30°C

3. H_3O^+

OH
$(CH_2)_5$
$CH=CHCO_2H$
77%

Villieras, J.[*]; Rambaud, M.; Graff, M.
                    Tetrahedron Lett, (1985), **26**, 53

$EtO-C\equiv CCH_2SiMe_3$

MeCHO, $TiCl_4$

-60 - 0°C

$CH_3CH=C(CH_2SiMe_3)(CO_2Et)$

70%

Pornet, J.; Khouz, B.; Miginiac, L.
                    Tetrahedron Lett, (1985), **26**, 1861

(5   :   1)

90%

Bache, M.D.[*]; Bosch, E.   **Tetrahedron Lett**, (1986), **27**, 641

$nC_6H_{13}-\overset{\overset{OCO_2Me}{|}}{\underset{\underset{CH_3}{|}}{C}}-C\equiv CH$ $\xrightarrow[\substack{5 \text{ atm CO, 15h} \\ 40°C}]{\substack{Pd(dba)_3 \cdot CHCl_3 \\ PPh_3, \text{ MeOH}}}$ $\underset{nC_6H_{13}}{\overset{CH_3}{>}}C=C=CHCO_2Me$ 99%

Tsuji, J.[*]; Suguira, T.; Minami, I.
                    Tetrahedron Lett, (1986), **27**, 731

$\xrightarrow[\text{xylene, reflux}]{DBU, MgSO_4}$

52%

Carling, R.W.; Holmes, A.B.[*]   **JCS Chem Comm**, (1986), 325

PhCHO $\xrightarrow[\text{2. aq. } H_2SO_4]{\substack{1. \ BrCH_2CO_2Me, \ Na_2Te \\ DMF, \ THF, \ -20°C}}$ $PhCH=CHCO_2Me$

73%

Suzuki, H.[*]; Inouye, M.   **Chem Lett**, (1986), 403

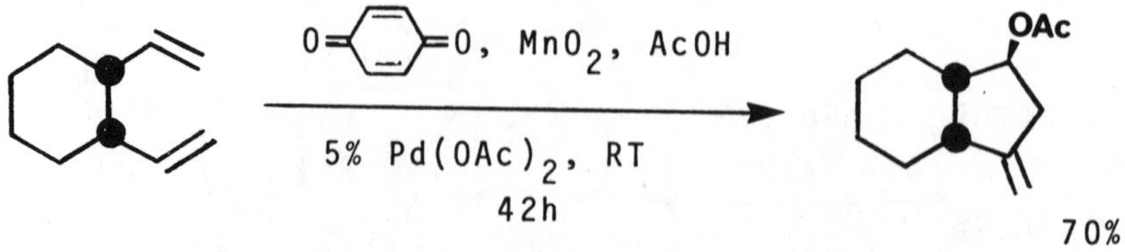
O=⟨⟩=O, MnO2, AcOH
5% Pd(OAc)2, RT
42h
OAc
70%

Antonsson, T.; Heumann, A.; Moberg, C.
JCS Chem Comm, (1986), 518

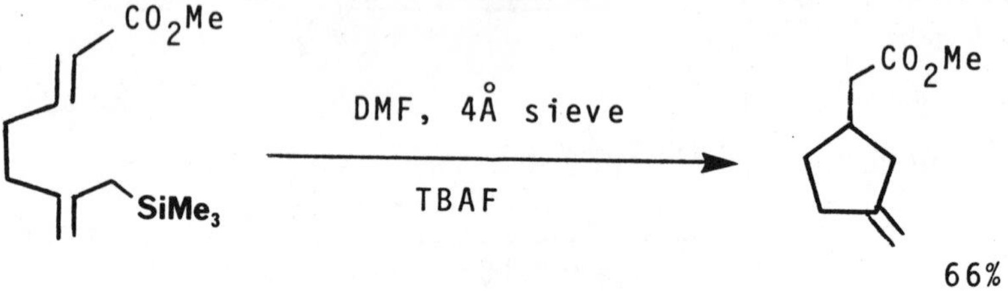
CO2Me
SiMe3
DMF, 4Å sieve
TBAF
CO2Me
66%

Majetich, G.*; Desmond Jr., R.W.; Soria, J.J.
J Org Chem, (1986), **51**, 1753

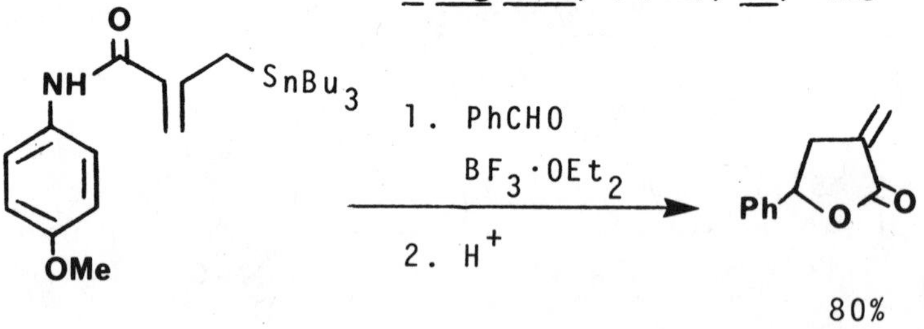
O
NH
SnBu3
OMe
1. PhCHO
BF3·OEt2
2. H+
Ph
O
O
80%

Tanaka, K.*; Yoda, H.; Isobe, Y.; Kaji, A.
J Org Chem, (1986), **51**, 1856

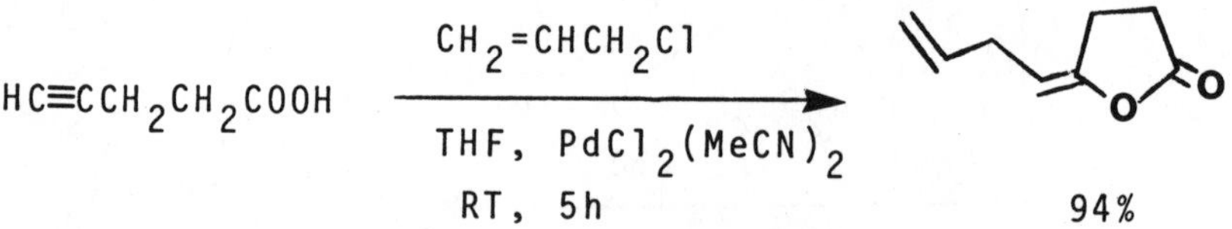
HC≡CCH2CH2COOH
CH2=CHCH2Cl
THF, PdCl2(MeCN)2
RT, 5h
O
O
94%

Yanagihara, N.; Lambert, C.; Iritani, K.; Utimoto, K.*; Nozaki, H.
J Am Chem Soc, (1986), **108**, 2753

$HC\equiv C\underline{n}Bu$ $\xrightarrow[\substack{1\% \ Ru(\eta^5COD)_2}]{\substack{EtCOOH, \ PBu_3 \\ PhCH_3, \ 80°C, \ 4h}}$

72% + 2%

Mitsudo, T.[*]; Hori, Y.; Yamakawa, Y.; Watanabe, Y.[*]
Tetrahedron Lett, (1986), **27**, 2125

$\underline{n}PrCHO$ $\xrightarrow[\substack{100°C, \ 2.5h}]{\substack{BrCH_2CO_2Et, \ \underline{n}Bu_3Sb}}$ $\underline{n}PrCH=CHCO_2Et$

92%

Huang, Y.; Shen, Y.; Chen, C.
Tetrahedron Lett, (1986), **27**, 2903

$BnOCC=CHiPr$
          $|$
         CH_3 (with O above first C)

$\xrightarrow[\substack{(-)-ephedrine \\ -78°C \quad (28\% \ ee \ R)}]{\substack{h\nu, \ CH_2Cl_2}}$

$BnOCCHCH=CMe_2$
         $|$
        CH_3 (with O above first C)

71%

Piva, O.; Henin, F.; Muzart, J.; Pete, J.-P.[*]
Tetrahedron Lett, (1986), **27**, 3001

$\xrightarrow[\substack{10 \ min}]{\substack{PdCl_2(PhCN)_2}}$

95%

Auburn, P.R.; Whelan, J.; Bosnich, B.[*]
Organometallics, (1986), **5**, 1533

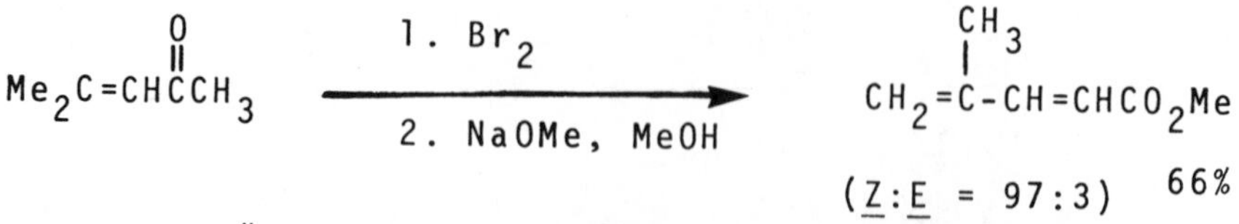
Me2C=CHCCH3
O
1. Br2
2. NaOMe, MeOH
CH3
CH2=C-CH=CHCO2Me
(Z:E = 97:3)
66%

Engler, T.A.*; Falter, W. **Tetrahedron Lett**, (1986), 27, 4115

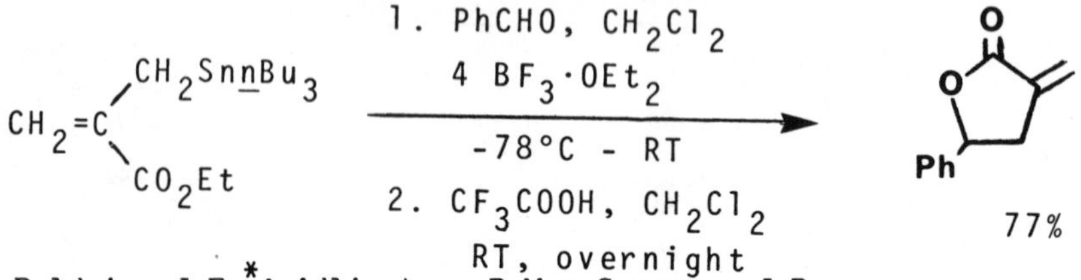
CH2SnnBu3
CH2=C
CO2Et
1. PhCHO, CH2Cl2
4 BF3·OEt2
-78°C - RT
2. CF3COOH, CH2Cl2
RT, overnight
O
O
Ph
77%

Baldwin, J.E.*; Adlington, R.M.; Sweeney, J.B.
Tetrahedron Lett, (1986), 27, 5423

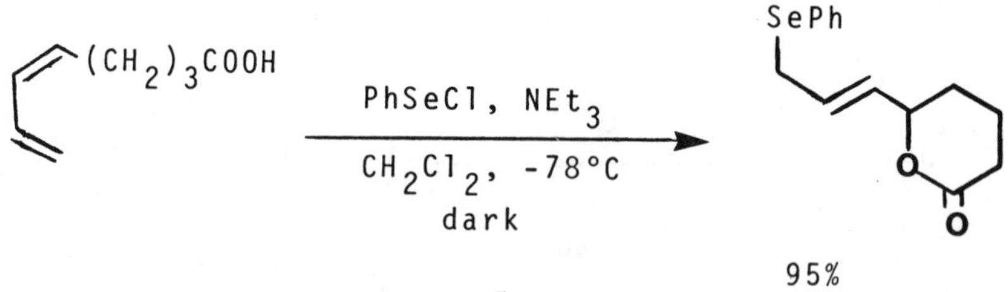
(CH2)3COOH
PhSeCl, NEt3
CH2Cl2, -78°C
dark
SePh
O
O
95%

Huckstep, M.R.; Taylor, R.J.K.*; Caton, M.P.L.
Tetrahedron Lett, (1986), 27, 5919

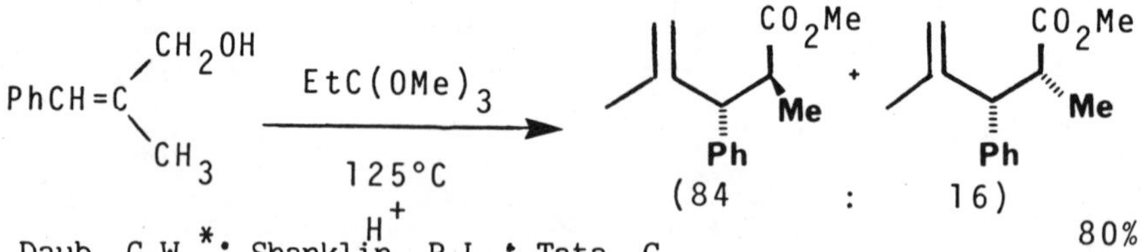
CH2OH
PhCH=C
CH3
EtC(OMe)3
125°C
H+
CO2Me
Me
Ph
CO2Me
Me
Ph
(84 : 16)
80%

Daub, G.W.*; Shanklin, P.L.; Tata, C.
J Org Chem, (1986), 51, 3402

$$\text{HO} \overset{O}{\underset{}{\diagdown}} \text{O} \quad \overset{Me_3Si}{\diagup}$$

1. LiN(TMS)$_2$
2. Me$_3$SiCl

3. CH$_2$N$_2$
4. KH

$$\longrightarrow \quad CH_3CH=CHCH=CHCO_2Me$$

83%

Sato, T.; Tsunekawa, H.; Kohama, H.; Fujisawa, T.[*]
Chem Lett, (1986), 1553

Review: "Synthesis of α-Methylene-γ-Lactones"

Sarma, J.C.; Sharma, R.P.   **Heterocycles**, (1986), **24**, 441

Related Methods: Protection of Aldehydes (Section 60A);
Protection of Ketones (Section 180A)

Also via: Acetylenic esters (Section 306); Olefinic acids
(Section 322; β-hydroxyesters (Section 327)

SECTION 363: Ether, Epoxide, Thioether - Ether, Epoxide, Thioether

See Section 60A (Protection of Aldehydes) ad Section 180A
(Protection of Ketones) for reactions involving formation of
Acetals and Ketals.

tBuOOSiMe$_3$

VO(acac)$_2$
PO(OSiMe$_3$)$_3$

(46 : 54)

78%

Hujama, T.[*]; Obayashi, M.   **Tetrahedron Lett**, (1983), **24**, 395

Mouzin, G.; Cousse, H.; Rieu, J.-P.; Duflos, A.
                    Synthesis, (1983), 117

Ley, S.V.*; Lygo, B.   **Tetrahedron Lett**, (1984), **25**, 113

Murai, T.; Kato, S.; Murai, S.*; Toki, T.; Suzuki, S.; Sonoda, N.
                    J Am Chem Soc, (1984), **106**, 6093

Amouroux, R.*   **Heterocycles**, (1984), **22**, 1489

SECTION 364: Ether, Epoxide, Thioether - Halide, Sulfonate

1. NaBr, 18-crown-6
   CH_2Cl_2, RT, 5 min

2. mcpba, 0°C
   5 min   80%

Srebnik, M.; Mechoulam, R.[*]   **JCS Chem Comm**, (1984), 1070

I_2, NaHCO$_3$

CH_3CN, 0°C

80%

Williams, D.R.[*]; Grote, J.; Harigaya, Y.
                    Tetrahedron Lett, (1984), **25**, 5231

1. TiCl$_4$, -45°C
   CH_2Cl_2

2. MeOH, 3N HCl

95%
(97% ee)

Melany, M.L.; Lock, G.A.; Thompson, D.W.
                    J Org Chem, (1985), **50**, 3925
Winstead, R.C.; Simpson, T.H.; Lock, G.A.; Schiavelli, M.D.;
Thompson, D.W.[*]
                    J Org Chem, (1986), **51**, 275
Bunnelle, W.H.[*]; Seamon, D.W.; Mohler, D.L.; Ball, T.F.;
Thompson, D.W.[*]

                    Tetrahedron Lett, (1984), **25**, 2653

CH$_3$OCH$_2$COOH

1. SOCl$_2$, 120°C

2. AlCl$_3$, 80°C

CH$_3$OCH$_2$Cl

78%

Stadlwieser, J.[*]   **Synthesis**, (1985), 490

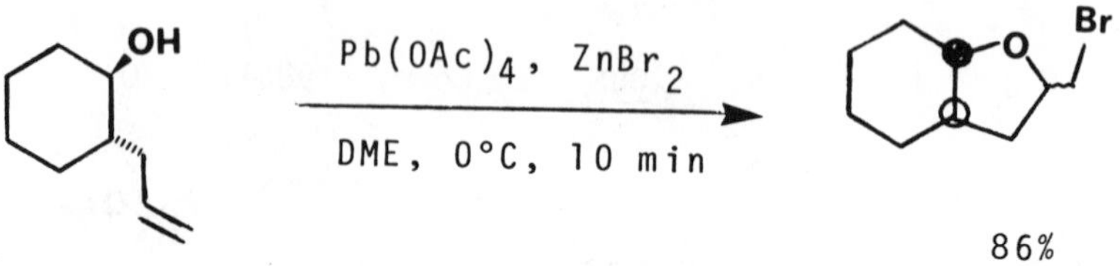

Pb(OAc)$_4$, ZnBr$_2$

DME, 0°C, 10 min

86%

Motohashi, S.[*]; Satomi, M.; Fujimoto, Y.[*]; Tatsuno, T.
Heterocycles, (1985), **23**, 2035

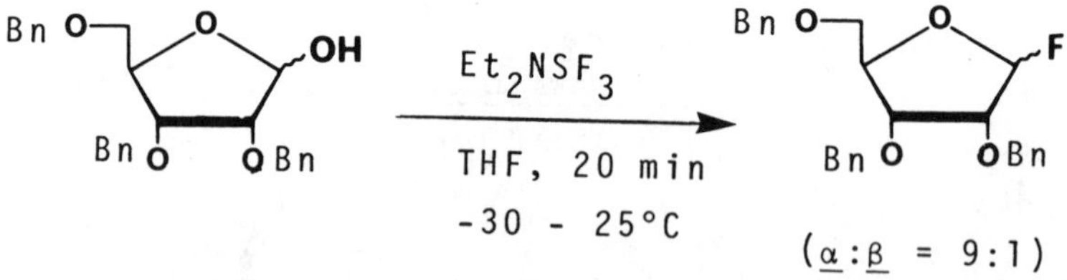

Et$_2$NSF$_3$

THF, 20 min

-30 - 25°C

(α:β = 9:1)

90%

Posner, G.H.[*]; Haines, S.R.[*]   **Tetrahedron Lett**, (1985), **26**, 5

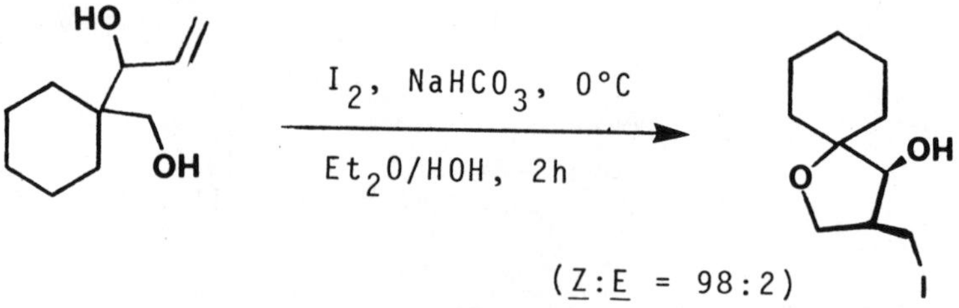

I$_2$, NaHCO$_3$, 0°C

Et$_2$O/HOH, 2h

98%

(Z:E = 98:2)

Tamaru, Y.; Kawamura, S.; Yoshida, Z.[*]
Tetrahedron Lett, (1985), **26**, 2885

$$\text{AgF, CH}_3\text{CN}$$
$$\xrightarrow{}$$
$$\text{PhSCl, RT}$$

57%

Purrington, S.T.[*]; Correa, I.D.   **J Org Chem**, (1986), **51**, 1080

Review: "Organic Synthesis with α-Chlorosulfides"

Dilworth, B.M.; McKervey, M.A.   **Tetrahedron**, (1986), **42**, 3731

SECTION 365: Ether, Epoxide, Thioether - Ketone

$$\xrightarrow[\text{PhH}]{h\nu}$$

69%

Carless, H.A.J.[*]; Fekarurhobo, G.K.
                    Tetrahedron Lett, (1983), **24**, 107

$$\xrightarrow[\text{2. H}_3\text{O}^+]{\text{1. Ni(R)}}$$

42%

Curran, D.P.[*]; Singleton, D.H.
                    Tetrahedron Lett, (1983), **24**, 2079

$$\xrightarrow[\substack{\text{2. TolSO}_2\text{-SnPr} \\ -80°\text{C, 16h}}]{\text{1. LDA, THF, } -80°\text{C}}$$

83%

Scholz, D.[*]   **Liebigs Ann Chem**, (1983), 259, 264
                    Monatsh Chem, (1983), **114**, 655

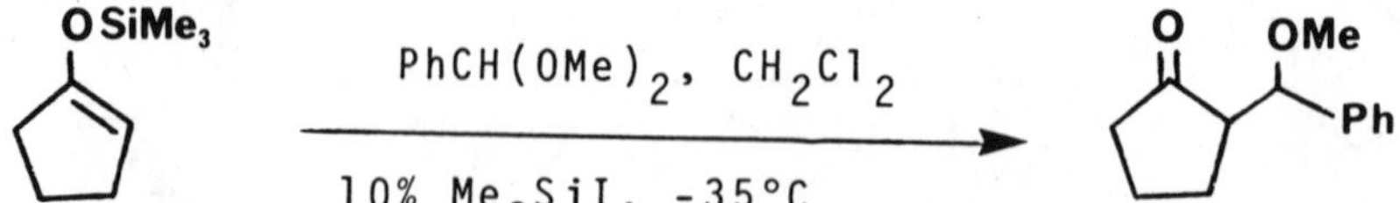

$$PhCH(OMe)_2, CH_2Cl_2$$

$$10\% \ Me_3SiI, \ -35°C$$

87%

(erythro:threo = 99:1)

Sakurai, H.[*]; Sasaki, K.; Hosomi, A.
Bull Chem Soc Jpn, (1983), 56, 3195

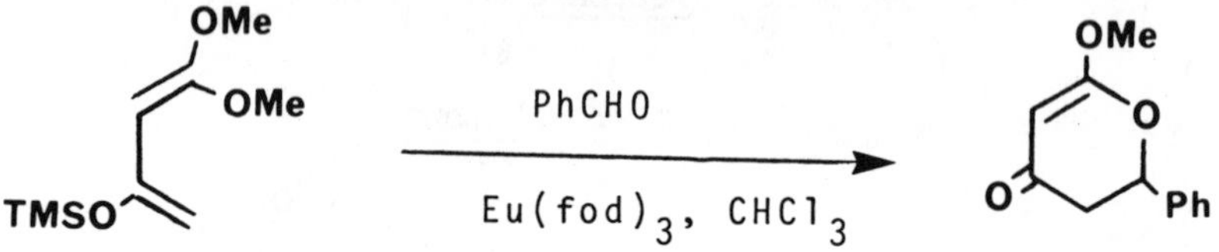

PhCHO

$$Eu(fod)_3, \ CHCl_3$$

85%

Danishefsky, S.[*]; Harvey, D.F.; Quallich, G.; Uang, B.J.
J Org Chem, (1984), 49, 392

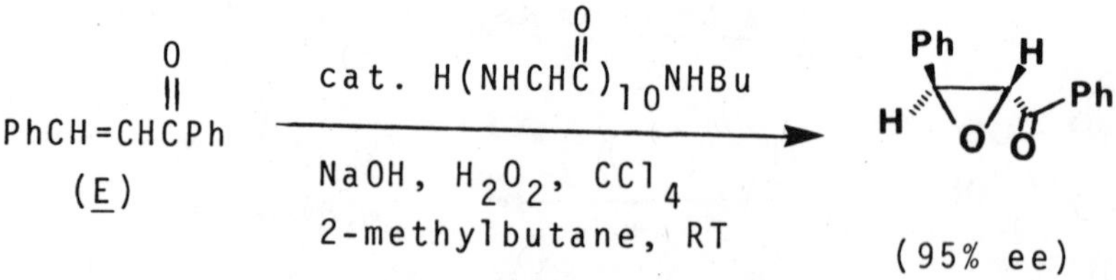

$$PhCH=CHCPh$$
(E)

cat. $H(NHCHC)_{10}NHBu$

$$NaOH, H_2O_2, CCl_4$$
2-methylbutane, RT

76%

(95% ee)

Banfi, S.; Colonna, S.[*]; Molinari, H.; Julia, S.[*]; Guixer, J.
Tetrahedron, (1984), 40, 5207

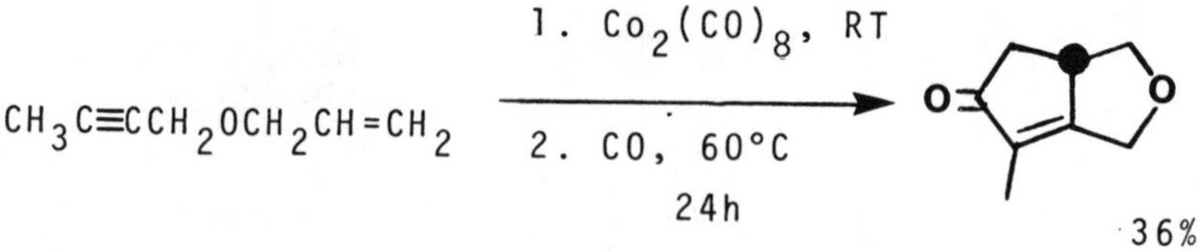

$$CH_3C≡CCH_2OCH_2CH=CH_2$$

1. $Co_2(CO)_8$, RT

2. CO, 60°C
24h

36%

Billington, D.C.[*]; Willison, D.
Tetrahedron Lett, (1984), 25, 4041

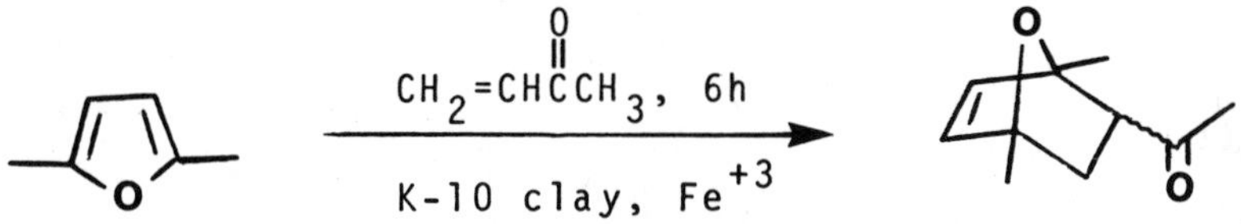

$$CH_2=CHCCH_3, \text{ 6h}$$

K-10 clay, Fe^{+3}

65%

(<u>exo</u>:<u>endo</u> = 0.6:1)

Laszlo, P.[*]; Lucchetti, J. **Tetrahedron Lett**, (1984), **25**, 4387

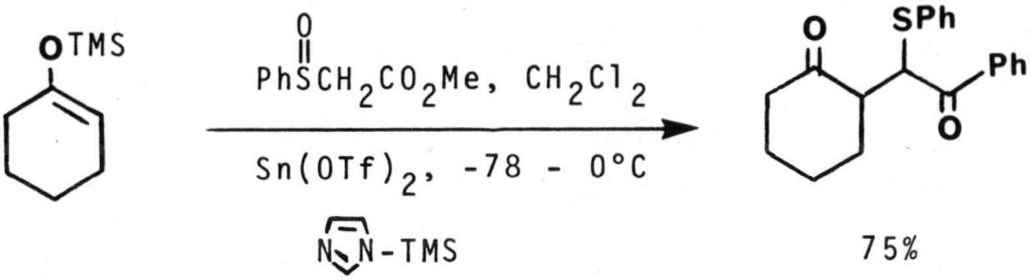

$$PhSCH_2CO_2Me, CH_2Cl_2$$

$Sn(OTf)_2$, -78 - 0°C

N N-TMS

75%

Shimizu, M.; Akiyama, T.; Mukaiyama, T.
                    Chem Lett, (1984), 1531

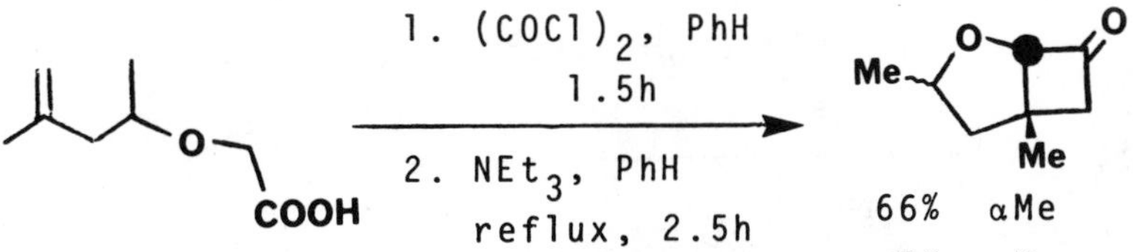

1. $(COCl)_2$, PhH
   1.5h

2. NEt_3, PhH
   reflux, 2.5h

66% αMe

7%   βMe

Snider, B.B.[*]; Hui, R.A.H.F.   **J Org Chem**, (1985), **50**, 5167
Snider, B.B.[*]; Hui, R.A.H.F.; Kulkarni, Y.S.
                    J Am Chem Soc, (1985), **107**, 2194

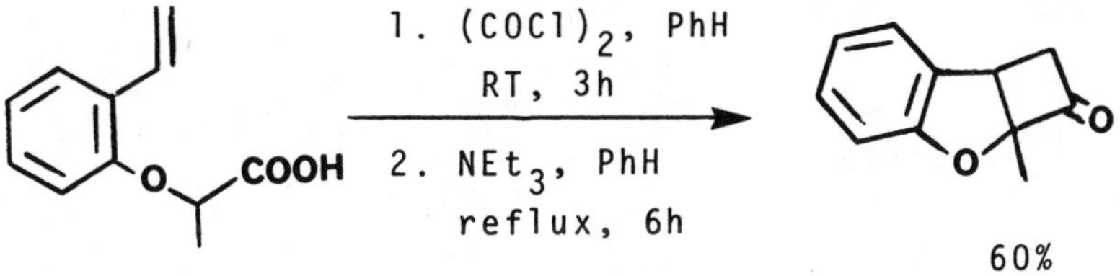

1. $(COCl)_2$, PhH
   RT, 3h

2. NEt_3, PhH
   reflux, 6h

60%

Brady, W.T.[*]; Giang, Y.F.   **J Org Chem**, (1985), **50**, 5177

$$CH_3\overset{O}{\overset{\|}{C}}CH_3 \xrightarrow[\underline{n}C_5H_{11}S^-Na^+]{CH_2O} \underline{n}C_5H_{11}SCH_2CH_2\overset{O}{\overset{\|}{C}}CH_3$$

61%

Sabirov, S.S.; Gnevasheva, L.M.; Ismailov, M.I.; Isobaev, M.D.
J Org Chem USSR, (1984), **20**, 1239

1. KH, THF, HMPA
   -11 - 25°C
2. gl. AcOH, HOH

Cohen, T.[*]; Yu, L.-C.; Daniewski, W.M.   69%
J Org Chem, (1985), **50**, 4596

1. SnCl$_2$·2H$_2$O, EtOH, RT
2. ice
3. pH 8
4. levulinic acid, 80°C
   0.1N HCl, 3.5h   79%

Varma, R.S.; Kabalka, G.W.[*]   **Syn Commun**, (1985), **15**, 443

PhCH(OMe)$_2$, CH$_2$Cl$_2$
-78°C   86%

(syn:anti = 88:12)

Mukaiyama, T.; Iwakiri, H.   **Chem Lett**, (1985), 1363

Naruse, Y.; Ukai, J.; Ikeda, N.; Yamamoto, H.[*]
        Chem Lett, (1985), 1451

Kobayashi, S.; Mukaiyama, T.   **Chem Lett**, (1986), 221

Takeda, T.[*]; Kaneko, Y.; Fujiwara, T.
        Tetrahedron Lett, (1986), **27**, 3029

Miyake, H.[*]; Yamamura, K.   **Bull Chem Soc Jpn**, (1986), **59**, 89

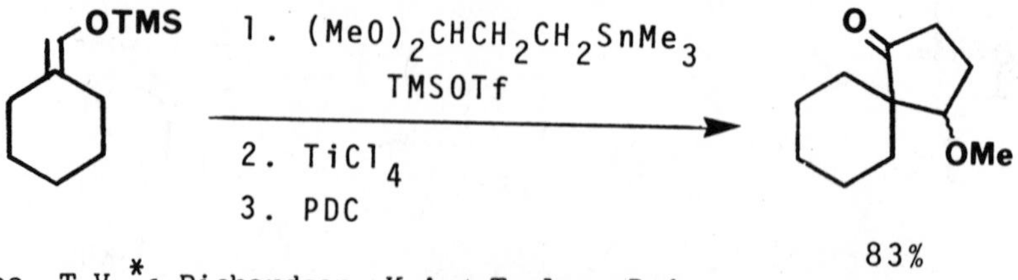

1. $(MeO)_2CHCH_2CH_2SnMe_3$
   TMSOTf

2. $TiCl_4$

3. PDC

83%

Lee, T.V.[*]; Richardson, K.A.; Taylor, D.A.
Tetrahedron Lett, (1986), **27**, 5021

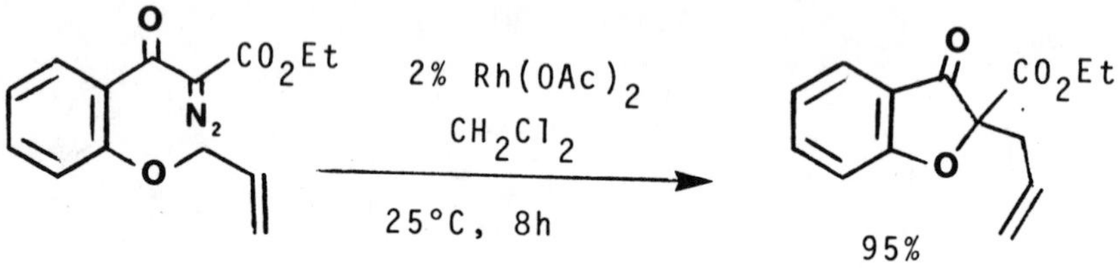

2% $Rh(OAc)_2$

CH_2Cl_2

25°C, 8h

95%

Pirrung, M.C.[*]; Werner, J.A.   **J Am Chem Soc**, (1986), **108**, 6060

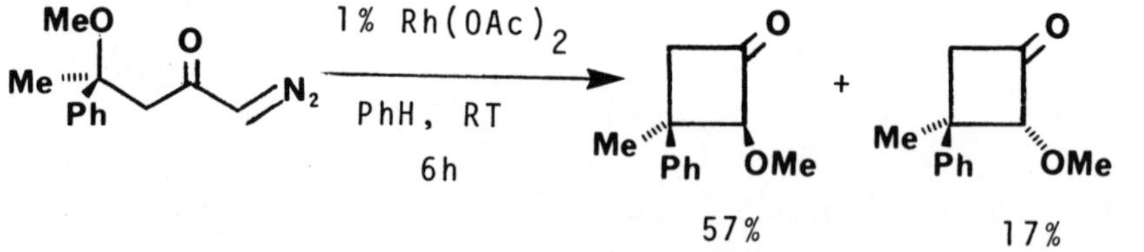

1% $Rh(OAc)_2$

PhH, RT

6h

57%   +   17%

Roskamp, E.J.; Johnson, C.R.[*]   **J Am Chem Soc**, (1986), **108**, 6062

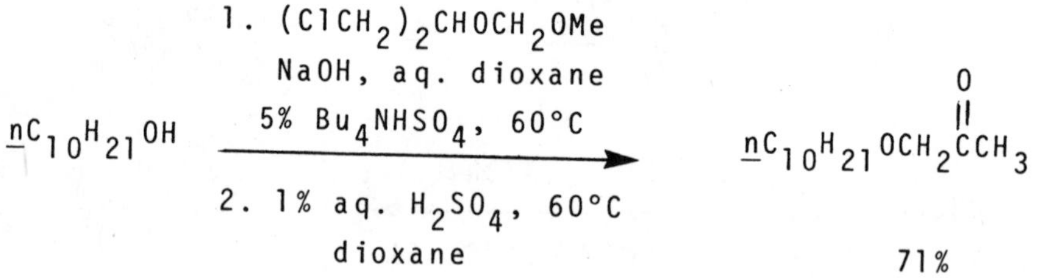

1. $(ClCH_2)_2CHOCH_2OMe$
   NaOH, aq. dioxane
   5% Bu_4NHSO_4, 60°C

$\underline{n}C_{10}H_{21}OH$ → $\underline{n}C_{10}H_{21}OCH_2\overset{O}{\overset{\|}{C}}CH_3$

2. 1% aq. H_2SO_4, 60°C
   dioxane

71%

Gu, X.-P.; Ikeda, I.; Komada, S.; Masuyama, A.; Okahara, M.[*]
J Org Chem, (1986), **51**, 5425

SECTION 366: Ether, Epoxide, Thioether - Nitrile

$nC_{11}H_{23}$ dioxane (Me, Me)
$\xrightarrow[\text{CH}_2\text{Cl}_2]{\text{Me}_3\text{SiCN} \quad \text{TiCl}_3}$

$\underset{\text{HO}}{\overset{nC_{11}H_{23}}{O\cdots C\text{-CN}}}$Me   + $\underset{\text{HO}}{\overset{nC_{11}H_{23}}{O\text{-}C\text{-CN}}}$Me

(95 : 5)   98%

Elliott, J.D.[*]; Choi, V.M.F.; Johnson, W.S.
J Org Chem, (1983), 48, 2294

$PhCH_2CN$
$\xrightarrow[\text{Bu}_4\text{NBr, 20°C, 1h}]{\text{iPrCHO, 50\% NaOH}}$
$\underset{CN}{\overset{O}{Ph\text{-}C\text{-}CHiPr}}$ 80%

Makosza, M.[*]; Kwast, A.; Kwast, E.; Jonczyk, A.
J Org Chem, (1985), 50, 3722

$CH_3CH(OMe)_2$
$\xrightarrow[\text{25°C, 15 min}]{\text{TMS-CN, ZnI}_2}$
$\underset{}{\overset{CN}{CH_3CHOMe}}$ 63%

Kirchmeyer, S.; Mertens, A.; Arvanaghi, M.; Olah, G.A.[*]
Synthesis, (1983), 498

$\underset{}{\overset{\text{MeO OMe}}{\bigcirc}}$
$\xrightarrow[\text{2. aq. NaHCO}_3]{\begin{array}{l}\text{1. } \underline{t}\text{BuN}\equiv\text{C, TiCl}_4 \\ \text{CH}_2\text{Cl}_2, -78 - 0°C\end{array}}$
$\underset{}{\overset{\text{MeO CN}}{\bigcirc}}$ 92%

Ito, Y.[*]; Imai, H.; Segoe, K.; Saegusa, T.[*]
Chem Lett, (1984), 937

Stork, G.[*]; Sher, P.M.   **J Am Chem Soc**, (1986), **108**, 303
                            J Am Chem Soc, (1983), **105**, 6765

SECTION 367: Ether, Epoxide, Thioether - Olefin

Enol ethers are found in this section.

van Schaik, T.A.M.; Henzen, A.V.; van der Gen, A.[*]
                    Tetrahedron Lett, (1983), **24**, 1303

Miller, R.D.[*]; McKean, D.R.   **Tetrahedron Lett**, (1983), **24**, 2619

Ousset, J.B.; Mioskowski, C.[*]; Solladie, G.
                    Tetrahedron Lett, (1983), **24**, 4419

Hartner Jr., F.W.; Schwartz, J.; Clift, S.M.
J Am Chem Soc, (1983), 105, 640

Stanton, S.A.; Felman, S.W.; Parkhurst, C.S.; Godleski, S.A.[*]
J Am Chem Soc, (1983), 105, 1964

R = OTMS, R' = H   82%

Bednarski, M.; Danishefsky, S.[*]
J Am Chem Soc, (1983), 105, 3716

R = H, R' = Me   10 KBar, 20h   81% (E:Z = 1:1)

Jurczak, J.; Golebiowski, A.; Bauer, T.   Synthesis, (1985), 928

Semmelhack, M.F.[*]; Tamura, R.   J Am Chem Soc, (1983), 105, 6750

CH$_2$=C=C(OMe)(CMe$_2$OAc)

$\xrightarrow[\text{2\% Pd(PPh}_3)_4]{\text{PhZnCl, THF}}$

CH$_2$=C(Ph)-C=CMe$_2$(OMe)

95%

Kleijn, H.; Westmijze, H.; Meijer, J.; Vermeer, P.
Rec Trav Chim Pays Bas, (1983), **102**, 378

Et,Me-C(O)-CHSiMe$_2$Ph

$\xrightarrow[\text{CH}_2\text{Cl}_2, \ 5 \ \text{min}]{\text{BF}_3 \cdot \text{OEt}_2, \ -78°C}$

Et,Me-C=CHOSiMe$_2$Ph

68%

($\underline{E}$:$\underline{Z}$ = 4:96)

Fleming, I.[*]; Newton, T.W.   **JCS Perkin I**, (1984), 119

$\xrightarrow[\substack{\text{TMS-Cl}, \ -78°C \\ \text{THF}}]{\substack{\text{tBu} \\ \text{LiN}-\underline{\text{toctyl}}}}$

OTMS   OTMS

+

(97.5 : 2.5)

Corey, E.J.[*]; Gross, A.W.   **Tetrahedron Lett**, (1984), **25**, 495

$\xrightarrow[\substack{\text{Co}_2(\text{CO})_8, \ 120°C \\ 50 \ \text{atm CO}}]{\text{HSiEt}_2\text{Me, PPh}_3}$

OSiMeEt$_2$

OSiMeEt$_2$

88%

Chatani, N.; Furukawa, H.; Kato, T.; Murai, S.[*]; Sonoda, N.
J Am Chem Soc, (1984), **106**, 430

1. Hg(OAc)$_2$, gl. AcOH
   RT, 30 min

2. NaBH$_4$, NaOH

63%

Larock, R.C.[*]; Harrison, L.W.   **J Am Chem Soc**, (1984), **106**, 4218

TMS-C≡CCH$_2$OH

1. iBuMgBr, Et$_2$O, 6h
   Cp$_2$TiCl$_2$, 0 - 25°C

2. $\underline{n}$C$_6$H$_{13}$CCH$_3$
         ‖
         O

3. BF$_3$·OEt$_2$, CH$_2$Cl$_2$

$\underline{n}$C$_6$H$_{13}$

TMS

90%

Sato, F.[*]; Kanbara, H.; Tanaka, Y.
              Tetrahedron Lett, (1984), **25**, 5063

O
‖
PhS$\underline{n}$Bu

1. 3 LDA, THF, -10°C

2. TMS-Cl, -10 - 25°C

3. aq. NaHCO$_3$

PhS
   \
    C=CHEt
   /
TMS

80%

($\underline{E}$:$\underline{Z}$ = 2:1)

Miller, R.O.[*]; Hässig, R.   **Tetrahedron Lett**, (1984), **25**, 5351

1. $\underline{n}$BuLi, THF
   HMPA

2. $\underline{n}$C$_9$H$_{19}$CHO

$\underline{n}$C$_9$H$_{19}$

80%

Ousset, J.B.; Mioskowski, C.[*]; Yang, Y.-L.; Falck, J.R.[*]
              Tetrahedron Lett, (1984), **25**, 5903

1. LiCHPh(OMe), THF, -78°C

2. CS_2, RT - reflux

3. MeI, reflux

4. $\underline{n}Bu_3SnH$, AIBN
   reflux, PhH

62%

($\underline{Z}$:$\underline{E}$ = 4:6)

Vatele, J.-M.[*]   **Tetrahedron Lett**, (1984), **25**, 5997

pTsOH·HOH

PhH, 18h

reflux

(Dean Stark)

67%

Nickon, A.[*]; Rodriguez, A.D.; Shirhatti, V.; Ganguly, R.
J Org Chem, (1985), **50**, 4218

e^-, 60°C, NaOH

Et_4NOTs, MeOH

cat.

87%

cat. = chloropyridine cobaloxime (III)

Torii, S.[*]; Inokuchi, T.; Yukawa, T.
J Org Chem, (1985), **50**, 5875

OAc
|
$CH_3CH=CHCH\underline{n}Pr$

$\underline{n}Bu_3SnOPh$, THF

$Pd(PPh_3)_4$, RT

82%

OPh
|
$CH_3CH=CHCH\underline{n}Pr$

(19)

+

PhO
|
$CH_3CHCH=CH\underline{n}Pr$ (81)

Keinan, E.[*]; Sahai, M.; Roth, Z.; Nudelman, A.[*]; Herzig, J.
J Org Chem, (1985), **50**, 3558

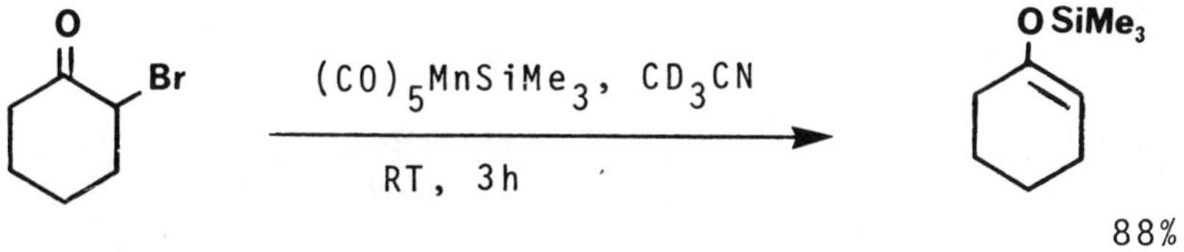

88%

Marsi, M.; Brinkman, K.C.; Lisensky, C.A.; Vaughn, G.D.;
Gladysz, J.A.

J Org Chem, (1985), 50, 3396

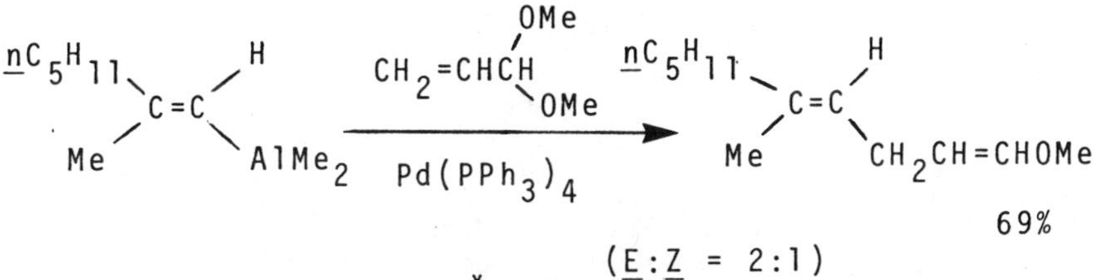

69%

$(\underline{E}:\underline{Z} = 2:1)$

Chatterjee, S.; Negishi, E.[*]   J Org Chem, (1985), 50, 3406

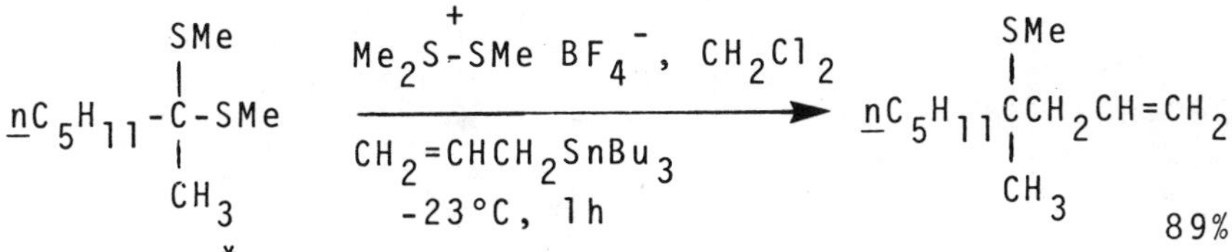

89%

Trost, B.M.[*]; Sato, T.   J Am Chem Soc, (1985), 107, 719

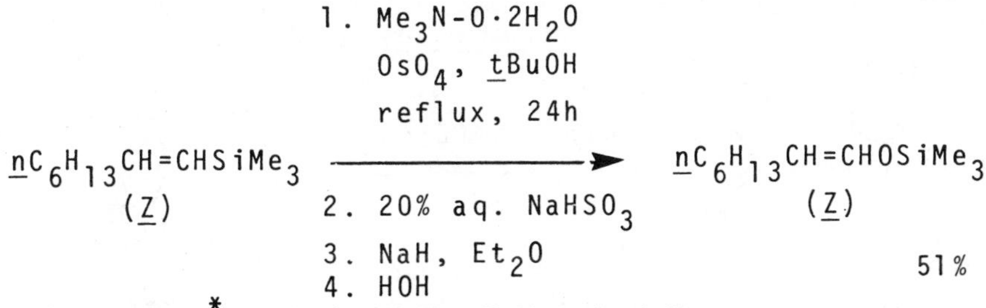

1. $Me_3N-O\cdot2H_2O$
   OsO_4, $\underline{t}BuOH$
   reflux, 24h

$\underline{n}C_6H_{13}CH=CHSiMe_3$ ⟶ $\underline{n}C_6H_{13}CH=CHOSiMe_3$
   $(\underline{Z})$ $(\underline{Z})$

2. 20% aq. $NaHSO_3$

3. NaH, Et_2O 51%
4. HOH

Hudrlik, P.F.[*]; Hudrlik, A.M.; Kulkarni, A.K.
                    J Am Chem Soc, (1985), 107, 4260

$$CH_2=CHCH(OEt)_2 \xrightarrow[\text{NiCl}_2(\text{dppp}), \ 3h]{\text{PhMgBr, Et}_2O, \ 0°C} EtOCH=CHCH_2Ph$$

60%

$$(\underline{E}:\underline{Z} = 97:3)$$

Sugimura, H.; Takei, H.*   **Chem Lett**, (1985), 351

$$BrCH_2CO_2Et \xrightarrow[\substack{\text{Pd(PPh}_3)_4, \ 80°C \\ \text{tetralin, 20h}}]{\underline{n}Bu_3SnSn\underline{n}Bu_3, \ TMS-Cl} CH_2=C\begin{smallmatrix} Et \\ \\ OSiMe_3 \end{smallmatrix}$$

69%

Kosugi, M.*; Koshiba, M.; Sano, H.; Migita, T.*
Bull Chem Soc Jpn, (1985), **58**, 1075

$$\underline{n}BuC\equiv CCH_2SiMe_3 \xrightarrow[\substack{50°C \\ 1\%}]{\underline{n}Bu_3SnCH_2CH=CMe_2, \ THF} CH_2=C=C\begin{smallmatrix} nBu \\ \\ CHOEt \\ iBu \end{smallmatrix}$$

76%

Pornet, J.; Miginiac, L.*; Jaworski, K.; Rondrianielina, B.
Organometallics, (1985), **4**, 333

$$Ph_3Sn(CO)_3Co=C\begin{smallmatrix} OMe \\ \\ Ph \end{smallmatrix} \xrightarrow[\substack{PhH, \ 50°C \\ 6h}]{2.5 \ EtC\equiv CEt} $$

93%

Wulff, W.D.*; Gilbertson, S.R.; Springer, J.P.
J Am Chem Soc, (1986), **108**, 520

Me◁⋯⟨cyclohexene ring⟩⋯OCSMe (C=S)
          → Pd(PPh$_3$)$_4$ / CHCl$_3$, 25°C, 1h → Me◁⋯⟨cyclohexene ring⟩⋯SH

95%

Auburn, P.R.; Whelan, J.; Bosnich, B.
       JCS Chem Comm, (1986), 146

Me$_3$Si\
      C=CHnBu
Br/

1. I(CH$_2$)$_3$OMEM
2. SnCl$_4$, CH$_2$Cl$_2$
   -15°C, 12h

→ ⟨tetrahydropyran ring with =n-Bu⟩

71%

Overman, L.E.[*]; Castañeda, A.; Blumenkopff, T.A.
       J Am Chem Soc, (1986), **108**, 1303

BnCN

1. nBuLi, THF, -85°C
2. [PhSMe/NCS/CCl$_4$], THF
3. nBuLi, -85 - -30°C
4. HOH

PhSCH=CHPh

52%

Harada, T.; Karasawa, A.; Oku, A.[*]   **J Org Chem**, (1986), **51**, 842

MeCH=CHCH$_2$SiMe$_3$
(**E**)

$\xrightarrow[\substack{\text{TiCl}_4 \\ \text{CH}_2\text{Cl}_2, \text{ 6h} \\ -78 - 0°C}]{\text{tBuCH(OMe)$_2$}}$

t-Bu⟨—CH(OMe)—CH(Me)—CH=CH$_2$⟩ + t-Bu⟨—CH(OMe)—CH(Me)—CH=CH$_2$⟩

(96 : 4)
63%

Hosomi, A.[*]; Ando, M.; Sakurai, H.[*]   **Chem Lett**, (1986), 365

$\underline{n}C_7H_{15}CH(OMe)_2$ $\xrightarrow[\begin{array}{c}CH_2Cl_2, \ 30 \ min\end{array}]{\begin{array}{c}CH_2=CHCH_2SiMe_3 \\ \hline K10 \ clay, \ 0°C\end{array}}$ $\underset{90\%}{\underline{n}C_7H_{15}\overset{\overset{\displaystyle OMe}{|}}{C}HCH_2CH=CH_2}$

Kawai, M.; Onka, M.*; Izumi, Y.   **Chem Lett**, (1986), **381**

$\xrightarrow[\begin{array}{c}PhH, \ 25°C, \ 36h\end{array}]{\underline{t}BuCH_2O)_2I_2W{\Longleftarrow}}$

85%

Aguero, A.; Kress, J.; Osborn, J.A.   **JCS Chem Comm**, (1986), 531

$\underset{MeSCH_2CPh}{\overset{\overset{\displaystyle O}{\|}}{}}$ $\xrightarrow[\begin{array}{c}28°C, \ 4h\end{array}]{\begin{array}{c}\diagup\!\diagdown\!\diagup\!\diagdown\!\!/\!/ \ , \ h\nu \\ \hline CH_2Cl_2\end{array}}$

70%   21%

Vedejs, E.*; Eberlein, T.H.; Mazur, D.J.; McClure, C.K.; Perry,
D.A.; Ruggeri, R.; Schwartz, E.; Stults, J.S.; Varie, D.L.;
Wilde, R.G.; Wittenberger, S.
                    J Org Chem, (1986), **51**, 1556

$\xrightarrow[\begin{array}{c}LiClO_4\end{array}]{\begin{array}{c}e^-, \ cat., \ MeOH\end{array}}$

70%

cat. = chloro(pyridino)-bis-(dimethylglyoximato)
       cobalt (III)

Bhandal, H.; Pattenden, G.*; Russell, J.J.
                    Tetrahedron Lett, (1986), **27**, 2299

$$\text{TMS-Cl, Zn, Et}_2\text{O}$$
$$40°C, \text{ THF, 30h}$$
))))

88%

Boudjouk, P.[*]; So, J.H.　**Syn Commun**, (1986), **16**, 775

PhCH=CHCH$_2$SiMe$_3$
　　　　$\xrightarrow[\substack{\text{C electrodes} \\ \text{MeCN, Et}_4\text{NOTs}}]{e^-, \text{MeOH}}$

OMe
|
PhCHCH=CH$_2$　(63)

+　　　　　76%

PhCH=CHCH$_2$OMe (37)

Yoshida, J.[*]; Murata, T.; Isoe, S.
　　　　Tetrahedron Lett, (1986), **27**, 3373

$$\text{EtSSiMe}_3$$
$$3\% \text{ dppe, THF}$$
$$1\% \text{ Pd}_2(\text{dba})_3 \cdot \text{CHCl}_3$$
$$\text{RT, 96h}$$

95%

Trost, B.M.[*]; Scanlan, T.S.　**Tetrahedron Lett**, (1986), **27**, 4141

PhCH=CHCHO
　　　$\xrightarrow[\substack{\text{TMS-CH}_2\text{C=CH}_2, 25\% \text{ PPh}_3 \\ \text{CH}_2\text{OAc}}]{\substack{5\% \text{ Pd(OAc)}_2, \text{ THF} \\ 20\% \underline{n}\text{Bu}_3\text{SnOAc, reflux}}}$

89%

Trost, B.M.[*]; King, S.A.　**Tetrahedron Lett**, (1986), **27**, 5971

Review: "Silyl Enol Ethers In Synthesis"

Brownbridge, P.　**Synthesis**, (1983), 1, 85

Related Methods: Protection of Ketones (Section 180A)

SECTION 368: Halide, Sulfonate - Halide, Sulfonate

Halocyclopropanations are found in Section 74F (Alkyls from Olefins).

PPh$_3$, CCl$_4$

reflux, 2d

74%

Croft, A.P.; Bartsch, R.A.[*]   **J Org Chem**, (1983), **48**, 3353

4 MoO$_2$Br$_2$(MeCN)$_2$

MeCN, 25°C, 6h

70%

Arzoumanian, H.[*]; Krentzien, H.; Lai, R.; Metzger, J.;
Petrignoni, J.-F.

J Organomet Chem, (1983), **243**, 175

Me$_2$C=CH$_2$

17% PhI(OH)OTs, 0°C
─────────────────────
CH$_2$Cl$_2$, 36h

OTs
|
Me$_2$CCH$_2$OTs

54%

Rebrovic, L.; Koser, G.F.   **J Org Chem**, (1984), **49**, 2462

PrCH=CHnPr
(**E**)

Mn(OAc)$_3$, CH$_3$CCl (O)
─────────────────────
AcOH, 0°C, 5h

Cl
|
nPrCHCHnPr
|
Cl

(meso:dl = 11:1)   71%

Donnelly, K.D.; Fristad, W.E.[*]; Gellerman, B.J.; Peterson, J.R.;
Selle, B.J.

Tetrahedron Lett, (1984), **25**, 607

$$\underline{n}C_5H_{11}CH=CHCH_3 \xrightarrow[\text{reflux, } Bu_4NCl]{\begin{array}{l}\text{1. PhSeCl, MeCN}\\ \text{2. } Cl_2, CCl_4\\ \text{3. } H_2O_2\end{array}} \underline{n}C_5H_{11}\overset{\displaystyle Cl}{\underset{\displaystyle Cl}{CHCHCH_3}}$$

(Z)   (erythro)   87%

Morella, A.M.; Ward, A.D.[*] **Tetrahedron Lett**, (1984), **25**, 1197

$$\xrightarrow[\text{reflux, 30 min}]{\begin{array}{l}\text{1. } TeCl_4, CCl_4\\ \text{2. } \underline{t}BuOOH, \text{ dioxane}\end{array}}$$

90%

Uemura, S.[*]; Fukuzawa, S.   **J Organomet Chem**, (1984), **268**, 223

$$\xrightarrow[CH_2Cl_2, \text{ RT}]{Me_2NBr, BF_3 \cdot OEt_2}$$

66%

Heasley, G.E.[*]; Janes, J.M.; Stark, S.R.; Robinson, B.L.;
Heasley, V.L.; Shellhamer, D.F.
                    Tetrahedron Lett, (1985), **26**, 1811

$$\xrightarrow[0°C, \text{ 8h}]{BBr_3, CH_2Cl_2}$$

77%

Napolitano, E.[*]; Fiaschi, R.; Mastorilli, E.
                    Synthesis, (1986), 122

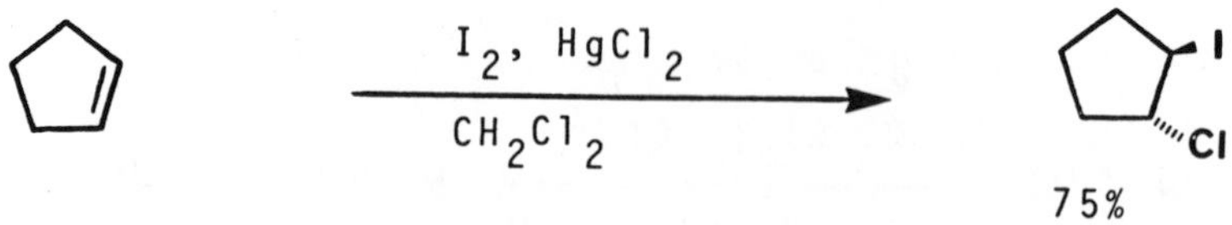

75%

Barluenga, J.[*]; Martinez-Gallo, J.M.; Najera, C.; Yus, M.
J Chem Res, (1986), 274

SECTION 369: Halide, Sulfonate - Ketone

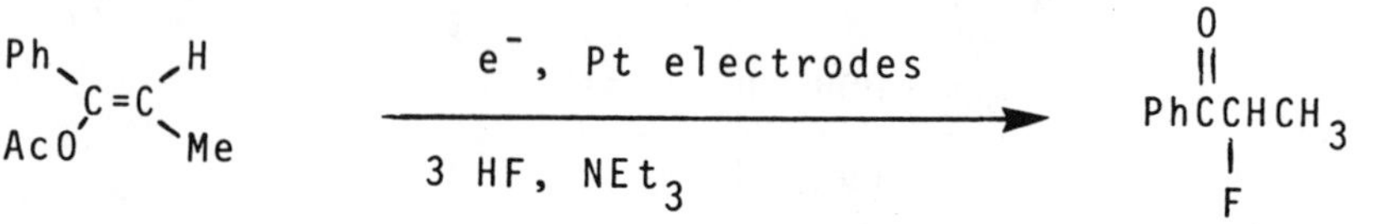

Laurent, E.; Tardivel, R.; Thiebault, H.   63%
Tetrahedron Lett, (1983), 24, 903

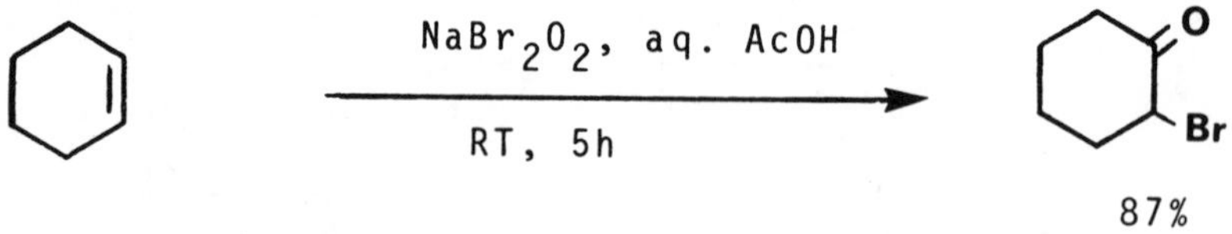

87%

Kageyama, T.[*]; Tobito, Y.; Katoh, A.; Ueno, Y.[*]; Okawara, M.
Chem Lett, (1983), 1481

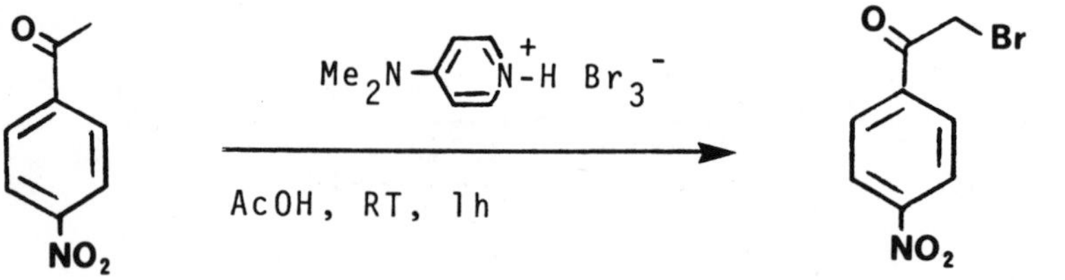

86%

Arrieta, A.; Ganboa, I.; Palomo, C.
Syn Commun, (1984), 14, 939

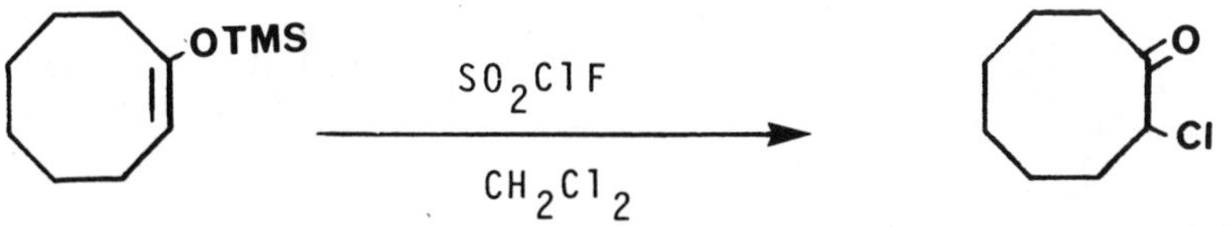
OTMS
SO_2ClF
CH_2Cl_2
O
Cl

86%

Olah, G.A.*; Ohannesian, L.; Arvanaghi, M.; Prakash, G.K.S.
J Org Chem, (1984), 49, 2032

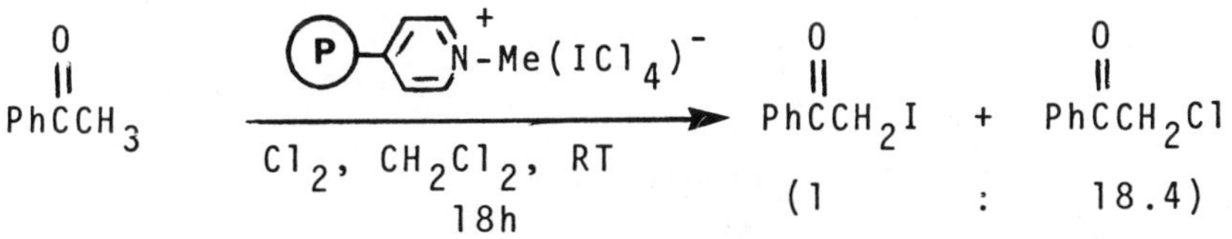
PhCCH3
P
N-Me(ICl4)⁻
Cl2, CH2Cl2, RT
18h
PhCCH2I + PhCCH2Cl
(1 : 18.4)

97%

Šket, B.; Zupan, M.* Tetrahedron, (1984), 40, 2865

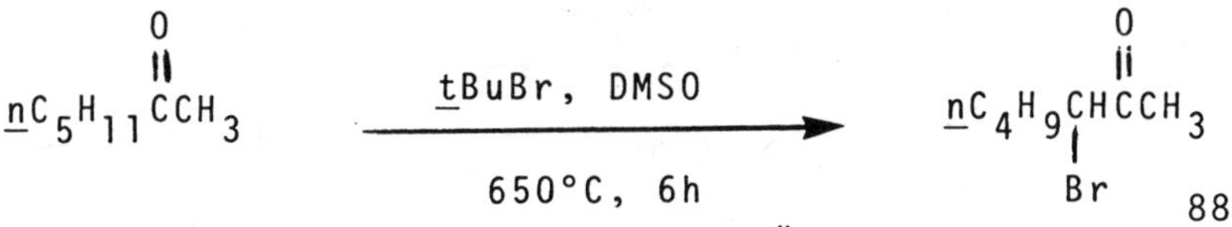
nC5H11CCH3
tBuBr, DMSO
650°C, 6h
nC4H9CHCCH3
Br

88%

Armani, E.; Dossena, A.; Marchelli, R.*; Casnati, G.
Tetrahedron, (1984), 40, 2035

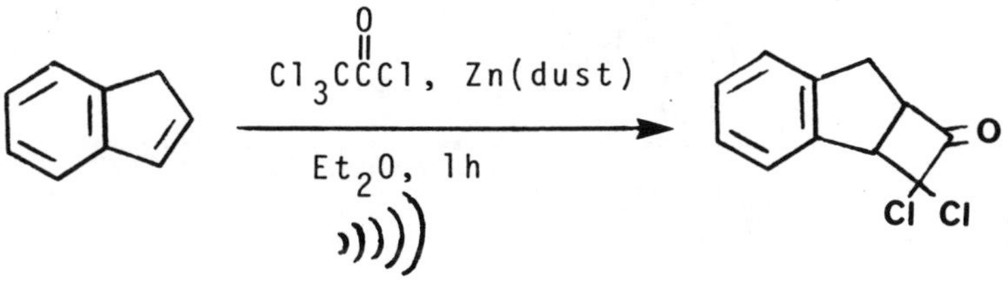
Cl3CCCl, Zn(dust)
Et2O, 1h
O
Cl Cl

80%

Mehta, G.*; Rao, H.S.P. Syn Commun, (1985), 15, 991

1. [LDA, CH_2Br_2], -90°C

2. nBuLi

$PhCH_2CH_2CO_2Et$ $\xrightarrow{\hspace{3cm}}$ $PhCH_2CH_2CCH_2Br$

3. CH_3CCl, EtOH
        $\overset{\|}{O}$ -78°C

75%

Kowalski, C.J.[*]; Haque, M.S.   J Org Chem, (1985), 50, 5140

$PhCCH_3$ (with C=O)

1. LDA, THF, -78°C - RT

2. CH_3CF, -78°C
        $\overset{\|}{O}$

3. $Na_2S_2O_3$

$PhCCH_2F$ (with C=O)

75%

Rozen, S.[*]; Brand, M.   Synthesis, (1985), 665

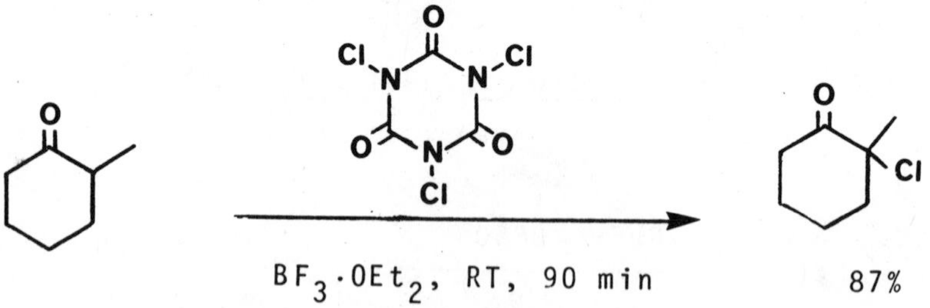

$BF_3 \cdot OEt_2$, RT, 90 min

87%

Hiegel, G.A.[*]; Peyton, K.B.   Syn Commun, (1985), 15, 385

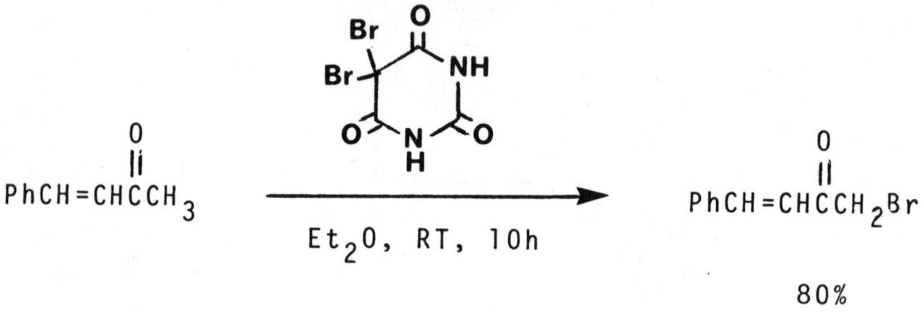

$PhCH=CHCCH_3$ (with C=O)

$\xrightarrow{\hspace{3cm}}$

Et_2O, RT, 10h

$PhCH=CHCCH_2Br$ (with C=O)

80%

Grundke, G.; Keese, W.; Rimpler, M.[*]
                Chem Ber, (1985), 118, 4288

MeO$_2$C   CO$_2$Me
O   O
Ph   ethyl
(RR)

$\xrightarrow[\text{cat. HBr, 15°C}]{\text{Br}_2, \text{CCl}_4}$

MeO$_2$C   CO$_2$Me
O   O
Ph
Br   94%
(S:R$_*$ = 93:7)

Castaldi, G.[*]; Cavicchioli, S.; Giordano, C.[*]; Uggeri, F.
Angew Chem Int Ed Engl, (1986), **25**, 259

Me$_2$C=CHCH$_2$OH

$\xrightarrow[\substack{\text{CH}_3\text{NO}_2, \text{BF}_3 \cdot \text{OEt}_2 \\ 0°C, 30 \text{ min}}]{\substack{\text{Ph} \\ \text{TMSO} } C=CHCl}$

$$\text{PhCCHCH}_2\text{CH}_2\text{CCH}_3$$
O   O
Cl
60%

Poirier, J.-M.; Hennequin, L.; Fomani, M.
Bull Chem Soc Fr, (1986), II436

O TMS

$\xrightarrow[\text{CH}_2\text{Cl}_2, \text{RT, 7h}]{}$ (⟨ ⟩N-F OTs$^-$)

O
F
87%

(also for o,p fluorination of phenols)

Umemoto, T.[*]; Kawada, K.; Tomita, K.
Tetrahedron Lett, (1986), **27**, 4465

nC$_5$H$_{11}$C≡CTMS

$\xrightarrow[\substack{2. \text{Me}_3\text{N-O} \\ 3. \text{NBS}}]{1. \text{BH}_3\text{-Me}_2\text{S}}$

O
nC$_5$H$_{11}$CHCSiMe$_3$
Br
61%

Page, P.C.B.[*]; Rosenthal, S.
Tetrahedron Lett, (1986), **27**, 5421

Bellesia, F.; Ghelfi, F.; Grandi, R.; Pagnoni, U.M.[*]
J Chem Res, (1986), 428

Review: "α-Fluoro Carbonyl Compounds"

Rozen, S.; Filler, R.   **Tetrahedron**, (1985), **41**, 1111

SECTION 370: Halide, Sulfonate - Nitrile

Kaiser, D.A.; Kaye, P.T.   **Syn Commun**, (1984), **14**, 883

Kiyooka, S.[*]; Fujiyama, R.; Kawaguchi, K.
Chem Lett, (1984), 1979

SECTION 371: Halide, Sulfonate - Olefin

$\underline{n}C_8H_{17}C\equiv CH$

1. ![benzo-dioxaborole]BH, 70°C
2. HOH
3. NCS, aq. NaBr, THF

$BrCH=CH\underline{n}C_8H_{17}$
(Z)
63%

Kabalka, G.W.[*]; Sastry, K.A.R.; Knapp, F.F.; Srivastava, P.C.
Syn Commun, (1983), **13**, 1027

$Br(CH_2)_5Br$

HMPT, 5 min
195 - 220°C

$Br(CH_2)_3CH=CH_2$
57%

Kraus, G.A.[*]; Landgrebe, K. **Synthesis**, (1984), 885

NaOH, I_2, 0°C
1h
81%

1. Br_2, -25°C
2. NaOMe, MeOH
-25°C - RT
80%

Brown, H.C.[*]; Samayaji, V. **Synthesis**, (1984), 919

Cl
|
$Me_3SiCHCH=CH_2$

$PhCH(OMe)_2$, 10h
$BF_3\cdot OEt_2$, CH_2Cl_2
0°C - RT

OMe
|
$PhCHCH_2CH=CHCl$
85%
($\underline{Z}:\underline{E}$ = 85:15)

Hosomi, A.[*]; Ando, M.; Sakurai, H.[*] **Chem Lett**, (1984), 1385

$\underline{n}C_6H_{13}$ \
C=CH

$\xrightarrow[\substack{\text{Ni}(PPh_3)_4 \\ PhH/hexane}]{\substack{\underline{E}-ClCH=CHCl \\ 20°C, 6h}}$

$\underline{n}C_6H_{13}CH=CHCH=CHCl$ \
$(\underline{E},\underline{E})$

80%

Ratovelomanana, V.; Linstrumelle, G. \
**Tetrahedron Lett**, (1984), **25**, 6001

$HC\equiv CH$
$\xrightarrow{\substack{1.\ \underline{n}BuLi/CuBr \cdot SMe_2,\ THF \\ -25°C \\ 2.\ I_2,\ THF,\ -60\ -\ -10°C \\ 3.\ NaHSO_3}}$
$\underline{n}BuCH=CHI$ \
$(\underline{Z})$

75%

Alexakis, A.; Cahiez, G.; Normant, J.F.[*] \
**Org Syn**, (1984), **62**, 1

$PhCH=C=CH_2$
$\xrightarrow[\substack{CuCl_2,\ MeCN,\ 0°C \\ 3d}]{\text{cat. } PdCl_2(MeCN)_2}$

CH_2Cl \
$|$ \
$PhCH=C-C=CHPh$ \
$|$ \
Cl

$(\underline{EE}:\underline{ZZ} = 2.8:1)52\%$

Hegedus, L.S.[*]; Kambe, N.; Ishii, Y.; Mori, A. \
**J Org Chem**, (1985), **50**, 2240

1.

$\xrightarrow[\substack{2.\ KOH,\ EtOH}]{\substack{CH_2Cl_2,\ -78°C\ -\ RT}}$

46%

Hudrlik, P.F.[*]; Kulkarni, A.K.**Tetrahedron**, (1985), **41**, 1179

$$(OH)_2BCH=CHnC_8H_{17} \xrightarrow[\substack{2. \text{ aq. } Na_2SO_3 \\ reflux, 3h}]{\substack{1. \ Cl_2, \ CH_2Cl_2 \\ 0-20°C}} ClCH=CHnC_8H_{17}$$

(E)   (Z)   81%

Kunda, S.A.; Smith, T.L.; Hylarides, M.D.; Kabalka, G.W.[*]
Tetrahedron Lett, (1985), **26,** 279

$$nC_6H_{13}C\equiv CH \xrightarrow[\substack{2. \ TolSO_2N(Cl)Na \cdot xH_2O \\ NaBr, \ NaOAc, \ 0°C \\ aq. \ THF}]{\substack{1. \ BBr_3, \ CH_2Cl_2 \\ -78°C - RT}}}$$

$$\begin{array}{c} nC_6H_{13} \\ \diagdown \\ \diagup C=CHBr \\ Br \end{array}$$

84%

Hara, S.; Kato, T.; Shimizu, H.; Suzuki, A.[*]
Tetrahedron Lett, (1985), **26,** 1065

+
$Mo(CO)_2Cp$
PF_6^-

1. MeMgBr
2. $Ph_3C^+PF_6^-$
3. ArMgBr
4. I_2, MeCN
20°C

Ar = ⊰⟨○⟩—OMe

(2   :   1)

72%

Pearson, A.J.[*]; Khan, Md.N.I.
Tetrahedron Lett, (1985), **26,** 1407

$$nC_8H_{17}MgCl \xrightarrow[\substack{5\% \ Pd(PPh_3)_4, \ 6h \\ Et_2O/PhH}]{5 \ ClCH=CCl_2, \ 20°C} nC_8H_{17}CH=CCl_2$$

65%

Ratovelomanana, V.; Linstrumelle, G.; Normant, J.-F.[*]
Tetrahedron Lett, (1985), **26,** 2575

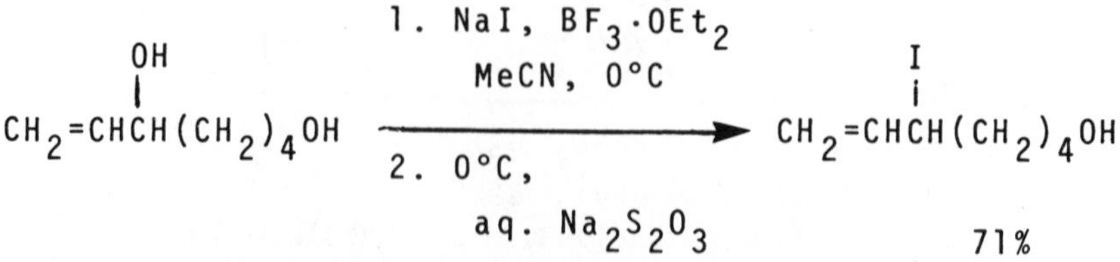

1. NaI, $BF_3 \cdot OEt_2$
   MeCN, 0°C

$CH_2=CHCH(CH_2)_4OH$ $\xrightarrow{\hspace{3cm}}$ $CH_2=CHCH(CH_2)_4OH$

2. 0°C,
   aq. $Na_2S_2O_3$

(OH above first structure, I above product)

71%

Vankar, Y.D.[*]; Rao, C.T.   **Tetrahedron Lett**, (1985), **26**, 2717

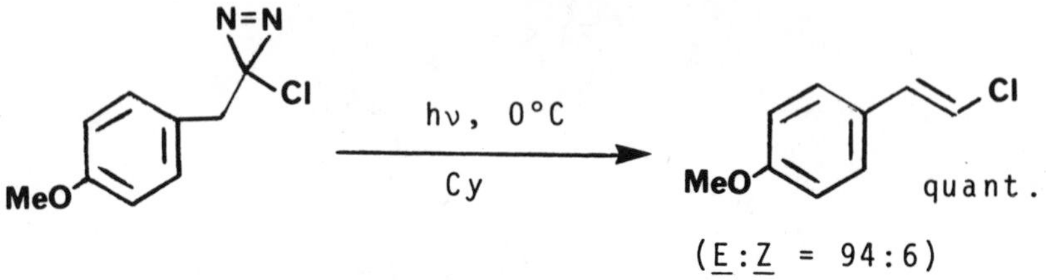

$\xrightarrow[Cy]{h\nu, \ 0°C}$

(E:Z = 94:6)

Tomioka, H.[*]; Hayashi, N.; Inoue, N.; Izawa, Y.
Tetrahedron Lett, (1985), **26**, 1651

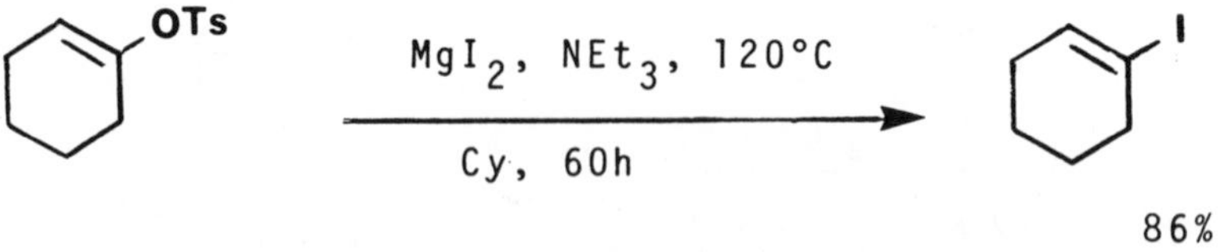

$\xrightarrow[Cy, \ 60h]{MgI_2, \ NEt_3, \ 120°C}$

86%

García Martínez, A.[*]; Martinez Alvarez, R.; García Fraile, A.
Synthesis, (1986), 222

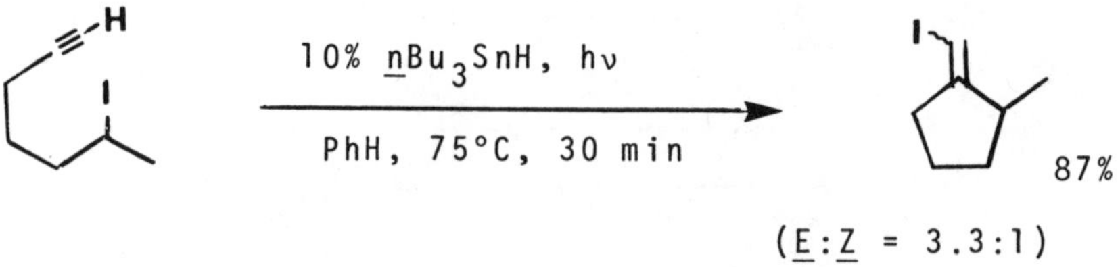

$\xrightarrow[PhH, \ 75°C, \ 30 \ min]{10\% \ \underline{n}Bu_3SnH, \ h\nu}$

87%

(E:Z = 3.3:1)

Curran, D.P.[*]; Chen, M.-H.; Kim, D.
J Am Chem Soc, (1986), **108**, 2489

4 CrCl$_2$, DMF/HOH

pH 3.5, 0°C

16h

88%

Wolf, R.; Steckhan, E.[*]   **JCS Perkin I**, (1986), 733

$\underline{n}C_6H_{13}C{\equiv}CBr$

1. $\underline{n}PrBHBr \cdot SMe_2$, 0°C

   Et_2O

2. NaOMe, MeOH, RT

3. Br_2, CH_2Cl_2, -40°C

4. NaOMe, MeOH, -40°C

71%

Brown, H.C.[*]; Bhat, N.G.; Rajagopalan, S.
          Synthesis, (1986), 480

$Et-C{\equiv}C-Et$

$I(Py)_2 \cdot BF_4$, NaH

CH_2Cl_2, NaF, -40°C

1.5h

50%

Barluenga, J.[*]; Rodríguez, M.A.; González, J.M.; Campos, P.J.;
Asensio, G.

          Tetrahedron Lett, (1986), **27**, 3303

2 $\underline{n}BuC{\equiv}CH$

1. BBr_3, CH_2Cl_2
   -78°C - RT

2. Me_2S, 0°C

3. $EtC{\equiv}CEt$, Dibal-H
   -78°C - 0°C

4. I_2, AcOK, THF

74%

Hyuga, S.; Takinami, S.; Hara, S.; Suzuki, A.[*]
          Chem Lett, (1986), 459

CHO

$$\xrightarrow[\text{THF, 1h}]{CHI_3, CrCl_2, 0°C}$$

78%

$(\underline{E}:\underline{Z} = 89:11)$

Takai, K.*; Nitta, K.; Utimoto, K.
J Am Chem Soc, (1986), **108**, 7408

SECTION 372: Ketone - Ketone

$$\underset{\text{PhCCl}}{\overset{O}{\underset{||}{}}}$$

1. PhCHO, ZnCl$_2$
2. PPh$_3$
3. NaN(TMS)$_2$

$$\xrightarrow{\hspace{3cm}}$$

$$\underset{\text{PhC-CPh}}{\overset{O \quad O}{\underset{|| \quad ||}{}}}$$

92%

Anders, E.*; Gassner, T.
Angew Chem Int Ed Engl, (1983), **22**, 619

$$\xrightarrow[-80°C]{\underline{t}BuOK, THF}$$

85%

Nakashita, Y.; Hesse, M.   **Helv Chim Acta**, (1983), **66**, 845

$$\xrightarrow[-15°C]{Cu(BF_4)_2, Et_2O}$$

$$\underset{\text{PhC(CH}_2)_4\text{CPh}}{\overset{O \qquad O}{\underset{|| \qquad\quad ||}{}}}$$

78%

Ryu, I.; Ando, M.; Ogawa, A.; Murai, S.*; Sonoda, N.
J Am Chem Soc, (1983), **105**, 7192

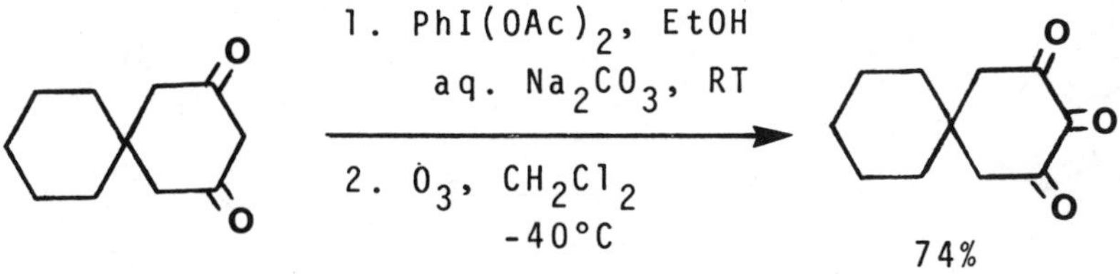

1. $PhI(OAc)_2$, EtOH
   aq. Na_2CO_3, RT

2. O_3, CH_2Cl_2
   $-40°C$

74%

Schank, K.*; Lick, C.   **Synthesis**, (1983), 392

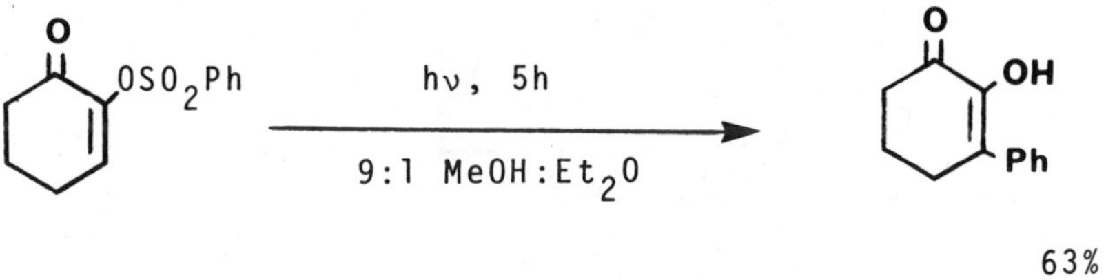

OSO_2Ph

$h\nu$, 5h

9:1 MeOH:Et_2O

OH

Ph

63%

Feigenbaum, A.*; Pete, J.-P.; Scholler, D.
                    J Org Chem, (1984), **49**, 2355

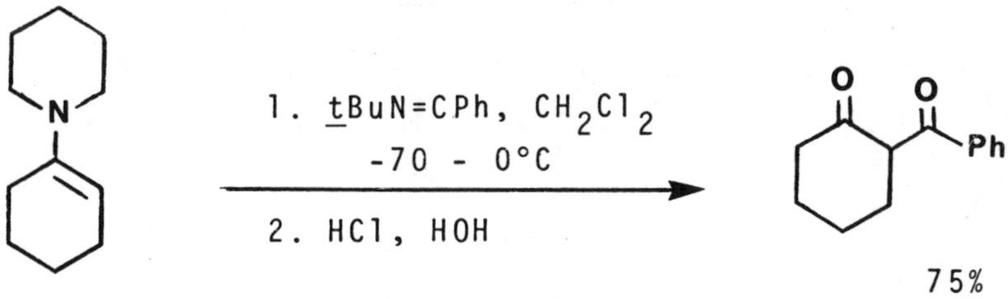

1. $\underline{t}BuN=CPh$, CH_2Cl_2
   $-70 - 0°C$

2. HCl, HOH

Ph

75%

Baudoux, D.; Fuks, R.*   **Bull Chem Soc Belg**, (1984), **93**, 1009

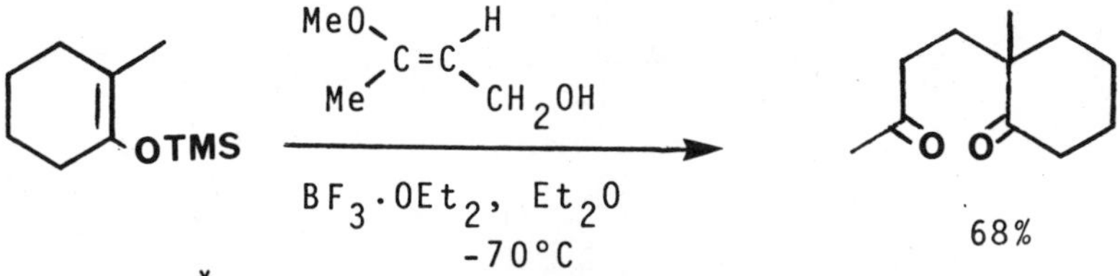

MeO, H
    C=C
Me CH_2OH

$BF_3 \cdot OEt_2$, Et_2O
$-70°C$

OTMS

68%

Duhamel, P.*; Poirier, J.-M.; Tavel, G.
                    Tetrahedron Lett, (1984), **25**, 43

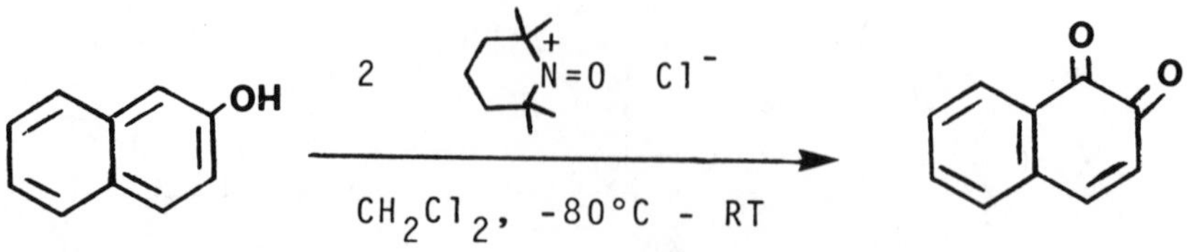

Hunter, D.H.[*]; Barton, D.H.R.; Motherwell, W.J.
                    Tetrahedron **Lett**, (1984), **25**, 603

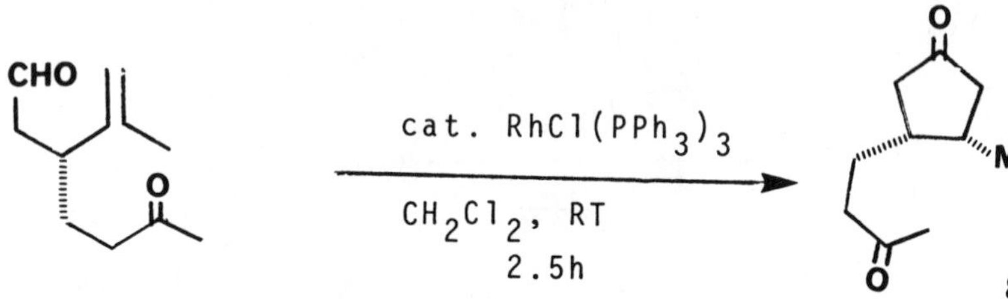

95%

Sakai, K.[*]; Ishiguro, Y.; Funakoshi, K.; Ueno, K.; Suemune, H.
                    Tetrahedron **Lett**, (1984), **25**, 961

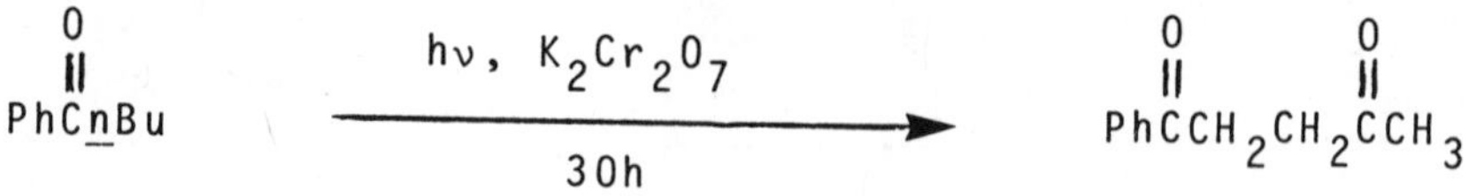

84%

$$\underset{\text{PhCnBu}}{\overset{O}{\overset{||}{}}} \xrightarrow[\text{30h}]{h\nu,\ K_2Cr_2O_7} \underset{\text{PhCCH}_2\text{CH}_2\text{CCH}_3}{\overset{O\quad\quad O}{\overset{||\quad\quad ||}{}}}$$

71%

Mitani, M.[*]; Tamada, M.; Uehara, S.; Koyama, K.
                    Tetrahedron **Lett**, (1984), **25**, 2805

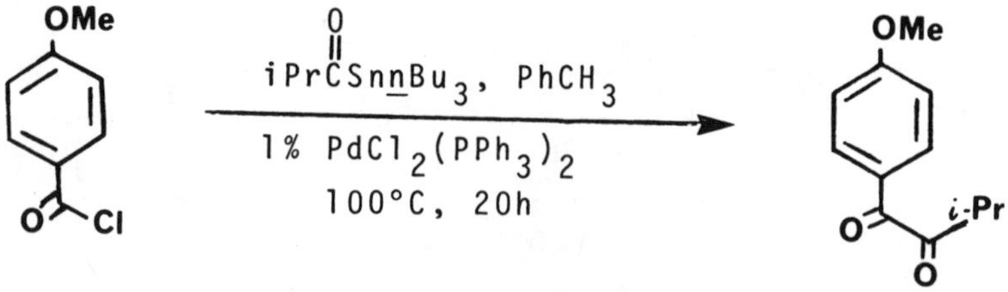

Verlhac, J.-B.; Chanson, E.; Jousseaume, B.; Quintard, J.-P.
                    Tetrahedron **Lett**, (1985), **26**, 6075

93%

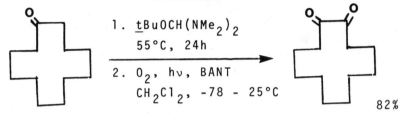

1. MeCH=CHCCH$_3$, TiCl$_4$
   Ti(OiPr)$_4$, CH$_2$Cl$_2$
   -95°C

2. aq. K$_2$CO$_3$

72%

Huffman, J.W.[*]; Potnis, S.M.; Satish, A.V.
J Org Chem, (1985), 50, 4266

1. tBuOCH(NMe$_2$)$_2$
   55°C, 24h

2. O$_2$, hν, BANT
   CH$_2$Cl$_2$, -78 - 25°C

82%

BANT = bis-acenaphthalene thiophene

Wasserman, H.H.[*]; Ives, J.L.   J Org Chem, (1985), 50, 3573

nBu$_2$(CN)CuLi$_2$/CO

THF, -110°C
90 min

86%

n-Bu

Seyferth, D.[*]; Hui, R.C.   J Am Chem Soc, (1985), 107, 4551

1. ⬡NEt

Sn(OTf)$_2$

CH$_2$Cl$_2$, -45°C

2. CH$_2$=CHCCH$_3$, -45°C
        ‖
        O

Me$_2$CHCCH$_2$CH=CH$_2$ → Me$_2$CHCCH=CH(CH$_2$)$_3$CCH$_3$
                                        (E)   55%

Stevens, R.V.; Mukaiyama, T.   Chem Lett, (1985), 851

$$PhCH=CHCCH_3 \xrightarrow[\begin{array}{c} TrClO_4, \ CH_2Cl_2 \\ 2. \ aq. \ NaHCO_3 \end{array}]{\begin{array}{c} 1. \ CH_2=C \stackrel{Ph}{\diagdown} OTMS, \ -78°C \end{array}} CH_3CCH_2CHCH_2CPh$$

92%

Kobayashi, S.; Murakami, M.; Mukaiyama, T.
Chem Lett, (1985), 953

$$\xrightarrow[\begin{array}{c} PhCl, \ reflux \\ 24h \end{array}]{}$$

81%

Gonzalez, A.; Güell, F.; Marquet, J.; Moreno-Mañas, M.[*]
Tetrahedron Lett, (1985), **26**, 3735

1. NaNH$_2$, tBuONa
2. MeSSMe
3. CuCl$_2$·CuO, acetone
   HOH, 35°C, 30 min

80%

Carré, M.C.; Caubere, P.[*]   **Tetrahedron Lett**, (1985), **26**, 3103
Carré, M.C.; Ndebeka, G.; Riondel, A.; Bourgasser, P.; Caubere, P.[*]

Tetrahedron Lett, (1984), **25**, 1551

$$(CO)_5Cr=C \stackrel{OMe}{\diagdown} CH_3 \xrightarrow[\begin{array}{c} MeCN, \ 70°C \\ 2. \ AcOH/HOH/EtOH \end{array}]{1. \ HC\equiv C(CH_2)_3CH=CH_2}$$

32%

Wulff, W.D.[*]; Kaesler, R.W.   **Organometallics**, (1985), **4**, 1461

$$\begin{array}{c} O \\ \parallel \\ PhCCHPh \\ \mid \\ Br \end{array} \quad \xrightarrow[\text{PhH, reflux, 2h}]{\text{Amberlyst A-26(NO}_3^-)} \quad \begin{array}{c} O \\ \parallel \\ PhC-CPh \\ \parallel \\ O \end{array}$$

80%

Cainelli, G.[*]; Manescalchi, F.; Plessi, L.
Gazz Chim Ital, (1986), **116**, 163

$$PhC \equiv CPh \quad \xrightarrow[\text{Hg(OAc)}_2, \ 40°C, \ 2h]{\text{(HMPA)MoO(O}_2)_2, \ DCE} \quad \begin{array}{c} O \\ \parallel \\ PhC-CPh \\ \parallel \\ O \end{array}$$

80%

Ballistreri, F.P.; Failla, S.; Tomaselli, G.A.[*]; Curci, R.
Tetrahedron Lett, (1986), **27**, 5139

SECTION 373: Ketone - Nitrile

1. $LiCMe_2CN$, CO, THF
   -78°C
2. CF_3COOH, -78°C
3. CAN, aq. acetone

85%

Semmelhack, M.F.[*]; Herndon, J.W.; Liu, J.K.
Organometallics, (1983), **2**, 1885

1. Li(TMS)CHCN
   THF, -78°C
2. CsF, aq. MeCN

96%

Tomioka, K.; Koga, K.   **Tetrahedron Lett**, (1984), **25**, 1599

$$CH_2=C \overset{CN}{\underset{NMePh}{\diagdown}} \quad \xrightarrow[\substack{2.\ BnBr \\ 3.\ H_3O^+}]{\substack{1.\ Li \overset{CH_3}{\underset{CN}{\diagup}CH} }} \quad BnCCH_2CH_2CHCH_3 \overset{O \quad\quad CN}{\underset{}{}}$$

92%

Ahlbrecht, H.*; Ibe, M.   **Synthesis**, (1985), 421

$$\xrightarrow[NaCN]{Na_2S_2O_8}$$

30%

Nikishin, G.I.*; Troyansky, E.I.; Misintsev, V.V.; Molokanov, A.N.; Ogibin, Y.N.

Tetrahedron Lett, (1986), **27**, 4215

SECTION 374: Ketone - Olefin

For the oxidation of allylic alcohols to olefinic ketones, see Section 168 (Ketones from Alcohols and Phenols).

For the oxidation of allylic methylene groups (C=C-CH$_2$ → C=C-CO), see Section 170 (Ketones from Alkyls ad Methylenes).

For the alkylation of olefinic ketones, see also Section 177 (Ketones from Ketones), and Section 74E (Alkyls from Olefins) for conjugate alkylations.

$$\xrightarrow[\substack{2.\ NEt_3,\ RT,\ overnight}]{\substack{1.\ PhSCl,\ Et_2O,\ 0°C}}$$

75%

McKervey, M.A.*; Ratananukul, P.

Tetrahedron Lett, (1983), **24**, 117

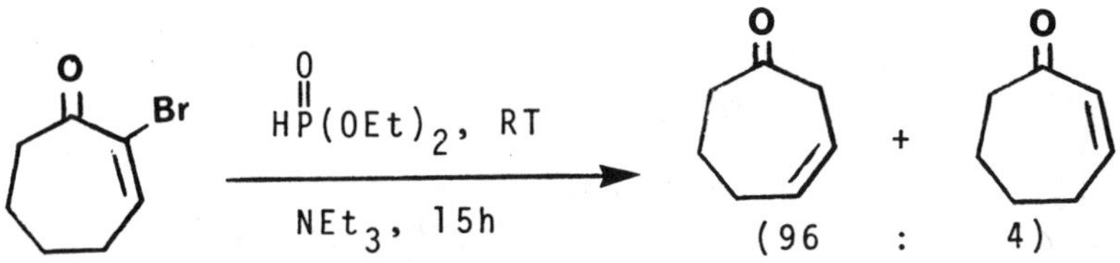

(96 : 4)

66%

Hirao, T.*; Masunga, T.; Hayashi, K.; Ohshiro, Y.*; Agawa, T.
Tetrahedron **Lett,** (1983), **24**, 399

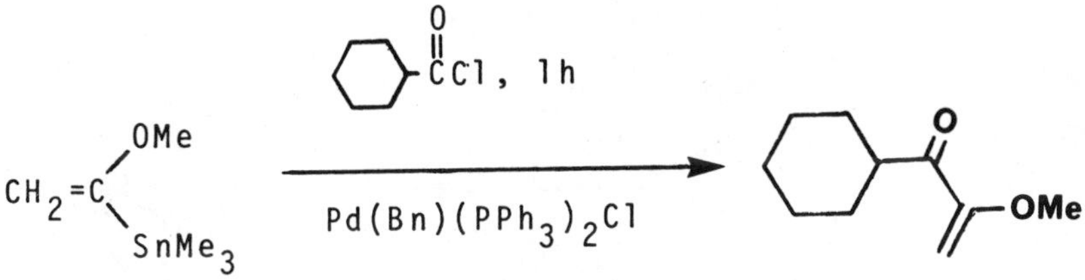

86%

Soderquist, J.A.; Leong, W.W.H.
Tetrahedron **Lett,** (1983), **24**, 2301

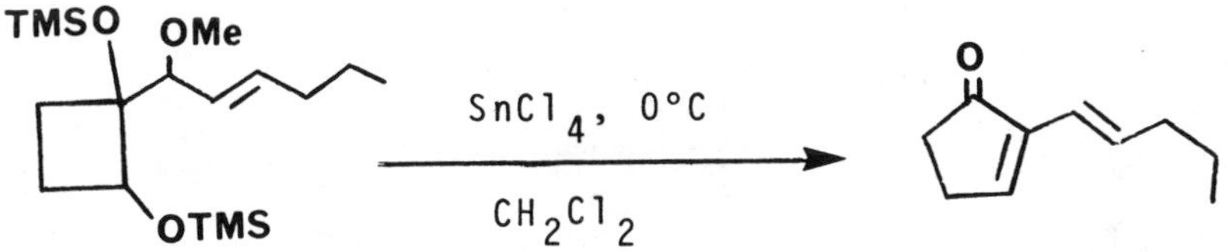

60%

Nakamura, E.; Shimada, J.; Kuwajima, I.*
JCS **Chem** **Comm,** (1983), 498

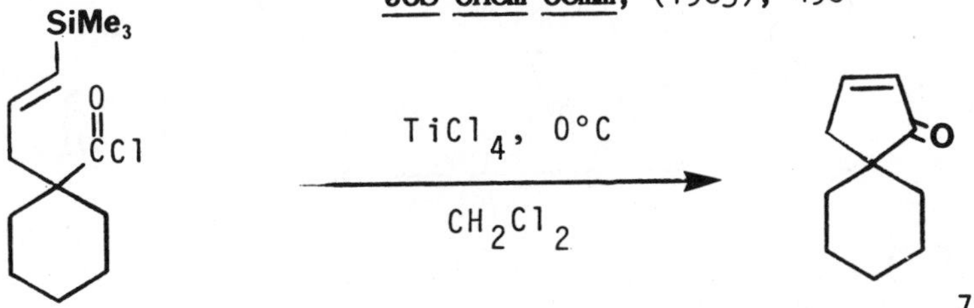

70%

Nakamura, E.; Fukuzaki, K.; Kuwajima, I.*
JCS **Chem** **Comm,** (1983), 499

$$
\begin{array}{c}
\text{OMe} \\
| \\
\text{MeC(CH}_2)_3\text{CO}_2\text{Me} \\
| \\
\text{OMe}
\end{array}
\quad
\xrightarrow[\substack{120°C \\ 24h}]{\text{EtCOOH}}
\quad
\begin{array}{c}
\overset{\overset{\displaystyle O}{\|}}{\text{CH}_2\text{=CHCHCH}_2\text{C(CH}_2)_3\text{CO}_2\text{Me}} \\
| \\
\text{Ph} \qquad + \qquad (12)
\end{array}
$$

$$
\begin{array}{c}
\overset{\overset{\displaystyle O}{\|}}{\text{CH}_2\text{=CHCHCHCCH}_3} \\
\text{69\%} \qquad | \quad | \text{CH}_2\text{CO}_2\text{Me} \quad (88) \\
\text{Ph}
\end{array}
$$

Daub, G.W.*; Lunt, S.R.   **Tetrahedron** **Lett**, (1983), **24**, 4397

$$\xrightarrow[\substack{2\ 30\%\ H_2O_2,\ RT \\ 5h}]{1\%\ RuCl_3 \cdot H_2O}$$

90%

Ito, S.; Aihara, K.; Matsumoto, M.*
Tetrahedron **Lett**, (1983), **24**, 5249

$$\xrightarrow[\text{MeCN}]{2\ [ClCu^{II}O\text{-}Cu^{II}Cl]}$$

80%

Capdevielle, P.; Maumy, M.   **Tetrahedron** **Lett**, (1983), **24**, 5611

$$\xrightarrow[\substack{\text{MeCN, reflux} \\ 1h}]{1\%\ Pd(OAc)_2,\ dppe}$$

78%

Tsuji, J.*; Minami, I.; Shimizu, I.
Tetrahedron **Lett**, (1983), **24**, 5635

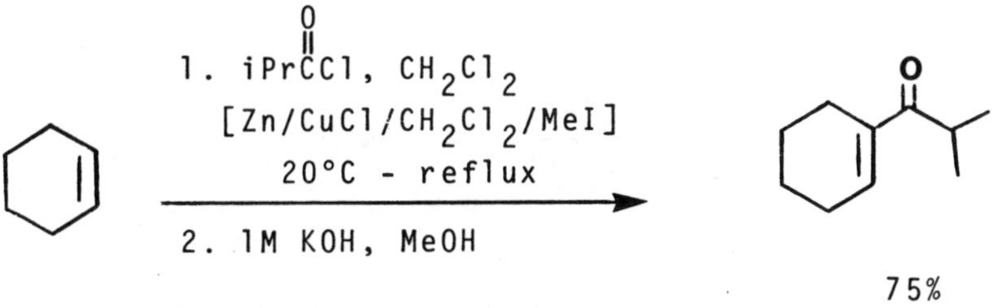

1. iPrCCl, CH_2Cl_2
   [$Zn/CuCl/CH_2Cl_2/MeI$]
   20°C - reflux

2. 1M KOH, MeOH

75%

Shono, T.[*]; Nishiguchi, I.; Sasaki, M.; Ikeda, H.; Kurita, M.
J Org Chem, (1983), **48**, 2503

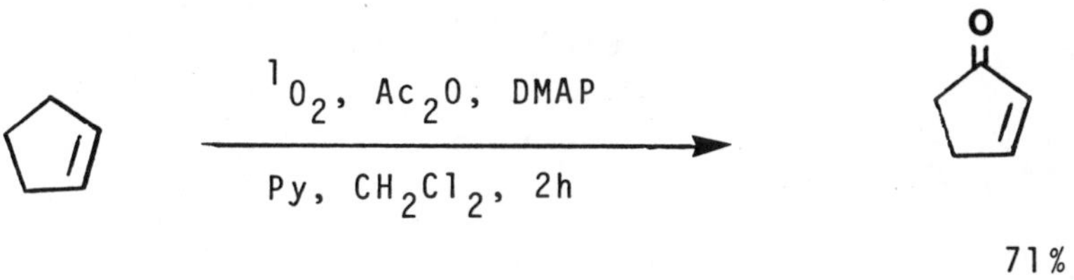

$^1O_2$, Ac_2O, DMAP

Py, CH_2Cl_2, 2h

71%

Mihelich, E.D.[*]; Eickhoff, D.J.   **J Org Chem**, (1983), **48**, 4135

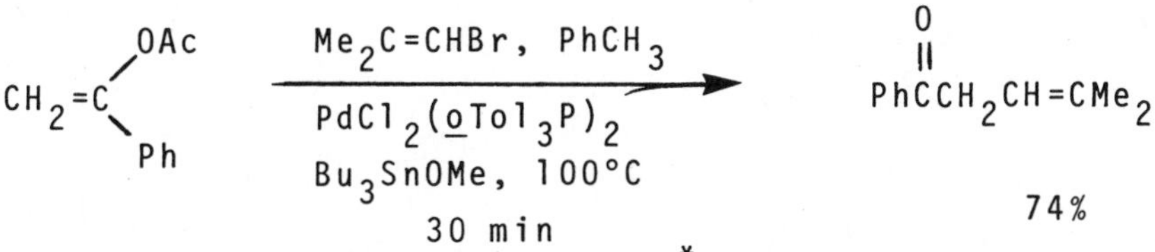

$CH_2=C$ OAc / Ph

$Me_2C=CHBr$, $PhCH_3$

$PdCl_2(\underline{o}Tol_3P)_2$
Bu_3SnOMe, 100°C
30 min

$PhCCH_2CH=CMe_2$

74%

Kosugi, M.; Hagiwara, I.; Migita, T.[*]   **Chem Lett**, (1983), 839

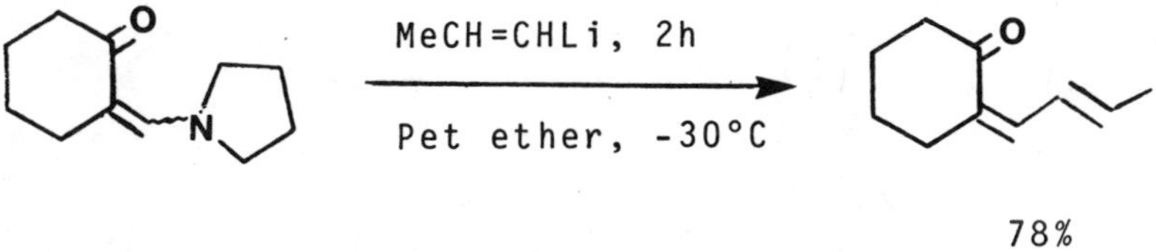

$MeCH=CHLi$, 2h

Pet ether, -30°C

78%

Mukaiyama, T.[*]; Ohsumi, T.   **Chem Lett**, (1983), 875

$$CH_2=C=CHOMe \quad \xrightarrow[\substack{3.\ Tol-I,\ RT,\ 8h \\ PdCl_2(PPh_3)_2,\ Dibal-H \\ 4.\ aq.\ H^+}]{\substack{1.\ \underline{n}BuLi \\ 2.\ ZnCl_2,\ THF}} \quad CH_2=CH\overset{\overset{\displaystyle O}{\|}}{C}-Tol$$

66%

Russell, C.E.; Hegedus, L.S.[*]   J Am Chem Soc, (1983), 105, 943

$$\xrightarrow[\substack{RT\ -\ reflux \\ CH_2Cl_2}]{2\ EtAlCl_2,\ 3h}$$

97%

Trost, B.M.[*]; Adams, B.R.   J Am Chem Soc, (1983), 105, 4849

$$\xrightarrow[\substack{NEt_3,\ THF,\ 60°C \\ 24h}]{Pd(PPh_3)_4,\ CO}$$

54%

Negishi, E.[*]; Miller, J.A.   J Am Chem Soc, (1983), 105, 6761

$$MeCH=CH\overset{\overset{\displaystyle O}{\|}}{C}Cl \quad \xrightarrow[\substack{AlCl_3,\ 50°C,\ 30\ min}]{CH_2=CHSiMe_3,\ CCl_4}$$

63%

Kjeldsen, G.; Knudsen, J.S.; Ravn-Petersen, L.S.; Torssell, K.B.G.[*]

Tetrahedron, (1983), 39, 2237

Trost, B.M.[*]; Self, C.R.   **J Org Chem**, (1984), **49**, 468

Wilson, S.R.[*]; Price, M.F.   **J Org Chem**, (1984), **49**, 722

Rautenstrauch, V.[*]   **J Org Chem**, (1984), **49**, 950

Hegedus, L.S.[*]; Perry, R.J.   **J Org Chem**, (1984), **49**, 2570

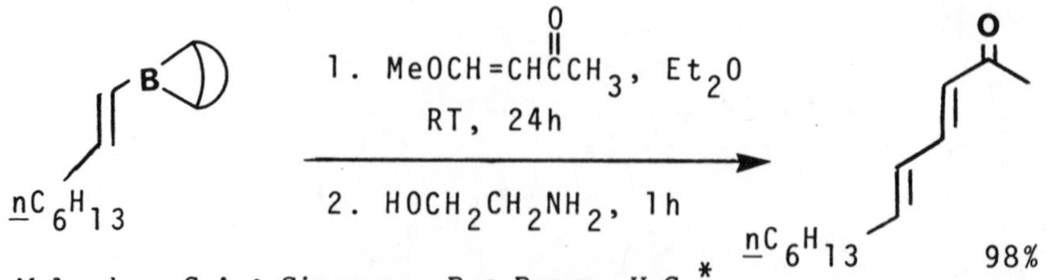
B
nC6H13
1. MeOCH=CHCCH3, Et2O
O
RT, 24h
2. HOCH2CH2NH2, 1h
O
nC6H13
98%

Molander, G.A.; Singarem, B.; Brown, H.C.*
J Org Chem, (1984), 49, 5024

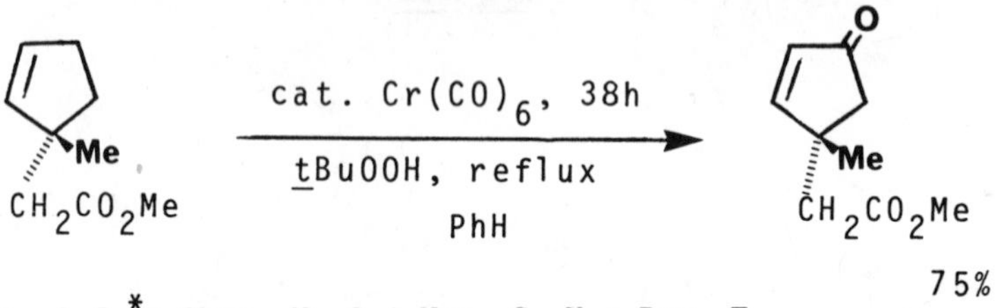
Me
CH2CO2Me
cat. Cr(CO)6, 38h
tBuOOH, reflux
PhH
O
Me
CH2CO2Me
75%

Pearson, A.J.*; Chen, Y.-S.; Hsu, S.-Y.; Ray, T.
Tetrahedron Lett, (1984), 25, 1235

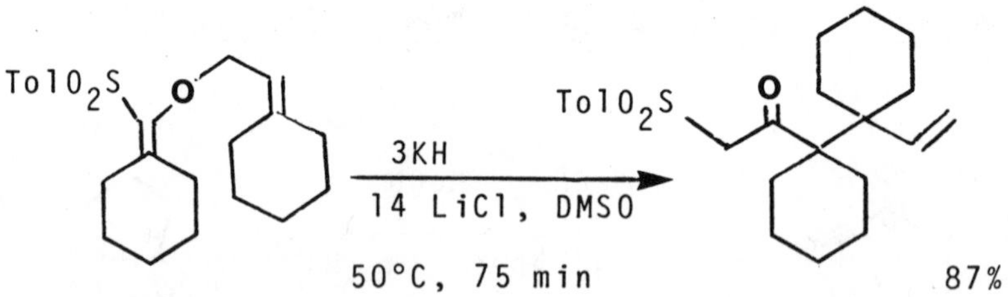
TolO2S
O
3KH
14 LiCl, DMSO
50°C, 75 min
TolO2S
O
87%

Denmark, S.E.*; Harmata, M.A.
Tetrahedron Lett, (1984), 25, 1543

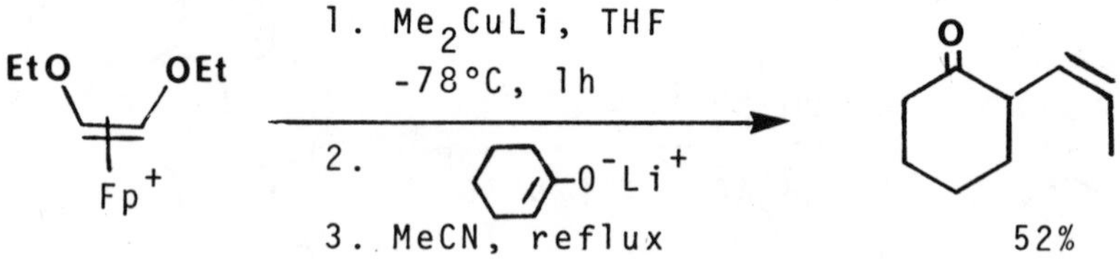
EtO
OEt
Fp+
1. Me2CuLi, THF
-78°C, 1h
2.
O⁻Li+
3. MeCN, reflux
O
52%

Marsi, M.; Rosenblum, M.* J Am Chem Soc, (1984), 106, 7264

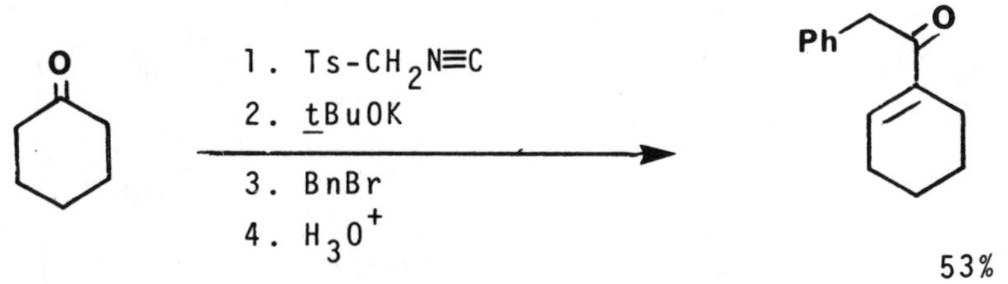

1. Ts-CH$_2$N≡C
2. $\underline{t}$BuOK
3. BnBr
4. H$_3$O$^+$

53%

Moskal, J.; van Leusen, A.M.

Tetrahedron Lett, (1984), 25, 2585

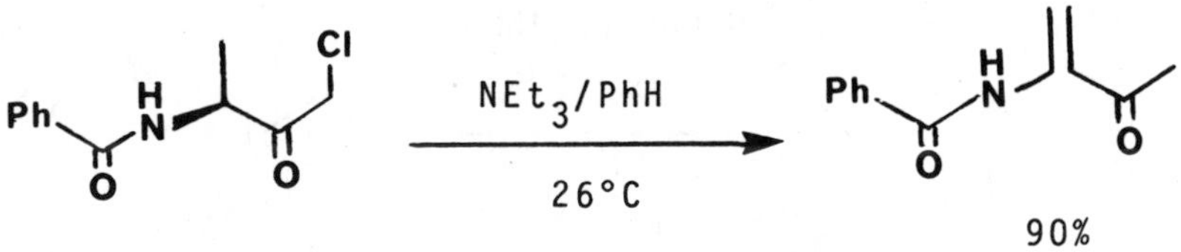

NEt$_3$/PhH

26°C

90%

Gordon, E.M.[*]; Pluščec, J.; Delaney, N.G.; Natarajan, S.;
Sundeen, J.

Tetrahedron Lett, (1984), 25, 3277

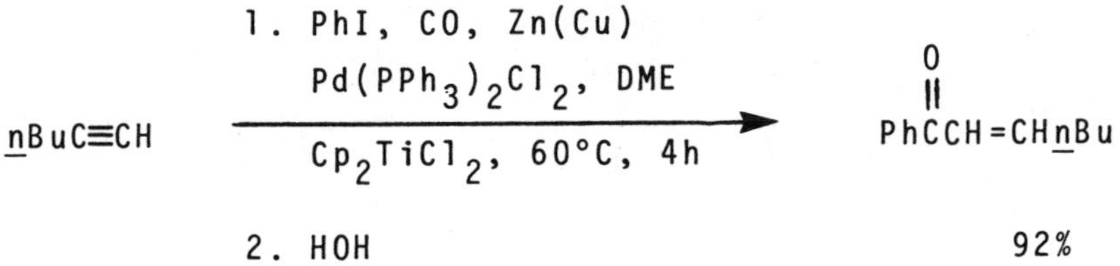

$\underline{n}$BuC≡CH

1. PhI, CO, Zn(Cu)
   Pd(PPh$_3$)$_2$Cl$_2$, DME
   Cp$_2$TiCl$_2$, 60°C, 4h

2. HOH

PhCCH=CH$\underline{n}$Bu

92%

Tamaru, Y.; Ochiai, H.; Yoshida, Z.[*]

Tetrahedron Lett, (1984), 25, 3861

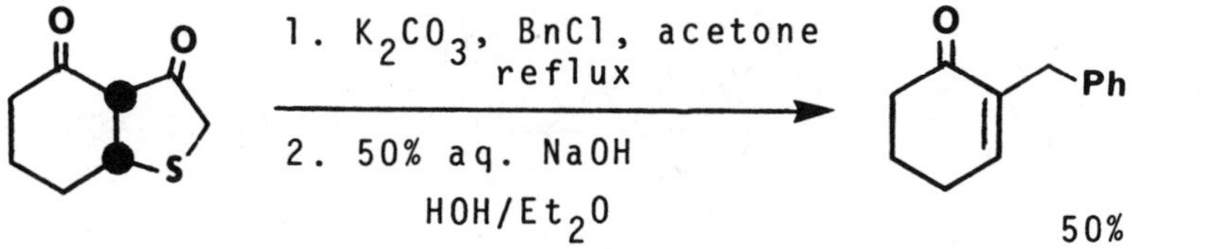

1. K$_2$CO$_3$, BnCl, acetone
   reflux

2. 50% aq. NaOH
   HOH/Et$_2$O

50%

Beraldi, P.G.; Barco, A.; Benetti, S.; Pollini, G.P.; Zanirato,
V.

Tetrahedron Lett, (1984), 25, 4291

$$\text{2-(CO}_2\text{CH}_2\text{CH=CH}_2\text{) cyclohexanone} \quad \xrightarrow[\text{PhCH}_3, \text{ reflux}]{\text{Mo(CO)}_6} \quad \text{2-allylcyclohexanone}$$

85%

Tsuji, J.*; Minami, I.; Shimizu, I.   **Chem Lett**, (1984), 1721

$$\xrightarrow[\substack{\text{2. aq. NaHCO}_3 \\ \text{3. 1N HCl}}]{\substack{\text{1. Pb(OAc)}_4, \text{ PhH} \\ \text{reflux, 5 min}}}$$

69%

Nakatani, K.; Isoe, S.*   **Tetrahedron Lett**, (1984), **25**, 5335

$$\underset{\text{PhCH=CHCCH}_3}{\overset{O}{\|}} \quad \xrightarrow[\text{2. HOH, -40°C - RT}]{\substack{\text{1. Me}_3\text{SiCH}_2\text{CH=CH}_2, -40°C \\ \text{TiCl}_4, \text{ CH}_2\text{Cl}_2, 30 \text{ min}}} \quad \underset{\substack{| \\ \text{CH}_2\text{CH=CH}_2}}{\overset{O}{\underset{\|}{\text{PhCHCH}_2\text{CCH}_3}}}$$

80%

Sakurai, H.*; Hosomi, A.; Hayashi, J.   **Org Syn**, (1984), **62**, 86

$$\text{Me}_2\text{C=CHCH}_2\text{Cl} \quad \xrightarrow[\substack{\text{1% PdCl}_2(\text{MeCN})_2 \\ \text{CHCl}_3, 48h}]{\substack{\text{Me}_3\text{SnCH}_2\text{CH=CH}_2, 25°C \\ 6 \text{ atm CO, PPh}_3}} \quad \underset{\|}{\overset{O}{\text{Me}_2\text{C=CHCH}_2\text{CCH}_2\text{CH=CH}_2}}$$

54%

Merrifield, J.H.; Godschalx, J.P.; Stille, J.K.*
Organometallics, (1984), **3**, 1108

$PhC\equiv CPh$ $\xrightarrow[\text{PhCH}_3,\ 100°C]{(CO)_5W=C\begin{smallmatrix}OMe\\Me\end{smallmatrix}}$ $\underset{\underset{Ph}{\big|}}{PhCH=CCCH_3}$ (O above)

64%

Macomber, D.W.[*] **Organometallics**, (1984), **3**, 1589

$\xrightarrow[\text{PhCHO, RT, 1h}]{CeI_3,\ THF}$

90%

Fukuzawa, S.[*]; Fujinami, T.; Sakai, S.
                    JCS Chem Comm, (1985), 777

$\underset{nC_6H_{13}CCl}{\overset{O}{\|}}$
1. $\underline{n}Bu_3Sn-C\equiv C-SiMe_3$, DCE
   $PdCl_2(PPh_3)_2$, reflux
2. THF, RT, 6h
3. 620°C, 14 Torr

56%

Ackroyd, J.; Karpf, M.; Dreiding, A.S.[*]
                    Helv Chim Acta, (1985), **68**, 338

$\underset{CH_3CCH=CH_2}{\overset{O}{\|}}$
1. PhSe-**B**
   $PhCH_3$, 25°C, 2h
2. iPrCHO
3. Py, H_2O_2, 0 - 30°C

$\underset{\overset{CH_2}{\|}}{\overset{O\quad OH}{\overset{\|}{CH_3C-C-CHiPr}}}$

71%

Leonard, W.R.; Livinghouse, T.[*]   **J Org Chem**, (1985), **50**, 730

$$CH_2=C \begin{smallmatrix} CH_3 \\ \\ CH_2OCOMe \\ \quad \overset{\|}{O} \end{smallmatrix} \quad \xrightarrow[\substack{Pd_2(dba)_3 \cdot CHCl_3 \\ 12 \ min}]{\substack{EtCCHCO_2Me, \ 30°C \\ \overset{|}{Me}}} \quad CH_2=CCH_2-\overset{Me}{\underset{Me}{C}}-\overset{CO_2Me}{\underset{O}{C}} Et$$

$$O$$

92%

Tsuji, J.*; Shimizu, I.; Minami, I.; Ohashi, Y.; Sugiura, T.;
Takahashi, K.

J Org Chem, (1985), **50**, 1523

$$\begin{smallmatrix} OH \\ | \\ \underline{n}BuCHCH=CH_2 \end{smallmatrix} \quad \xrightarrow[\substack{2. \ H_3O^+ \\ 3. \ 2 \ KH, \ 120°C \\ 6h}]{\substack{1. \ ClCH=CHCO_2H, \ 2 \ NaH \\ THF, \ 65°C, \ 4h}} \quad \underline{n}BuCH=CHCH_2CH_2\overset{\overset{O}{\|}}{C}CH_3$$

58%

Büchi, G.*; Vogel, D.E.   **J Org Chem**, (1985), **50**, 4664

$$\xrightarrow[\substack{HgCl_2, \ RT \\ \\ 2. \ CO, \ 0°C, \ 1h}]{1. \ Cl_2ZrCp_2, \ Mg}$$

62%

Negishi, E.*; Holmes, S.J.; Tour, J.M.; Miller, J.A.
J Am Chem Soc, (1985), **107**, 2568

$$\xrightarrow[\substack{THF, \ -25°C, \ 15h}]{CH_2=CHCH_2Ti(NEt_2)_3} \quad PhC\overset{\overset{O}{\|}}{C}CH_2CH=CH_2$$

95%

Reetz, M.T.*; Wenderoth, B.; Urz, R.
Chem Ber, (1985), **118**, 348

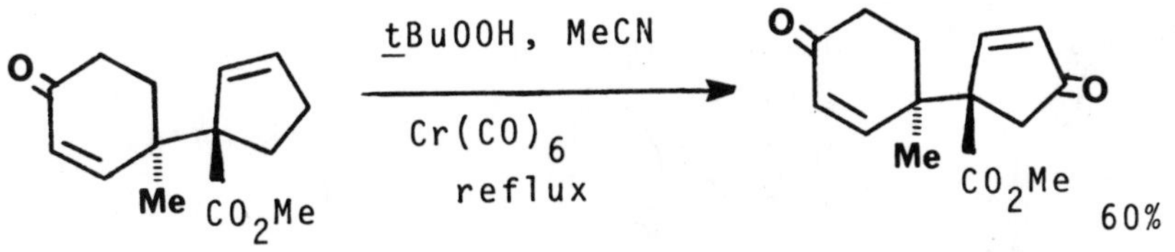

tBuOOH, MeCN

Cr(CO)$_6$
reflux

60%

Pearson, A.J.[*]; Chen, Y.-S.; Han, G.R.; Hsu, S.-Y.; Ray, T.
JCS Perkin I, (1985), 267

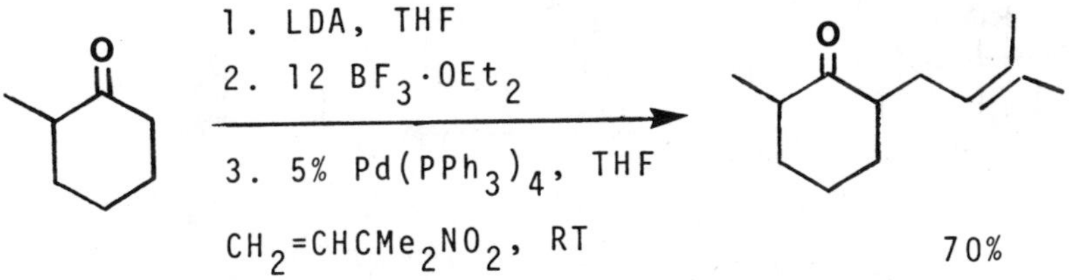

1. LDA, THF
2. 12 BF$_3$·OEt$_2$
3. 5% Pd(PPh$_3$)$_4$, THF

CH$_2$=CHCMe$_2$NO$_2$, RT

70%

Ono, N.[*]; Hamamoto, I.; Kaji, A.[*]
Bull Chem Soc Jpn, (1985), 58, 1863

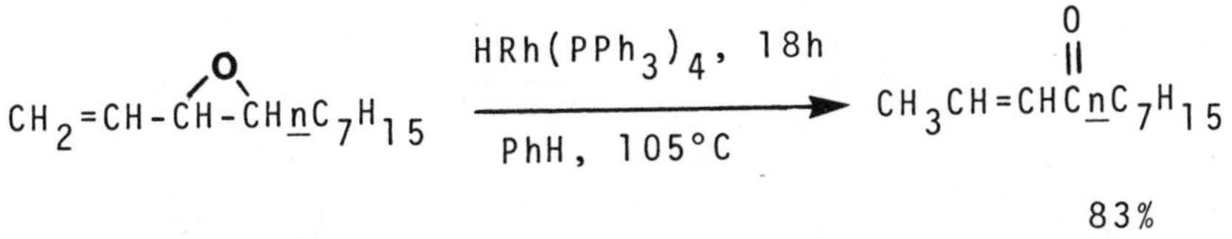

CH$_2$=CH-CH-CHnC$_7$H$_{15}$

HRh(PPh$_3$)$_4$, 18h

PhH, 105°C

CH$_3$CH=CHCnC$_7$H$_{15}$

83%

Sato, S.; Matsuda, I.[*]; Izumi, Y.
Tetrahedron Lett, (1985), 26, 1527

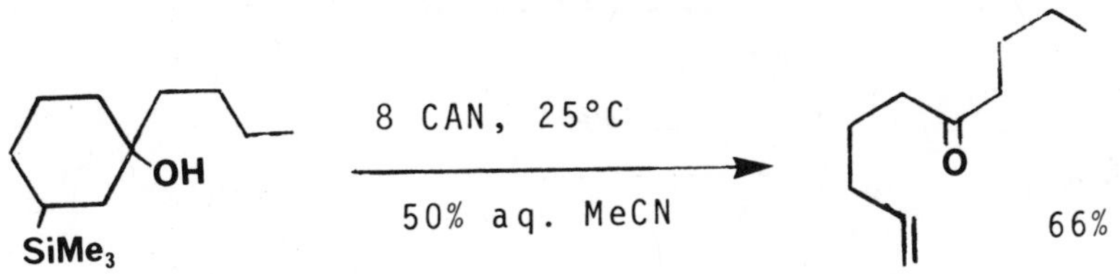

8 CAN, 25°C

50% aq. MeCN

66%

Wilson, S.R.[*]; Zucker, P.A.; Kim, C.-W.; Villa, C.A.
Tetrahedron Lett, (1985), 26, 1969

$$\underset{\underset{O}{\overset{O}{\parallel}}}{nPrCH-CHCCH_3} \xrightarrow[\text{2. } BF_3 \cdot OEt_2]{\substack{\text{1. 2 } nBu_3SnCH_2Li \\ \text{THF, } -78°C - RT}} nPrCH=CHCH_2\overset{O}{\overset{\parallel}{C}}CH_3$$

76%

Sato, T.[*]; Kikuchi, T.; Sootome, N.; Murayama, E.
Tetrahedron Lett, (1985), 26, 2205

$$\xrightarrow[\substack{\text{THF, reflux} \\ \text{2.5d}}]{\substack{2 \; O=\!\!\bigcirc\!\!=O \\ PdCL_2(PhCN)_2}}$$

67%

Clark, G.R.[*]; Thiensathit, S.
Tetrahedron Lett, (1985), 26, 2503

$$\underset{\underset{OEt}{|}}{nBuC\equiv CCH-\overset{OEt}{\underset{|}{C}}-CH_3} \xrightarrow[\substack{\text{2. 4 EtMgBr, THF} \\ \text{CuBr, } -30°C \\ \text{3. SiO}_2}]{\substack{\text{1. MeSO}_2Cl, NEt_3 \\ CH_2Cl_2, 0°C}} \underset{Et}{\overset{nBu}{>}}C=C=CHC\overset{O}{\overset{\parallel}{C}}CH_3$$

68%

Bernard, D.; Doritheau, A.[*]
Tetrahedron Lett, (1985), 26, 4923

$$CH_3\overset{O}{\overset{\parallel}{C}}(CH_2)_8CHO \xrightarrow[\text{THF, 25°C, 48h}]{CH_2I_2-Zn-Me_3Al} CH_3\overset{O}{\overset{\parallel}{C}}(CH_2)_8CH=CH_2$$

96%

Okazoe, T.; Hibino, J.; Takai, K.[*]; Nozaki, H.
Tetrahedron Lett, (1985), 26, 5581

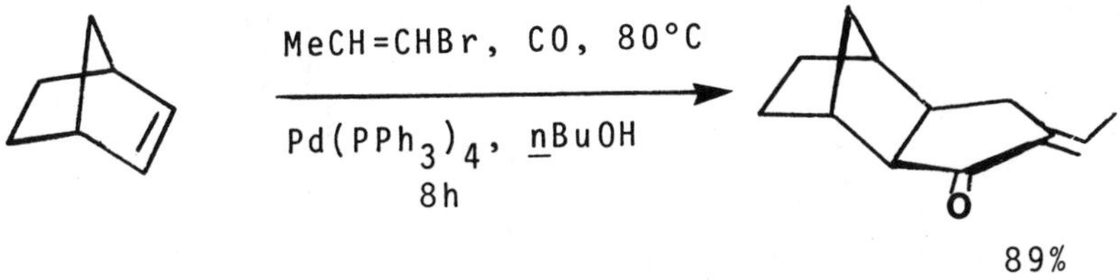

MeCH=CHBr, CO, 80°C

Pd(PPh$_3$)$_4$, nBuOH
8h

89%

Amari, E.; Catellani, M.; Chiusoli, G.P.
J Organomet Chem, (1985), **285**, 383

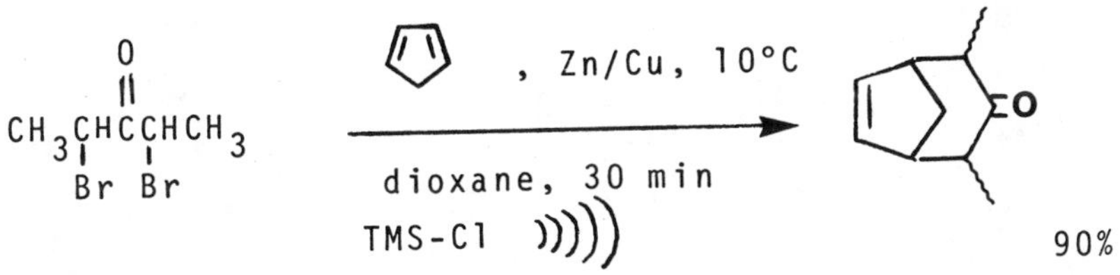

O
‖
CH$_3$CHCCHCH$_3$
   | |
  Br   Br

, Zn/Cu, 10°C

dioxane, 30 min
TMS-Cl)))))

90%

Joshi, N.N.; Hoffmann, H.M.R.*
Tetrahedron Lett, (1986), **27**, 687

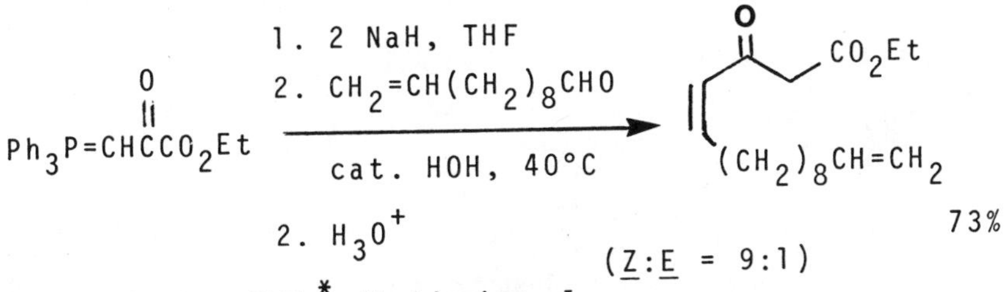

1. 2 NaH, THF
2. CH$_2$=CH(CH$_2$)$_8$CHO

O
‖
Ph$_3$P=CHCCO$_2$Et

cat. HOH, 40°C

2. H$_3$O$^+$

73%

($\underline{Z}$:$\underline{E}$ = 9:1)

Pietrusiewicz, K.M.*; Monkiewicz, J.
Tetrahedron Lett, (1986), **27**, 739

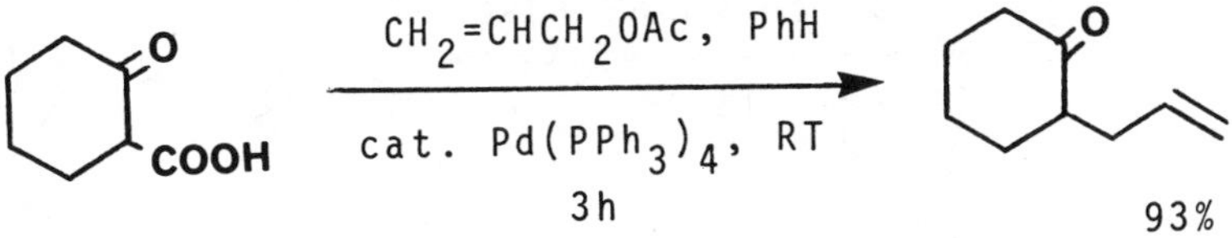

CH$_2$=CHCH$_2$OAc, PhH

cat. Pd(PPh$_3$)$_4$, RT
3h

93%

Tsuda, T.*; Okada, M.; Nishi, S.; Saegusa, T.*
J Org Chem, (1986), **51**, 421

PhCHCH=CHPh $\xrightarrow[\text{TiC}_3\text{H}_5\text{PdCl/L, 40°C}]{\text{NaCH(Ac)}_2\text{, THF, 19h}}$ PhCHCH=CHPh
| |
OAc $\text{CH}_3\overset{\text{O}}{\underset{\text{||}}{\text{C}}}\text{CHC}\overset{\text{O}}{\underset{\text{||}}{\text{C}}}\text{CH}_3$

$$L = \text{Ph}_2\text{P} \quad \text{Fe} \quad \begin{matrix}\text{Me}\\ \text{CHNMe}\\ \text{CH(CH}_2\text{OH)}_2 \text{ (90% ee S)}\\ \text{PPh}_2\end{matrix}$$ 97%

Hayashi, T.*; Yamamoto, A.; Hagihara, T.; Ito, Y.
                          Tetrahedron Lett, (1986), 27, 191

HC≡C, $\text{Co}_2\text{(CO)}_6$ ... OMe ... Me ... HO $\xrightarrow[\text{60°C, 4h}]{\text{SiO}_2\text{, sealed tube}}$ [product]   74%

Smit, W.A.*; Gybin, A.S.; Shashkov, A.S.; Strychkov, Y.T.;
Kyz'mina, L.G.; Mikaelian, G.S.; Caple, R.*; Swanson, E.D.
            Tetrahedron Lett, (1986), 27, 1241, 1245

$$\text{CH}_2=\text{CH(CH}_2)_3\overset{\text{OH}}{\underset{|}{\text{C}}}\text{HCH=CH}_2 \quad \xrightarrow[\substack{\text{RuH}_2\text{(PPh}_3)_4\\ \text{PhCH}_3\text{, reflux}}]{\text{CH}_2=\text{CHCH}_2\text{OCO}_2\text{Me}} \quad \text{CH}_2=\text{CH(CH}_2)_3\overset{\text{O}}{\overset{\text{||}}{\text{C}}}\text{CH=CH}_2$$ 95%

Minami, I.; Yamada, M.; Tsuji, J.*
                          Tetrahedron Lett, (1986), 27, 1805

1. KH(I$_2$ pretreated)
   18-crown-6, THF
   ────────────────
   reflux, 2h
2. EtOH, -78°C

(improved procedure)   75%
Macdonald, T.L.*; Natalie Jr., K.J.; Prasad, G.; Sawyer, J.S.
                    J Org Chem, (1986), 51, 1124

$$MeCH=CHCH_2OH \xrightarrow[\text{RT, 4h}]{PhCCHN_2, \ BF_3 \cdot OEt_2} PhCCHCH=CH_2$$

with $\overset{O}{\overset{\|}{\underset{}{}}}$ on the PhCCHN_2 group, product $PhCCHCH=CH_2$ with $\overset{O}{\overset{\|}{}}$ and Me substituent   83%

Kachinsky, J.L.C.; Salomon, R.G.[*]   **J Org Chem**, (1986), **51**, 1393

$$\xrightarrow[\text{Me}_3\text{SiI, CH}_2\text{Cl}_2 \atop 4h]{\underline{n}C_5H_{11}CHO, \ 0°C}$$

63%

Hatanaka, Y.; Kuwajima, I.[*]   **J Org Chem**, (1986), **51**, 1932

1. $CH_3\overset{O}{\overset{\|}{C}}CH_2CH_2CHPhCOOH$
   PhH, reflux, 16h
2. LDA, THF, -78°C
3. MeI
4. Red-Al
5. H^+

42%

(99.8% ee)

Meyers, A.I.[*]; Lefker, B.A.; Wanner, K.Th.; Aitken, R.A.
      J Org Chem, (1986), **51**, 1936

$$\xrightarrow[\text{PPh}_3, \ \text{CH}_3\text{CN} \atop 25°C, \ 20 \ min]{Pd_2(dba)_3 \cdot CHCl_3}$$

67%

Tsuji, J.[*]; Nisar, M.; Minami, I.
      Tetrahedron Lett, (1986), **27**, 2483

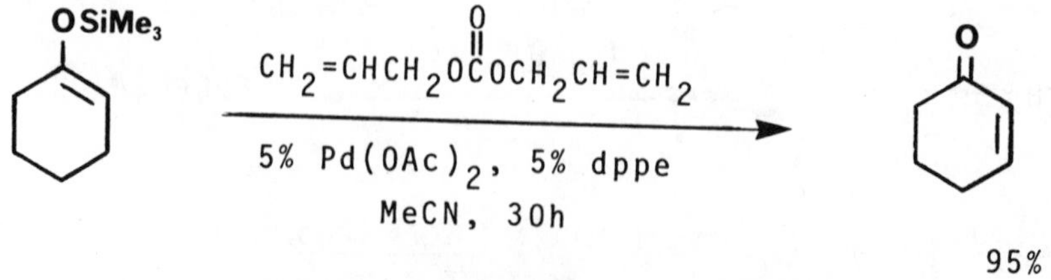

OSiMe₃

$$CH_2=CHCH_2OCOCH_2CH=CH_2$$

$\xrightarrow{\text{5\% Pd(OAc)}_2, \text{ 5\% dppe}}$
MeCN, 30h

→ (cyclohexenone)

95%

Minami, I.; Takahashi, K.; Shimizu, I.; Kimura, T.
Tetrahedron, (1986), **42**, 2971

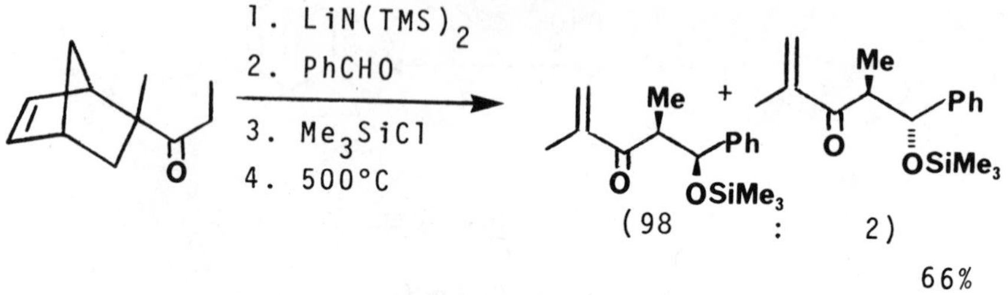

1. LiN(TMS)₂
2. PhCHO
3. Me₃SiCl
4. 500°C

→

Me (+) Me / Ph

(98 : 2)

66%

Bloch, R.; Gilbert, L.   **Tetrahedron Lett**, (1986), **27**, 3511

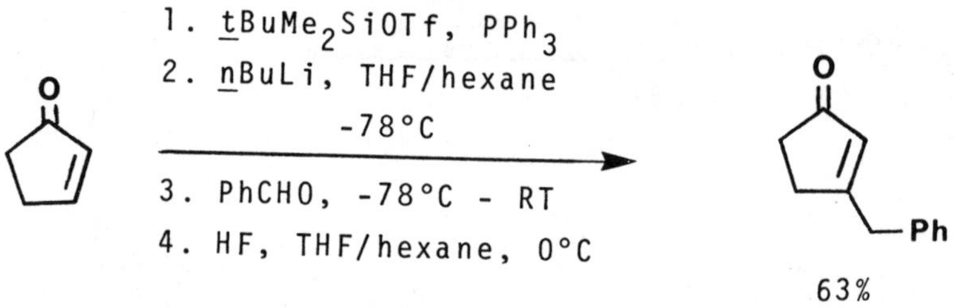

1. tBuMe₂SiOTf, PPh₃
2. nBuLi, THF/hexane
   -78°C
3. PhCHO, -78°C - RT
4. HF, THF/hexane, 0°C

→ —Ph

63%

Kozikowski, A.P.[*]; Jung, S.H.   **J Org Chem**, (1986), **51**, 3400

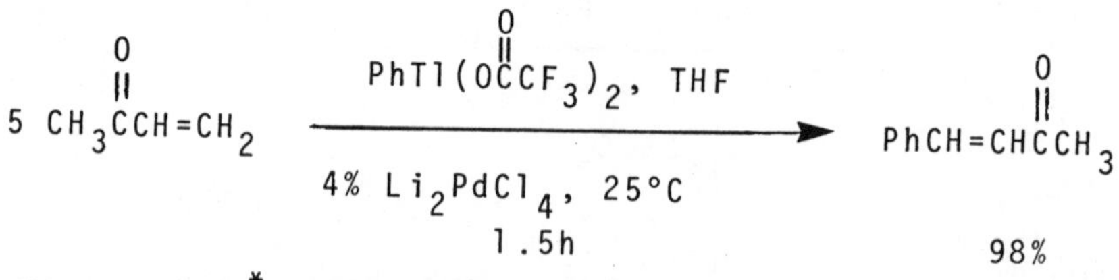

$$5 \ CH_3CCH=CH_2$$

$\xrightarrow{\text{PhTl(OCCF}_3)_2, \text{ THF}}$
4% Li₂PdCl₄, 25°C
1.5h

$$PhCH=CHCCH_3$$

98%

Kjonaas, R.A.[*]   **J Org Chem**, (1986), **51**, 3708

Reviews:

"Chromium (II) Reagents: Reduction of α-Acetylenic
Ketones to _trans_-Enones"

Smith III, A.B.[*]; Levenberg, P.A.; Suits, J.Z.
                        Synthesis, (1986), 184

"Oxidation of Alkenes: Metal Induced Formation of an Allylic
Carbon-Oxygen Bond"

Muzart, J.[*]   **Bull Chem Soc Fr**, (1986), II65

SECTION 375: Nitrile - Nitrile

1. LDA, PhH, 5°C, 15 min

2. [structure: CH₂SCN / Cl]

3. HOH
4. 10% aq. NaOH

91%

Davis, W.A.; Cava, M.P.[*]   **J Org Chem**, (1983), **48**, 2774

NaCH(CN)$_2$, CuI

HMPA, 120°C   59%

Suzuki, H.[*]; Kobayashi, T.; Osuka, A.[*]   **Chem Lett**, (1983), 589

SECTION 376: Nitrile - Olefin

1. TMS-PF, CH$_2$Cl$_2$

2. NEt$_3$

3. aq. NaHCO$_3$ 89%

Nishiyama, H.; Sakuta, K.; Osaka, N.; Itoh, K.[*]
                        Tetrahedron Lett, (1983), **24**, 4021

CHO

1. $\underset{TMS}{\overset{Me}{>}} C=C=NSiMe_3$

   $TiCl_4$, $Ti(OiPr)_4$

   CH_2Cl_2, -78°C

2. aq. Na_2CO_3

3. $BF_3 \cdot OEt_2$, CH_2Cl_2 ($\underline{E}:\underline{Z}$ = 9:1)

66%

Okada, H.; Matsuda, I.*; Izumi, Y.   **Chem Lett**, (1983), 97

$Fe(CO)_3$

1. $LiCMe_2CN$, THF/HMPA
   -78 - 25°C

2. CF_3COOH, -78°C
   30 min

CN

85%

Semmelhack, M.F.*; Herndon, J.W.
                    Organometallics, (1983), **2**, 363

1. CO_2, $Ca(OCl)_2$
   CH_2Cl_2, HOH, 10°C

2. NaCN

3. Bu_4NI, HOH, 110°C

4. $\underline{t}BuOK$, THF, RT

CN

58%

Suzuki, S.*; Fujita, Y.; Nishida, T.
                Syn Commun, (1984), **14**, 817

1. $(EtO)_2\overset{O}{\overset{\|}{P}}CN$, LiCN
   THF, RT, 5 min

2. $BF_3 \cdot OEt_2$, PhH
   RT, 2h

CN

80%

Harusawa, S.; Yoneda, R.; Kurihara, T.*; Hamada, Y.; Shioiri, T.
                Tetrahedron Lett, (1984), **25**, 427

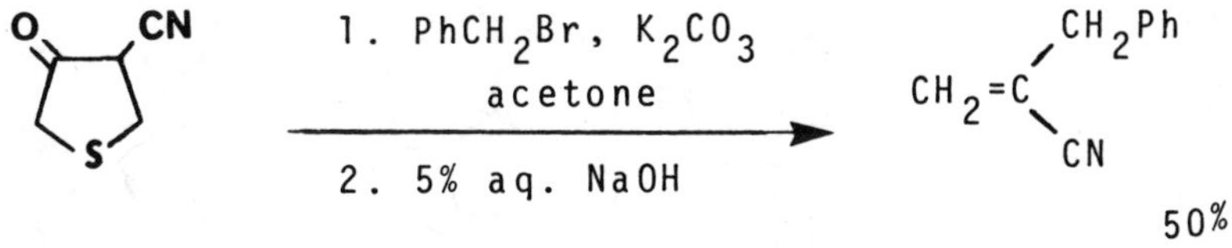
O
CN
S
1. PhCH2Br, K2CO3
acetone
2. 5% aq. NaOH
CH2Ph
CH2=C
CN
50%

Baraldi, P.G.; Pollini, G.P.; Zanirato, V.; Barco, A.; Benetti, S.
Synthesis, (1985), 969

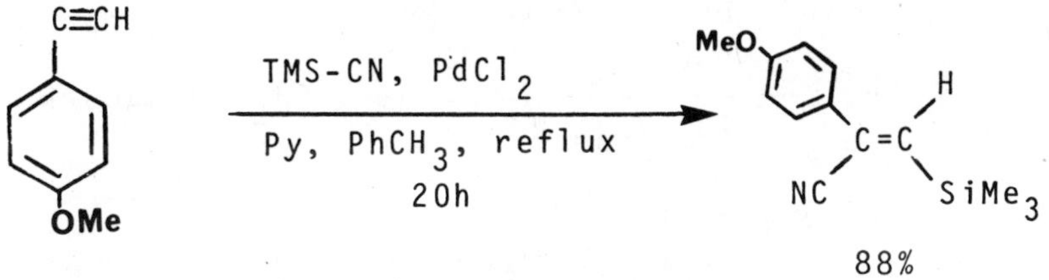
C≡CH
OMe
TMS-CN, PdCl2
Py, PhCH3, reflux
20h
MeO
H
C=C
NC
SiMe3
88%

Chatani, N.*; Hanafusa, T. **JCS Chem Comm**, (1985), 838

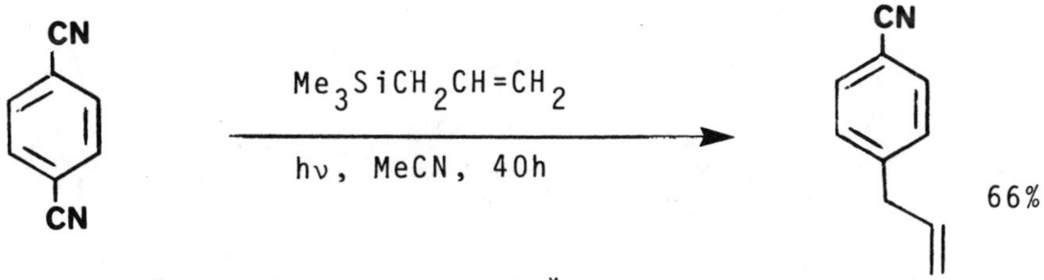
CN
CN
Me3SiCH2CH=CH2
hν, MeCN, 40h
CN
66%

Mizuno, K.*; Ikeda, M.; Otsuji, Y.*
Tetrahedron Lett, (1985), **26**, 461

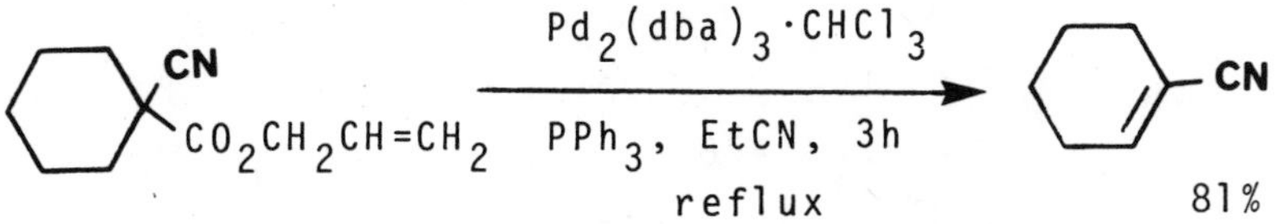
CN
CO2CH2CH=CH2
Pd2(dba)3·CHCl3
PPh3, EtCN, 3h
reflux
CN
81%

Minami, I.; Yuhara, M.; Shimizu, I.; Tsuji, J.*
JCS Chem Comm, (1986), 118

$$TMS\text{-}CN, RT \xrightarrow{SnCl_4, 30\ min} CH_2Cl_2$$

(2   :   8)   65%

Miyake, H.[*]; Yamamura, K.   **Tetrahedron Lett**, (1986), **27**, 3025

$$CH_2=CHCH_2SiPh_3 \xrightarrow[\text{2. Zn, THF, 1h}]{\begin{array}{c}1.\ PhCH=CHNO_2,\ TiCl_4 \\ CH_2Cl_2,\ -15°C\end{array}} CH_2=CHCH_2\overset{\overset{\displaystyle CN}{|}}{C}HPh$$

69%

Uno, H.[*]; Fujiki, S.; Suzuki, H.[*]
Bull Chem Soc Jpn, (1986), **59**, 1267

SECTION 377: Olefin - Olefin

$$\xrightarrow[\begin{array}{l}2.\ \underline{n}C_5H_{11}I \\ 3.\ 650°C,\ 0.1\ Torr\end{array}]{\begin{array}{l}1.\ \underline{n}BuLi,\ THF \\ \quad -78°C\end{array}} CH_2=CHCH=CH\underline{n}C_5H_{11}$$

90%

Bloch, R.; Abecassis, J.; Hasson, D.
Can J Chem, (1984), **62**, 2019
Bloch, R.; Abecassis, J.   **Tetrahedron Lett**, (1983), **24**, 1247

$$CH_2=CHOEt \xrightarrow[\begin{array}{l}3.\ ICH=CH\underline{n}C_5H_{11} \\ \quad 5\%\ Pd(PPh_3)_4\end{array}]{\begin{array}{l}1.\ 2\ \underline{t}BuLi,\ 2\ TMEDA \\ \quad hexane,\ -78°C \\ 2.\ ZnCl_2,\ THF,\ -78\text{-}0°C\end{array}} CH_2=\overset{\overset{\displaystyle OEt}{|}}{C}-CH=CH\underline{n}C_5H_{11}$$

(**E**)

74%

Negishi, E.[*]; Luo, F.-T.   **J Org Chem**, (1983), **48**, 1560

$$\text{TMSO(CH}_2)_6\text{C}{\equiv}\text{CH} \xrightarrow[\substack{\text{3. EtCH=CHI} \\ \text{Pd(PPh}_3)_4}]{\substack{\text{1.} \\ \text{2. aq. NaOH}}} \text{HO(CH}_2)_6\text{CH=CHCH=CHEt}$$

in PhCH$_3$, 120°C   74%   ($\underline{EE}$:$\underline{ZE}$ = 99:1)
in THF , 65°C   74%   ($\underline{ZE}$:$\underline{EE}$ = 99:1)

Cassani, G.; Massardo, P.; Piccardi, P.
Tetrahedron Lett, (1983), **24**, 2513

$$\underline{n}\text{C}_8\text{H}_{17}\text{CH=CH(CH}_2)_7\text{CO}_2\text{Me} \xrightarrow[\substack{\text{WCl}_6, \ 80°C \\ \text{SnMe}_4, \ 18h \ 68\%}]{} \begin{array}{l} \text{MeO}_2\text{(CH}_2)_7\text{CH=CH} \\ \quad\quad\quad\quad\quad\quad\ | \\ \quad\quad\quad\quad\quad (\text{CH}_2)_{10} \\ \quad\quad\quad\quad\quad\quad\ | \\ \underline{n}\text{C}_8\text{H}_{11}\text{CH=CH} \end{array}$$

Villemin, D.*   **Tetrahedron Lett**, (1983), **24**, 2855

54%

(77% ee)

Sakane, S.; Fujiwara, J.; Maruoka, K.; Yamamoto, H.*
J Am Chem Soc, (1983), **105**, 6154

$$\xrightarrow[\substack{\text{Pd(acac)}_2, \ \text{PPh}_3 \\ \text{dioxane, reflux} \\ 70h}]{\substack{\text{OH} \\ | \\ \text{CH}_2\text{=CHCH}\underline{n}\text{C}_5\text{H}_{11}}}$$

68%

Moreno-Mañas, M.*; Trius, A.
Bull Chem Soc Jpn, (1983), **56**, 2154

1. $BrCH_2SO_2Br$, hν, -15°C
   CH_2Cl_2, 2h

2. NEt_3, CH_2Cl_2, reflux

3. $\underline{t}BuOK$, THF/$\underline{t}BuOH$
   0°C

74%

Block, E.*; Aslam, M.   J Am Chem Soc, (1983), 105, 6165
Block, E.*; Eswarakrishnan, V.; Gebreyes, K.
                        Tetrahedron Lett, (1984), 25, 5469

$Me_2C=CH)_2CuLi$

Et_2O/THF

5% $Pd(PPh_3)_4$ (97 : 3)   94%

Jabri, N.*; Alexakis, A.; Normant, J.F.
                        Bull Chem Soc Fr, (1983), II321

$C≡C(CH_2)_3CH=CH_2$ 1.isooctane
                      $CpCo(CO)_6$
$C≡C-TMS$ reflux, 97h

2.$CuCl_2·HOH$
  NEt_3, MeCN

SiMe₃   71%

Sternberg, E.D.; Vollhardt, K.P.C.*
                        J Org Chem, (1984), 49, 1564

$CH_2=CHCH=CHMe$, NEt_3

$Pd(OAc)_2$, 100°C

$P(\underline{o}Tol)_3$, 15h

CH=CHCH=CHMe

71%

($\underline{EE}$:$\underline{EZ}$ = 8:2)

CO_2Me

Mitsudo, T.; Fischetti, W.; Heck, R.F.*
                        J Org Chem, (1984), 49, 1640

$$CH_2=C=CHCH_2O\overset{O}{\overset{\|}{P}}(OEt)_2 \xrightarrow[\text{THF, }-25°C]{\underline{n}C_7H_{15}MgBr}$$

$$\underset{CH_2=C-CH=CH_2}{\overset{\underline{n}C_7H_{15}}{|}}$$

80%

Djahanbini, D.[*]; Cozes, B.; Gore, J.
Tetrahedron, (1984), **40**, 3645
Tetrahedron Lett, (1984), **25**, 203

K-10 clay
Fe^{III}, 0°C
tbutylphenol
1h   (4   :   1)   77%

Laszlo, P.[*]; Lucchetti, J.   **Tetrahedron Lett**, (1984), **25**, 1567

C≡CMe
|
$(CH_2)_4$
|
C≡CMe

$$\xrightarrow[\text{Ph}_2\text{PMe, }-20°C]{Cp_2TiCl_2,\ Na(Hg)}$$

80%

Nugent, W.A.[*]; Calabrese, J.C.
J Am Chem Soc, (1984), **106**, 6422

1. Ca(sand), tBuOH
2. EDA, nBuNH_2
THF, 24h

90%

Benkeser, R.A.[*]; Laugal, J.A.; Rappa, A.
Tetrahedron Lett, (1984), **25**, 2089

$\underline{n}C_6H_{13}C{\equiv}CH$ $\xrightarrow{\begin{array}{c}1.\ MeCH{=}CHCH_2AliBu_2\\ Cl_2ZrCp_2,\ DCE\\ \underline{\qquad 60°C \qquad}\\ 2.\ H_3O^+\end{array}}$

(75   :   25)

60%

Miller, J.A.; Negishi, E.*   **Tetrahedron Lett**, (1984), **25**, 5863

$PhCH{=}CHCH_2OAc$
(**E**)
$\xrightarrow{\begin{array}{c}1.\ 5\%\ Pd(PPh_3)_4,\ RT\\ PPh_3,\ NaBr,\ THF\\ MeOH\\ \underline{\qquad\qquad}\\ 2.\ \underline{n}BuLi\\ 3.\ PhCHO\\ -17°C{-}RT\end{array}}$ $PhCH{=}CHCH_2CH{=}CHPh$

82%

(**EZ**:**EE** = 79:21)

Tsukahara, Y.; Kinoshita, H.*; Inomata, K.; Kotake, H.
                Bull Chem Soc Jpn, (1984), **57**, 3013

$EtOCH_2C{\equiv}CCH_2OEt$ $\xrightarrow[Et_2O,\ 35°C]{LiAlH_4,\ MgBr_2}$ $CH_2{=}C{=}CHCH_2OEt$

57%

Barbot, F.; Dauphin, B.; Miginiac, P.*   **Synthesis**, (1985), 768

$HOCH_2C{\equiv}CCH_2OH$ $\xrightarrow{\begin{array}{c}1.\ Py,\ (EtO)_2\overset{\overset{\displaystyle O}{\|}}{P}Cl\\ 0°C,\ 2h\\ \underline{\qquad\qquad}\\ 2.\ \underline{n}BuMgBr,\ CuI\\ THF,\ 0°C\ {-}\ RT\\ 12h\end{array}}$

69%

Araki, S.; Ohmura, M.; Butsugan, Y.*   **Synthesis**, (1985), 963

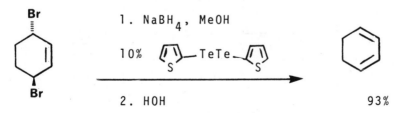

1. $Ph_3P=CHCH_2CH_2NMe_2$
   $PhCH_3$, 23°C

2. mcpba, $CHCl_3$

3. 50°C, 4h

65%

Corey, E.J.[*]; Desai, M.C. **Tetrahedron Lett,** (1985), **26**, 5747

1. $CF_3\overset{\overset{O}{\|}}{C})_2O$, CH_2Cl_2
   RT, 3h

2. $CH_2=CHnC_5H_{11}$, 0°C
   CF_3COOH, 90 min

3. mcpba, CH_2Cl_2 4. 140°C

$CH_2=CHCH=CHnC_5H_{11}$

63%

Ishibashi, H.[*]; Komatsu, H.; Maruyama, K.; Ikeda, M.
Tetrahedron Lett, (1985), **26**, 5791

1. tBuLi, tBuOK
   pentane/hexane

2. 2 $nC_{12}H_{25}Br$
   THF, RT, 1h

$nC_{12}H_{25}$

$nC_{12}H_{25}$

74%

Gordon III, B.[*]; Blumenthal, M.; Mera, A.E.; Kumpf, R.J.
J Org Chem, (1985), **50**, 1540

1. $NaBH_4$, MeOH

10% (thienyl)—TeTe—(thienyl)

2. HOH

93%

Engman, L.[*]; Byström, S.L. **J Org Chem**, (1985), **50**, 3170

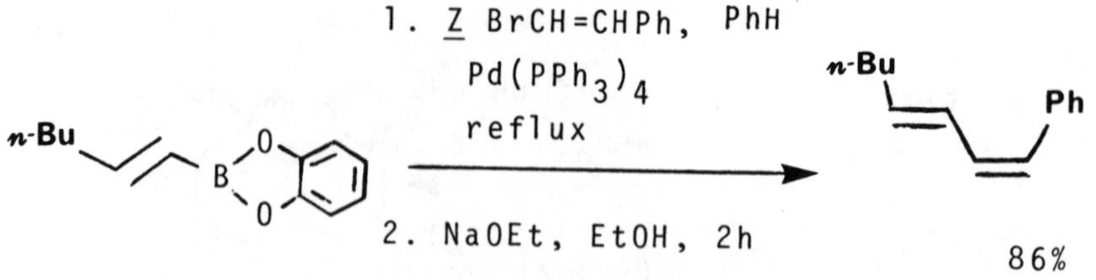

1. $\underline{Z}$ BrCH=CHPh, PhH
   Pd(PPh$_3$)$_4$
   reflux

2. NaOEt, EtOH, 2h

86%

Miyaura, N.; Yamada, K.; Suginome, H.; Suzuki, A.[*]
J Am Chem Soc, (1985), 107, 972

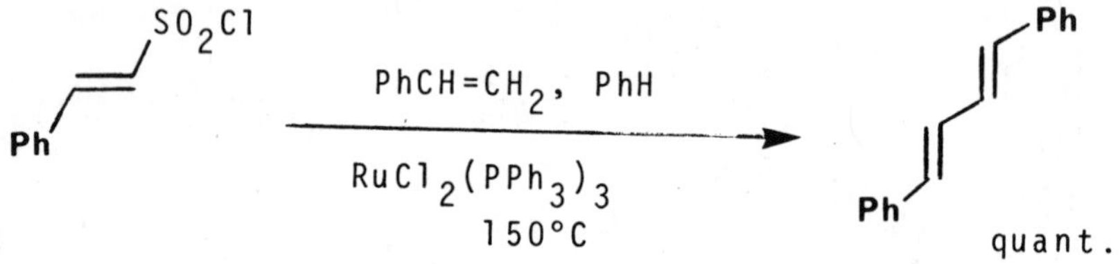

PhCH=CH$_2$, PhH

RuCl$_2$(PPh$_3$)$_3$
150°C

quant.

Kamigata, N.[*]; Ozaki, J.; Kobayashi, M.   Chem Lett, (1985), 705

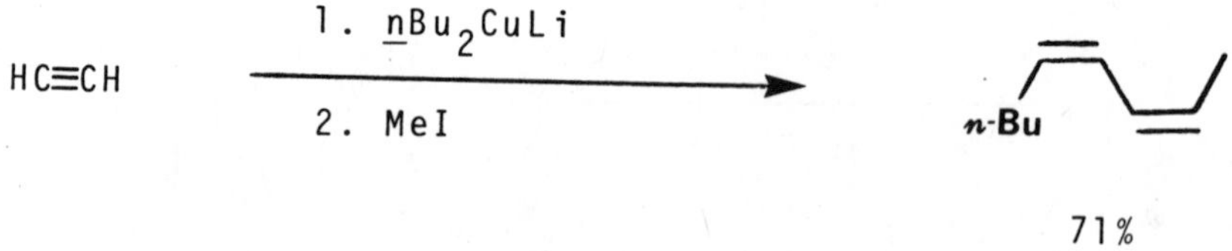

HC≡CH

1. $\underline{n}$Bu$_2$CuLi

2. MeI

71%

Furber, M.; Taylor, R.J.K.[*]; Burford, S.C.
Tetrahedron Lett, (1985), 26, 3285

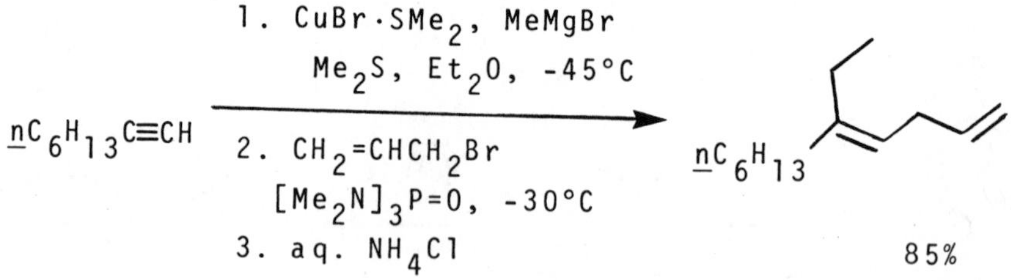

$\underline{n}$C$_6$H$_{13}$C≡CH

1. CuBr·SMe$_2$, MeMgBr
   Me$_2$S, Et$_2$O, -45°C

2. CH$_2$=CHCH$_2$Br
   [Me$_2$N]$_3$P=O, -30°C

3. aq. NH$_4$Cl

$\underline{n}$C$_6$H$_{13}$

85%

Iyer, R.S.; Helquist, P.[*]   Org Syn, (1985), 64, 1

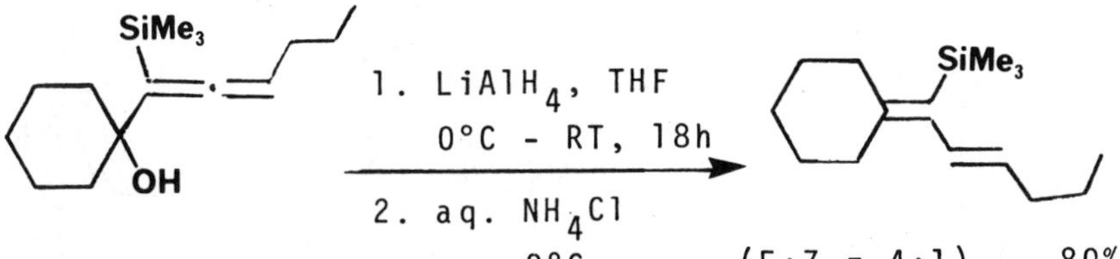
SiMe3
OH
1. LiAlH4, THF
0°C - RT, 18h
2. aq. NH4Cl
0°C
SiMe3

($\underline{E}$:$\underline{Z}$ = 4:1) 80%

Wang, K.K.*; Nikam, S.S.; Marcano, M.M.
Tetrahedron Lett, (1986), **27**, 1123

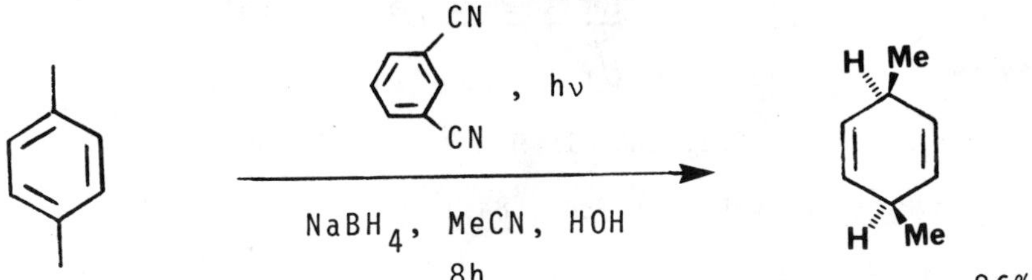
CN
, hν
CN
NaBH4, MeCN, HOH
8h
H Me
H Me

86%

Epling, G.A.*; Florio, E. **Tetrahedron Lett**, (1986), **27**, 1469

$$PhCH{=}CHCH_2OH \xrightarrow[\text{Pd(acac)}_2\text{, reflux, 111h}]{\text{PhCHO, dioxane}} PhCH{=}CHCH{=}CHPh$$

53%

Moreno-Mañas, M.*; Ortuño, R.M.; Prat, M.; Galán, M.A.
Syn Commun, (1986), **16**, 1003

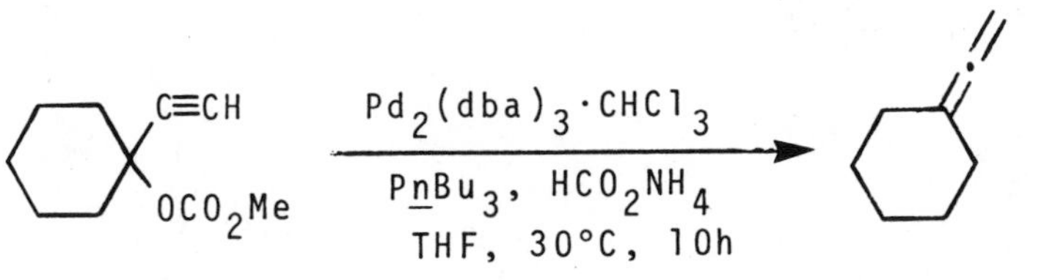
C≡CH
OCO2Me
Pd2(dba)3·CHCl3
PnBu3, HCO2NH4
THF, 30°C, 10h

66%

Tsuji, J.*; Sugiura, T.; Yuhara, M.; Minami, I.
JCS Chem Comm, (1986), 922

$$2 \text{ PhCH=CHCH}_2\text{Cl} \xrightarrow[\substack{\text{PdCl}_2, \text{ PPh}_3, \text{ DMF} \\ \text{Et}_4\text{NOTs, 24h}}]{e^-, \underline{n}\text{Bu}_3\text{SnCl, 50°C}}$$

92%

Yoshida, J.; Funahashi, H.; Iwasaki, H.; Kawabata, N.[*]
                    Tetrahedron Lett, (1986), **27**, 4469

Reviews:

"Catalysis of the Cope and Claisen Rearrangements"

Lutz, R.P.[*]   **Chem Rev**, (1984), **84**, 205

"Mercury (II)and Palladium (II) Catalyzed [3,3]-Sigmatropic
  Rearrangements"

Overman, L.E.[*]   **Angew Chem Int Ed Engl**, (1984), **23**, 579

"Asymmetric Diels-Alder and Ene Reactions in Organic Synthesis"

Oppolzer, W.[*]   **Angew Chem Int Ed Engl**, (1984), **23**, 876

"Acetylene Equivalents in Cycloaddition Reactions"

DeLucchi, O., Modena, G.   **Tetrahedron**, (1984), **40**, 2585

"η^3-Allylpalladium Compounds"

Jolly, P.W.[*]   **Angew Chem Int Ed Engl**, (1985), **24**, 283

AUTHOR INDEX